Lecture Notes in Physics

373

C.B. de Loore (Ed.)

Late Stages of Stellar Evolution Computational Methods in Astrophysical Hydrodynamics

Proceedings of the Astrophysics School II
Organized by the European Astrophysics
Doctoral Network at Ponte de Lima
Portugal, 11–23 September 1989

Springer-Verlag
Berlin Heidelberg GmbH

Editor

C. B. de Loore
Astrophysical Institute, Vrije Universiteit Brussel
Pleinlaan 2, B-1050 Brussels, Belgium

This book was processed by the authors using the TEX macro package
from Springer-Verlag Berlin Heidelberg GmbH.

ISBN 978-3-662-14501-2 ISBN 978-3-540-46921-6 (eBook)
DOI 10.1007/978-3-540-46921-6

Originally published by Springer-Verlag Berlin Heidelberg New York in 1991
Softcover reprint of the hardcover 1st edition 1991

2153/3140-543210 – Printed on acid-free paper

Preface

Scientific progress is heavily dependent upon increased collaboration, strong mobility of scientists and an adequate use of large research facilities available through multinational or bilateral cooperation. This is especially true for astronomy, since it is such a fast developing science,from both theoretical and observational viewpoints. These considerations led to the foundation in 1986 of the European Astrophysics Doctoral Network, which today federates 21 departments of astronomy in European universities, all having a graduate programme in astrophysics. The Network decided to organize each year a summer school at the predoctoral level, gathering European students at the beginning of their thesis research period and placing them in interaction with the international community of scientists, represented by the lecturers, and among themselves. Such a school offers them a broad exposure to major fields of astronomy at an early stage of their own research, in order to deepen their scienitific education, and helps them to gain maximum advantage from the possibilities offered by the international, in particular European, collaboration. These possibilities include access to the large observing capabilities provided by space- and ground-based European instruments. The 1989 school dealt with two parallel topics: "Theory and Observations of Late Stages of Stellar Evolution" and "Computational Methods in Astrophysical Hydrodynamics". These topics were chosen to cover areas where considerable theoretical progress has been made and is expected in the coming years, given the new observational tools Europe will have at its disposal: the Space Telescope, the Infrared Space Observatory, the Very Large Telescope, the IRAM radio-interferometer, and the Southern Millimetric Telescope. Eight lecturers delivered 50 hours of lectures during ten working days over two weeks. A careful planning and ordering of topics ensured that the lectures are accessible to research students of one year's standing or less. Workshops were organized at the end of the day, and at night the students gave a short presentation of their current work to other students and lecturers. The time available for these presentations was approximately 12 hours, distributed over two weeks. Moreover a set of personal computers was put at the disposal of students and lecturers, allowing demonstrations to illustrate the theoretical courses and giving the students the opportunity to build and to run programs as practical exercises.

Graduate studies in Europe are arranged in many diverse ways: some countries have, or can afford to have, a large number of graduate courses, while others have none, or, in some cases, lack the minimum geographic concentration of students necessary. The opportunity to have courses understandable by students with the minimum requirement of a solid education in physics therefore appears to be a valuable and appreciated objective. This volume contains all lectures presented at the Astrophysics School II, with the exception of the course " Radiative Hydrodynamics". We take the opportunity here to express our gratitude to the European bodies in Brussels, and mainly the ERASMUS programme. Very favourable conditions were offered by the local authorities at

Ponte de Lima in Portugal (Câmara Municipal de Ponte de Lima), namely free use of the town assembly hall for the lectures, and the secretariat, including electricity, telephone, cleaning, free transportation from Porto's international airport and town centre to Ponte de Lima and back, and even local secretarial help, to the local tourist authority for arranging special accomodation and for a closing countryside tour with folk dances and dinner, as well as a local gift for the participants. We should like to acknowledge the support received from the following institutions and Foundations: the Carl Zeiss Foundation, Germany; the Centre National de Recherche Scientifique, France; the CNR, Italy; the European Council's European Programme for Postgraduate Studies; The European Economic Community's ERASMUS Programme, ICP 89/F/0415-13; the European Southern Observatory; the Fundação Calouste Gulbenkian, Portugal; the Instituto Nacional de Investigação Cientifica,Portugal; the Instituto Nacional de Investigaçao Cientifica e Tecnológica, Portugal; the Leids Kerkhoven-Bosscha Fonds, Netherlands; Rank Xerox, Portugal;the Science and Engineering Research Council, UK; Sweden; Switzerland and all national laboratories that individually supported some of the students and lecturers. The school was attended by 59 students, representing 13 different countries. There was much stimulating interaction among the students and with the lecturers: the spirit as well as the language in the school was truly European.

This publication was made possible thanks to the interest of Springer Verlag, through Prof. W. Beiglböck.

Brussels, 1 December 1990 Bert de Loore

Contents

Part I Theory and Observation of Late Stages of Stellar Evolution

Stellar Structure and Stellar Evolution........3
by Bert C. de Loore

Supernovae: Observations, Theory, Models and Nucleosynthesis........97
by Ewald Muller

Circumstellar Envelopes of Late Type Stars........161
by Alain Omont

Evolution of Close Binary Systems: Application to X-Ray Binaries........213
by Gertjan J. Savonije

Part II Computational Methods in Astrophysical Hydrodynamics

An Introduction to Computational Methods in Hydrodynamics........259
by Willy Benz

Basic Hydrodynamics........313
by Jean Heyvaerts

Convection and Turbulence in Stars........355
by Jean-Paul Zahn

PARTICIPANTS

Aguilera, Immaculada Univ. de Granada, Spain
Angonin, Marie-Christine Observatoire de Meudon, France
Argyropoulos, Stylianos University of Athens, Greece
Barrio, David Univ. College London, United Kingdom
Boffin, Henri Université Libre de Bruxelles,Belgium
Caetano, Armindo Observ. Astr. de Lisboa, Portugal
Cananzi, Katia Observatoire de Marseille, France
Chatzichristou, Helène Observatoire de Meudon, France
Colomer, Francisco Univ. de Valencia, Spain
David,. Pedro Observatoire de Meudon, France
Devaney, Nicholas Dunsink Observatory, Ireland
Ferrara, Andrea Università de Firenze, Italy
Figueredo, Jorge Universidade do Porto, Portugal
Gallagher, Fergus Mullard Radio Astr. Obs., United Kingdom
Gebhardt, Ute Ruhr-Univ. Bochum, Germany
Gomez, Christina Obs. Astr." Ramon Aller", Spain
Groenewegen, Martin Inst.Anton Pannekoek,the Netherlands
Grün, Karl Technische Universität Wien, Austria
Guérin, Frédéric Observatoire de Meudon, France
Heemskerk, Martin Inst.Anton Pannekoek, the Netherlands
Herzig, Klaus Universität Göttingen, Germany
Hourdin,Frédéric Ecole Normale Supérieure, France
Joncour, Isabelle Institut d'Astrophysique, Paris, France
Keong, Ng Yuen Sterrewacht Leiden, the Netherlands
Koninckx, Jean-Paul Sterrek. Inst. Utrecht, the Netherlands
Kozma, Cecilia Stockholm Observatory, Sweden
Kruss, Andrea Universität Kiel, Germany
Kunz, Mathias Universität Tubingen, Germany
Lapuente, Pilar Univ. de Barcelona, Spain
Lebre, Agnes Universite de Montpellier, France
Li, Janle University of Sussex, United Kingdom
Lopes, Ilidio Universidade do Porto, Portugal
Ludwig, Hans-Gunther Universität Kiel, Germany
Malbet, Fabien Institut d'Astrophysique, Paris, France
Marti-Puig, Jose Univ. de Valencia, Spain
Mathias, Philippe CSNSM Campus Orsay, France
Mavromatakis, Fotis University of Crete, Greece
Mendes, Pedro Inst. Sup. Tecnico Lisboa, Portugal
Monteiro, Màrio Universidade do Porto, Portugal
Morin, Stephane Observatoire de Marseille, France
Pajot, Philippe CEN Saclay, France
Pavlakis, Kostas University of Crete, Greece
Pietrini, Paola Università de Firenze, Italy
Pijpers, Frank Sterrewacht Leiden, the Netherlands

iPont, Jordi Jose Univ. Polit. de Catalunya, Spain
Pujol, Francesco Univ. de Barcelona, Spain
Raiteri, Claudia ISAS, Italy
Reynolds, Alistair University of St Andrews, United Kingdom
Rokaki, Evlabia Institut d'Astrophysique, Paris, France
Rola, Claudio Universidade do Porto, Portugal
Ruffert, Maximilian Max Planck Institut, Garching, Germany
Schaller, GérardObservatoire de Geneve, Switzerland
Teodorani, Massimo Unioversità di Bologna, Italy
Theuns, Tom Univ. Antwerp, RUCA,Belgium
Thiebaut, Eric Observatoire de Meudon, France
Tringham, Roger Manchester University, United Kingdom
Vozikis, Christos University of Thessaloniki, Greece
Wisotzki, Lutz Hamburger Sternwarte, Germany

Camiel, Bert de Loore

Astrophysical Institute, Vrije Universiteit Brussel, Belgium
and
Astrophysics Research Group, University of Antwerp (RUCA)

PART I

Theory and Observation of Late Stages of Stellar Evolution

Stellar Structure and Stellar Evolution

Camiel, Bert de Loore

Astrophysical Institute, Vrije Universiteit Brussel, Brussels, Belgium and
Astrophysics Research Group, University of Antwerp (RUCA), Belgium

1 Introduction. Stellar parameters from observations

1.1 Magnitudes and luminosities

The physical properties of stars are usually expressed by their luminosity L, i.e. the total radiated energy from the surface of the star, their mass M, their radius R and the effective temperature T_{eff}. L, M and R can be expressed in solar units $L_\odot$, $M_\odot$, $R_\odot$:

$L_\odot = 3.89\ 10^{33}$erg s^{-1}

$M_\odot = 1.989\ 10^{33}$g

$R_\odot = 6.960\ 10^{10}$ cm

T_{eff} is related to the radius R by the Stefan-Boltzmann law

$L = 4\ \pi R^2\ \sigma\ T_{eff}{}^4$

with σ the Stefan-Boltzmann constant. In the cgs system $\sigma = 5.67\ 10^{-5}$ erg cm^{-2}deg^{-4}s^{-1}. The solar effective temperature is: $T_{eff\odot} = 5800$K.

For the various types of stars these parameters cover a wide range, especially the luminosity :

$-6 < logL/L_\odot < +6$

$0.08 < M/M_\odot < 100$

$0.001 < R/R_\odot < 1500$

$3.3 < \log T_{eff} < 5$

The luminosity of a star can be defined as the total power of the star required to sustain the energy outflow from a large surface surrounding the star. This energy flux consists of a flux of photons, neutrinos and of mass loss. Photons constitute the traditional luminosity; this form of the energy is generally the most important, since photons usually produce the dominant luminosity; moreover this form is the only one that is generally observable. Neutrinos are produced by nuclear reactions in the stellar interior and by weak interactions of photons with matter. Neutrinos are difficult to detect; they have very large free paths, and react seldom with other particles. The estimated brightness of a star is its apparent magnitude. It is dependent on the real brightness of the star and on its distance. The apparent magnitudes of two stars, m_1 and m_2, are related to their brightnesses b_1 and b_2 as follows

$$\frac{b_1}{b_2} = 100^{\frac{1}{5}(m_2 - m_1)} = 2.512^{m_2 - m_1} \tag{1.1}$$

$$m_2 - m_1 = 2.5\log\frac{b_1}{b_2} \tag{1.2}$$

The conversion of the observed energy flux into the photon luminosity of the star requires an estimate of the distance to the star. For an absolute comparison of intrinsic brightness the stellar magnitude is transformed to the magnitude the star should have when placed at a distance of 10 pc; this is called the absolute magnitude M.

From the inverse square law for energy and distance follows that the the ratio of the absolute brightness B and the apparent brightness b, for a star at a distance r in parsec, is given by

$$\frac{B}{b} = (\frac{r}{10})^2 \tag{1.3}$$

Using Eq.1.2 the relation between apparent magnitude m, absolute magnitude M and the distance d in pc can be expressed as

$$M = m + 5 - 5\log d \tag{1.4}$$

If we are interested in observations related to the visual part of the spectrum we use the visual apparent and absolute magnitudes m_v and M_v.

$M_v = m_v + 5 - 5 \log d$

In order to obtain the total energy flux, the visual magnitude has to be corrected for the contributions of the non-visual parts of the flux, i.e. the flux in the ultraviolet, and in the infrared. The bolometric magnitude M_b is a magnitude scale describing the stellar energy flux, integrated over all wavelengths.

The bolometric correction BC is the correction to be applied to the visual magnitude M_v to find M_b.

$M_b = M_v + BC$

It is evident that BC is always negative. The absolute luminosity is the total energy emitted by the star in all wavelengths. The luminosity can be linked to the bolometric magnitude, by using the sun as calibration.

For the sun:

$M_v = + 4.83$ BC = - 0.08

$M_b = + 4.75$ $L_\odot = 3.83\ 10^{33} \text{erg s}^{-1}$

For stars :

$M_b = 4.75 - 2.5 \log L/L_\odot$

or, using the Stefan-Boltzmann's law $L = 4\pi R^2 \sigma T_{eff}{}^4$

$\log L = - 15.04 + 2 \log R + 4 \log T_{eff}$

where L and R are expressed in solar units.

Stellar magnitudes can be measured at different wavelengths, by using filters. One of the oldest and best known systems is the UBV-photometry system of Johnson. The U filter registers the radiation in the ultraviolet, centred around 365 nm, with a bandwidth of 68 nm, the B filter is centred around 440 nm, with a bandwidth of 98 nm, the V filter (visual) is centred around 550 nm with a bandwidth of 89 nm. Magnitude differences between two filters, e.g. U - B or B - V are called "colours".

For a blue star we have B - V < 0, U - B < 0 and for a red star B - V > 0. For a yellow star we have B - V > 0, U - B > 0.

1.2 Spectral types

Although stellar spectra are different for different stars, it is possible to distinguish a limited number of spectral classes, from O-type stars (blue, hot stars) to M-type stars (red, cool stars).The table shows only the most general characteristics. This repartition in classes can be refined, by measuring the intensity of the spectral lines and by using the ratio of the intensity of various lines.

Table 1.1. The characteristics of the various spectral classes

spectral class	absorption lines of the most important ions
O	H(faint), He, He^+,C^{++},N^{++},O^{++},Si^{+++}
B	H(stronger),He, C^+, N^+, O^+, Si^+, Si^{++},Mg^+
A	H(very strong), O, Si^+, Mg^+, Ca^+, Ti^+, Fe,
F	H(strong), Ca^+, Cr, Cr^+,Fe, Fe^+,Sr^+
G	H(weaker), Ca+ (strong), Fe, Fe^+, Cr, Cr^+
K	H(weak), Ca, Ca^+, Fe, Cr, TiO bands
M	H(very weak), Ca, Fe, Cr, TiO bands stronger

In this way each class can be subdivided into 10 types, B0, B1, B2, ..., A0, A1, A2, ... The hottest stars are of type O3.
Stars of the same spectral type can show large differences in absolute magnitude. The faintest stars are called dwarfs, the brightest stars are giants, and the most luminous supergiants. This has led to a repartition in luminosity classes, I for the most luminous supergiants to V for main sequence stars, and even to VI for subdwarfs.This is a twodimensional classification of stars.

1.3 Stellar temperatures

The radiation of a black body can be described by the Planck function, $B_\lambda(T)$. πB_λ is the radiated energy emitted per unit of time, per surface unit, per wavelength interval

$$\pi B_\lambda(T) = \frac{\pi c_1}{\lambda^5}(e^{c_2/\lambda T} - 1)^{-1} \tag{1.1}$$

with $c_1 = 1.191\ 10^{-5}$ erg cm^2 s^{-1}
$c_2 = 1.439$ cm deg

The Planck curves for various temperatures depicted in Figure 1.1 reveal that for shorter wavelengths the intensity of the radiation increases with the temperature, and that the peak of the emission shifts to lower wavelengths with increasing temperature (Wien's shift law). The integral of the Planck function is

$$\int_0^\infty \pi B_\lambda(T) d\lambda = \sigma T^4 \tag{1.2}$$

with $\sigma = 5.67\ 10^{-5}$ ergcm^{-2} deg^{-4}

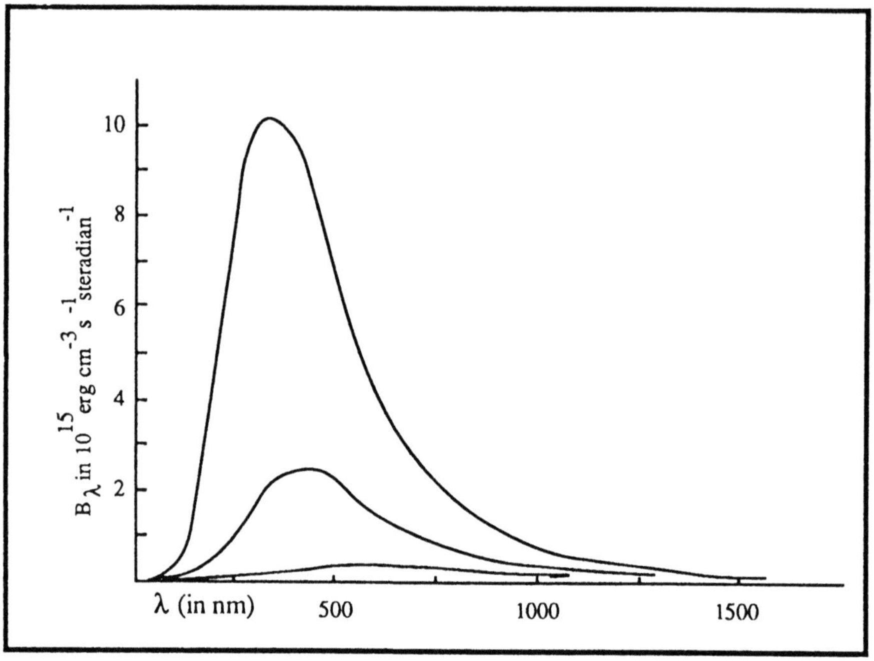

Fig. 1.1. Planck curves, i.e. the black body radiation B as a function of 1 for various temperatures. The intensities are higher for larger temperatures, and the peak of the intensity shifts to lower wavelengths for higher temperatures (Wien's shift law)

A widely used temperature definition is the *colour temperature*.

The colour of a star, B-V, can be measured; it can also be derived from the Planck curve. The temperature which has to be used in the Planck energy distribution, to give B-V as observed , is the colour temperature. The definition has the advantage, that it is independent on the distance of the star, since B-V represents a ratio. However it is not a very good representation of the real temperature at the stellar surface, since the energy distribution, affected by absorption lines, strongly deviates from a Planck distribution.

Another way to proceed is to determine the ratio of the line strengths of the same ions of the same element, and to calculate the ionization fraction e.g. He^+/He, for O-stars, or Fe^+/Fe for stars of spectral type G to M. Comparison of of the measured ionization fractions with the calculated ones for various temperatures, leads to the *ionization temperature*.

The most sensible temperature is the *effective temperature*. It is the temperature of an object of the same dimensions as the star and with the same bolometric magnitude, producing Planck radiation, in other words, the temperature which satisfies

$L_{bol} = 4\pi R^2 \sigma T_{eff}^4$

For the solar case, L = 3.89 10^{33} ergs-1, and R =6.96 10^{10} cm, this gives T_{eff} = 5770K. For other stars the effective temperatures have to be derived indirectly by comparison of the observed energy distribution with calculated model atmospheres, or by comparison of the observed absorption lines with predictions of the occurrence of these lines and their strengths by these model atmospheres.

Models, computed for various values of the flux, hence for various values of T_{eff}, show how the temperature, pressure and ionization rates increase with stellar depth. For each model the spectrum, including the absorption lines, is calculated, and this can be compared with observations.

Table 1.2. gives an overview of B-V, T_{eff}, BC L, R and M for various spectral types and luminosity classes.

1.4 The Hertzsprung- Russell diagram

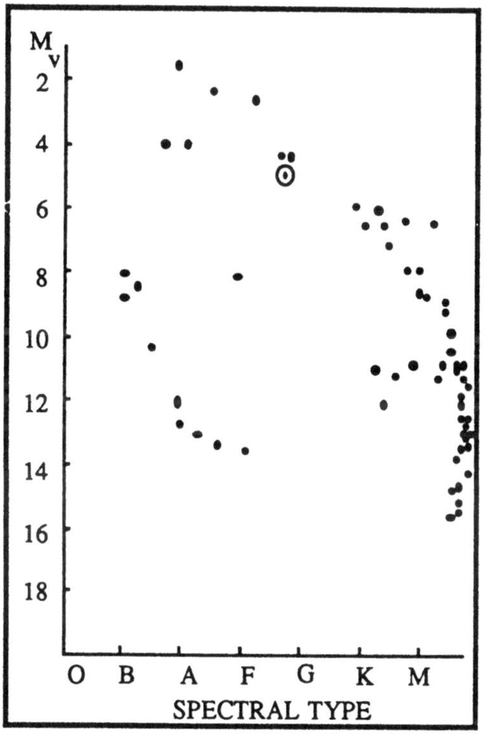

Fig. 1.2. The nearest stars in the HRD

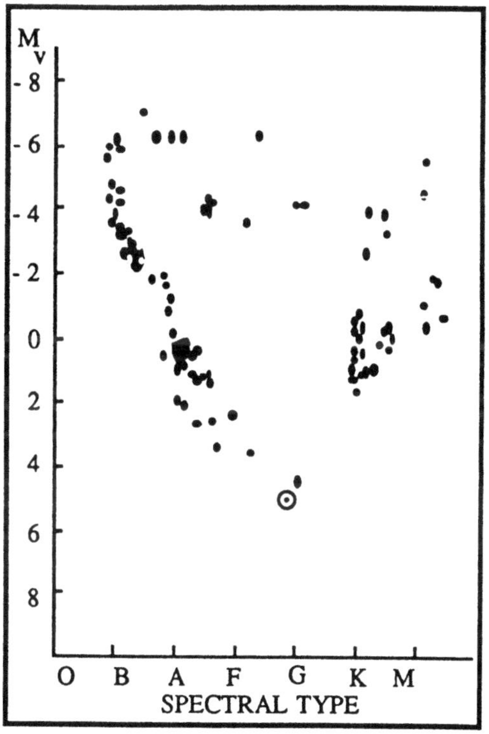

Fig. 1.3. The brightest stars in the HRD

A diagram where the magnitude of stars M_v is positioned versus their spectral type is called a Hertzsprung-Russell diagram. It is a fundamental tool for the study of stellar evolution. Figure 1.2 shows a large number of stars of known distance. The stars are not

Table 1.2. Stellar characteristics for the various types and luminosity classes (adapted from Allen, 1963)

Spec.type	T_{eff}	B-V	M_v	B.C.	M_b	log $L/L_\odot$	$R/R_\odot$	$M/M_\odot$
Main sequence stars								
O5 V	40.000	-0.35	-5.8	-4.0	-10	5.90	19.3	40
B0 V	28.000	-0.31	-4.0	-2.8	-6.8	4.62	9.0	18
B5 V	15.500	-0.16	-1.1	-1.5	-2.6	2.94	4.3	6.5
A0 V	9.900	0.00	+0.6	-0.5	+0.1	1.86	3.01	3.2
A5 V	8.500	+0.13	+1.9	-0.2	+1.7	1.22	2.0	2.1
F0 V	7.400	+0.27	+2.7	-0.1	+2.6	0.86	1.7	1.7
F5 V	6.580	+0.42	+3.4	+0.0	+3.4	0.54	1.5	1.3
G0 V	6.030	+0.58	+4.3	-0.0	+4.3	0.18	1.2	1.1
G5 V	5.520	+0.70	+5.1	-0.1	+5.0	-0.10	1.0	0.93
K0 V	4.900	+0.98	+6.0	-0.2	+5.8	-0.42	0.89	0.78
K5 V	4.130	+1.10	+7.3	-0.6	+6.7	-0.78	0.83	0.69
M0 V	3.480	+1.45	+9.0	-1.2	+7.8	-1.22	0.70	0.47
M5 V	2.800	+1.63	+12.1	-2.3	+9.8	-2.10	0.43	0.21
M8 V	2.400	+1.8	+16			-3.0	0.11	0.10
Giants								
G0 III	5.600	+0.65	+1.1	-0.0	+1.1	1.46	6	2.5
G6 III	5.000	+0.85	+0.7	-0.2	+0.5	1.70	10	3
K0 III	4.500	+1.07	+0.5	-0.5	+0.0	1.90	15	4
K50III	3.800	+1.41	-0.2	-0.8	-1.0	2.30	34	5
M0 III	3.200	+1.60	-0.3	-1.5	-1.8	2.62	69	6
M5 III	2.700?	+1.85	-0.8	-2.8	-3.6	3.34	200	?
Supergiants								
B0 I	30.000	-0.25	-6.4	-3	-9	5.5	22	36
A0 I	12.000	0.00	-6.2	-0.5	-7	4.7	54	18
F0 I	7.000	+0.25	-6	-0.1	-6	4.3	100	13
G0 I	5.700	+0.70	-6	-0.1	-6	4.3	150	13
K0 I	4.100	+1.39	-5	-0.7	-6	4.3	290	13
M0 I	3.000?	+1.94	-5	-2	-7	4.7	860?	18

uniformly spread in the diagram, but are positioned along a narrow band, from luminous, blue OB-star at the upper left corner to faint red cool M stars, at the lower right corner.

Figure 1.2 shows the HRD for the nearest stars (d < 6.5 pc) and Figure 1.3 shows the HRD for the brightest stars (V <2.60). The striking difference between the two diagrams is due to selection effects. Most stars in the Galaxy are intrinsiqually weak. When we select all stars within a given distance of the sun, they will all have a large absolute magnitude $M_v \geq 6$; the sun is in this sample one of the brigthest stars. The selection of the brightest stars contains a large number of intrinsiqually bright stars which are far away. This means that this sample contains many giants and supergiants. Figure 1.4 shows the HRD of stars with known distance. The figure reveals that most stars (about 95 band, from the upper left (luminous, blue, massive OB stars) to the lower right (weak, red, low mass M stars). This band is called the main sequence. The stars of the main sequence are of luminosity class V or IV. Besides the main sequence band two other regions may be seen, the domain of the red supergiants (class I, $M_v \geq -5$), in the upper right part of the HRD, and the domain of the white dwarfs, at the bottom of the diagram. Finally one remarks the giant branch, going up to the right, and comprising the red giants (classes III and II), and below the main sequence the domain of the subdwarfs, of luminosity class VI. Figure 1.5 shows the location of these various groups.

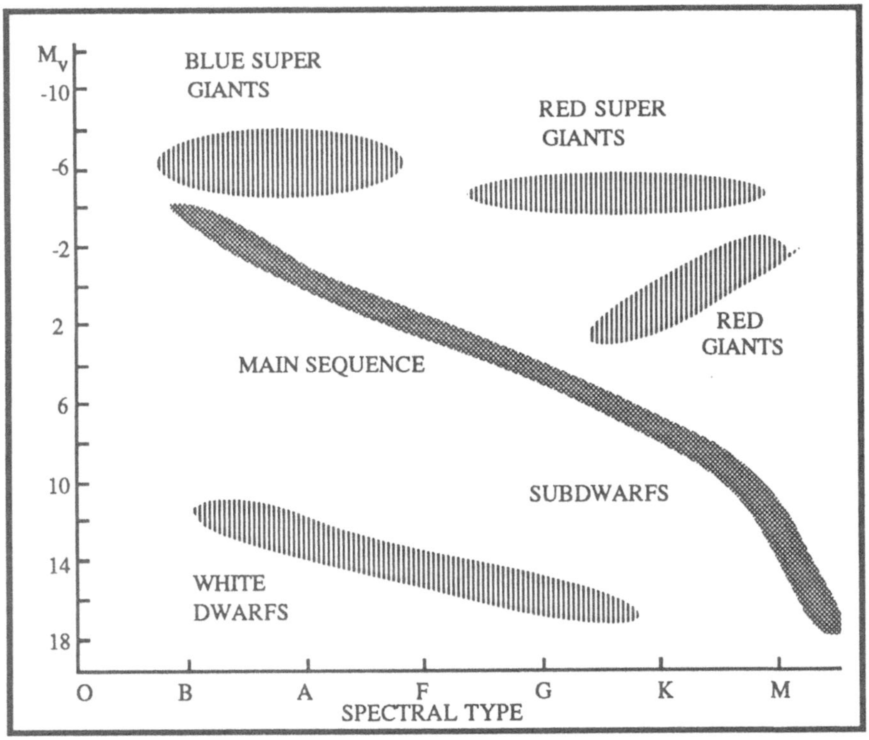

Fig. 1.4. A schematic view of the HRD, and the location of various groups : main sequence stars, red giants, blue and red supergiants, white dwarfs.

An alternative form of the HRD is the colour-magnitude diagram, where the absolute visual magnitude M_v is depicted versus the colour index B-V. A third form is the luminosity-temperature diagram, showing M_b (or L) versus T_{eff}. For this form of the HRD the bolometric correction is required; the determination of T_{eff} is more difficult

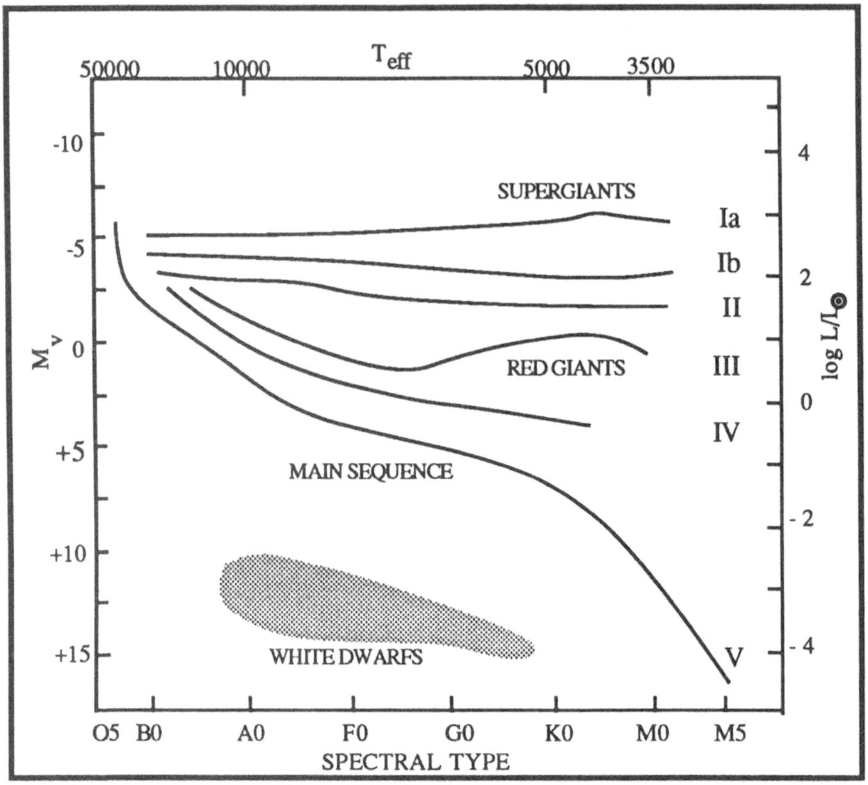

Fig. 1.5. The HRD and the position of the different luminosity classes.

than the spectral type or the colors. For the study of the stellar evolution the L-Teff diagram is more suitable, since in this way the total energy production, and the radius are given, the parameters characterizing the evolution of a star of given mass. Figure 1.6 shows the L-Teff diagram. The HRD is an important tool for the presentation of the theoretical stellar models, and for the comparison of computed models with observations.

1.5 Open clusters - globular clusters - Stellar populations

The material out of which our Galaxy was formed was probably mainly hydrogen, a smaller fraction of helium, and a still smaller amount of other elements. These fractions are generally indicated as X,- the mass fraction of hydrogen-, Y, -the mass fraction of helium-, and Z, -the fraction of heavy elements. Obviously $X + Y + Z = 1$. In the early stages the Galaxy was spherical, and started collapsing afterwards to a flat disk. The first stars, formed out of a mixture of hydrogen and helium, $Z \sim 0$, had a more or less spherical distribution about the galactic center. It is believed that these stars formed in globular clusters. They retained their spherical distribution about the center of the

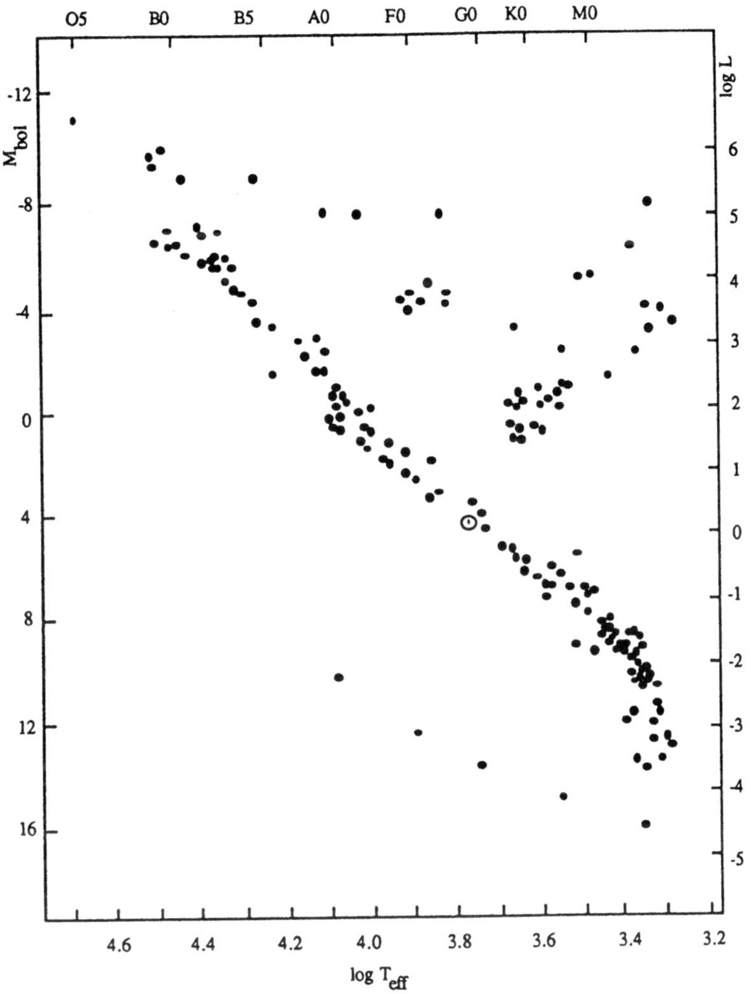

Fig. 1.6. The Hertzsprung-Russell diagram : M_{bol}, and log L versus log T_{eff} and the spectral type

Galaxy. The oldest objects of our Galaxy are thus these metal poor stars, spherically distributed, called population II stars.

During stelllar evolution part of the stellar material may be restituted to the interstellar medium, by stellar wind mass loss from massive stars during the hydrogen burning phase, by lower mass stars during advanced phases.

Massive stars undergo a supernova explosion where most of the stellar matter is blown away, and only a minor part, if any, of the star is left as a neutron star or a black hole. In this way the interstellar matter is enriched in heavy elements. Out of this material new stars are formed, which evolve, and on their turn contribute to the enrichment of the insterstellar medium. In this way the Z-value increased gradually from very low values ≈ 0 to 0.02 or 0.03, by the nuclear burning in successive generations of stars.

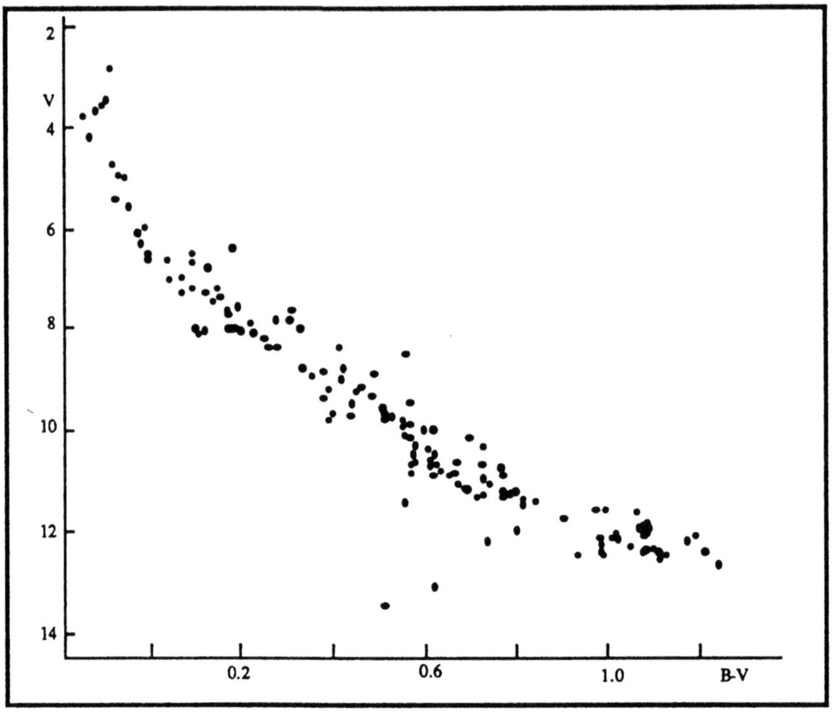

Fig. 1.7. The colour-magnitude diagram of the young open cluster, the Pleiades (after Mitchell and Johnson, 1957)

The youngest objects were formed recently out of material within the galactic disk. Hence these stars must have a larger metallicity. These stars are population I stars. A typical star cluster consisting of this type of stars is called an open or galactic cluster.

About 2000 of these open clusters are known, each containing some 100 to 1000 stars, showing no concentration to the center of the cluster. A special type of open cluster containing the most luminous 0- and B stars of the main sequence, is called an association. They can be detected by the fact that they have not a random distribution at the sky, but that they have a tendency to be in the vicinity of other 0 stars. In this way they form a group, together with B stars and Wolf-Rayet stars, covering an area with a dimension of the order of 100 pc. It is generally believed that these stars share a common origin. All the stars of open clusters are at the same distance, which means that for making the HRD or the CMD of a cluster we only need the apparent bolometric magnitude (m_v+ BC) ; the diagram found in this way has the same form as the diagram of the absolute magnitudes ; it is only shifted down over a distance m_v - M_v. The most remote open clusters known until now are at a distance of 5 kpc. Open clusters and associations are found to be associated with clouds of interstellar gas. From the expansion rate and the actual size of the cluster its age can be derived. Some clusters are younger than 10^8 yrs,

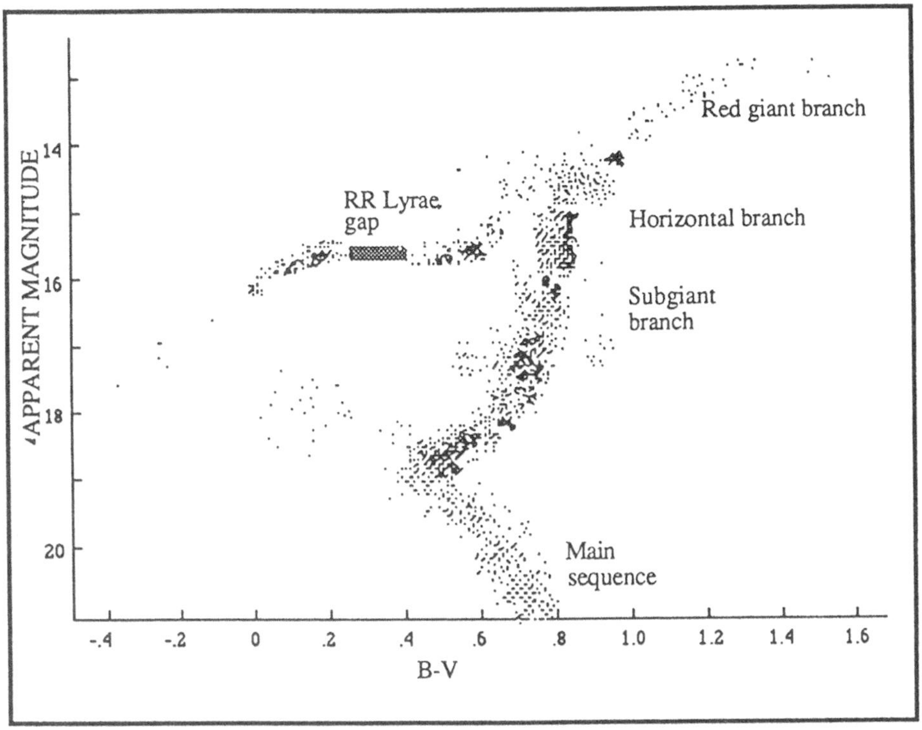

Fig. 1.8. The color-magnitude diagram of the old globular cluster M3 (after Johnson and Sandage, 1956)

some associations are not older than a few million years. Figure 1.7 shows the CMD of the young open cluster the Pleiades.

The main characteristics of CMD of open clusters are a long main sequence extending at the upper end to domain of the blue stars, with B - V in the range - 0.4 to 0. The lower end is limited by the observational possibilities. The main sequence may continue much farther to the bottom than shown in the diagram. Open clusters contain generally a few red giants or supergiants, of the same magnitude as the top of the main sequence. Clusters with a short main sequence, i.e. not rising beyond B - V = + 0.4, contain a well populated red giant branch.

Globular clusters contain 10^5 to 10^7 stars. They have a spherical form, their size is 5 to 30 pc. The mean stellar density is 100 to 1000 stars per cubic parsec. Figure 1.8 shows the CMD of the globular cluster M3. The main characteristics are that the main sequence is short (extending to B - V $\sim$ + 0.4, the red giant branch is well occupied, and that a horizontal branch is present, B - V $\sim$ 0 to $\sim$ 0.3. In the horizontal branch a gap is present, at B - V $\sim$ 0.3, called the Hertzsprung gap. In this gap we find the RR Lyrae variables, with M_v from 0.5 to 1.2, hence also the horizontal branch has this magnitude.

Blue stars, O, B and A stars, are found mainly in the galactic plane, in the spiral arms while red stars have a much loose spatial distribution. This effect was discovered by Baade in 1944. He distinguished two stellar populations : Population I stars consist mainly of O, B and A stars, concentrated in the galactic plane, and associated with gas and dust regions. Typical population I objects are the open clusters. Population II stars are cool stars, outside the plane, and outside gas and dust regions. Typical population II objects are the globular clusters. The population II stars are metal poor stars. The metallicity Z ranges between 0.0003 and 0.003. Population II stars were found to be much older than population I stars. It turned out that in between the pop. I and pop. II also an intermediate population may be defined.

1.6 Associations

In the early fifties was derived from observations (Ambartsumian,1949; Blaauw, 1952) that O and B stars are always found in dilute groups of some 50 members. It is only by measuring radial velocities and proper motions, that can be seen that these stars belong together; the groups are not conspicuous against the background stars. The elements of the groups are drifting away from one or more centers. The stars of these groups belong mechanically to each other, they are unlike clusters, not gravitationally bound. Open and galactic clusters occupy small areas at the sky : an open cluster has typically a diameter of 3 pc, a globular cluster 60 pc, the dimension of an association is larger than 100 pc. From the expension velocity and the direction of motion of the stars the time lapse, required for the stars to move from the center of the association to their actual position, can be computed. The derived times are found to be of the order of 4 to 30 million years.

Also an other type of associations exists, the T-associations. T Tauri stars are also found in nebular regions, hence T Tauri stars can also form associations; they are called T-associations. Observations reveal that each OB association contains T Tauri stars, but that not every T association contains OB stars. In the HRD they are located at the right side of the main sequence. Probably they are the progenitors of stars of later spectral type than OB at the main sequence.

1.7 Age determination of clusters

All stars of a given stellar cluster are at the same distance. The knowledge of this distance allows the derivation of the absolute visual magnitude M_v , and the bolometric magnitude M_b from the apparent visual magnitude m_v.

The lifetime of a star is determined by two factors : its total energy fuel content, and the amount of energy emitted per unit of time, i.e. its mass and its luminosity. The total lifefime of a star t^* can be scaled to that of the sun, which is estimated at 10^{10} years.

$t_* = t_\odot \times M/L = 10^{10} \times M/L$

Using as approximate mass-luminosity relation

$L = M^{10/3}$

we can express the stellar lifetime as a function of the stellar mass M

$t_* = 10^{10} \times M^{-7/3}$

$\log t_* = 10 - 7/3 \log M$

An estimate of stellar lifetimes is given in Table 1.3. The table gives the relation: spectral type, mass (as given in table 1.2.) and the age in years. Starting from the assumption that all stars of a given cluster were formed at the same moment - (which is not so sure) it is possible to derive the age of the cluster. In a forming condensation, stars of all masses are produced, covering the whole range from the smallest stars to the

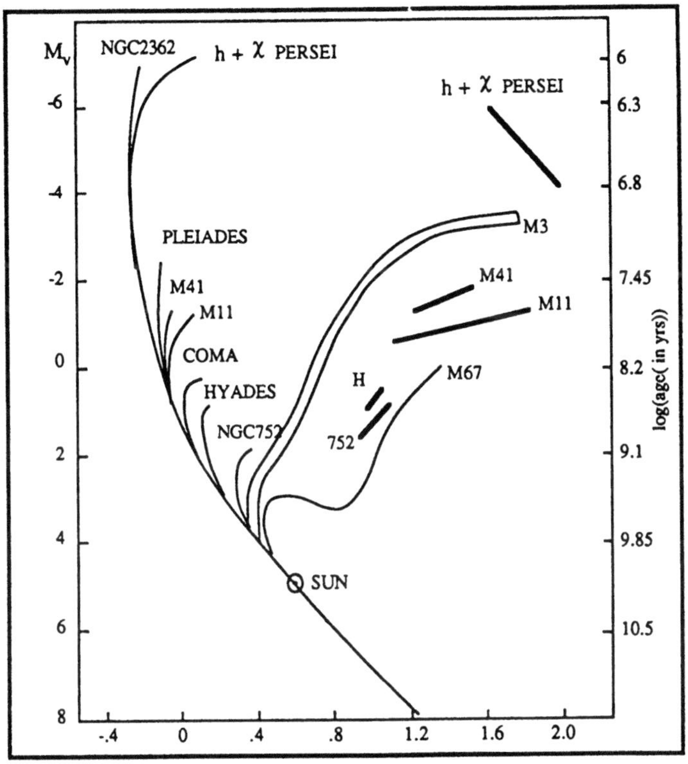

Fig. 1.9. A composite HR diagram for a number of galactic clusters, showing an overlay of several individual diagrams. The further down the turnoff points at the main sequence the older the cluster. Also a globular cluster (M3) is included for comparison. The tracks are interrupted in the middle part of the diagram; this is the Hertzsprung gap, where relatively few stars are found, indicating that in this region the stars evolve rapidly. At the right axis an age schale is added, allowing an estimate of the ages of the various clusters. h and χ Persei are the youngest objects shown, M67 the oldest open cluster (after Sandage, 1957).

most massive ones. Hence at the moment of formation, all spectral types of the main sequence, ranging from class M (low mass stars) to class 0 (50 - $80M_{\odot}$) were present. The evolutionary times increase with decreasing mass; for the most massive stars, the lifetimes are of the order of million years, for low mass stars they are of the order of billion years. Hence during the evolution of the cluster first the 0 stars will disappear after 10^6- 10^7 years, later on the B stars, than the A stars (7.5 10^8 yrs), the F stars (2 - 5 10^9 yrs) ....

This means that the main sequence becomes shorter during the evolution of the cluster since the stars leave the main sequence and evolve towards the right part of the HRD. The more massive stars finish their main sequence stage early, and they leave the main sequence before the lower mass stars. Hence the age of the cluster is determined by the

Table 1.3. Approximate lifetimes of stars in years (t_{max}) of different masses

	M in $M_{\odot}$	t_{max}
O5	50	10^6
B0	18	10^7
B5	6	$1.2\ 10^8$
A0	3.2	$6.5\ 10^8$
A5	2.4	$1.3\ 10^9$
F0	1.7	$2.8\ 10^9$
F5	1.4	$5.3\ 10^9$
F8	1.2	$6.5\ 10^9$
G2	1	10^{10}
G0	0.9	$1.15\ 10^{10}$
K0	0.78	$1.8\ 10^{10}$
M0	0.42	$8\ 10^{10}$
M5	0.27	$2\ 10^{10}$

age of the most massive star at the main sequence. A cluster containing as most massive stars, stars of 18 Mo, hence of spectral type B0, has an age of 10 million years, as may be read from table 1.3 The ages of clusters may be derived from a composite HRD as shown in Figure 1.9. At the right axis, the ages corresponding with the luminosity of the stars at the top of the main sequence are indicated. The figure reveals that the open cluster h and χ Persei has an age of $\approx 1.510^7$ yrs, the Pleiades $\approx 6.10^7$ yrs, the Hyades $\approx 8.10^8$ yrs, M 67 $\approx 6.10^9$ yrs. It should be noted that the diagram of Figure 1.9 is not suited for the age determination of globular clusters, since for these groups of population II-stars the chemical composition, hence the evolution and the lifetimes, are different from that of population I stars. Population II stars of a given spectral type have a slightly smaller mass than pop I stars of the same type. However, as a rough guess the age of globular clusters is $\approx 10^{10}$ years; they are the oldest objects of an Galaxy.

2 The equations of stellar structure

The physical state of the stellar interior depends on ***position*** and on ***time***: the star has a spatial structure at any time, but also has a (generally slow) time evolution. Presently, we are mainly concerned with the space dependence, and we assume that all time dependence is very small. We consider the star as spherically symmetric, not deformed by rotation; furthermore we neglect possible magnetic fields. In this case, the structure can be described by four physical principles:

- the equation of motion of the mass layers
- conservation of mass
- conservation of energy
- the energy transport mechanism.

These principles yield five differential equations describing the stellar structure.

The parameters used to describe the physical conditions inside stars:

- r: for spherical models all parameters depend only on the distance r from the star's center. The total radius of the star is denoted by R. As basic unit the radius of the sun, $R_\odot$, is used

- ρ: the density of the stellar matter in g/cm^3;

- M_r: the total mass, contained in a sphere with radius r. We will use the mass of the sun, $M_\odot$, as basic unit. In fact, M_r is a density parameter, since from the definition of Mr, we have

$$M_r = \int_0^r 4\pi r'^2 \rho \, dr' \tag{2.1}$$

The total stellar mass, M_R, is denoted by M;

- T: the temperature of the stellar material;

- P: the total pressure in the stellar interior.

The pressure, exerted by the particles and the photons of the stellar plasma, is related to the density of the material, its temperature and its chemical composition. This relation is the equation of state (Section 3);

- L_r: The parameter L_r denotes the amountof energy passing across a sphere with radius r. The energy at the stellar surface, L_R, is denoted by L. We will use the total luminosity of the sun, $L_\odot$, as basic unit.

We analyse the physical principles, governing the structure of a star, and derive the structure equations as functions of the distance r. Afterwards we will transform them into coordinates, more suitable for the computation of models.

We can use the distance r from the centre to the surface as the spatial coordinate. For the evolution in time we have to know also the dependence of the stellar parameters on the time t. If r and t are considered as independent variables we call the treatment Eulerian (in the sense of classical hydrodynamics). The space coordinates are purely geometric and e.g. $\partial x/\partial t$ and dx/dt are identically 0. To this has to be added a space-time description for the field of the involved variables, e.g. the velocity field v=v(r,t). We define the function M(r,t) as the mass contained in a sphere with radius r at time t. The variation of M with respect to r and t is then

$$dM = 4\pi r^2 \rho dr - 4\pi r^2 \rho v dt$$

The meaning of the two terms in the right hand side is the following:

first term : the mass inside a spherical shell of thickness dr; it gives the variation of M(r,t) due to a variation of r at constant t, hence

$$\frac{\partial m}{\partial r} = 4\pi r^2 \rho$$

second term: the spherically symmetric mass outflow of the sphere with radius r is constant owing to a radial velocity v in the time lapse dt, hence

$$\frac{\partial m}{\partial t} = 4\pi r^2 \rho v$$

For each of the partial derivatives the other independent variable has to be considered as constant.

2.2 Conservation of mass

From the definition (2.1) of M_r, we obtain by differentiation:

$$\frac{dM_r}{dr} = 4\pi r^2 \rho \tag{2.2}$$

We can derive the equation for the velocity of a layer at M_r from this equation.The velocity of the layer is $v = \partial r/\partial t$, hence, using (2.2):

$$v = \frac{dr}{dt} = \int_0^{M_r} \frac{\partial}{\partial t}\left(\frac{1}{\rho r^2}\right)\frac{dM_r}{4\pi} = \frac{-1}{4\pi}\int_0^{M_r} \frac{1}{r^4\rho^2}\left(2\rho r\frac{\partial r}{\partial t} + r^2\frac{\partial r}{\partial t}\right) dM_r \tag{2.3}$$

$$\frac{dv}{dM_r} = \frac{-1}{4\pi r^4 \rho^2}\left(2\rho r v + r^2\frac{\partial \rho}{\partial t}\right)$$

$$\frac{dv}{dM_r} = \frac{-dr}{dM_r}\left(2\frac{v}{r} + \frac{\partial ln\rho}{\partial t}\right)$$

By derivation we find

$$\frac{d\frac{v}{r}}{dM_r} = \frac{1}{r}\frac{dv}{dM_r} - \frac{v}{r^2}\frac{dr}{dM_r}$$

Applying this to the previous equation we obtain

$$\frac{d\frac{v}{r}}{dM_r} = \frac{-dr}{dM_r}\frac{1}{r}\left(2\frac{v}{r} + \frac{\partial ln\rho}{\partial t}\right) - \frac{dr}{dM_r}\frac{1}{r}\frac{v}{r}$$

or

$$\frac{d\frac{v}{r}}{dM_r} = -\frac{dr}{dM_r}\frac{1}{r}\left[3\frac{v}{r} + \left(\frac{\partial ln\rho}{\partial t}\right)_{M_r}\right] \tag{2.4}$$

This is a special form of the continuity equation of hydrodynamics: variations in the velocity of the layers are caused by the changing density inside the star.

2.3 The equation of motion

Equation (2.4) describes only the relation between the velocities of the matter inside the star and the variation of the density. We have to determine what forces act on an element of matter inside a star. We consider a small cylinder of stellar matter with a section dS, a height dr, and situated at a distance r from the center (Figure 2.1). Its motion (inside the star) is governed by Newton's laws. Since we only consider radial motions, we may write

$$F = m\frac{\partial v}{\partial t} \quad \text{with} \quad m = \rho\, dS\, dr \tag{2.5}$$

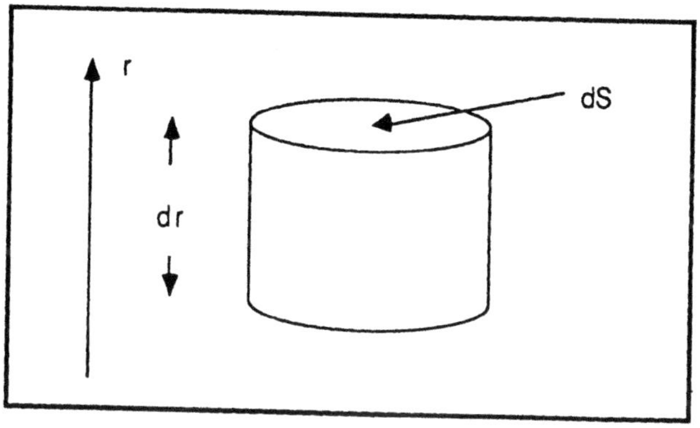

Fig. 2.1.

where F is the external force, exerted on the cylinder, and m is its mass. Now there are two external forces: the gravitation and the pressure. The gravitational force on the cylinder is

$$F_g = g\, m = G\frac{M_r}{r^2}m \tag{2.6}$$

The pressure force is the difference between the pressure on the bottom surface and the top surface of the cylinder:

$$\begin{aligned} F_p &= P(r)dS - P(r+dr)dS \\ &= -\frac{dP}{dr}dS\, dr \end{aligned} \tag{2.7}$$

Combination of (2.5) and (2.7) leads to

$$\frac{dP}{dr} = -G\frac{M_r\rho}{r^2} - \rho r\left[\left(\frac{\partial \frac{v}{r}}{\partial t}\right)_{M_r} + \left(\frac{v}{r}\right)^2\right] \tag{2.8}$$

In many circumstances of normal stellar evolution, the velocities of the layers are very small, and the second term in (2.8) is so small that it may be neglected; the star is in hydrostatic equilibrium. In advanced burning stages however that the 'hydrodynamical' terms become important. They have also to be included for phases where the stellar parameters change very rapidly, e.g. for pulsations.

2.4 Conservation of energy

L_r is the total amount of energy leaving a sphere with radius r per unit of time. For thermal equilibrium is reqiuired that the amount of energy, leaving the sphere equals the amount of energy, produced in the whole sphere,.
Denoting by ϵ the net energy production per unit of mass and per unit of time, and denoting by Q the heat content of the stellar matter, per gram, we can write the equilibrium condition as:

$$L_r = \int_0^r 4\pi r'^2 \rho \left(\varepsilon - \frac{dQ}{dt} + v \cdot g \right) dr' \tag{2.9}$$

with v the velocity of the layer and g the local gravity. Since dQ = T dS (S is the entropy) we can also write:

$$\frac{dL_r}{dr} = 4\pi r^2 \rho \left[\varepsilon - T \left(\frac{dS}{dt} \right)_{M_r} - \frac{v}{r} \frac{GM_r}{r} \right] \tag{2.10}$$

The first of the three terms of the right hand side represents the net production of internal energy ϵ. The second term is the work exerted by the layers. For expanding layers this term is negative since energy is needed. In the case of contraction, this term is positive, and provides the star with an extra source of energy. However, in stable phases of stellar evolution this term is negligibly small. It will only become important if the production of internal energy vanishes. The last term is the work done against gravity.

2.5 Transport of energy

Energy can be transported across a star in various ways: by radiation, by conduction and by convection. Although radiation and conduction are two separate transfer mechanisms, they may be described by the same type of equation, which we will call the equation of radiative transfer. If convection is present, a different transfer equation has to be used, based on the fact that a convective region is nearly adiabatic, i.e. that there is no exchange of heat between the convective cells and the surrounding medium.

2.5.1 Radiative energy transport

The fact itself that stars radiate implies that their interior must be hotter than their surface. The rate of energy transport is determined by the temperature gradient inside the star, and by the 'transparency' of the stellar material. Only these two factors determine the flow of radiation.

Each point inside the star is surrounded by particles, moving at large velocities, and by radiation. Continuously, it sees photons, emitted by atoms, travelling only a few centimeters, and then being absorbed by other atoms. Light seems to reach the point from all directions with the same intensity: it gives the impression that the radiation field is isotropic. We can easily estimate the degree of anisotropy of the radiation field. Inside the star, we can consider the radiation, emitted by a small amount of matter as black body radiation. The emitted flux is :

$$H = \sigma T^4 \tag{2.11}$$

with σ the constant of Stefan-Boltzmann. Consider now two small amounts of matter, one of them at a distance l further away from the center, and at a temperature T - dT. There will be a small difference in the emitted radiation flux, dH, causing anisotropy in the radiation field. Since dT<<T, we have:

$$\frac{dH}{H} = 4\frac{dT}{T} = 4\frac{1}{T}\frac{dT}{dr}l \tag{2.12}$$

It can be easily verified that the typical temperature gradient inside a star is approximately 2.2 10^{-4} K/cm (dT/dr $=T_c R_\odot$-in the solar case: T_c=1.5 10^7K; R=7 10^{10}), so dH/H=1.2$10^{-10}$ l(cm). We find the degree of anisotropy by inserting for l the mean free path of the photons, i.e. the distance they can travel before they are absorbed by an atom. The study of the absorption of radiation in stars reveals that the mean free path of photons is of the order of centimeters. Hence the deviation from isotropy is extremely small. However, it is exactly this small deviation from isotropy that causes the outward radiation flux: since outside, where it is a little cooler, a little less radiation is emitted, a net outward flux originates, transporting energy towards the surface.

In order to derive an exact expression for the radiative energy transport, we have to look at the radiation field in more detail. The radiation field may be described by a function I(r,θ) giving the radiation intensity (the energy per unit surface per unit solid angle) at a distance r, inclined to r by an angle θ. We may now determine the radiation balance within a solid angle ω, in a small cylinder with length dl and section dS, inside the star (Figure 2.2).

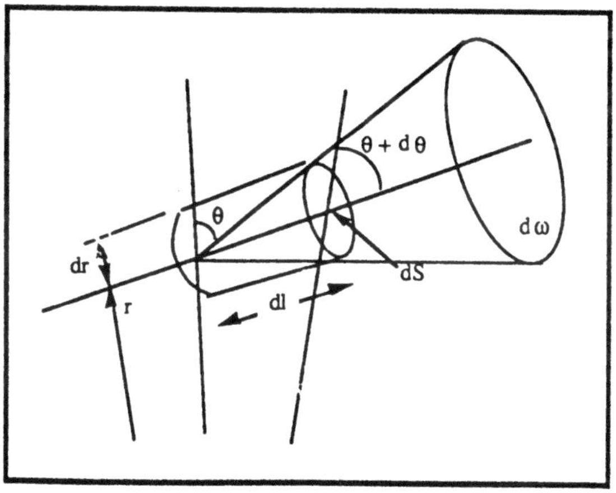

Fig. 2.2. The radiation balance in a cylinder dl,ds, within a solid angle ω

gain in radiation at the bottom surface:

$$I(r,\theta)d\omega\ dS \tag{2.13}$$

loss at the top:

$$-I(r+dr,\theta+d\theta)d\omega\ dS \tag{2.14}$$

dr = dl cos θ and r dθ = - dl sin θ, hence the sum of (2.13) and (2.14) is

$$\left(-\frac{\partial I}{\partial r}+\frac{\partial I}{\partial \theta}\frac{\tan\theta}{r}\right) dr\ d\omega\ dS \tag{2.15}$$

Inside the cylinder, a fraction $\kappa\rho$dl of the radiation is absorbed by the matter, with κ the absorption coefficient per gram.

The dimension of κ is 1/cm/g since $\kappa\rho$ dl must be dimensionless. The total loss of radiation by absorption thus amounts to I $d\omega dS\kappa\rho$ dl. However, there is a gain in radiation in the cylinder by the emission of radiation by an amount j, isotropic in all directions. Within the solid angle $d\omega$ this represents a gain in radiation of

$$j\ \rho\ dS\ dl\ \frac{d\omega}{4\pi} \tag{2.16}$$

From the combination of (2.15) and (2.16) the equation of radiative transfer follows:

$$\frac{\partial I}{\partial r}\cos\theta-\frac{\partial I}{\partial \theta}\frac{\sin\theta}{r}+\kappa\rho I-\frac{j\rho}{4\pi}=0 \tag{2.17}$$

The second term in (2.17) may be neglected since in stellar interiors the radiation field is nearly isotropic. Multiplication of this equation by cos θ/ c and integration over all solid angles gives:

$$\frac{\partial}{\partial \rho}\left(\frac{1}{c}\int I\cos^2\theta\ d\omega\right)+\frac{\kappa\rho}{c}\int I\ \cos\theta\ d\omega=0 \tag{2.18}$$

In this equation one recognizes the radiation pressure

$$P_{\mathrm{rad}}=-\int I\cos^2\theta\frac{d\theta}{c}$$

and the radiation flux F_{rad} (the energy per unit time per unit of surface):

$$F_{\mathrm{rad}}=\int I\cos\theta d\theta$$

$$\frac{dP_{\mathrm{rad}}}{dr}+\frac{\kappa\rho}{c}F_{\mathrm{rad}}=0 \tag{2.19}$$

In thermodynamic equilibrium the radiation pressure is given by (see Section 3):

$$P_{\mathrm{rad}}=\frac{1}{3}aT^4 \tag{2.20}$$

The radiation flux F_{rad} is related to L_r by

$$L_r=4\pi r^2F_{\mathrm{rad}}$$

so that finally equation (2.19) may be transformed into

$$\frac{1}{3}a\frac{dT^4}{dr}+\frac{\kappa\rho}{c}\frac{L_r}{4\pi r^2}=0 \tag{2.21}$$

Since L_r is known from the equation of energy conservation, we may convert (2.21) into an equation for the temperature gradient:

$$\frac{dT}{dr}=-\frac{3}{16\pi ac}\frac{\kappa\rho}{T^3}\frac{L_r}{r^2} \tag{2.22}$$

We recognize the physical principles of radiative transport in this equation: if other quantities remaining the same, L_r becomes larger, the temperature gradient becomes larger; if the stellar material is more opaque (κ larger), a larger temperature gradient is required to transport the same amount of energy (i.e. at constant L_r).

2.5.2 Convective energy transport

If in the medium inside a star convective motions occur, the temperature gradient of the medium is largely influenced by the temperature gradient inside the convection cells. It is generally assumed that convective cells during their motion, do not exchange heat with their surroundings: they are adiabatic. We may write the temperature gradient of the medium conveniently as

$$\frac{dT}{dr}=\nabla'_c\frac{T}{P}\frac{dP}{dr} \tag{2.23}$$

where

$$\nabla'_c=\left(\frac{dlnT}{dlnP}\right)_{\text{convective}} \tag{2.24}$$

is a thermodynamic quantity which is in most cases near the adiabatic value.

2.6 Transformations of the equations

For the spherical symmetric case it turns out that the use of a Lagrangian coordinate instead of r is more adequate. In that case the spatial coordinate of a given mass element does not vary in time.

As coordinate can be chosen the mass M_r. To any mass element the value M_r is assigned, i.e. the mass contained in a concentric sphere at the moment t_o.

The new independent variables are then M_r and t; all other variables depend on those two. At the centre we have M_r=0, and at the surface M_r=M, the total mass of the star.

We can easily transform the partial derivatives from the case (r,t) to the case (m,t). For any function depending on two variables r,t, where one of the two , e.g. r is replaced by a new variable m, the partial derivatives with respect to the new variables are given by

$$\frac{\partial}{\partial m}=\frac{\partial}{\partial r}.\frac{\partial r}{\partial m}$$

$$\left(\frac{\partial}{\partial t}\right)_m = \frac{\partial}{\partial r}\cdot\left(\frac{\partial r}{\partial t}\right)_m + \left(\frac{\partial}{\partial t}\right)_r$$

Applying this to equation 2.2 gives:

$$\frac{\partial r}{\partial m} = \frac{1}{4\pi r^2 \rho}$$

This equation replaces 2.2 of the Eulerian description in the Lagrangian description. By substituting this into the general transformation equation we find a transformation rule for the equations (2.4), (2.8), (2.10) and (2.21)-(2.22):

$$\frac{\partial}{\partial m} = \frac{1}{4\pi r^2 \rho}\frac{\partial}{\partial r}$$

2.6.1 The relative mass coordinate

We transform the structure equations (2.1), (2.4), (2.8), (2.10) and (2.21)-(2.22) into a more suitable form. We define a new independent relative mass parameter μ as

$$\mu = 1 - \frac{M_r}{M} \tag{2.25}$$

with M the total mass of the star. The use of this parameter, - 0 at the surface and 1 at the center-, will avoid difficulties with the numerical integration of the atmosphere: near the surface, we are forced to take extremely small integration steps, because insufficient accuracy of M_r will cause problems. For any function ψ of space and

$$D\Psi \equiv \left(\frac{d\Psi}{dt}\right)_{M_r} = (1-\mu)\frac{\dot{M}}{M}\left(\frac{d\Psi}{d\mu}\right)_t + \left(\frac{d\Psi}{dt}\right)_\mu \tag{2.26}$$

where

$$\dot{M} = \frac{dM}{dt} \tag{2.27}$$

is the mass loss rate (negative if mass is lost, positive if mass is gained). Remark that D• is a linear operator. Using (2.25) and (2.26), we can transform the structure equations into

$$\frac{dr}{d\mu} = -\frac{M}{4\pi r^2 \rho} \tag{2.28a}$$

From (2.2) we find

$$r = \int_0^{M_r} \frac{dM_r}{4\pi r^2 \rho}$$

The velocity of a layer at M_r is then

$$v = \frac{\partial r}{\partial t} = \int_0^{M_r} \frac{\partial}{\partial t}\left(\frac{1}{r^2\rho}\right)\frac{dM_r}{4\pi} = \frac{-1}{4\pi}\int_0^{M_r} \frac{1}{r^4\rho^2}\left(2\rho r\frac{\partial r}{\partial t} + r^2\frac{\partial \rho}{\partial t}\right) dM_r$$

$$\frac{dv}{dM_r} = \frac{-1}{4\pi r^4 \rho^2} \left(2\rho r v + r^2 \frac{\partial \rho}{\partial t}\right)$$

$$\frac{d\frac{v}{r}}{dM_r} = \frac{1}{r}\frac{dv}{dM_r} - \frac{v}{r^2}\frac{dr}{dM_r}$$

$$\frac{d\frac{v}{r}}{d\mu} = -\frac{dr}{d\mu}\frac{1}{r}\left(3\frac{v}{r} + Dln\rho\right) \tag{2.28b}$$

$$\frac{dL_r}{d\mu} = -M\left(\varepsilon - TDS - \frac{v}{r}\frac{GM(1-\mu)}{r}\right) \tag{2.28c}$$

$$\frac{dP}{d\mu} = G\frac{M^2(1-\mu)}{4\pi r^4} + \frac{M}{4\pi r}\left[\left(\frac{v}{r}\right)^2 + D\frac{v}{r}\right] \tag{2.28d}$$

$$\frac{dT}{d\mu} = \frac{3}{64\pi^2 ac}\frac{\kappa M}{T^3}\frac{L_r}{r^4}(\text{radiative}) \tag{2.28e}$$

$$\frac{dT}{d\mu} = \nabla_c' \frac{T}{P}\frac{dP}{d\mu}(\text{convective}) \tag{2.28f}$$

2.6.2 The use of any thermodynamic quantity

It is useful in computations of stellar structure to use ***another thermodynamic quantity***, say α, ***instead of the pressure***. Depending on the situation this quantity could be e.g. the density, the electron density or the so-called 'degeneracy parameter'. The reason for introducing such a quantity is purely technical. If we retain the pressure, we have to compute the density from the pressure and the temperature. In some circumstances, such as e.g. when the gas is partially ionized, this may be cumbersome . In this case, it would be more appropriate to take the electron density as fundamental parameter, since it occurs directly in the Saha equation. If the pressure or the density were chosen, iterations on the Saha equations are needed to obtain an electron density, consistent with the pressure and the ionization equilibrium. We will not specify what quantity α will represent, but suppose that the pressure only depends on α and T (and, of course, on the chemical composition). The general relation between P, T and α can be expressed as

$$\frac{dP}{d\mu} = \frac{\partial P}{\partial \alpha}\frac{d\alpha}{d\mu} + \frac{\partial P}{\partial T}\frac{dT}{d\mu} + \sum_j \frac{\partial P}{\partial X_j}\frac{dX_j}{d\mu} \tag{2.29}$$

with X_j the composition parameters.

Using equations (2.28) and (2.29) the equation for $d\alpha/d\mu$ can be derived. In 2.28f the substitution

$$\frac{dT}{d\mu} = \nabla_c T \frac{d\alpha}{d\mu} \tag{2.30}$$

with

$$\nabla_c = \left(\frac{dlnT}{d\alpha}\right)_{\text{convective}} \tag{2.31}$$

has to be applied.

Elimination of $d\alpha/d\mu$ from equation (2.30), using the equation for $d\alpha/d\mu$ derived before leads then to the following full set of equations :

$$\frac{dr}{d\mu} = -\frac{M}{4\pi r^2 \rho} \tag{2.32a}$$

$$\frac{d\frac{v}{r}}{d\mu} = -\frac{dr}{d\mu}\frac{1}{r}\left(3\frac{v}{r} + Dln\rho\right) \tag{2.32b}$$

$$\frac{dL_r}{d\mu} = -M\left(\varepsilon - TDS - \frac{v}{r}\frac{GM(1-\mu)}{r}\right) \tag{2.32c}$$

$$\frac{d\alpha}{d\mu} = \frac{G\frac{M^2(1-\mu)}{4\pi r^4} + \frac{M}{4\pi r}\left(\frac{v^2}{r^2} + D\frac{v}{r}\right)\frac{P}{T}\frac{dT}{d\mu} - \sum_j \frac{\partial P}{\partial X_j}\frac{dX_j}{d\mu}}{\partial P \partial \alpha} \tag{2.32d}$$

$$\frac{d\alpha}{d\mu} = \frac{G\frac{M^2(1-\mu)}{4\pi r^4} + \frac{M}{4\pi r}\left(\frac{v^2}{r^2} + D\frac{v}{r}\right)\frac{P}{T}\frac{dT}{d\mu} - \sum_j \frac{\partial P}{\partial X_j}\frac{dX_j}{d\mu}}{\partial P \partial \alpha} \qquad (2.32d)$$

$$\frac{dT}{d\mu} = \frac{3}{64\pi^2 ac}\frac{\kappa M}{T^3}\frac{L_r}{r^4}(\text{radiative}) \tag{2.32e}$$

$$\frac{dT}{d\mu} = \nabla_c T \frac{G\frac{M^2(1-\mu)}{4\pi r^4} + \frac{M}{4\pi r}\left(\frac{v^2}{r^2} + D\frac{v}{r}\right) - \sum_j \frac{\partial P}{\partial X_j}\frac{dX_j}{d\mu}}{\frac{\partial P}{\partial \alpha} + T\nabla_c \frac{\partial P}{\partial T}}(\text{convective}) \tag{2.32f}$$

Finally for practical computations, it may be more suitable to use ln r instead of r, L_r/L_m where L_m is a suitable normalization, ln T instead of T. This will ensure a more stable integration since the variation of the logarithms throughout the star is much smaller than the real parameters .

2.7 The boundary conditions

In the center (μ=1) we have the trivial boundary conditions r(μ=1)=0 and L_r (μ=1)=0. In a small area near the center we can consider the density to be constant and equal to the central density ρ_c. We then have:

$$M_r = \frac{4}{3}\pi r^3 \rho_c \tag{2.33}$$

giving immediately

$$lnr = \frac{1}{3}ln\left[\frac{3M}{4\pi}(1-\mu)\right] - \frac{1}{3}ln\rho_c \tag{2.34}$$

so we have for v/r, in (and near) the center :

$$\frac{v}{r} = \frac{dlnr}{dt} = -\frac{1}{3}\frac{\partial ln\rho_c}{\partial t} = -\frac{1}{3}Dln\rho_c \tag{2.35}$$

At the very outer boundary of the star we can use the Eddington approximation for the temperature. This gives

$$T^4 = \frac{L}{8\pi\sigma r^2} \tag{2.36}$$

with L the total luminosity of the star.

The outer boundary condition for α depends on the exact meaning of α. An appropriate rule is to determine α at the boundary both for the temperature, given by (2.41), and for a density of 10^{-12} g cm^{-3}. Summarizing, the boundary conditions for the five structure equations are:

$$r(\mu = 1) = 0 \tag{2.37a}$$

$$\frac{v}{r}(\mu = 1) = -\frac{1}{3}Dln\rho_c \tag{2.37b}$$

$$L_r(\mu = 1) = 0 \tag{2.37c}$$

$$\alpha(\mu = 0) = \alpha(T(\mu = 0), \rho = 10^{-12} g\ cm^{-3}) \tag{2.37d}$$

$$T(\mu = 0) = \left(\frac{L}{8\eta\sigma r^2}\right)^{\frac{1}{4}} \tag{2.37e}$$

2.8 Development near the center

In the center of the star r = 0. This yields an infinite value for $dr/d\alpha$ and poses computational problems for the other derivatives. Therefore the parameters near the center ($\mu = 1$) have to be developed in series. The integration of the stellar structure can then start at a point, near the center. The relation between the parameters at this point and the central conditions is given by the series developments. We see that the value of the radius r near the center is given by (2.33) or (2.34), and that (2.35) gives the velocity near the center. The luminosity near the center can be readily obtained by integration of (2.32c), assuming $\epsilon = \epsilon_c$, $T = T_c$, $DS = DS_c$:

$$L_r = (1-\mu)M(\varepsilon_c - T_c DS_c)r \tag{2.38}$$

For the determination of the temperature near the center we substitute (2.38) into (2.22), giving for the radiative temperature gradient

$$\frac{dT}{dr} = -\frac{1}{4ac}\frac{\kappa_c \rho_c^2}{T_c^3}(\varepsilon_c - T_c DS_c)r \tag{2.39}$$

Integration of (2.39) leads to an expression for T near the center.
If the center is convective we obtain from (2.30)

$$T - T_c = \nabla_c T_c(\alpha - \alpha_c) \tag{2.40}$$

where α_c is the central value of α.

In order to determine α near the center we use the fundamental relation (2.29):

$$dP = \frac{\partial P}{\partial \alpha}d\alpha + \frac{\partial P}{\partial T}dT + \sum_j \frac{\partial P}{\partial X_j}dX_j \tag{2.41}$$

The equation of motion (2.8), developed near the center, gives

$$P = P_c - \frac{2}{3}\pi G \rho_c^2 r^2 \qquad (1-\mu << 1) \tag{2.42}$$

Substitution of (2.42) into (2.41) and taking $dP = P - P_c$, $d\alpha = \alpha - \alpha_c$ and $dT = T - T_c$, we obtain

$$\alpha = \alpha_c - \frac{\frac{2}{3}\pi G\rho_c^2 r^2 - \frac{\partial P}{\partial X_j}(T-T_c) - \sum_j \frac{\partial P}{\partial X_j}(X_j - X_{jc})}{\partial P \partial \alpha_c} \qquad (1-\mu << 1) \tag{2.43}$$

If the center is radiative, $T - T_c$ is given by (2.39). If the center is convective, we have to eliminate $T - T_c$ between (2.43) and (2.40). This leads to

$$\alpha = \alpha_c - \frac{\frac{2}{3}\pi G\rho_c^2 r^2 + \sum_j \frac{\partial P}{\partial X_j}(X_j - X_{jc})}{\frac{\partial P}{\partial \alpha_c} + \mathrm{T}\nabla_c \frac{\partial P}{\partial \mathrm{T}_c}} \qquad (1-\mu << 1) \tag{2.44}$$

Using the central values of the thermodynamic quantities and equations (2.34), (2.35), (2.38), (2.39), (2.40), (2.43), (2.44) we can determine the parameters near the center, avoiding numerical problems.

In principle, the mathematical problem of stellar structure is now completely determined, since we assume the energy generation rate, the opacity and the equation of state to be known functions. The problem consists of a coupled system of five, non-linear differential equations. Obviously, no analytical solutions exist for the complete problem, so we have to solve it numerically. Since the boundary conditions are given at both sides of the integration interval, the problem is not a simple initial value problem but rather a boundary value problem.

3 The equation of state

3.1 Introduction

The equation of state of stellar matter gives the relation between the pressure, the density, the temperature and the chemical composition of the matter. It is always assumed that the environment is in local thermodynamic equilibrium. This ensures that all atomic processes (such as ionization and excitation of atoms) are in equilibrium. This assumption facilitates the computation of the equation of state. It is only in the very outer atmosphere that this condition is not satisfied. However, these layers are in most cases not important for the computation of the overall structure of the stars. The equation of state without local thermodynamic equilibrium has only to be considered if one is e.g. interested in the formation of spectral lines in stellar atmospheres.

A general expression for the equation of state is

$$f(P, \rho, T, X_i) = 0 \tag{3.1}$$

where f symbolizes the relation to befound. The chemical composition is normally expressed by the mass fractions X_i of all elements i: of one gram of stellar matter, X_i gram is of species i; the sum of all mass fractions is one. Another possibility is to adopt the number fractions $Y_i = X_i/A_i$with A_i the atomic weight of species i in a.m.u.. These quantities are related to the number density of species i (N_i: the number of atoms i per unit volume):

$$N_i = \rho N_0 Y_i = \rho N_0 \frac{X_i}{A_i} \tag{3.2}$$

N_o is Avogadro's number.

The gas inside a star consists of a mixture of molecules, atoms and electrons. At temperatures, higher than a few times 10^4 K, the atoms have lost part of their electrons (or all their electrons); they are (partly) ionized. Molecules are only present in the outer layers of the star where temperatures of a few thousand K prevail. They are formed from neutral atoms such as hydrogen (H_2), carbon and oxygen (e.g. CO) and from heavier species (e.g. TiO). The free electrons present between the ions come from the ionization of the atoms, although under extreme circumstances (very high temperatures and very low densities) electrons and positrons may be created spontaneously from radiation. Finally, the atoms and electrons are embedded in an intense radiation field, being very nearly isotropic.

The classical way to compute the equation of state is to evaluate the pressure, given the density and the temperature and the chemical composition. According to the values of the density and the pressure, different regions can be distinguished. In the low-temperature region one has partial ionization. In the high-density region, where the electrons are very close to each other and quantum effects occur, one has to use expressions for the equation of state of degenerate electrons. In the high-temperature region, one has to consider the relativistic motion of the free electrons. Furthermore, in some circumstances, these effects have to be combined in order to obtain precise results. Therefore, the computation of a "complete" equation of state can be complicated.

Although the problem can be avoided by using pre-computed tables it may be worthwhile to compute the equation of state explicitly since the use of these pre-computed results requires still much storage memory, and the interpolation remains demanding, especially in zones of partial ionization.

One can replace each of these by other thermodynamical quantities, as long as they are mutually independent. In practice, we can choose the parameters in such a way that the equation of state is easy to calculate.

The pressure integral can be computed directly, taking into account the momentum distribution of the different particles of our gas, as given by the laws of statistical mechanics. Another fundamental thermodynamic quantity will be chosen, related to the momentum distribution of the electrons. This is the reason why in the derivation of the equations of stellar structure, the pressure has been substituted by "any thermodynamic quantity". In this way, the main physical effects, -ionization, degeneration and relativistic motion-, can be treated simultaneously and consistently.

3.2 The pressure integral

Pressure of gas (or radiation)is caused by the collisions of particles with a surface in the medium. As a result of these collisions, momentum is transferred to the surface, exerting a force per unit surface called pressure. From these mechanical considerations can be derived:

$$P = \int_0^\infty dp \int_0^{\frac{\pi}{2}} d\theta \; \Delta p(p,\theta) \cdot f(p,\theta) \tag{3.3}$$

where Δp (p,θ) is the momentum, transferred by the collision of one particle with momentum p, colliding at an angle θ with the normal, and f(p,θ) is the number of particles with momentum p striking one square centimeter of the surface per second at an angle θ.

Newton's law prescribes

$$\Delta p(p,\theta) = 2p \;\cos\theta \tag{3.4}$$

Let n(p,θ) be the number density of particles having a momentum p and moving at an angle θ. In one second a quantity v(p) n(p,θ) strikes the surface with a velocity v(p) at an angle θ. If the beam has a section of one square centimeter it strikes 1/cosθ cm^2 of the surface. Therefore

$$f(p,\theta)dp\; d\theta = \cos\theta \cdot v(p) \cdot n(p,\theta)dp\; d\theta \tag{3.5}$$

Assuming the velocities as isotropic the fraction of particles moving in a given direction θ is

$$\frac{n(p,\theta)}{n(p)}d\theta = \frac{1}{2}\sin\theta\, d\theta \tag{3.6}$$

where n(p) is the moment distribution of the particles. Combination of (3.3) - (3.6) and integrating over θ, leads to

$$P = \frac{1}{3}\int_0^\infty p\, v(p)n(p)dp \tag{3.7}$$

Hence the pressure integral depends on the relation between velocity and moment v(p), and on the moment distribution n(p) for the considered particles .
The pressure is additive. We will consider three contributions to the total pressure: the pressure of atoms and ions, P_i, the pressure of the electrons, P_e, and the pressure of the radiation, P_r.

3.2.1 The ion pressure

With the exception of the extreme conditions where nuclear matter density is reached, the ions satisfy a Maxwellian moment distribution:

$$n_i(p)dp = \frac{N_i 4\pi p^2}{(2\pi m_i kT)^{\frac{3}{2}}}\exp - \frac{p^2}{2m_i\, kT}dp \tag{3.8}$$

where N_i is the ion density and m is the mass of an ion. Since ions are not relativistic we have v_i=pi/m_i.

3.2.2 The electron pressure

Since electrons are fermions, having a half integral spin, they obey Fermi-Dirac statistics. The momentum distribution is then:

$$n_e(p)dp = \frac{8\pi p^2}{h^3}\left[1 + \exp\left(-\eta + \frac{E}{kT}\right)\right]^{-1} dp \tag{3.9}$$

where E is the kinetic energy of the electron (excluding the rest energy) and η is the so-called degeneracy parameter. The relativistic relation between E and p is

$$E = m_e c^2\left[\sqrt{1 + \left(\frac{p}{m_e c}\right)^2} - 1\right] \tag{3.10}$$

where m_e is the electron mass. The relativistic velocity is given by

$$v(p) = \frac{p}{m_e\sqrt{1 + \left(\frac{p}{m_e c}\right)^2}} \tag{3.11}$$

Substituting (3.9) - (3.11) into (3.7) gives

$$P_e = \frac{16\pi\sqrt{2}}{3h^3} m_e^4 c^5 \beta^{\frac{5}{2}} \int_0^\infty \frac{x^{\frac{3}{2}}\left(1+\frac{1}{2}\beta x\right)^{\frac{3}{2}}}{1+\exp(-\eta+x)} dx \tag{3.12}$$

where

$$\beta = \frac{kT}{m_e c^2} = \frac{T}{5.93 \cdot 10^9 K} \tag{3.13}$$

The electron pressure can be expressed in terms of the so-called Fermi-Dirac integrals

$$F_k(\eta,\beta) = \int_0^\infty \frac{x^k\left(1+\frac{1}{2}\beta x\right)^{\frac{1}{2}}}{1+\exp(-\eta+x)} dx \qquad (k > -1) \tag{3.14}$$

as

$$P_e = \frac{16\pi\sqrt{2}}{3h^3} m_e^4 c^5 \beta^{\frac{5}{2}} \left[F_{\frac{3}{2}}(\eta,\beta) + \frac{\beta}{2} F_{\frac{5}{2}}(\eta,\beta)\right] \tag{3.15}$$

3.2.3 The radiation pressure

Since the moment of a photon is hν/c we can transform (3.7) into

$$P_{rad} = \frac{1}{3}\int_0^\infty h\nu\, n(\nu)\, d\nu \tag{3.16}$$

where n(ν) gives the frequency distribution of the photons. Since photons obey Einstein-Bose statistics, the frequency distribution is given by

$$n(\nu)d\nu = \frac{8\pi\nu^2}{c^3}\left(\exp\frac{h\nu}{kT} - 1\right)^{-1} d\nu \tag{3.17}$$

The radiation pressure is

$$\begin{aligned} P_r &= \frac{1}{3}\frac{8\pi^5 k^4}{15 c^3 h^2} T^4 \\ &= \frac{1}{3} a T^4 \end{aligned} \tag{3.18}$$

with a the Stefan-Boltzmann constant.

3.3 The computation of the equation of state

The equation of state will be calculated as a function of the parameter η, the "degeneracy parameter", the temperature T and the chemical composition. In this respect the degeneracy parameter replaces the density, which is commonly used as input parameter for the equation of state. The Fermi-Dirac integrals and the electron density and pressure can be calculated directly, without the use of tables of these quantities as functions of the total matter density. Since η and T are known, the electron pressure follows immediately from (3.15). The electron density $N_e = \int n_e(p)\, dp$ can be evaluated, using (3.9) and (3.10), as

$$N_e = \frac{8\pi\sqrt{2}}{h^3} m_e^3 c^3 \beta^{\frac{3}{2}} \left[F_{\frac{1}{2}}(\eta,\beta) + \beta F_{\frac{3}{2}}(\eta,\beta) \right] \tag{3.19}$$

From the density of free electrons N_e and the chemical composition we can derive the ion density and the ion pressure. The radiation pressure can be computed from (3.18).

3.3.1 Computation of the Fermi-Dirac integrals

In order to compute the electron density and pressure we have to calculate the Fermi-Dirac integrals $F_{1/2}$, $F_{3/2}$, and $F_{5/2}$.
We must approximate these integrals by appropriate analytic functions, since no general analytic expressions exist. An accurate method for approximating Fermi-Dirac integrals has been given by Eggleton et al. (1973). This approximation assures the correct asymptotical behavior of the integrals.
We define new parameters f and g, given by

$$\eta = 2\sqrt{1+f} + ln\frac{\sqrt{1+f}-1}{\sqrt{1+f}+1} \tag{3.20a}$$

$$g = \beta\sqrt{1+f} \tag{3.20b}$$

A Fermi-Dirac integral can then be approximated as

$$F = f\frac{(1+f)^k}{(1+g)^{4.5}}\sigma(f,g) \tag{3.21a}$$

with

$$\sigma(f,g) = \sum_{i=0}^{5}\sum_{j=0}^{5} E_{ij} f^i g^j \tag{3.21b}$$

The values of k and E_ij for the Fermi-Dirac integrals $F_1/2$, $F_3/2$ and $F_5/2$ are given in Table 3.1. The approximation (3.21) gives the desired Fermi-Dirac integrals with a precision of 0.01asymptotic behavior for very small and very large values of h and b. It can therefore be used under all circumstances. The value of f can be derived easily by Newton-Raphson iteration on equation (3.20a), taking into account that

$$\frac{d\eta}{df} = \frac{1+f}{f} \tag{3.22}$$

3.3.2 The electrons

$$A = \frac{\pi m_e^4 c^5}{3h^3} = 6.002 \cdot 10^{22} \text{dynes cm}^{-2} \tag{3.23a}$$

$$B = \frac{8\pi m_e^3 c^3}{3h^3 N_0} = 9.736 \cdot 10^5 \text{g cm}^{-3} \tag{3.23b}$$

$$\beta = \frac{kT}{m_e c^2} = \frac{T}{5.93 \cdot 10^9 K} \tag{3.23c}$$

Then we have:

$$N_e = 3\sqrt{2}N_0B\ \beta^{\frac{3}{2}}\left[F_{\frac{1}{2}}(\eta,\beta) + F_{\frac{3}{2}}(\eta,\beta)\right]\ \mathrm{cm}^{-3} \tag{3.24}$$

$$\mathrm{P_e} = 16\sqrt{2}\mathrm{A}\beta^{\frac{5}{2}}\left[F_{\frac{3}{2}}(\eta,\beta) + \frac{\beta}{2}F_{\frac{5}{2}}(\eta,\beta)\right]\ \mathrm{dyne\ cm}^{-2} \tag{3.25}$$

Table 3.1. The values of k and Eij in expression (3.21) for the Fermi-Dirac integrals F1/2, F3/2, F5/2.

	$F_{1/2}$	k=-5.25				
j:	0	1	2	3	4	5
i: 0	1.63710	7.96385	15.6594	14.6385	7.15296	1.30621
1	8.35455	41.8681	62.6050	91.7005	3.9250	6.62459
2	17.1100	80.2592	194.782	120.956	76.2612	13.2063
3	17.3033	82.9480	167.599	136.456	77.3268	13.4056
4	9.06674	43.8384	82.1991	81.6230	37.1594	6.82456
5	1.88562	9.04591	17.3355	16.3376	7.75433	1.41421
	$F_{3/2}$	k=-4.75				
j:	0	1	2	3	4	5
i: 0	2.45564	12.5856	25.5680	25.9059	13.0570	2.61243
1	12.0793	65.5606	79.0484	171.473	57.5098	12.5341
2	23.7606	109.787	347.651	128.293	131.356	23.4115
3	23.4927	118.143	218.565	243.864	112.979	22.3412
4	11.4072	56.3159	110.673	108.809	53.1110	10.3111
5	2.26274	10.9893	21.3098	20.5527	9.89668	1.88562
	$F_{5/2}$	k=-4.25				
j:	0	1	2	3	4	5
i:0	6.13911	33.0168	69.2682	73.1489	37.8497	7.83728
1	28.1289	152.536	291.015	343.308	163.896	34.5195
2	51.0794	257.900	596.375	517.350	296.866	58.6793
3	45.3982	232.350	468.267	486.634	241.567	49.0211
4	19.7141	98.7694	197.655	197.191	98.6103	19.6140
5	3.23249	15.8025	30.8619	30.0576	14.6142	2.82843

3.3.3 Computation of the density

We denote by N_e the number density of electrons, coming from the ionization of ions. It is equal to the density of free electrons N_e, eventually corrected for the presence of electron-positron pairs. ν_i is the average number of free-ionization electrons, contributed by one ion of type i. We may then write the identity

$$\sum_i N_i \nu_i = \overline{N}_e \tag{3.26}$$

Substitution of (3.2) into (3.26) gives:

$$\rho = \frac{\overline{N}_e}{N_0} \left(\sum_i \nu_i \frac{X_i}{A_i} \right)^{-1} \equiv \frac{\overline{N}_e}{N_0} \mu_e \tag{3.27}$$

μ_e is called the mean molecular weight per free electron. It is the only composition-dependent parameter in the computation of the matter density. For complete ionization ν_i = Z_i, and

$$\frac{1}{\mu_e} = \sum_i Z_i \frac{X_i}{A_i} \tag{3.28}$$

Z_i/A_i= 1/2 for $Z_i > 1$; we have for *complete* ionization , with $X_H$the mass fraction of hydrogen

$$\mu_e = \frac{2}{1 + X_H} \tag{3.29}$$

In order to evaluate ν_i, the distribution of the ions over the different ionization states is required. Let us denote by N_r the number density of r times ionized ions of type i. The ratio of r+1 times ionized to r times ionized ions is given by the Saha equation:

$$\Psi_r \equiv \frac{N_r + 1}{N_r} = e^{-\eta} \frac{B_r + 1}{B_r} \exp - \frac{L_r}{kT} \qquad (0 \leq r \leq Z_i) \tag{3.30}$$

where I_r is the energy to ionize the r times ionized ion once more. B_r is called the partition function and may be written as

$$B_r = \sum_j g_{r,j} \exp - \frac{\phi_{r,j}}{kT} \tag{3.31}$$

where the sum extends over all excited levels, bound to the r times ionized ion. In (3.31) $g_{r,j}$ is the statistical weight of level j and Φ_r, j is its energy above the ground state. From the ratios Ψ_r we can compute the ratio of the number density of r times ionized ions to the number density of neutral atoms:

$$\beta_r' = \frac{N_r}{N_{r=0}} = \prod_{j=0}^{r-1} \Psi_r \tag{3.32}$$

$$\beta_0' = 1 \qquad (0 \leq r \leq Z_i)$$

These ratios can be normalized to 1 to give the ratio of r times ionized ions to the total ion density:

$$\beta_r = \frac{\beta_r'}{\sum_{j=0}^{Z_i} \beta_j'} \tag{3.33}$$

The number of ionization electrons per ion is now given by

Table 3.2. Statistical weight of the ground level and ionization potentials of some elements (from Moore, 1970)

r	H	He	C	N	O
r :0	2 13.595	1 24.580	3 11.264	4 14.540	3 13.614
1		2 54.403	2 24.376	3 29.605	4 35.146
2		1	1 47.864	2 47.426	3 54.934
3			2 64.476	1 77.450	2 77.394
4			1 391.99	2 97.863	1 113.87
5			2 489.84	1 551.93	2 138.08
6			1	2 666.83	1 739.11
7				1	2 871.12
8					1

$$\nu_i = \sum_{r=0}^{z_i} r\beta_r \tag{3.34}$$

In many cases it is reasonable to restrict the sum (3.31) to the ground level, so that $B_r = g_{r,0}$. In this case the evaluation of Ψ_r is straightforward. The statistical weight of the ground levels and the ionization potentials of some elements are given in Table 3.2. The total number density of the ions is given by (cf.(3.2))

$$N_{\text{ion}} = \rho N_0 \sum_i \frac{X_i}{A_i} \equiv \rho N_0 \mu_i^{-1} \tag{3.35}$$

where μ_i is the mean molecular weight per ion. The ion pressure is immediately given by

$$P_i = N_{\text{ion}} kT \tag{3.36}$$

3.4 Corrections to the equation of state

The simple equation of state, derived in 3.3 needs several corrections to account for realistic situations. We will consider three corrections: the formation of electron-positron pairs at high temperatures, coulomb interactions between ions and the depression of the continuum and the formation of molecules.

3.4.1 The formation of electron-positron pairs

At temperatures higher than a few times 10^9K a considerable fraction of the photons in the radiation field has energies larger than $2m_ec$ (m_e is the rest mass of the electron). Under these conditions, electron-positron (e±) pairs are created spontaneously. Since these particles annihilate very rapidly, an equilibrium is reached between photons and e± pairs:

$$e^+ + e^- \leftrightarrow \gamma \tag{3.37}$$

This reaction adds particles to the gas and therefore influences the pressure. For more details see e.g. de Loore and Doom, 1990.

3.4.2 Coulomb interactions

The shielding by electrons.

A partially ionized gas consists of ions, with a positive charge and of free electrons, moving between the ions. Although these electrons are principally unbound to the ions, they are attracted by them, and "cluster" around the ions. In this way an "ion sphere" is formed, consisting of an ion, surrounded by free electrons. This effect lowers the Coulomb force of the ion: an electron in the neighborhood of an ion only feels a Coulomb interaction, shielded by the presence of the electrons in the ion sphere. This effect has important consequences for the ionization.

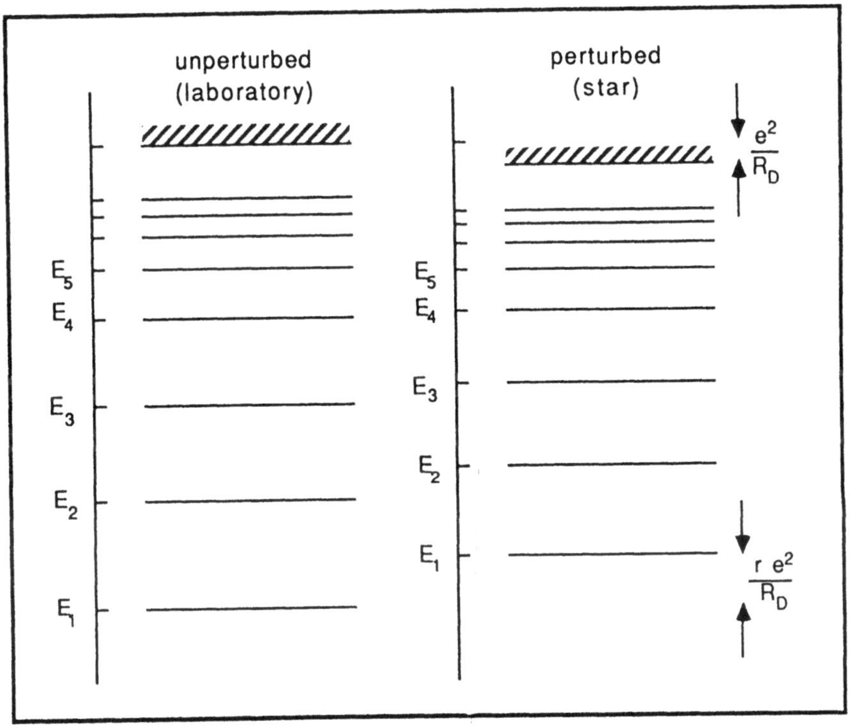

Fig. 3.1. Energy levels and continuum depression in a "perturbed" ion.

We may study this effect in the framework of the Debye-Hckel model (see e.g. Eddington, 1926; Landau, Lifschitz, 1958) assuming that the energy of the ions, due to the Coulomb interaction, is much smaller than their thermal kinetic energy. Consider an atom of charge Z, having lost r electrons (Fig. 3.1). In the laboratory, we measure the energy levels E_1, E_2,..., while $E = 0$ corresponds to the continuum, i.e. a free electron.

Now consider a free electron in the neighborhood of the ion. Its potential energy due to the interaction with the ion is zero. But due to the interaction with other charges in the ion sphere the electron acquires an extra potential energy of

$$\phi_{e,f} = -\frac{e^2}{R_D} \tag{3.38}$$

R_D, is called the Debye radius. It is a typical scale of the ion sphere, and is given by

$$R_D = \left(\frac{kT}{4\pi e^2 \rho N_0 \sigma}\right)^{\frac{1}{2}} = 8.9 \cdot 10^{-12}\sqrt{\frac{T}{\rho\sigma}}\text{cm} \tag{3.39a}$$

where

$$\sigma = (\nu_i + \nu_i^2)\frac{X_i}{A_i} \tag{3.39b}$$

The extra potential $\phi_{e,f}$ makes that a bound electron already becomes free from the ion when it acquires an energy E $= \phi_{e,f}$. We say that the continuum is depressed by an amount e^2/R_D.

Interaction with free electrons.

We consider the interaction of an electron with the free electrons of a gas. We suppose that the free electrons are uniformly distributed throughout the gas, i.e. that they do not cluster around the ions. Around each nucleus of charge eZ_i we consider a small sphere with radius ri, containing Z_i electrons, bound or free. If ν_i of these electrons are free, they possess a total electrostatic energy of

$$E_i = \frac{3}{5}\frac{v_i^2 e^2}{r_i} \tag{3.40}$$

We can average (3.40) over all elements i to obtain the electrostatic energy of an average free electron:

$$E_0 = \frac{1}{\overline{N}_e}\sum_i N_i E_i \tag{3.41}$$

The value of the radius r_i can be found from the constraint that the sphere must be electrically neutral:

$$\frac{4}{3}\pi r_i^3 N_{el} = Z_i \tag{3.42}$$

where

$$N_{el} = \rho N_0 \sum_i Z_i \frac{X_i}{A_i} \tag{3.43b}$$

$$\frac{1}{r_i} = \left(\frac{4\pi\rho N_0}{3}\right)^{\frac{1}{3}} Z_i^{-\frac{1}{3}} \sum_i Z_i \frac{X_i}{A_i} \tag{3.44}$$

and we find for E_O, using (2)

$$E_0 = \frac{3}{5}e^2\left(\frac{4\pi N_0^4}{3}\right)^{\frac{1}{3}}\frac{\rho^{\frac{1}{3}}}{\overline{N}_e}\sigma$$

$$= 7.09 \cdot 10^{24}\frac{\rho^{\frac{1}{3}}}{\overline{N}_e}\sigma eV \qquad (3.45a)$$

if ρ is in g cm-3 where

$$\sigma = \left(\sum_i Z_i \frac{X_i}{A_i}\right)^{\frac{1}{3}} \sum_i \nu_i^2 Z_i^{-\frac{1}{3}} \frac{X_i}{A_i} \qquad (3.45b)$$

The first factor of (3.45b) is $\mu_e{}^1/3$ in the case of complete ionization and can be approximated using (3.29)

The modified ionization potential.

Taking into account the Coulomb interactions,the ionization potential I_r, measured in the laboratory, has to be replaced by an effective ionization potential I_r' in the Saha equation. I_r' is given by

$$I_r' = I_r - \frac{(r+1)e^2}{R_D} - E_0 \qquad (3.46)$$

where R_D is given by (3.39) and E_0 is given by (3.45). If I_r' is negative, the atom is considered to be at least r times ionized. 3.3.2.

3.3.2 The formation of molecules

In most conditions in stellar atmospheres the formation of molecules has a negligible effect on the equation of state. If k_i atoms of type i and k_j atoms of type j combine into one molecule of type l, then the effect is that the number densities of atoms i and j decrease by k_i and k_j respectively and that one particle l is added each time a molecule is formed. If the number density of the molecules of type l is m_l then the total n_ion density nion should be corrected by an amount

$$\Delta n_{\text{ion}} = m_1(1 - k_i - k_j) \qquad (3.47)$$

The dissociation equilibrium of molecules is usually expressed in the form

$$\frac{n_{i,n}^{k_i} \cdot n_{j,n}^{k_j}}{m_l} = K_{ijl}(T) \qquad (3.48)$$

where n_i, n is the number density of the neutral atom of type i and K_{ijl} (T) is the dissociation rate. The equations of type (3.48) have thus to be combined with the Saha equation to find the molecule density. Convenient analytical approximations to $K_5 ijl$ have been computed by Mihalas (1967) and are based on results by Stanger (quoted by

Aller 1963), Yamashita (1962) and Vardya (1961,1964). The interpolation formula has the form

$$\log K_{ijl} = a + b\theta + c\theta^2 + d\theta^3 \tag{3.49a}$$

where

$$\theta = \frac{5040K}{T} \tag{3.49b}$$

Values of the coefficients a-d for some molecules are given in Table 3.3.
If many molecules are taken into account, a specialiteration scheme is required to compute all molecular densities. In most cases concerning stellar structure, only the formation of the H_2 molecule has to be taken into account.

Table 3.3. Coefficients in the analytical fit to log K_{ijl} (Mihalas, 1967)

molecule	a	b	c	d
H_2	12.5335	-4.9252	0.0562	-0.0033
C_2	12.5996	-6.5040	0.0230	0
N_2	13.1946	-10.0934	0.0211	0
O_2	13.1969	-5.3838	0.0238	0
CH	12.0854	-3.8428	0.0240	0
CN	12.9750	-8.2930	-0.0093	0
O	13.7429	-11.4432	0.0036	0
OH	12.2363	-4.6829	0.0051	0
CO_2	15.5100	-5.3500	0	0
H_2O	12.4900	-5.0900	0	0

3.5 Asymptotic expansions of the equation of state

Approximate equations of state may be derived for special values of η and β . These yield simple expressions and show part of the underlying physics more clearly. We will consider the case of the "perfect gas", non-degenerate and non-relativistic. This corresponds to the situation inside many stars and is characterized by $\eta \ll 0$ and $\beta \ll 1$. Further, we consider extreme degeneracy, a situation encountered at very high densities and corresponding to $\eta \gg 0$ and $\beta \ll 1$. The last case considered is extreme relativistic degeneracy when density and temperature are very high. This corresponds to $\eta \gg 0$ and $\beta \gg$. We will just give the equations for each case; for the derivation of the equations see e.g. De Loore and Doom (1990).

3.5.1 No degeneracy, non-relativistic

If $\eta \ll 0$ and $\beta \ll$, the gas is said to be non-degenerate and non-relativistic, and we can derive simple expressions for the equation of state.

$$N_e = 10\sqrt{2\pi}\,\pi \left(\frac{m_e c}{h}\right)^3 e^{\eta} \beta^{\frac{3}{2}} \tag{3.50}$$

$$P_e = 10\sqrt{2\pi}\pi\left(\frac{m_e c}{h}\right)^5 \frac{h^2}{m} e^{\eta}\beta^{\frac{5}{2}} \tag{3.51}$$

The total gas pressure is:

$$\begin{aligned} P_g &= P_i + P_e \\ &= N_i kT + n_e kT \\ &= \sum_j n_j kT + n_e kT \end{aligned} \tag{3.52}$$

where n_j is the number density of species j.

$$\begin{aligned} P_g &= N_0 k\rho\left[\sum_j (1+\nu_j)\frac{X_j}{A_j}\right]T \\ &= \frac{\Re}{\mu}\rho T \end{aligned} \tag{3.53}$$

where $= N_0 k$ is the gas constant and

$$\mu = \left[\sum_j (1+\nu_j)\frac{X_j}{A_j}\right]^{-1} \tag{3.54}$$

is called the mean molecular weight. Equation (3.54) is the law of perfect gasses for a mixture of different species.

3.5.2 Large degeneracy, non-relativistic

If $\eta \gg 0$ and $\beta \gg 1$, we have

$$N_e = 2\sqrt{2}N_0 B\eta^{\frac{3}{2}}\beta^{\frac{3}{2}} \tag{3.55}$$

$$\begin{aligned} P_e &= \frac{32\sqrt{2}}{5}A\eta^{\frac{5}{2}}\beta^{\frac{5}{2}} \\ &= \frac{2}{5}\eta N_e kT \end{aligned} \tag{3.56}$$

The density is

$$\rho = 2\sqrt{2}B\mu_e\eta^{\frac{3}{2}}\beta^{\frac{3}{2}} \tag{3.57}$$

The gas pressure is :

$$P_g = 6\sqrt{2}A\left(\mu_e \sum_j \frac{X_j}{A_j} + \frac{2}{5}\eta\right)\eta^{\frac{3}{2}}\beta^{\frac{5}{2}}$$

$$= \left(\mu_e \sum_j \frac{X_j}{Aj} + \frac{2}{5}\eta\right) N_e kT \tag{3.58}$$

The first term between brackets is related to the ion pressure, the second one to the electron pressure and is proportional to η. This implies that if η becomes large enough, the electron pressure will be much larger than the ion pressure, since the first term is of the order unity. This effect is called electron degeneracy.
Electrons, being fermions, must obey the Pauli exclusion principle, i.e. each cell in the phase space of volume $1/h^3$ can only be occupied by 2 electrons. If the electron density becomes very large all cells corresponding to lower momenta are occupied. If more electrons are added, they are forced to occupy still higher momenta, thereby raising the electron pressure rapidly.

The physical picture of degeneracy implies that the moment distribution of the electrons should only be determined by the electron density. Expressing P_e as a function of ρ/μ_e one obtains

$$P_e = 1.004 \cdot 10^{13}\left(\frac{\rho}{\mu_e}\right)^{\frac{5}{3}} \text{dynes cm}^{-2} \tag{3.59}$$

3.5.2 Large degeneracy, relativistic

If $\eta >> 0$ but β is not small, the Fermi-Dirac integrals become

$$F_k(\eta,\beta) = \int_0^{\eta} y^k \left(1 + \frac{\beta}{2}y\right)^{\frac{1}{2}} dy \tag{3.60}$$

$$N_e = 8N_0 B x^3 (x^2+1)^{\frac{3}{2}} \tag{3.61}$$

$$P_e = 2A\left[3\sinh^{-1}x + x\sqrt{x^2+1}(2x^2+1)\cdot(8x^4+8x^2-3)\right] \tag{3.62}$$

This result shows that the electron density and pressure depend only on the product $\eta\beta$ and that the electron pressure only depends on the electron density.

4 Nuclear reactions - nuclear burning in stellar interiors

4.1 Introduction

Nuclear reactions in the stellar interior produce the energy to make the star shine and transform elements into others, like chemical reactions transform molecules into other molecules.

Examples:

- $^{12}C + ^{16}O \rightarrow ^{24}Mg + ^{4}He$

One ^{12}C particle and one ^{16}O particle react and produce a compound nucleus of ^{28}Si; the latter one breaks up into a ^{24}Mg nucleus and a ^{4}He nucleus. It breaks up through the "α channel" (a ^{4}He nucleus is also called an alpha particle).

- $^{13}C + ^{1}H \rightarrow ^{14}N + \gamma$

This reaction terminates through the "γ channel". The compound nitrogen nucleus decays to its ground state and emits a γ photon.

- $^{25}Mg + ^{1}H\ ^{26}Al^{*} \rightarrow ^{26}Mg + e+ + \nu$

The compound nucleus of ^{26}Al does not decay to its ground state but transforms into ^{26}Mg through a beta decay.

At the beginning of its nuclear lifetime a star converts hydrogen into helium. Hydrogen "burning" reactions occur already at low temperatures ($\approx 10^7$K) because the Coulomb barriers are then rather smal. When all hydrogen in the interior is consumed, the temperatures in the stellar interior rise above a hundred million Kelvin and helium burning reactions are initiated. Stars can use the ashes of nuclear burning as fuel for a subsequent burning phase: helium is converted into carbon, oxygen and, to a lesser extent into neon. When helium is exhausted in the stellar interior the temperature rises to roughly 10^9K and carbon, oxygen and neon get involved in fusion reactions (carbon burning, neon burning and oxygen burning). The main final product of these reactions is silicon. In an ultimate phase silicon is transformed into elements of the iron group. Since this kind of fusion requires energy rather than liberating it, iron cannot be used as a nuclear fuel for fusion reactions. With the formation of iron the nuclear burning lifetime of a star comes to an end.

4.2 The equations for the nuclear species

We follow the time evolution of an element, which is produced and destroyed by nuclear reactions. The equations governing this evolution follow immediately from the definition of the cross section for nuclear reactions.

Consider an element i, involved in a number of reactions.

The element is *destroyed* by nuclear reactions involving n_i nuclei of type i and n_j nuclei of type j per reaction.

The element is *produced* by nuclear reactions involving n_k particles k and n_l particles l and producing n_{ijk} particles of type i per reaction.

According to the definition of the cross section we can write for the number density N_i of the particles i:

$$\frac{dN_i}{dt} = -\sum_j n_{ij} \frac{N_i^{n_{ij}} N_j^{n_j}}{n_{ij}! n_j!} < \sigma v >_{ij} + \sum_{k,l} n_{ijk} \frac{N_k^{n_k} N_l^{n_l}}{n_k! n_l!} < \sigma v >_{kl} \tag{4.1}$$

with $<\sigma v>_{ij}$ the averaged cross section for the considered reaction . The first summation in (4.1) extends over all nuclear reactions destroying the element i, while the second summation extends over all reactions producing the element i.

4.3 Cross sections for nuclear reactions

4.3.1 The reaction rate

Consider a nuclear reaction of the type A+b → C+d, where nuclei of type A collide with particles b, producing nuclei of type C and particles of type d. We suppose that in one reaction, respectively n_A, n_b, n_C and n_d particles of types A, b, C and d are involved. For such a reaction, the cross section is defined as

$$\sigma(\text{cm}^2) = \frac{\text{number of reactions/nucleus A/unit of time}}{\text{number of incoming particles b/cm}^2\text{/unit of time}} \tag{4.2}$$

If we only consider particles of types A and b, having a relative velocity v, we can write the number of reactions per cm^3 as

$$r = v\ \sigma(v) N_A N_B \tag{4.3}$$

where N_A and N_b are the number densities of the particles A and b. We must now take into account the velocity distribution n(v). If n(v) is normalized as

$$\int_0^\infty n(v) dv = 1 \tag{4.4}$$

we can deduce from (4.3):

$$r(\text{reactions cm}^{-3}) = N_A N_B \int_0^\infty v\ \sigma(v)\ n(v)\ dv$$

$$= N_A N_B < \sigma v > \tag{4.5}$$

This expression is deduced from the assumption that the particles A and b are not identical, since $N_A \cdot N_b$ is the number of independent pairs of particles (A,b). If both particles are identical, the number of independent pairs of particles is only $N_A \cdot N_b/2$. Generally, if more than one particle of type A or b is involved, we must replace (4.5) by

$$\text{r}(\text{reactions cm}^{-3}) = \frac{N_A}{n_A!} \frac{N_b}{n_b!} < \sigma v > \tag{4.6}$$

Both the particles of type A and b are part of a gas, so they have a velocity distribution $n_1(\mathbf{v}_1)$ for A and $n_2(\mathbf{v}_2)$ for b. Since both particles are ions, they are non-degenerate and their velocity distribution is the Maxwell-Boltzmann distribution:

$$n_i(\mathbf{v}_i)d^3\mathbf{v}_i = \left(\frac{m_i}{2\pi kT}\right)^{\frac{3}{2}} \exp\left(-\frac{v_i^2 m_i}{2kT}\right) d^3\mathbf{v}_i \tag{4.7a}$$

where

$$\mathbf{v}_i(v_{ix}, v_{iy}, v_{iz}) \tag{4.7b}$$

$$d^3v_i = dv_{ix}dv_{iy}dv_{iz} \tag{4.7c}$$

Since both velocity distributions are independent the probability that a particle A has a velocity around $\mathbf{v}_1$ and particle b has a velocity around $\mathbf{v}_2$ is simply

$$n_1(\mathbf{v}_1)n_2(\mathbf{v}_2)d^3\mathbf{v}_1 d^3\mathbf{v}_2 \tag{4.8}$$

Since mass is conserved with great accuracy in nuclear reactions (only a very small amount of the mass is transformed into energy), distribution $\mathbf{v}$ is of no importance. Moreover, the velocities $\mathbf{v}$ are distributed isotropically, so we can replace d^3v by $4\pi v^2 dv$. We then obtain for $<\sigma v>$:

$$< \sigma v >= 4\pi \left(\frac{\mu}{2kT}\right)^{\frac{3}{2}} \int_0^\infty v^3\sigma(v)\exp\left(-\frac{\mu v^2}{2kT}\right) dv \tag{4.9}$$

or, using the kinetic energy $E = \frac{1}{2}\mu v^2$:

$$< \sigma v >= \left(\frac{8}{\pi}\right)^{\frac{1}{2}} \frac{1}{\mu^{\frac{1}{2}}(kT)^{\frac{3}{2}}} \int_0^\infty \sigma(E)E\exp\left(-\frac{E}{kT}\right) dE \tag{4.10}$$

This gives the desired expression for the reaction rate. In order to calculate it explicitly, we need further information on the dependence of the cross section on the energy. This information can only be obtained by studying the atomic nucleus more closely.

It is more appropriate to transform the equation (4.1) into a form using the number fractions Y_i of the elements. Since

$$N_i = \rho N_A \frac{X_i}{A_i} \equiv \rho N_A Y_i$$

where X_i is the mass fraction, we can immediately write for (4.1):

$$\frac{dY_i}{dt} = -\sum_j \rho^{n_{ij}+n_j-1} n_{ij} \frac{Y_i^{n_{ij}} Y_j^{n_j}}{n_{ij}!n_j!} N_A^{n_{ij}+n_j-1} < \sigma v >_{ij}$$

$$+\sum_{k,l} \rho^{n_k+n_l-1} n_{ijk} \frac{Y_k^{n_k} Y_l^{n_l}}{n_k!n_l!} N_A^{n_k+n_l-1} < \sigma v >_{kl} \tag{4.11}$$

which is the most general form of the reaction equation. In most cases the reaction equation is much simpler, since in many nuclear reactions only one particle of each species is involved, or at most two or three particles of only one type. Tables of reaction rates (Fowler et al. 1975, Harris et al. 1983) give the quantity

$$N_A,^{n_k+n_l-1} = r_{kl}$$

as functions of the temperature.

We can easily derive a general expression for the nuclear energy generation, by using (4.11). Consider an extremely simple and imaginary reaction

C → D

M_C and M_D are the masses of the nuclei C and D. The total energy produced per reaction, Q, is clearly equal to the difference in mass, converted into energy:

$Q = -(M_D - M_C)\,c^2$

Since the number of nucleons, A, is conserved in the reaction, we can equally write

$$Q = -(\varepsilon_D - \varepsilon_D)c^2 \tag{4.12}$$

where ϵ_i (= M_i - A_i) is called the mass excess of the nucleus i. It is a measure of the binding energy between nucleons. The mass excess for a number of nuclei is given in Table 5.1. We can interpret (5.3) as if the reaction C → D adds an energy $\epsilon_D c^2$ and removes an energy $\epsilon_C c^2$. Generally we can thus write for the energy generation

$$\varepsilon \text{ (power per unit mass)} = -N_A \sum_i \frac{dY_i}{dt} \varepsilon_i c^2 \tag{4.13}$$

where the summation extends over all elements involved in nuclear reactions.

$$\varepsilon \text{ (erg/g/s)} = -9.647\ 10^{17} \sum_i \frac{dY_i}{dt} \varepsilon_i (MeV) \tag{4.14}$$

This way of computing the energy generation rate is more stable than summing up the contributions of the different reactions. The nuclear energy generation should be corrected for the production of neutrinos. A β unstable nucleus can decay by emitting an electron (β^- decay) or a positron (β^+ decay) and a neutrino. The neutrino immediately escapes from the star (except in extremely dense conditions) and the kinetic energy the neutrino acquired during the β decay is therefore lost. The energy, lost in the neutrino, $Q\nu$, depends on the total energy, $Q\text{-}\beta$, which is liberated by the β decay (Fowler et al. 1975):

$$Q_v = -0.511 \frac{(1 - x^2) \cdot (1 - \frac{1}{4}x - \frac{1}{9}x^2)}{x} MeV \tag{4.15a}$$

with

$$x = (1.957 Q_{\beta^-}(MeV) + 1)^{-1} \text{ for } \beta^- \text{ decay} \tag{4.15b}$$

Table 4.1. The atomic mass excess in MeV, normalized to 0 for the ^{12}C nucleus (Wapstra and Gove 1971)

n 8.07169		
^{1}H 7.28922	^{2}H 13.13627	
^{3}He 14.93173	^{4}He 2.42494	
^{7}Li 14.9086		
^{7}Be 15.7703		
^{8}B 22.9223		
^{12}C 0.	^{13}C 3.12527	
^{14}N 2.86382	^{15}N 0.1018	
^{16}O -4.73668	^{17}O -0.8074	^{18}O -0.78250
^{19}F -1.4861		
^{20}Ne -7.0417	^{21}Ne -5.7312	^{22}Ne -8.0251
^{23}Na -9.5290		
^{24}Mg -13.9313	^{25}Mg -13.1945	^{26}Mg-16.2134
^{27}Al-17.1950		
^{28}Si-21.4911		

4.4 Hydrogen burning

During the first stage of nuclear burning in stellar interiors hydrogen is being converted into helium. This occurs at temperatures of the order of ten million Kelvin, and mainly through the proton-proton chain and the CNO cycle.

4.4.1 The proton-proton chain

The complete proton-proton (or p-p) chain consists of the reactions

$$1.\ 2\,{}^{1}\mathrm{H} \rightarrow {}^{2}\mathrm{H} + e^{+} + \nu \qquad (4.16a)$$

$$2.\ {}^{2}\mathrm{H} + {}^{1}\mathrm{H} \rightarrow {}^{3}\mathrm{He} \qquad (4.16b)$$

$$3.\ 2\,{}^{3}\mathrm{He} \rightarrow {}^{4}\mathrm{He} + 2\,{}^{1}\mathrm{H} \quad \text{, the ppI chain} \qquad (4.16c)$$

$$4.\ {}^{3}\mathrm{He} + {}^{4}\mathrm{He} \rightarrow {}^{7}\mathrm{Be} \qquad (4.16d)$$

$$5.\ {}^{7}\mathrm{BeH} + e^{-} \rightarrow {}^{7}\mathrm{Li} + \nu \qquad (4.16e)$$

5. $^{7}\mathrm{BeH} + e^{-} \rightarrow^{7} \mathrm{Li} + \nu$ (4.16e)

6. $^{7}\mathrm{Li} +^{1} \mathrm{H} \rightarrow 2\ ^{4}\mathrm{He}$, the ppII chain (4.16f)

7. $^{7}\mathrm{Be} +^{1} \mathrm{H} \rightarrow^{8} \mathrm{B}$ (4.16g)

8. $^{8}\mathrm{B} \rightarrow^{8} \mathrm{Be} + e^{+} + \nu$ (4.16h)

9. $^{8}\mathrm{Be} \rightarrow 2\ ^{4}\mathrm{He}$, the ppIII chain (4.16i)

The equation determining the abundance of deuterium (^{2}H) can easily be derived from the general form (4.11):

$$\frac{\mathrm{d}Y_2}{\mathrm{dt}} = \rho\left(Y_1^2\frac{r_1}{2} - Y_1Y_2r_2\right) \tag{4.17}$$

with Y_1 and Y_2 the number fractions of ^{1}H and ^{2}H (deuterium) and r_i the rate of reaction i in the pp chain (4.16). Assuming that Y_1, ρ, r_1 and r_2 are constant, we can solve (4.17), leading to

$$Y_2(t) = Y_{2,0}e^{-\frac{t}{\tau}} + Y_{2\infty}(1 - e^{-\frac{t}{\tau}}) \tag{4.18}$$

where $Y_2, 0$ is the initial mass fraction of deuterium and

$$Y_{2\infty} \equiv \frac{X_{2\infty}}{A_2} = Y_1\frac{r_1}{2r_2} = \lim_{t\rightarrow\infty} Y_2(t) \tag{4.19}$$

is the equilibrium number fraction. $X_{2\infty}$ is the equilibrium abundance of deuterium.

$$\tau = (\rho Y_1 r_2)^{-1} \tag{4.20}$$

is the typical time scale for deuterium to reach equilibrium. When the species reaches equilibrium, its derivative becomes zero, and the abundance does not change anymore. It should be noticed that this equilibrium abundance is a stationary abundance, and should not be confounded with nuclear equilibrium.

Fig. 4.1 shows the equilibrium abundance of deuterium; the time scale τ is shown as a function of the temperature. It can be seen that the equilibrium abundance of deuterium is extremely small: even for low temperatures we have an equilibrium abundance smaller than 10^{-16}. The equilibrium abundance is even smaller for high temperatures, which implies that deuterium is destroyed as the temperature rises. Also the time scale for reaching equilibrium is very small (in most cases less than one year), so we can assume that deuterium has reached its equilibrium value and that its time derivative is zero. In the convective cores of massive stars, where $\log T < 7.4$ the time scale for equilibrium is of the order of seconds, which is much smaller than the typical time scale for convective mixing. In this case deuterium is not homogeneously distributed in the stellar core, but has its local equilibrium abundance at any point.

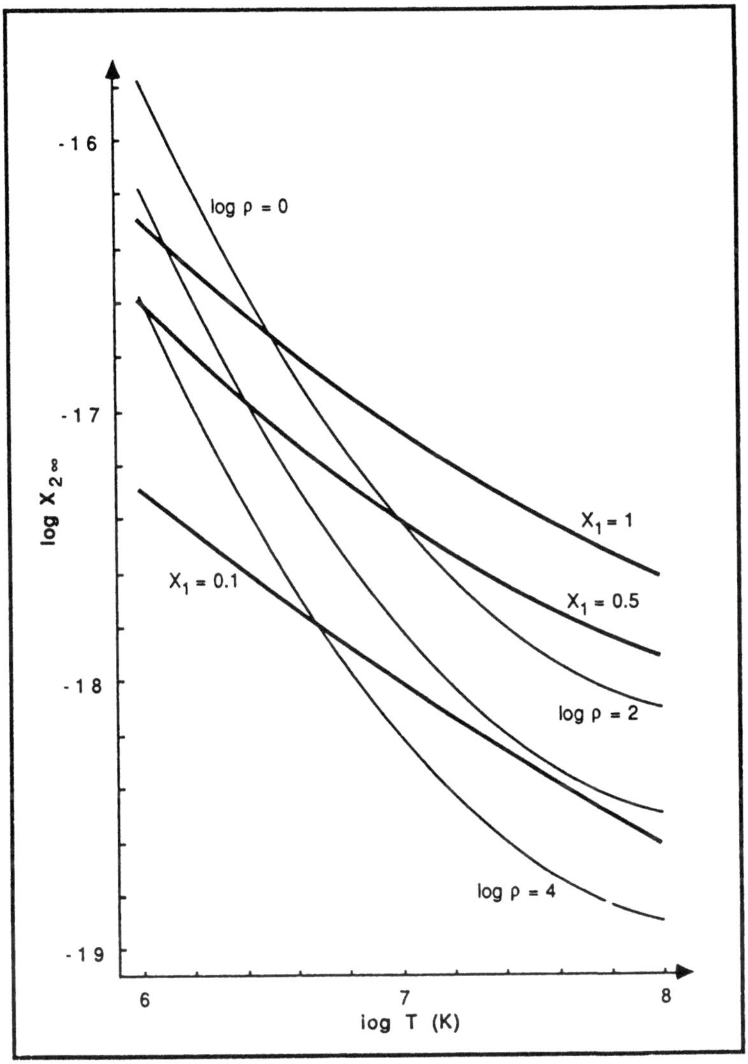

Fig. 4.1. The equilibrium abundances $X_{2\infty}$ (thick lines) and time scales (thin lines) of deuterium expressed as function of ρ. The time scale τ is linked to the density ρ as $\tau=(\rho\ Y_1\ r_2)^-1$. The reaction rates are from Fowler et al. (1975) and Harris et al. (1983).

The equation governing the abundance of ^{3}He is

$$\frac{dY_3}{dt} = \rho(Y_1 Y_2 r_2 - Y_3^2 r_3 - Y_3 Y_4 r_4) \tag{4.21}$$

with Y_3 and Y_4 the number fractions of ^3He and ^4He. If we assume that deuterium has reached equilibrium, we can put $dY_2/dt = 0$ and we can eliminate Y_2 from (4.21) using (4.17). This gives

$$\frac{dY_3}{dt} = a + bY_3 + cY_3^2 \tag{4.22a}$$

with

$$a = \rho y_1^2 \frac{r_1}{2} \tag{4.22b}$$

$$b = -\rho y_4 r_4 \tag{4.22c}$$

$$c = -\rho r_3 \tag{4.22d}$$

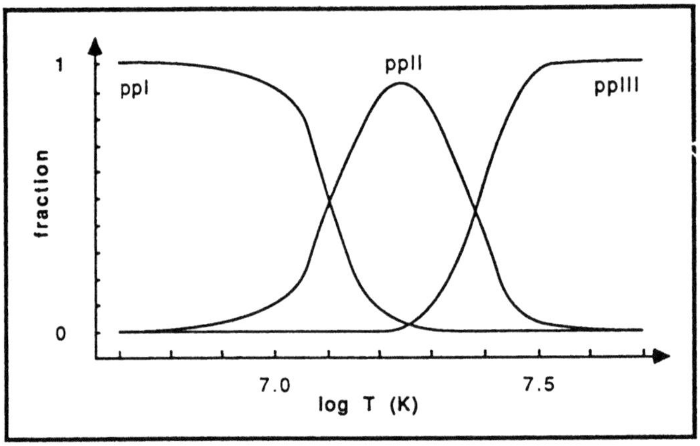

Fig. 4.2. The number of ^{4}He nuclei produced by the different pp chains.

The discriminant of (4.21) is

$$\Delta = \rho^2(Y_4^2 r_4^2 + 2Y_1^2 r_1 r_3) \tag{4.23}$$

and the solution is

$$Y_3(t) = \frac{1}{2c}\left\{\sqrt{\Delta}\tanh\left[\frac{\sqrt{\Delta}}{2}(t_0\text{-}t)\right]\text{-}b\right\} \tag{4.24}$$

where the parameter t_0 is related to the initial abundance of ^{3}He, $Y_{3,0}$:

$$t_0 = \frac{2}{\sqrt{D}}\tanh^{-1}\left(\frac{b+2c\,Y_{3,0}}{\sqrt{D}}\right) \tag{4.25}$$

The equilibrium abundance of ^{3}He is given by

$$Y_{3\infty} \equiv \frac{X_{3\infty}}{A_3} = \lim_{t\to\infty} Y_3(t) = \frac{\sqrt{Y_4^2 r_4^2 + 2Y_1^2 r_1 r_3} - Y_4 r_4}{2r_3} \tag{4.26}$$

As can be seen from (4.24) the typical time scale for reaching equilibrium is given by

$$t = \frac{2}{\sqrt{D}} \tag{4.27}$$

From the branching ratios between the different pp chains it is easy to derive what fraction of the ^{4}He nuclei is produced by any of the three pp chains.This is shown in Figure 4.2 for $X_1=X_4=0.5$ and assuming that all elements involved are in equilibrium.

4.4.2 The CNO cycle

Carbon is now used as a "catalyst" to convert hydrogen into helium. The complete reaction chain consists of 3 cycles: in the first, the CN cycle, only carbon and nitrogen are involved; in the other two cycles, the ONI and ONII cycles, oxygen isotopes are involved. The reactions are

1. $^{12}C+^1H \rightarrow ^{13}N$ (4.28a)

2. $^{13}N \rightarrow ^{13}C+e^+ +\nu$ (4.28b)

3. $^{13}C+^1H \rightarrow ^{14}N$ (4.28c)

4. $^{14}N+^1H \rightarrow ^{15}O$ (4.28d)

5. $^{15}O \rightarrow ^{15}N+e^+ +\nu$ (4.28e)

6. $^{15}N+^1H \rightarrow ^{12}C+^4He$ the CN cycle (4.28f)

7. $^{15}N+^1H \rightarrow ^{16}O$ (4.28g)

8. $^{16}O+^1H \rightarrow ^{17}F$ (4.28h)

9. $^{17}F \rightarrow ^{17}O+e^+ +\nu$ (4.28i)

10. $^{17}O+^{1}H \rightarrow ^{14}N+^{4}He$ the ONI cycle (4.28j)

11. $^{17}O+^{1}H \rightarrow ^{18}F$ (4.28k)

12. $^{18}F \rightarrow ^{18}O+e^{+}+\nu$ (4.28l)

13. $^{18}O+^{1}H \rightarrow ^{15}N+^{4}He$ the ONII cycle (4.28m)

The ONI cycle is entered through an alternative to the $^{15}N + ^{1}H \rightarrow ^{12}C + ^{4}He$ reaction: instead of a break up of the compound oxygen nucleus into a carbon and a helium nucleus, the reaction terminates through the γ channel. The ONII cycle is entered in the same way by the termination of the $^{17}O + ^{1}H$ reaction through the γ channel.

The β decay of all unstable nuclei (reactions 2, 5, 9 and 12) can be considered as immediate, so for stable phases of hydrogen burning some reactions can be written in a compressed form:

$$^{12}C + ^{1}H \rightarrow ^{13}C + e^{+} + \nu \tag{4.29a}$$

$$^{14}N + ^{1}H \rightarrow ^{15}N + e^{+} + \nu \tag{4.29b}$$

$$O + e^{+} + \nu \tag{4.29c}$$

$$O + e^{+} + n \tag{4.29d}$$

Since all reactions are single proton captures the equations describing the time evolution of the CNO abundances can easily be written in the form of a (singular) matrix:

$$\frac{d\mathbf{Y}}{dt} = M \cdot \mathbf{Y} \tag{4.30}$$

with

$$\mathbf{Y} = [\mathbf{Y}_{12}\ \mathbf{Y}_{13}\ \mathbf{Y}_{14}\ \mathbf{Y}_{15}\ \mathbf{Y}_{16}\ \mathbf{Y}_{17}\ \mathbf{Y}_{18}\ ] \tag{4.31}$$

Y_n is the number fraction of the species with atomic mass n (only stable nuclei are considered).

$$M = \rho Y_1 \begin{bmatrix} -r_1 & 0 & 0 & r_6 & 0 & 0 & 0 \\ r_1 & -r_3 & 0 & 0 & 0 & 0 & 0 \\ 0 & r_3 & -r_4 & 0 & 0 & r_{10} & 0 \\ 0 & 0 & r_4 & -r_6 - r_7 & 0 & 0 & r_{13} \\ 0 & 0 & 0 & r_7 & -r_8 & 0 & 0 \\ 0 & 0 & 0 & 0 & r_8 & -r_{10} - r_{11} & 0 \\ 0 & 0 & 0 & 0 & 0 & r_{11} & -r_{13} \end{bmatrix} \tag{4.32}$$

The equilibrium abundances of the CNO elements, $\mathbf{Y}\infty$, can be found by solving the equation

$$M \cdot \mathbf{Y} = 0 \tag{4.33}$$

for non-trivial solutions, requiring that the total number of CNO particles remains constant.

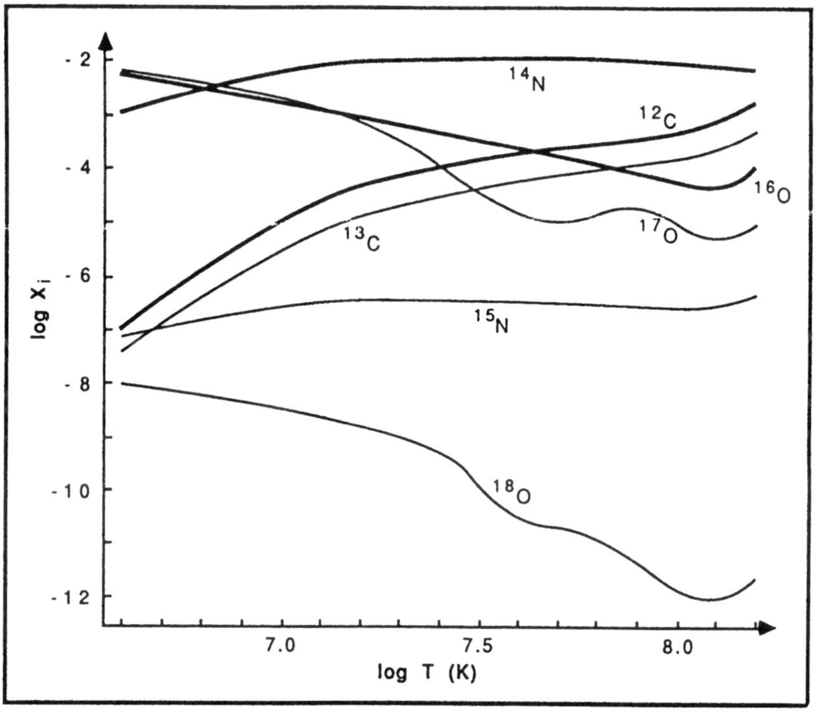

Fig. 4.3. The equilibrium abundances of the CNO cycle. The reaction rates are from Fowler et al. (1975).

The equilibrium abundances of the CNO elements are shown in Fig. 4.3; for the sum of the CNO mass fractions, the value for the solar atmosphere is taken, 7.4 10^{-4}. These equilibrium abundances can be compared to the abundances of carbon, nitrogen and oxygen in the solar photosphere: log X_{12} = -2.53, log X_{14} = -3.03 and log X_{16} = -2.17.

When the CNO cycle is in equilibrium, ^{14}N is the most abundant isotope, except at very low temperatures. At typical temperatures for hydrogen burning (7 < log T < 8) the largest part of carbon and oxygen is transformed into nitrogen. Furthermore at lower temperatures, more oxygen is left while at higher temperatures carbon prevails. Although the equilibrium abundances of ^{12}C and ^{13}C vary by four orders of magnitude, the ratio X_{12}/X_{13} remains remarkably constant at over the whole temperature range; the value of 3.14±0.10 is much smaller than the cosmic value which is ≈ 100. ^{15}N has an equilibrium abundance of 10^{-7} to 10^{-6} .The equilibrium abundance of ^{18}O is very small.

The set of differential equations (4.30) cannot be solved analytically and one cannot derive analytical expressions for the equilibrium time scales. They can only be found by solving the system of equations numerically, starting with e.g. the abundances in the solar photosphere, assuming a constant value for the temperature and the density and following the approach to equilibrium. Computations reveal that the CN cycle reaches equilibrium much faster that the ON cycles. First, ^{12}C is transformed largely into ^{14}N and ^{13}C is formed. Only afterwards part of the ^{16}O is transformed into ^{14}N and are the other oxygen isotopes formed. The equilibrium time scales for the CN and the ON cycles, computed numerically, are shown in Fig. 4.4 for a density of 1 g/cm^3. The figure shows that the equilibrium timescales are inversely proportional to the density. There is a large spread in the equilibrium time scales for different temperatures: for low temperatures, (e.g. in the central parts of the sun, log T < 7.2) they are larger than the Hubble time scale, so the CN cycle is definitely not in equilibrium. At high temperatures the equilibrium time scales are extremely short, implying that in massive stars the CNO cycle reaches its equilibrium in the early stages of hydrogen burning.

4.4.3 The energy generation

The pp chain as well as the CNO cycle transform four hydrogen nuclei into one helium nucleus; however the energy generated per helium nucleus produced is slightly different for both cases, a consequence of the fact that the neutrinos, produced in both reaction cycles, have slightly different energies. For each group of four transformed hydrogen nuclei the pp chain produces 27.622 MeV while the CNO cycle only produces 24.970 MeV.

When the pp and the CNO cycles are in equilibrium, all reactions proceed at the same rate. Then the total energy generation rate can be computed by multiplying the reaction rate of one reaction by the energy generation of one whole reaction chain. For the pp chain we take the initial reaction 2 $^1H \rightarrow {}^2H + e^+ + \nu$ and use half of the energy generated by one complete pp chain since two initial reactions are needed to initiate the production of one ^4He nucleus. For the CNO cycle we use the reaction $^{14}N + {}^1H \rightarrow {}^{15}O$. From the expressions for the reaction rates given by Fowler et al. (1975) one obtains, for temperatures below 10^8K:

$$\varepsilon_{pp}(\mathrm{erg/g/s}) = 2.81\,10^6 \frac{\rho X_1^2}{T_6^{\frac{2}{3}}} \exp\left(-\frac{38.8}{T_6^{\frac{1}{3}}}\right) \cdot (1+1.23\,10^{-2} T_6^{\frac{1}{3}} + 1.09\,10^{-2} T_6^{\frac{2}{3}} + 9.38\,10^{-4} T_6) \tag{4.34}$$

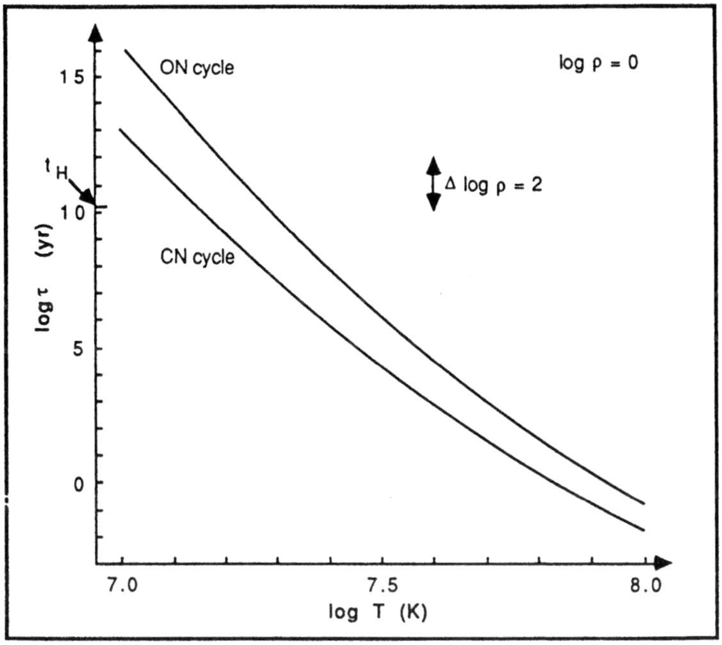

Fig. 4.4. The equilibrium timescales of the CN and the ON cycles for a density of 1 g/cm^3. The time scales are inversely proportional to the density.

$$\varepsilon_{CNO}(\text{erg/g/s}) = 8.71\ 10^{27} \frac{\rho X_1 X_{CNO}}{T_6^{\frac{2}{3}}} exp\left(-\frac{152.28}{T_6^{\frac{1}{3}}}\right) \cdot$$

$$(1 + 2.7\ 10_6^{\frac{1}{3}} - 7.78\ 10^{-3} T_6^{\frac{2}{3}} - 1.49\ 10^{-4} T_6 + 2.61\ 10^{-5} T_6^{\frac{4}{3}} + 1.27\ 10^{-6} T_6^{\frac{5}{3}}) \qquad (4.35)$$

X_1 is the mass fraction of hydrogen X_{CNO} is the total mass fraction of the CNO elements $T_6 = T / 10^6$K

4.5 Helium burning

When hydrogen at the center is exhausted , the internal temperature may rise beyond 10^8K and the transformation of helium into heavier species can start. The most important is the triple-alpha process:

$$3\ ^4\text{He} \rightarrow ^{12}\text{C} \qquad (4.36)$$

or, using a more compact notation:

$$^4\,\text{He}(2\alpha,\gamma)^{12}\text{C} \qquad (4.37)$$

Two alpha particles are captured by a helium-4 nucleus, a γ photon is released and a carbon-12 nucleus is produced. The triple-alpha process consists of two reactions. Two alpha particles interact and produce a ^{8}Be nucleus, which is highly unstable and disintegrates in 10^{-16} seconds again into two alpha particles. Obviously the unstable ^{8}Be nucleus reaches an equilibrium abundance in an extremely short time. A small part of the ^{8}Be in equilibrium reacts with another alpha particle to produce an exited compound nucleus of ^{12}C. This nucleus decays to its ground state, emitting a γ photon .The combined reaction is then terminated. The bulk of energy of the helium burning is released by the triple-alpha process .

Secondary helium burning reactions consist of the capture of alpha particles by heavier species. The main reactions are:

$$^{12}\mathrm{C}(\alpha,\gamma)^{16}\mathrm{O}(\alpha,\gamma)^{20}\mathrm{Ne}(\alpha,\gamma)^{24}\mathrm{Mg}(\alpha,\gamma)^{28}\mathrm{Si} \qquad \text{the "carbon" chain} \tag{4.38}$$

$$^{14}\mathrm{N}(\alpha,\gamma)^{18}\mathrm{F}(\beta^+)^{18}\mathrm{O}(\alpha,\gamma)^{22}\mathrm{Ne}(\alpha,\gamma)^{26}\mathrm{Mg}(\alpha,\gamma)^{30}\mathrm{Si} \qquad \text{the "nitrogen" chain} \tag{4.39}$$

Since the Coulomb barriers become higher as heavier species are involved the subsequent reactions in these chains proceed slower and slower. The last two elements of the carbon chain, Mg and Si, only become abundant for high temperatures. At lower temperatures, the rates of the last two reactions are not large enough to provoke a large flow towards the heavier elements. Between log T = 8.3 and 8.6 the neon producing reaction becomes active, while at still higher temperatures, the neon is rapidly converted into magnesium and silicon. Helium burning must proceed at temperatures larger than log T = 8.1 to 8.3 to burn helium within an acceptable time scale.

In the carbon chain no equilibrium is reached, because all reactions proceed on the same time scale as the triple-alpha process and because ^{12}2C is constantly produced.

On the other hand, one could say that the nitrogen chain reaches a quasi-equilibrium. At the beginning of helium burning the available ^{14}N, produced by the CNO cycle, is rapidly transformed into ^{18}O and further into ^{22}Ne. Thereafter, the ^{22}Ne is slowly converted into ^{26}Mg.

The heaviest isotopes are not only produced by alpha captures, but also by neutron captures. During helium burning, a number of secondary alpha capture reactions produce neutrons:

$$^{13}\mathrm{C}+^{4}\mathrm{He} \rightarrow^{16}\mathrm{O}+\nu \tag{4.40a}$$

$$^{18}\mathrm{O}+^{4}\mathrm{He} \rightarrow^{21}\mathrm{Ne}+\nu \tag{4.40b}$$

$$^{22}\mathrm{Ne}+^{4}\mathrm{Ne} \rightarrow^{25}\mathrm{Mg}+\nu \tag{4.40c}$$

$$^{25}\mathrm{Mg}+^{4}\mathrm{He} \rightarrow^{28}\mathrm{Si}+\nu \tag{4.40d}$$

$$^{26}\mathrm{Mg}+^{4}\mathrm{He}\rightarrow^{29}\mathrm{Si}+\nu \qquad (4.40e)$$

These neutrons can be captured by neon, magnesium and silicon isotopes to produce heavier isotopes of the same element.

Summarized: the main characteristics of helium burning are:
- carbon is produced by the triple-alpha process and is partially converted into heavier elements by the carbon chain
- at the beginning of the He-burning phase, ^{14}N is rapidly converted into ^{22}Ne, which is then slowly destroyed during the remaining time
- ^{13}C is rapidly transformed into ^{16}O, releasing a burst of neutrons at the beginning of helium burning
- neutrons are also produced by several secondary reactions, and may be captured by the heavy elements, producing heavier isotopes of the same element.

4.6 Carbon burning

As helium is exhausted in the interior of massive stars, the internal temperature rises and approaches 10^9K, allowing the fusion of carbon nuclei. The main reaction of the carbon burning phase is :

$$^{12}C \rightarrow {}^{23}Mg + n \qquad (4.41a)$$

$$^{20}Ne + \alpha \qquad (4.41b)$$

$$^{23}Na + p \qquad (4.41c)$$

Two carbon nuclei interact to produce an excited compound nucleus of ^{24}Mg, which may decay through the neutron channel (4.41a), the alpha channel (4.41b) or the proton channel (4.41c). The neutron channel is the least important since it requires energy (2.63 MeV) while the two other terminations of the carbon fusion reaction liberate energy (4.62 MeV and 2.24 MeV respectively)

The carbon fusion reaction may terminate through still other channels, such as the γ channel, or by releasing two α-particles; however the branching ratios to these channels are negligibly small.

Carbon fusion produces neutrons, protons and α particles. The (few) neutrons that are produced may be captured by heavy nuclei to produce heavier isotopes.

The α particles are captured by oxygen, neon and magnesium and lead to a rearrangement of the final carbon burning products:

$$4^{16}\mathrm{O}+^{4}\mathrm{He}\rightarrow^{20}\mathrm{Ne} \qquad (4.41d)$$

$$^{20}\mathrm{Ne}+^{4}\mathrm{He}\rightarrow^{24}\mathrm{Mg} \qquad (4.41e)$$

$$^{24}\mathrm{Mg} + ^{4}\mathrm{He} \rightarrow ^{27}\mathrm{Al} + \mathrm{p} \tag{4.41f}$$

The protons, produced by reactions (4.41c) and (4.41f), may may react with sodium, destroying in this way the bulk of the sodium, produced by the fusion of carbon.

$$^{23}Na + ^{1}H \rightarrow ^{20}Ne + \alpha \tag{4.41g}$$

The main products of carbon burning are ^{20}Ne and ^{24}Mg. Although oxygen is involved in an a capture reaction it largely survives carbon burning. The secondary reactions, including many more neutron-, proton- and alpha captures than given in (4.41) lead to the production of some ^{21}Ne, ^{22}Ne, ^{25}Mg, ^{26}Al (unstable), ^{29}Si, ^{30}Si and ^{31}P. Typically, less than 1% of ^{23}Na is left and some ^{27}Al is produced.

The overall nuclear energy generation rate can be computed easily by assuming that for each two carbon nuclei that disappear, one ^{20}Ne and one ^{24}Mg nucleus appear.

Energy generation :

Using the reaction rate for ^{12}C + ^{12}C given by Fowler et al. (1975) we obtain:

$$\varepsilon_C(\mathrm{erg/g/s}) = 5.63\ 10^{45} \frac{\rho Y_{12}^2 T_{9\alpha}^{\frac{5}{6}}}{T_9^{\frac{3}{2}}} \cdot \frac{\exp\left(-\frac{84.165}{T_{9\alpha}^{\frac{1}{3}}}\right)}{\exp(-0.01 T_{9\alpha}) + 5.56\ 10^{-3} \exp(1.685 T_{9\alpha}^{\frac{2}{3}})} \tag{4.42a}$$

where

$$T_9 = \frac{T}{10^9} K \tag{4.43b}$$

$$T_{9\alpha} = T_9(1 + 0.067 T_9) \tag{4.44c}$$

4.7 Neon burning

At the end of carbon burning the stellar interior is mainly composed of oxygen (left over from helium burning and having survived carbon burning), and neon and magnesium (products of carbon burning). One would expect that the next burning product is oxygen. However, oxygen is a double magic nucleus (Z = N = 8) and is therefore very stable (see Fig. 4.1). Before a temperature of 2 10^9K is reached and direct fusion of oxygen becomes possible, a new type of reaction intervenes. At temperatures of 1.2 - 1.8 10^9K, the ^{20}Ne spontaneously disintegrates into ^{16}O and an alpha particle. This process is called photodisintegration. The α-particle combines with the oxygen nucleus and restores the neon nucleus. After a short time lapse equilibrium is established, i.e. the neon nuclei photodisintegrate at the same rate at which they are formed. When the reaction has come to an equilibrium, the α particles start interacting with the ^{20}Ne to form ^{24}Mg.

The main reactions of neon burning are thus

$$1,2.\ ^{16}O+^{4}Ne \rightleftarrows\ ^{20}Ne \tag{4.45a}$$

$$3.^{20}Ne+^{4}He \rightarrow^{24} Mg \tag{4.45b}$$

The double arrow in (4.45a) shows that both the forward and the reversed reaction are active.
We analyse in somewhat more detail the reactions for the time evolution of ^{20}Ne, ^{24}Mg, ^{16}O and ^{4}He, the main species involved in neon burning. If we denote the rate of the forward reaction in (4.45a) by $r_1$and the photodisintegration rate by r_2, we have

$$\frac{dY_{20}}{dt} = \rho Y_{16} Y_4 r_1 - Y_{20} r_2 - \rho Y_4 Y_{20} r_3 \tag{4.46a}$$

$$\frac{dY_{16}}{dt} = -\rho Y_{16} Y_4 r_1 + Y_{20} r_2 \tag{4.46b}$$

$$\frac{dY_{24}}{dt} = \rho Y_{20} Y_4 r_3 \tag{4.46c}$$

$$\frac{dY_4}{dt} = Y_{20} r_2 - \rho Y_{16} Y_4 r_1 - \rho Y_{20} Y_4 r_3 \tag{4.46d}$$

The reactions producing and destroying the α particles are very rapid, so we may assume equilibrium. Furthermore, since the α-capture-rate on neon is much smaller than the capture rate on oxygen, we may neglect the third term in (4.46d) for the computation of the equilibrium abundance of the α particles. This leads to

$$Y_{4\infty} = \frac{1}{\rho} \frac{Y_{20}}{Y_4} \frac{r_2}{r_1} \tag{4.47}$$

The ratio r_2/r_1 is called the reverse ratio of the $^{16}O(\alpha,\gamma)$ reaction. It is a function of the temperature only, and is given in tables of nuclear reaction rates (Fowler et al. 1975, Harris et al. 1983).
If the α particles are in equilibrium, 4.46b becomes

$$\frac{dY_{16}}{dt} = \frac{dY_{24}}{dt} = -\frac{1}{2}\frac{dY_{20}}{dt} = \rho Y_{20} Y_4 r_3 \tag{4.48}$$

So neon burning can be summarized as

$$2\ ^{20}Ne \rightarrow^{16} O+^{24}Mg \tag{4.49}$$

Neon burning is not a direct fusion of neon into heavier elements, and therefore it would be more appropriate to call it a "melting" of neon into oxygen and magnesium.

The net transformation of two ^{20}Ne nuclei into one ^{16}O and one ^{24}Mg nucleus liberates 4.59 MeV.

Energy generation :

Using (4.47) and (4.48) and adopting the reaction rates of Fowler et al. (1975) we find for the energy generation rate by neon burning:

$$\varepsilon_{Ne}(\text{erg/g/s}) = \frac{Y_{20}^2}{Y_{16}} \left[1.32\ 10^{33} exp \left(-\frac{70.76}{T_9} \right) + 1.63\ 10^{33} T_9^2 \exp \left(-\frac{71.11}{T_9} \right) \right.$$

$$\left. + \theta^2\ 1.05\ 10^{31} \exp \left(-\frac{64.01}{T_9} \right) + \theta^2\ 8.00\ 10^{30} T_9^{\frac{5}{8}} exp \left(-\frac{64.27}{T_9} \right) \right. \qquad (4.50)$$

with $T_9 = T / 10^9$K. θ^2 is the reduced width of a resonance in the compound ^{24}Mg nucleus. For this factor, varying between 0 and 1, usually a value of 0.1 is adopted.

4.8 Oxygen burning

Temperatures of 2 10^9 K can be attaineded in stellarinteriors after neon "melting" and oxygen fusion starts. This reaction produces a compound nucleus of ^{32}S which can decay through four channels:

$$2\ ^{16}O \rightarrow {}^{31}S + n \qquad \text{(in 5\% of the cases)} \qquad (4.51a)$$

$$^{31}P + p \qquad \text{(in 56\% of the cases)} \qquad (4.51b)$$

$$^{30}P + {}^2H \qquad \text{(in 5\% of the cases)} \qquad (4.51c)$$

$$^{28}Si + \alpha \qquad \text{(in 34\% of the cases)} \qquad (4.51d)$$

All the reactions are exotherm, except the deuteron (^{2}H) channel which requires 2.41 MeV to terminate.

The fundamental reaction liberates neutrons, just as for carbon burning: protons and a particles are produced (the deuteron immediately disintegrates into a proton and a neutron). These particles interact with most of the heavy nuclei of the plasma and also here a complex network of secondary reactions is present, rearranging the final products of oxygen burning. The detailed formation of the elements by oxygen burning can only be followed using a large network of nuclear reactions, including all secondary reactions. ^{28}Si and ^{32}S are the main products of oxygen burning, representing some 90% of the mass at the end of the burning phase.

During oxygen burning the photodisintegration of heavy nuclei becomes important. Many nuclei heavier than iron are present in the stellar interior. Some of these come from the interstellar matter that has formed the star, and were already present when the star was formed. Others were formed by neutron captures during previous nuclear burning phases The photodisintegration reactions during oxygen burning "melt down" these heavy nuclei: they disintegrate, emitting α particles and decay into nuclei of the iron group (A ≈ 56). At the end of oxygen burning, almost no nuclei heavier than nickel remain in the deep interior of the star.

Another consequence of the large photodisintegration rate is the formation of quasi-equilibrium clusters, small groups of different nuclei coupled by nuclear reactions that are nearly balanced by their inverses. An example of such a cluster is:

$$^{28}Si + n_{\rightarrow,}^{\leftarrow}\, {}^{29}Si + p_{\rightarrow,}^{\leftarrow}\, {}^{30}P \tag{4.52}$$

All three of these nuclei are in equilibrium: they are produced at the same rate as they are destroyed, and the time derivative of their abundance is zero. Many other such small groups of nuclei are formed. Each group is internally in equilibrium, but a given group is not in equilibrium with another group. Stellar evolution proceeds at such a high rate that there is almost no time to establish equilibrium. For some reactions the equilibrium times that are required, are longer than the characteristic evolution time scale, determined e.g. by

$$\left(\frac{d\ ln\ T_c}{dt}\right)^{-1}$$

As the temperature rises towards the end of oxygen burning, more nuclei join the quasi-equilibrium clusters and small clusters merge into larger clusters. At the end of the burning phase nearly all nuclei heavier than ^{24}Mg and up to the iron group have joined into one large quasi-equilibrium cluster.
Energetically, oxygen burning can be considered as the direct fusion of two oxygen nuclei into one ^{32}S nucleus.

Energy generation :

The overall energy generation rate is (Fowler et al. 1975):

$$\varepsilon_O(\mathrm{erg/g/s}) = 2.88\ 10^{56}\,\frac{\rho Y_{16}^2 T_{9\alpha}^{\frac{5}{6}}}{T_9^{\frac{3}{2}}}\cdot\frac{\exp\left(-\frac{135.93}{T_{9\alpha}^{\frac{1}{3}}}\right)}{\exp\left(-0.032\ T_{9\alpha}^4\right)+3.89\ 10^{-4}\exp\left(2.659\ T_{9\alpha}^{\frac{2}{3}}\right)} \tag{4.53}$$

with $T_9 = T/10^9$K

$T_{9\alpha} = T_9/(1 + 0.067\ T_9)$

4.9 Silicon burning

At the end of oxygen burning a large amount of silicon has been produced in the stellar interior. In a final phase of nuclear burning this silicon is transformed into iron group-elements .

4.9.1 Silicon melting

Since the silicon nucleus has a very high Coulomb barrier a direct fusion reaction ^{28}Si + ^{28}Si does not occur. Instead, at a temperature of about 3.5 10^9K the ^{28}Si "melts down" by successive photodisintegration processes, releasing α particles. These α particles are captured by the nuclei between magnesium and nickel, leading to a gradual production of heavier nuclei. All nuclei are transformed into elements of the iron group.

The nuclear reactions governing silicon burning are very complicated. All nuclei between magnesium and iron can capture protons, neutrons and α particles. Each of these capture reactions on each isotope can evolve in three ways: an α capture reaction can evolve according to the γ, the proton or the neutron channel. In this way an enormous network of nuclear reactions is set up where all reactions are nearly balanced by their inverse, impossible to follow and to see how ^{28}Si builds up elements or the iron group.

4.9.2 Nuclear quasiequilibrium

Nevertheless, without solving the whole network we can have an approximated view of the situation. Near the end of oxygen burning quasi-equilibrium clusters appear, groups of bound by neutron, proton and a capture reactions and their inverses. At the onset of silicon burning these clusters have joined and all elements, from magnesium to the iron group elements, form one large quasi-equilibrium cluster. We can make use of the presence of equilibrium to determine the abundances of the elements: if an isotope is in equilibrium, its abundance is independent of the way it is synthesized.

The notion of nuclear equilibrium can be compared to the ionization equilibrium in atoms. An atom r times ionized can capture an electron and become a r-1 times ionized atom. The probability that this occurs (the "cross section") is a function of the structure of the atom and the temperature. The r-1 times ionized atom can capture another electron, or can capture a photon, emit an electron ("photodisintegrate") and become an r times ionized atom. Here, just as in the nuclear network, we recognize a complex system of reactions (electron capture), balanced by their inverse (ionization). However, we do not calculate the ionization balance by considering all ionization levels in the partly ionized atom together with all capture and decay probabilities of the states. Instead, we know that the captures and the decays are fast and therefore in equilibrium. We than appeal to a simpler statistical approach, giving the distribution of the different ionized atoms, independent of the way this distribution came to be: we use the Saha equation. This equation, giving the statistical equilibrium distribution of the atoms over their different ionization stages is only determined by the temperature and the ionization energies, i.e. the energies separating the different states.

If nuclear reactions and their inverses are in equilibrium, we can treat them in the same way. Consider as an example the reaction

$$^{28}\mathrm{Si} + {}^{4}\mathrm{He}_{\rightarrow}, {}^{\leftarrow 32}\mathrm{S} \tag{4.54}$$

If this reaction is in equilibrium, the leftward reaction is exactly balanced by the rightward reaction. Denoting the leftward reaction rate by r_1 and the photodisintegration rate by r_2, the equilibrium condition is

$$Y_{28} Y_4 r_1 = Y_{32} r_2 \tag{4.55}$$

and consequently

$$\frac{Y_{28}Y_4}{Y_{32}} = \frac{r_2}{r_1} \tag{4.56}$$

The term at the right-hand side is the "reverse ratio".

Since ions are not degenerate we can use for the right hand side of (4.56) the non-degenerate form of the Saha equation. The energies, separating the different "states" is here the energy produced by the forward reaction (and absorbed by the photodisintegration):

$$\frac{Y_{28}Y_4}{Y_{32}} = \frac{1}{\rho N_A}\frac{(2\pi\mu kT)^{\frac{3}{2}}}{h^3}\frac{G_{28}G_4}{G_{32}} exp - \frac{Q}{kT} \tag{4.57}$$

where μ is the reduced mass of the reacting isotopes, G_i are the partition functions and Q is the energy released by the reaction:

$$Q = \epsilon_{28} + \epsilon_4 - \epsilon_{32} \tag{4.58}$$

where ϵ_i is the mass excess of the considered nucleus (see Table 4.1). The partition functions are generally temperature-dependent and are found by integrating over all the excited states of the nucleus. The partition function for the neutron and the proton is 2, of the α particle 1 (no bound states). It can be shown that

$$\frac{Y_{28}Y_4}{Y_{32}} = F(\rho, \mathrm{T}_9, {}^{28}\mathrm{Si}, {}^{4}\mathrm{He}, {}^{32}\mathrm{S})$$

$$= 9.868\ 10^9 \frac{1}{\rho}\left(\frac{A_{28}A_4}{A_{32}}\right)^{\frac{3}{2}} T_9^{\frac{3}{2}} \frac{G_{28}G_4}{G_{32}} \exp\left(-\frac{11.605Q}{T_9}\right) \tag{4.59}$$

with

A_i the nuclear mass in a.m.u.
$T_9 = T / 10^9$K
Q in MeV
ρ in g/cm^3

This expression allows the determination of the abundance of ^{32}S:

$$Y_{32} = \frac{Y_{28}Y_4}{F(\rho, T_9, {}^{28}\mathrm{Si}, {}^{4}\mathrm{He}, {}^{32}\mathrm{S})}$$

This expression can be generalized for multiple captures of neutrons, protons and α particles by repeated application of 4.59. Since there are reactions that transform α particles into protons and neutrons ((α,p) and (α,n) reactions) the α particles are also in equilibrium.

4.9.3 Silicon burning

The basic reaction, triggering the melting of ^{28}Si and allowing the buildup of heavy nuclei is $^{24}Mg(\gamma,\alpha)^{20}Ne$. The destruction rate of ^{28}Si can approximately be determined by considering the following rearrangement reactions:

1. ^{28}Si is in quasi-equilibrium with ^{24}Mg by photodisintegration and α capture.
2. ^{24}Mg is in equilibrium with ^{28}Si and disintegrates into ^{20}Ne with a rate r.
3. ^{28}Si melts down and for each two ^{28}Si nuclei that disappear one ^{56}Ni nucleus is formed. Furthermore, ^{28}Si and ^{56}Ni are in equilibrium through the whole quasi-equilibrium cluster of silicon burning.

We can find a simplified expression for the time evolution of ^{28}Si by considering the ratio of the number fraction of ^{56}Ni to the number fraction of ^{28}Si (Y_{28}/Y_{56}). This yields

$$\frac{dY_{28}}{dt} = 9.55\ 10^{10} Y_{28} \left(\frac{Y_{28}}{Y_{56}}\right)^{\frac{1}{7}} T_9^{\frac{3}{2}} \exp\left(-\frac{142.06}{T_9}\right) r_1 \tag{4.60}$$

where r_1 is the rate of the reaction $^{20}Ne(\alpha,\gamma)^{24}Mg$. Remark that the destruction rate of ^{28}Si is independent of the density.

Energy generation :

The energy generation rate can easily be found easily by considering silicon burning as the direct fusion of two ^{28}Si nuclei into one ^{56}Ni nucleus. Neglecting the term $(Y_{28}/Y_{56})^{1/7}$ which is approximately unity we find, after introducing the reaction rate of Fowler et al. (1975):

$$\varepsilon_{Si}(\mathrm{erg/g/s}) = Y_{28}\left[2.59\ 10^{33} \exp\left(-\frac{157.93}{T_9}\right) + 3.20\ 10^{33} T_9^2 \exp\left(-\frac{158.29}{T_9}\right)\right] \tag{4.61}$$

4.9.4 Nuclear statistical equilibrium

Near the end of silicon burning *all* nuclei from helium to the iron group join into one equilibrium group. The abundance of an element ^AZ, with Z protons and N = A-Z neutrons as functions of the abundance of the free neutrons and the free protons, in the same way as was done for silicon burning. We consider a nucleus A_Z as formed by the "capture" of Z protons and N neutrons:

$$Y_A = \frac{1}{2}(1.974\ 10^{10})^{1-A} \rho^{1-A} A^{\frac{3}{2}} G(^AZ,T) T_9^{\frac{3}{2}(1-A)} \exp\left(\frac{11.605Q}{T_9}\right) Y_n^N Y_p^Z \tag{4.62a}$$

with

$$Q(MeV) = 8.07144A - 0.78245Z - \varepsilon_A \tag{4.62b}$$

The abundances of the free neutrons and protons can then be derived from the condition that the sum of all mass fractions must be one and from the definition of the neutron excess. Calculations of abundances of the elements in nuclear statistical equilibrium reveal that the abundances are influenced considerably by the value of the neutron

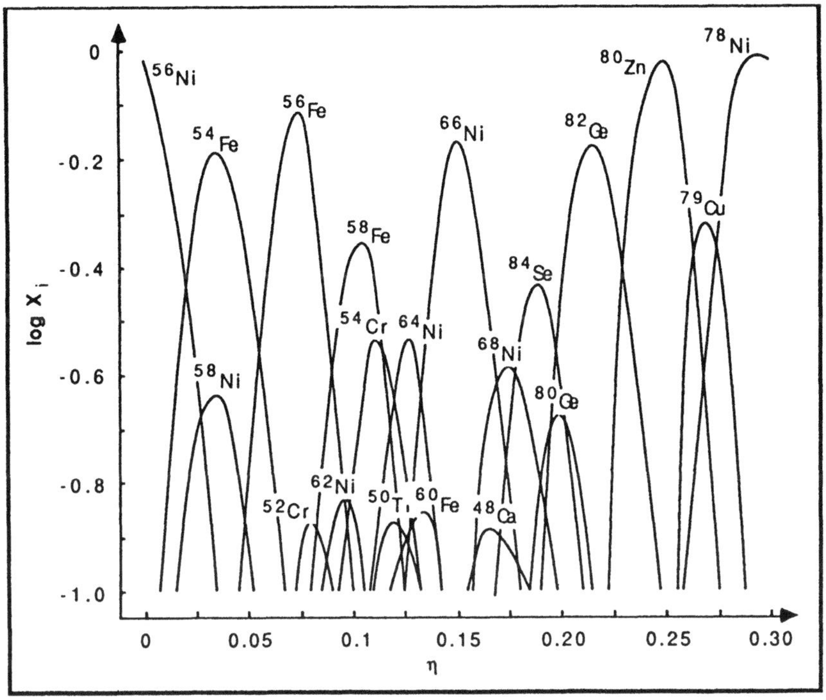

Fig. 4.5. The mass fractions of elements in nuclear statistical equilibrium as a function of the neutron excess for T = 3.5 10^9K and r = 10^7 g/cm^3 (Hartmann et al. 1985).

excess. If this excess rises, the plasma contains more neutrons and the equilibrium shifts towards neutron rich isotopes. The distribution of the heavy elements in nuclear statistical equilibrium is shown in Fig. 4.5 as a function of the neutron excess for a typical temperature and density for silicon burning. One sees that for a neutron excess of zero, mainly ^{56}Ni is formed. If the neutron excess becomes larger, we find first ^{54}Fe and then more neutron-rich isotopes: ^{56}Fe, ^{58}Fe, then ^{66}Ni and even ^{78}Ni for a neutron excess of 0.3. The neutron excess in a real star during silicon burning will thus be a dominating factor in the determination of the final outcome of silicon burning.

4.9.4 The end of silicon burning

The exact end products of silicon burning depend on the neutron excess and also on the temperature and the density. ^{56}Ni is a very important isotope: in about three months it decays into ^{56}Fe, which is the common iron isotope, found abundantly on earth.

We may assume that the two isotopes are linked by the equilibrium rearrangement reaction

$$^{54}\mathrm{Fe}{+}2\ ^{1}\mathrm{H} \rightarrow ^{56}\mathrm{Ni} \tag{4.63}$$

Hence the equilibrium ratio is, assuming that the partition functions of both nuclei are equal:

$$\frac{Y_{54}}{Y_{56}} Y_p^2 = \frac{1}{\rho^2} 3.69\ 10^{20} T_9^3 \exp\left(-\frac{141.95}{T_9}\right) \tag{4.64}$$

We assume now that the plasma consists of a mixture of ^{56}Ni nuclei and ^{54}Fe nuclei accompanied by two protons, an extremely simplifying assumption. This implies that $Y_p = 2Y^{56}$ and a neutron excess of zero.

We can easily derive from (4.64) the condition for which the mass fraction of ^{54}Fe is equal to the mass fraction of ^{56}Ni.

The division line between the formation of ^{56}Ni and the presence of ^{54}Fe + 2 ^{1}H is shown in Fig. 4.6. For lower temperatures and high densities (as in normal hydrostatic silicon burning) ^{56}Ni is dominant (see also Fig.4.5). For higher temperatures, as expected in explosive silicon burning, ^{54}Fe is dominant. This shows that the common iron is synthesized during silicon burning at moderate temperatures and high densities.

We are left with a very important question. At the end of silicon burning all elements from helium up to the iron group nuclei form one large equilibrium cluster. We have always assumed that this nuclear statistical equilibrium favors the formation of heavier nuclei, i.e. that the equilibrium tends towards the elements of the iron group. But we can imagine that under some circumstances, the equilibrium changes its sense and switches over towards the low end of the equilibrium cluster so that heavy nuclei disintegrate slightly faster than they can capture particles.

We consider the equilibrium condition between the two ends of the cluster: ^{4}He and ^{56}Ni:

$$14\ ^{4}\mathrm{He} \longleftrightarrow ^{56}\mathrm{Ni} \tag{4.64}$$

If we neglect the nuclear partition functions we find for the equilibrium ratio of helium and nickel:

$$\frac{Y_4^{14}}{Y_{56}} = 8.83\ 10^{139} \rho^{-13} T_9^{\frac{39}{2}} \exp\left(-\frac{1019.4}{T_9}\right) \tag{4.65}$$

The mass fractions of helium and nickel are both equal to 1/2 for

$$\rho(\mathrm{g/cm^3}) = 3.802\ 10^{11} T_9^{\frac{3}{2}} \exp\left(-\frac{78.42}{T_9}\right) \tag{4.66}$$

This division line is also shown in Fig. 4.6.

Above the boundary line, i.e. the isotopes cluster at the high end of the equilibrium cluster and the equilibrium can subsist.

Below the boundary line the heavy iron nuclei spontaneously disintegrate and the equilibrium tends towards the formation of α particles in great abundance.

Although this disintegration demands energy, it occurs spontaneously as the boundary, given by (4.66) is crossed. The sudden switch to the lower end of the equilibrium cluster absorbs an enormous amount of energy from the stellar interior and provokes the collapse of the star. With the disintegration of the nuclei of the iron group into α particles the nuclear life of the star has come to an end.

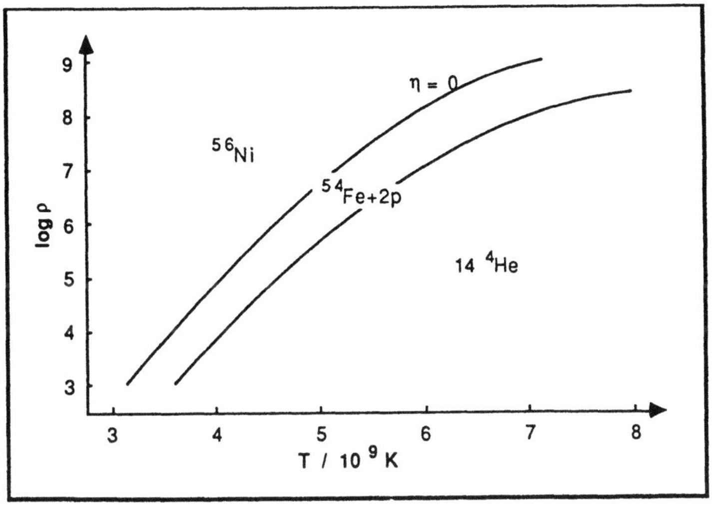

Fig. 4.6. The end products of silicon burning.

5 Numerical techniques

5.1 The solution of the structure equations

5.1.1 Introduction

The differential equations determining the stellar structure form a system of coupled, non-linear differential equations. The system cannot be solved analytically. Moreover, the equations present a boundary value problem, since boundary conditions for the stellar center and the surface are required.

Hence one has to apply numerical methods, implying that only an approximation of the real solution can be obtained in a discrete number of points inside the star. There is even no guarantee that the solution of the equations is unique. There is no general theorem in mathematics which states that a solution exists, and if so, that it is the only one.

A straightforward numerical method to solve the equations of stellar structure would be the following:

1. Integrate numerically the equations of stellar structure starting at the center, towards the surface and use some trial values of the parameters α and T at the center (α_c and T_c).
2. If the results of this integration - the values of the parameters at the surface - do not coincide with the boundary conditions at the surface, change α_c and T_c by some appropriate amounts.
3. Repeat the first two steps until the boundary conditions are fulfilled and convergence is reached.

This method fails in the first step, since the solution diverges near the surface. The outer boundary conditions for a star essentially mean that the temperature and the pressure tend to zero simultaneously. Since these quantities are very sensitive to the central conditions it is not simple to find values of the central parameters so that the surface temperature and pressure vanish simultaneously. Moreover, if the pressure reaches zero before the temperature does, or vice versa, severe numerical problems occur in the evaluation of the derivatives of the structure equations.

Reversing the sense of the integration does not help to solve the problem. In this case, one has to choose the surface boundary conditions as initial values for an inward integration. Approaching the center, the radius and the luminosity must approach zero simultaneously. If this is not the case, because the surface boundary conditions were not exactly correct, again instabilities occur in the evaluation of the derivatives of the structure equations.

The correct method would thus be an outward integration starting at the center and an inward integration starting at the surface, ensuring that both numerical integrations are stable. This is the *Schwarzschild integration* (Schwarzschild 1958):

1. Integrate outward from the center, using as initial conditions trial values for the central parameters. Integrate up to a 'fitting point', about half way between the center and the surface.
2. Integrate inward from the surface, using trial values of the surface boundary conditions as initial values of the integration. Integrate down to the fitting point.
3. If the parameters at the fitting point, obtained from both integrations are not equal, correct the trial values, as well at the center as at the surface, by appropriate amounts.

- Repeat steps one to three until both integrations yield the same values for the parameters at the fitting point and convergence is reached.

In this method the numerical accuracy of the integrations and of the parameter fitting at the fitting point are chosen arbitrarily, so it is possible to construct very accurate solutions of the equations of stellar structure. However, this method is only successful for 'simple' stars, e.g. stars having a homogeneous composition, but it does not converge for 'complicated' stars, e.g. stars containing narrow zones of nuclear burning. In these cases it is found that the values of the parameters at the fitting point are too heavily dependent on the boundary conditions, causing instabilities in the computation of the appropriate corrections to the boundary conditions.

These problems can be avoided by using a totally different method of solution, introduced by Henyey et al. (1959, 1964), the *Henyey iteration method* (See Kippenhahn et al. 1967 for a detailed description).

The star is divided into a number of N mass shells with boundaries at mass $M_r = M_i$ $(1 \leq i \leq N)$, where $M_1 = 0$ and $M_N = M$ (Fig. 5.1). The structure equations are used to establish relations between the solution at the different mesh points M_i.

The relation between M_1 and M_2, the two mesh points at the center, is given by a series expansion of the structure equations near the center.

The relation between M_{N-1} and M_N, the two mesh points at the surface is determined by a direct inward integration of the structure equations, in the same way as in the Schwarzschild integration technique.

The relation between all other mesh points between the center and the surface is determined by replacing the equations of structure by difference equations.

As an example the first structure equation, at mesh points M_j and M_{j+1}, is replaced by the difference equation

$$\frac{r_{j+1} - r_j}{M_{j+1} - M_j} = \left[-\frac{M}{4\pi r^2 \rho} \right]_{j+\frac{1}{2}} \tag{5.1}$$

where the subscript j+1/2 refers to an average between the mesh points j and j+1. In most cases one takes

$$[\,]_j + \frac{1}{2} = \frac{1}{2}\left([\,]_j + [\,]_{j+1}\right) \tag{5.2}$$

The replacement of the structure equations by difference equations introduces an error: at the right hand side of (5.1) we should write other terms containing higher order derivatives of the structure equations. These higher order terms we have neglected make up the truncation error. It is the same kind of error that appears if we truncate an infinite series to a finite number of terms (in this case this is restricted to one term).

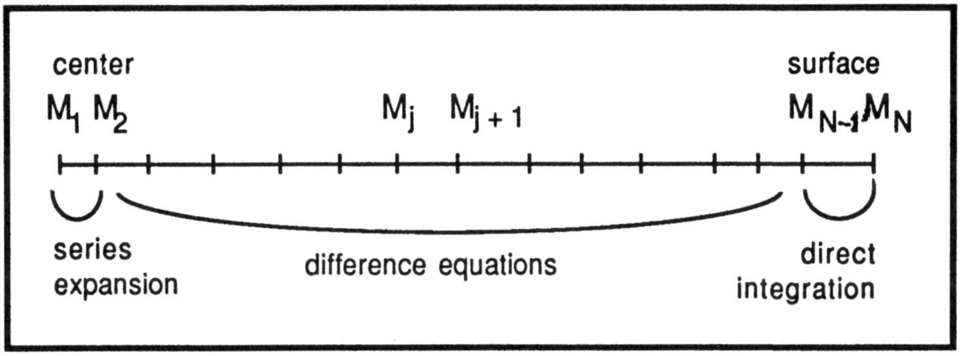

Fig. 5.1. Mesh points used in the Henyey iteration method

Equations of the type (5.1) are established for all structure equations and for all intermediate mesh points. For 5 structure equations, we have:

5 relations from the central series expansion

5 relations from the outer boundary conditions

5(N-3) difference equations.

Together with 5 boundary conditions one has in total 5N conditions for determining the 5N unknowns representing the solution (i.e. the values of the 5 parameters in all N mesh points). These 5N conditions can be written as 5(N-1) equations since five boundary conditions are fixed. These equations are solved for the parameters at all mesh points, using a generalized Newton iteration technique on the linearized equations. This involves the inversion of a 5(N-1) by 5(N-1) matrix which is only non-zero near the diagonal. In the Henyey iteration method this special form of the matrix is used to obtain an efficient inversion algorithm (see Kippenhahn et al. 1967).

The Henyey iteration method is very stable: small local errors, such as truncation errors, do not propagate to other mesh points and do not affect the general solution of the equations too much. Almost all present evolution codes use the Henyey iteration method to solve the structure equations. In most cases the generalized Newton iteration converges in *less that 10 iterations* on the difference equations. If a few tens of iterations are allowed for, the method converges in almost all possible and impossible situations. In this respect it is far superior to the Schwarzschild integration method.

However, the Henyey iteration method *can only be used if an approximate solution of the equations is available* in order to have good starting conditions for the Newton iteration on the equations. If an evolutionary time sequence of several models is computed, the solution is first approximated by an extrapolation from the previous model in the sequence. Only an approximation of the first model in the sequence, with the Schwarzschild integration technique has to be computed. In many cases this first model is homogeneous and the integration poses no problems.

The Henyey iteration method has a severe inconvenience: the control of the truncation error. If the difference equations are written in the form given by (5.1) and (5.2), the truncation error is

$$-\frac{(M_{j+1}-M_j)^3}{24}y''' \tag{5.3}$$

with y''' the third derivative of the parameter considered. Although the presence of the truncation error does not lead to spurious solutions, it makes up a deviation from the true solution even if the iterations have fully converged. The truncation error becomes smaller if the difference between adjacent mesh points is smaller, i.e. if more mesh points are used. If one uses a good repartition of some 100 to 200 mesh points over the star (a good 'zoning') it is possible to limit the relative error to 0.0001. But there exists no good criterion to determine where the zoning must be fine (many mesh points) and where it may be crude (less mesh points). If the zoning is not perfect, the truncation error may be much larger. Owing to the good stability of the method, these truncation errors are not distinguishable in the final solution. Moreover, in some evolutionary codes the difference equations have another form than those given by (5.1) and (5.2), implying in most cases a better stability but also a larger truncation error.

The advantages of the Schwarzschild integration method (accuracy) and the Henyey iteration method (stability) can be combined in an intermediate numerical method, proposed by Wilson (1981). The structure equations are not replaced by difference equations but are directly integrated. The method uses several fitting points instead of one. A generalized Newton iteration method is used to iterate on the values of the parameters at the fitting points, but the relations between the different fitting points are not given by difference equations. This method is generally known as the *Multiple Fitting* method or the *Multiple Shooting* method.

5.1.2 The multiple fitting method

We choose n_t fitting points, distributed over the star; fitting point 1 is close to the surface, fitting point n_t is located near the center of the star (Fig. 5.2). Among these points we choose a main fitting point, the point n_f. We further call the surface ($\mu = 0$) fitting point 0.

Assume that estimates of the parameters y_i ($1 \leq i \leq 5$: r, v/r, L_r, α and T) are known at the fitting points 1 to n_t, except at n_f. Further we have estimates of the central and surface boundary conditions (see Fig. 5.2).

- We start the inward integration from the surface toward fitting point 1, using the surface boundary conditions as initial values for the integration. This integration yields

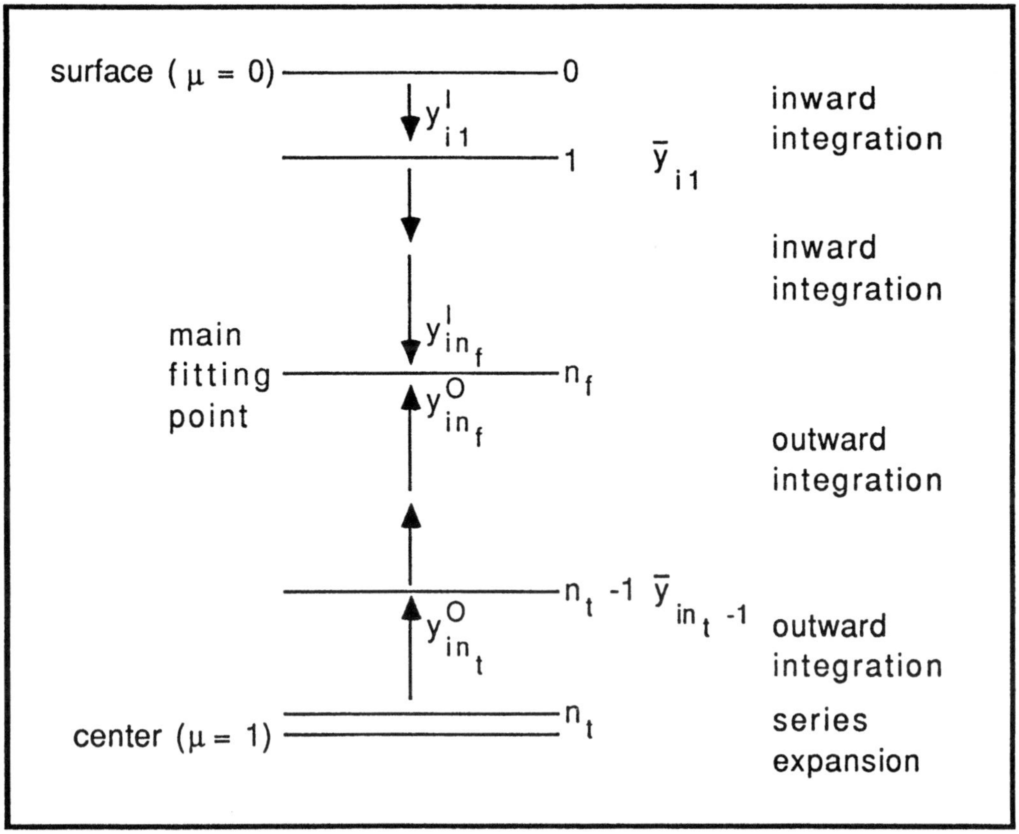

Fig. 5.2. The distribution of the fitting points over the star, at the top the stellar surface, at the bottom the stellar center, and the boundary conditions.

values y^I_{i1} at fitting point 1, where I stands for inward integration. The results of the inward integration, y^I_{i1} will in general be different from the estimates $\overline{y}_{i1}$ of the parameters at fitting point 1.

- We take $\overline{y}_{i1}$ as initial conditions and integrate from fitting point 1 to fitting point 2. This yields values y^I_{i2}, to be compared to the estimates $\overline{y}_{i2}$ at fitting point 2. We continue the integrations from fitting point j to j+1 until j+1 = n_f.

- Below the main fitting point we integrate outward.

- First we obtain new values at the fitting point n_t by the series expansion of the structure equations near the center.

- Next we integrate outward from fitting point j+1 to j (n_t-1 $\leq$ j $\leq$ n_f) using the estimates of the parameters at each fitting point as initial values. This yields values y^O_{ij}, to be compared to the estimates y^I_{ij} .

- At the main fitting point we have no estimates for y_i but we can compare y^O_{ij} to y^I_{ij}.

After a first series of integrations through the star the comparison between the integrated values of y_i and the estimates $\overline{y}_{i1}$ will yield differences. The estimates of the parameters at the fitting points and the free boundary conditions will have to be corrected by appropriate amounts. The integrations have to be repeated until the differences between the integrated values and the estimates are smaller than desired accuracies.

This method combines the advantages of the Schwarzschild and the Henyey methods. The numerical integrations between the fitting points can be carried out with any desired accuracy. Moreover, the main fitting point can be chosen in a way that all integrations proceed in the most stable way (inward near the surface, outward near the center). Besides, the method reduces to the Schwarzschild integration method for $n_t = 2$ and $n_f = 1$. In this case only the boundary conditions have to be specified since the first estimates of y_{i2} can be obtained from the series expansion near the center, and since no estimates are needed at the fitting point.

On the other hand the use of estimates at the fitting points and their iterative correction is essentially the Henyey iteration method. However, we do not iterate on the *difference equations* but on the *condition equations*; the iteration loop is repeated until the result of the integration equals the estimates. If the structure of the star becomes more complicated, extra fitting points can be inserted, taking advantage of the ability of the Henyey iteration method to solve complicated problems, but conserving the accuracy at the desired level. At the limit the method reduces to the Henyey iteration method if the fitting points are so closely spaced that each integration can be performed in one integration step.

Unknowns and equations. We will first consider the "Henyey iteration" part of the method, i.e. the method for obtaining the corrections on the estimates of the parameters, once all integrations throughout the star have been performed.

The complete solution of the structure equations in the fitting points is determined by the following parameters:

r_s: the radius at the surface;
$(v/r)_s$: the "velocity" at the surface;
L_s: the luminosity at the surface;
a_s: the value of α at the surface;
T_s: the temperature at the surface;
y, ¯sub $_{ij}$: the value of parameter i ($1 \leq i \leq 5$) at fitting point j ($1 \leq j \leq n_t, j \neq n_f$)
$(v/r)_c$: the central "velocity";
α_c: the value of α at the center;
T_c: the central temperature.

A total of $5n_t+3$ parameters t can be formed.
These parameters must satisfy the following conditions:

$$T_s^4 - \frac{L_s}{8\pi\sigma r_s^2} = 0 \tag{5.4a}$$

(this is the boundary condition 2.37e),

$$\alpha_s - f(T_s) = 0 \tag{5.4b}$$

(the boundary condition (2.37d); the exact form of f depends on the meaning of α),

$$\bar{y}_{ij} - y^I_{ij} = 0 \qquad (1 \le i \le 5,\ 1 \le j \le n_f) \tag{5.4c}$$

(the results of the inward integrations should be equal to the estimates at all fitting points above n_f),

$$y^I_{i\ n_f} - y^O_{i\ n_f} = 0 \qquad (1 \le i \le 5) \tag{5.4d}$$

(the inward and the outward integration toward the main fitting point should yield the same result),

$$\bar{y}_{ij} - y^O_{ij} = 0 \qquad (1 \le i \le 5,\ n_f < j < n_t) \tag{5.4e}$$

(the results of the outward integrations should beequal to the estimates at all fitting points below n_f),

$$\bar{y}_{i\ n_t} - s_i(\frac{v}{r_c}, \alpha_c, \underset{c}{\mathrm{T}}) = 0 \qquad (1 \le i \le 5) \tag{5.4f}$$

where s_i is the result of the series expansion near the center (the series expansion from the center to fitting point n_t should be equal to the estimates at that point), and finally

$$\frac{v}{r_c} - g(\alpha_c, \underset{c}{\mathrm{T}}) = 0 \tag{5.4g}$$

(the boundary condition (2.37b) where $g(\alpha_c,T_c) = -1/3\ D\ \ln\rho_c$).

In total we have $5n_t+3$ equations, completely determining the unknowns.

The solution method. The equations for the solution of the differential equations, form a system of $5n_t+3$ equations with $5n_t+3$ unknowns. Since the system is not linear , it cannot be solved by one simple matrix inversion. Instead, a kind of generalized Newton iteration technique is used.

Suppose we have to solve a system of N equations with N unknowns y_j, which can be written symbolically as

$$\eta_i(y) = 0 \qquad (1 \le i \le N) \tag{5.5}$$

The vector **y** depends on the N unknowns. If we insert an estimate of the solution, **y**,$^-$ into (5.5) the equations will not be fulfilled but we will obtain a "residue" $\Delta\ \eta_i$ for each equation:

$$\eta_i(\bar{y}) = \Delta\eta_i \qquad (1 \le i \le N) \tag{5.6}$$

It is clear that we want to make the residue zero by changing the estimates **y**,$^-$ in an appropriate manner. Suppose we want to change the estimates $\bar{y}_i$ by amounts dy_j in order to obtain $\eta_i(\bar{\mathbf{y}}+\mathbf{dy})$=0. The values of dy_j are then given to first order by differencing η_i:

$$d\eta_i = \sum_{j=1}^{N} \frac{\delta\eta_i}{\delta_{y_j}} \delta y_j = -\Delta\eta_i \qquad (1 \le i \le \mathrm{N}) \tag{5.7}$$

or

$$C \cdot \mathbf{dy} = -\Delta\eta \tag{5.8a}$$

where

$$C_{ij} = \frac{\partial \eta_i}{\partial y_j} \qquad (1 \leq i, j \leq \mathrm{N}) \tag{5.8b}$$

The corrections to the estimates of the solution $\mathbf{y}$ are determined by the solution of the linear system of equations (5.8). However, since the equations η_i are not linear, there are higher order terms that are neglected in (5.7), so the solution of (5.8) will only give an approximation of the corrections: we obtain a new estimate of the parameters $\mathbf{y}_j$: $\bar{y_j} + dy_j$.

If we insert the new estimates into the equations, the residues will probably be smaller, but not yet zero. Therefore the computation of the corrections has to be repeated until $|\Delta\ \eta_i|$ is small enough for all i.

In our problem the equations to be solved are the equations (5.4). If we note that $\mathrm{y}^{O}{}_{,ij}$ depends on the estimates in fitting point j+1 and that $\mathrm{y}^{I}{}_{,ij}$ depends on the estimates in fitting point j-1 we can easily write the equations (5.8) for our problem. The structure of the vector $\mathbf{dy}$ and the matrix C is given in Fig. 5.3. The order of the equations in the matrix C is the same as given in the previous section. The matrix C consists of a number of small submatrices. The submatrix A contains the information on the outer boundary conditions. The submatrices B and D contain the information on the inward and outward integrations respectively:

$$(B_j)_{ik} = -\frac{\partial y^I_{ij}}{\partial \bar{y}_{k\ j-1}} \qquad (1 \leq j \leq n_f,\ 1 \leq i, k \leq 5) \tag{5.9a}$$

$$(D_j)_{ik} = -\frac{\partial y^o_{ij}}{\partial \bar{y}_{k\ j+1}} \qquad (n_f \leq j < n_t, 1 \leq i, k \leq 5) \tag{5.9b}$$

The solution of this system of linear equations can be computed with standard techniques. C is a "band matrix" (only elements near the diagonal are non-zero) hence it can be stored efficiently in computer memory and it is appropriate to solve the system by decomposing C into a lower- and an upper diagonal matrix (LU decomposition).

The numerical integration. The numerical integration of the structure equations between the fitting points, in order to determine the coefficients of the B and D matrices, can in principle be performed with any standard method for the numerical integration of a system of differential equations. It is, however, desirable to have a control on the truncation error of the numerical method, since the final values of the integration must be computed with an error smaller than the tolerances $|\Delta\ \eta_i|$ we set on the condition equations.

This can be accomplished by using a so-called predictor-corrector method, where the predictor and the corrector formula have a truncation error of the same order (see e.g. Press et al. 1986).

In such a method, an estimate of the solution at the point x + dx is computed using the predictor. This predictor only uses values of the parameters and the derivatives in

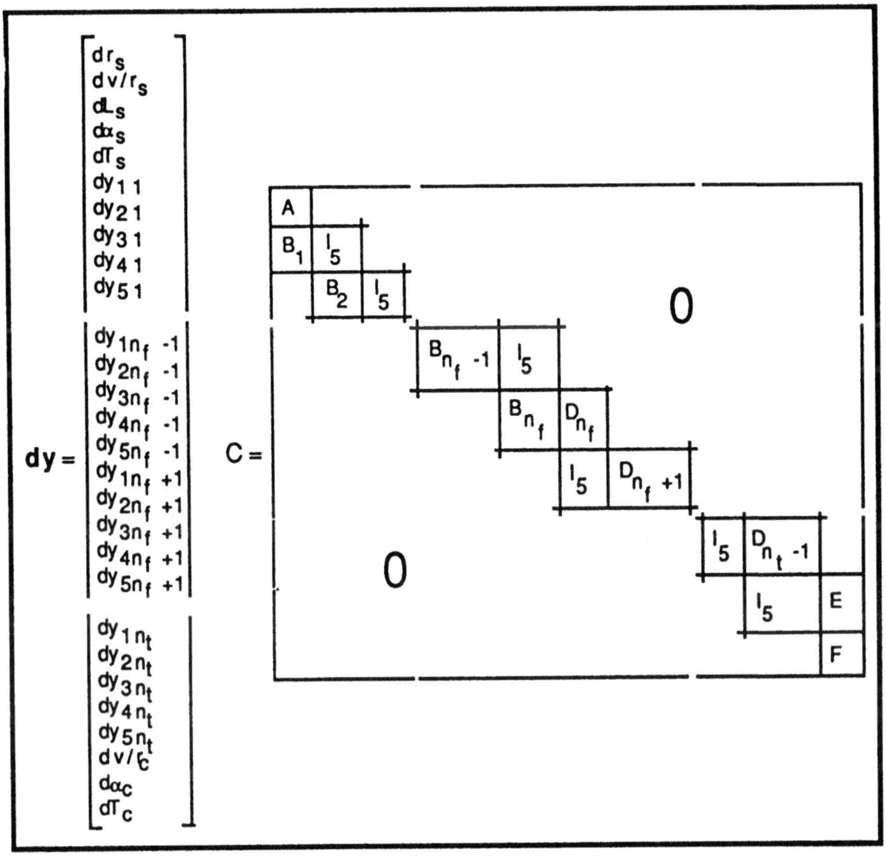

Fig. 5.3. The structure of the matrix C and the vector d_y.

previously computed points. This first estimate of the solution can be improved by applying a modifier. The ultimate numerical solution at x + dx is then computed by iteration on the corrector. This corrector is an implicit formula, i.e. it contains the derivatives of the parameters at x + dx, which must be evaluated at each iteration. The iteration on the corrector is continued until the desired accuracy is reached. The truncation error on the solution can then be estimated from the difference between the predictor and the corrector if both are of the same order.

As an example we can derive a very simple predictor-corrector algorithm for the solution of differential equations. Suppose we know the solution of the equation at a point x_o (solution y_o, derivative $y'_{,o}$ and we want to determine the solution at a point $x_1 = x_o + h$, where h is called the step size. Developing the solution into a Taylor series at x_o we find:

$$y_1 = y_0 + hy'_0 + \frac{h^2}{2} y''_0 + \tag{5.10}$$

We can derive from this series expansion a simple predictor formula:

$$y_1 = y_0 + hy_0' \tag{5.11}$$

This is called Euler's method. The truncation error is given by the first term in the series we have neglected:

$$\frac{h^2}{2}y'' \tag{5.12}$$

The truncation error is proportional to h^2 so we say that it is a first order method. The method is a predictor since no terms depending on the values of the parameters at x_1 appear.

The corrector for this method can be found by considering the Taylor series for y and y' in x_0:

$$y_1 = y_0 + hy_0'' + \frac{h^2}{2}y_0'' + \frac{h^3}{6}y_0''' + \dots \tag{5.13a}$$

$$y_1' = y_0' + hy_0'' + \frac{h^2}{2}y_0''' + \dots \tag{5.13b}$$

These Taylor expansions lead to the corrector formula, called the improved Euler method:

$$y_1 = y_0' + \frac{h}{2}(y_0' + y_1') + \frac{h^3}{24}y''' \tag{5.14}$$

The last term is the truncation error. The corrector is of second order.

We want to limit the truncation error to a strict minimum in order to obtain a solution which is as accurate as possible. This can be achieved by using a high order method in which the truncation error is proportional to h^4 or h^5. Generally a method is adopted where the values of the solutions and the derivatives are used in the points x, x - dx, x - 2dx, ... to obtain the solution at x + dx. Such methods (e.g. Adams-Moulton algorithms) are very accurate but they have as drawback that it is not easy to adjust the integration step dx to the desired accuracy.

Moreover, it is still necessary to evaluate the derivatives during each iteration on the corrector. In our case we have to compute the structure equations

to evaluate the equation of state

to calculate the energy generation rate

and to find the opacity

If many iterations and many integration steps are needed, this is very lengthy and it would be more suited to apply a method in which the number of times we have to evaluate the structure equations is limited, without losing accuracy.

These requirements can be fulfilled simultaneously by using the following method, which is based on the use of the second derivative of the solution and the Jacobian matrix.

Suppose we want to solve the system of first order differential equations

$$y_i' = f_i(y_j, x) \tag{5.15}$$

where the first derivative $y'_{,i}$ depends on the independent variable x and on all parameters y_j (but not on their derivatives). Suppose we know the solution and all derivatives at the

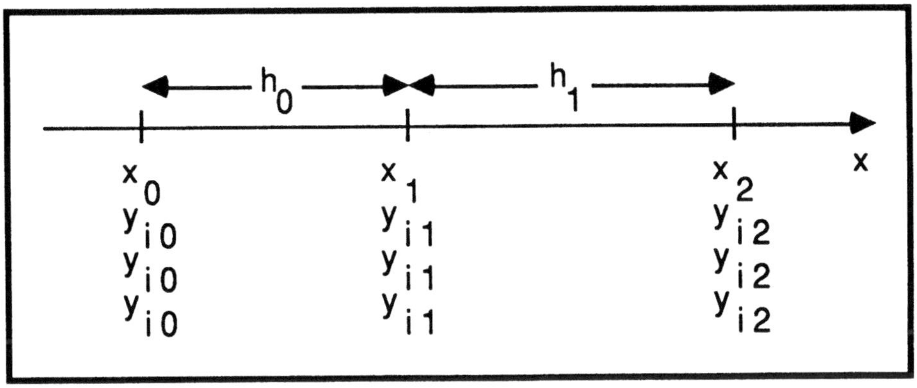

Fig. 5.4. From the solution and the derivatives at the points x_0 and $x_1=x_0+h$, the numerical solution at point $x_2= x_1+h$ is computed.

points x_0 and $x_1 = x_0 + h_0$ (Fig. 5.4). We now want to compute the numerical solution at the point $x_2 = x_1 + h_1$. Then a full fourth order predictor-corrector integration scheme is:

predictor:

$$\overline{y}_{i2} = h_{i1} + h_1(a_1 y'_{i0} + a_2 y'_{i1}) + h_1^2(a_3 y''_{i0} + a_4 y''_{i1}) + h_1^5 b y^V(\xi) \tag{5.16a}$$

where

$$a_1 = \alpha^2 + \frac{\alpha^3}{2} \tag{5.16b}$$

$$a_2 = 1 - a_1 \tag{5.16c}$$

$$a_3 = \frac{\alpha}{3} + \frac{\alpha^2}{4} \tag{5.16d}$$

$$a_4 = \frac{1}{2} + \frac{2}{3}\alpha + \frac{\alpha^2}{4} \tag{5.16e}$$

$$b = \frac{1}{720}\left(6 + \frac{15}{\alpha} + \frac{10}{\alpha^2}\right) \tag{5.16f}$$

and

$$\alpha = \frac{h_1}{h_0} \tag{5.16g}$$

$$x_0 < \zeta < x_2 \tag{5.16h}$$

modifier:

$$\overline{\overline{y}}_{j2} = \overline{y}_{i2} + \delta\gamma_{i1}(5\alpha^5 + 15\alpha^4 + 10\alpha^3) \tag{5.17}$$

where $\delta y_i 1$ is the truncation error on $y_i 1$;

corrector:

$$y_{i2} = y_{i1} + \frac{h_1}{2}(y''_{i2} + y''_{i1}) - \frac{h_1^2}{12}(y''_{i2} - y''_{i1}) + \frac{1}{720}h_1^5 y^V(\zeta) \tag{5.18}$$

The truncation error on $y_i 2$ is :

$$\delta y_{i2} = \frac{1}{720}h_1^5 y\hat{V}(\zeta) \approx \frac{\alpha^2(y_{i2} - \overline{y}_{i2})}{5\alpha^2 + 15\alpha + 10} \tag{5.19}$$

Since the truncation errors on the predictor and the corrector (the last term in (5.16a) and (5.18)) are proportional to h^5 the method is of fourth order. The purpose of the modifier is to bring the estimates of y_{i2} as close as possible to the solution of (5.12) in order to improve the convergence of the corrector.

One must iterate on the corrector until the results of two iterations are very close to each other. In principle, at each iteration one must re-evaluate $y'_{,i2}$ and $y''_{,i2}$ which are dependent on the result of the corrector. However, we will not re-evaluate the derivatives at each iteration. Since the values of the modifier, $\overline{\overline{y}}_{j2}$, are very close to the result of the corrector, y_{12}, we can safely develop the derivatives f_i in a series around $(x_2, \overline{\overline{y}}_{j2})$. Differencing f_i we find

$$df_i = \sum_{j=1}^{n} \frac{\partial f_i}{\partial y_j} dy_j + \frac{\partial f_i}{\partial x} dx \tag{5.20}$$

Let $\Delta y_j = y_{j2} - \overline{\overline{y}}_{j2}$ be the differences between the present values of the corrector and the modifier. Then the first derivative at the corrector values is

$$y'_{i2} = y'_{i2} + \sum_{j=1}^{n} \frac{\partial \overline{\overline{y}}'_{i2}}{\partial y_j} \Delta y_j \tag{5.21}$$

where $\overline{\overline{y}}_{j2} = f_i(x_2, \overline{\overline{y}}_{j2})$

The second derivatives are then given by

$$y''_{i2} = \sum_{j=1}^{n} \frac{\partial \overline{\overline{y}}'_{i2}}{\partial y_j} y'_{j2} + \frac{\partial \overline{\overline{y}}'_{i2}}{\partial x} \tag{5.22}$$

If we compute once the derivatives f_i and the partial derivatives of f_i for the parameter values given by the modifier, we can always evaluate the first and second derivatives using (5.21) and (5.22) without computing f_i again. The approximations used in (5.21) and (5.22) do not affect the truncation errors of the predictor and corrector.

The approximations (5.21) and (5.22) are also useful for the computation of the coefficients of the C matrix. In order to compute the submatrices B and D we must know the partial derivatives of the final values of the parameters at the end of the integration with respect to the initial conditions. These can be computed by disturbing the initial conditions and performing five parallel integrations, simultaneously with the main integration. The first and second derivatives, needed at each step in the parallel integrations, can also be computed using (5.21) and (5.22).

5.2 The evolution of the chemical abundances

The evolution of the abundances in the stellar interior is described by diffusion equations, containing the mixing of the elements by various processes such as convection and the transmutation of the elements by nuclear reactions. If we denote again by X_i the mass fraction of the element i, and by A_i its atomic weight, the evolution in space and time is described by the equation

$$\rho\frac{\partial X_i}{\partial t} = \text{div } \rho\sigma \text{ grad } X_i + \rho A_i f_i \tag{5.23}$$

where f_i represents the nuclear transformations and the first term describes all forms of mixing, depending on the value of the so-called 'diffusion coefficient '.

Since in our problem we only take into account radial terms, we can equally write

$$\text{div } \rho\sigma \text{ grad } X_i = \frac{1}{r^2}\frac{\partial}{\partial r}r^2\rho\sigma\frac{\partial X_i}{\partial r} \tag{5.24}$$

As in the structure equations, we will also use the relative mass coordinate $\mu = 1 - M_r/M$. Equation (5.23) can be transformed into

$$\frac{\partial X_i}{\partial t} = \frac{\partial}{\partial \mu}D\frac{\partial X_i}{\partial \mu} + A_i f_i \tag{5.25a}$$

with

$$D = \frac{16\pi^2\rho^2 r^4}{M^2}\sigma \tag{5.25b}$$

The finite element approach. The evolution of n chemical abundances X_i over an interval [0,1] in μ is described by n equations of the type (5.25). These equations are coupled by the terms of nuclear transmutation, f_i, since the nuclear reactions transform elements into others.

It is better to use the number fractions

$$Y^{(i)} \equiv \frac{X_i}{A_i} \qquad (1 \le i \le n) \tag{5.26}$$

instead of the mass fractions, since in all nuclear reactions the sum of the mass fractions is conserved.

The full system of equations to solve is then

$$\frac{\partial Y^{(i)}}{\partial t} = \frac{\partial}{\partial \mu}\left[D(\mu,t)\frac{\partial Y^{(i)}}{\partial \mu}\right] + f^{(i)}(Y,t) \qquad (1 \le i \le n) \tag{5.27a}$$

with as boundary conditions

$$\left(\frac{\delta Y^{(i)}}{\delta \mu}\right)_{\mu=0} = \left(\frac{\delta Y^{(i)}}{\delta \mu}\right)_{\mu=1} = 0 \qquad (1 \leq i \leq n) \tag{5.27b}$$

and the initial conditions

$$Y^{(i)} = Y^{(i)}_{(0)}(x) \qquad (1 \leq i \leq n) \tag{5.27c}$$

The total number of particles of one type is conserved by the diffusion.

A method to solve this type of parabolic partial differential equation is given in De Loore and Doom, 1990.

6 Successive burning stages and final stages of stellar evolution

6.1 Evolution of stars of different mass

After core hydrogen burning *low mass stars* develop a helium core where the electrons are degenerate.This degeneracy is lifted by the helium flash.

Intermediate mass stars ignite helium in non-degenerate conditions, but develop a highly electron-degenerate carbon-oxygen core after central helium exhaustion.

Massive stars evolve through all phases of nuclear burning without encountering degeneracy. The temperature can increase beyond 6 10^8K igniting central carbon burning in a quiescent way, and producing Ne and Mg.

The mass limits for this subdivision in low mass, intermediate and high mass stars depend on several assumptions adopted in the construction of the stellar models, such as mass-loss or overshooting. The treatment of convection in the interior stellar layers is particularly important in this respect, since the extent of convection in the stellar interior determines the mass of the different burning cores and the occurrence of electron degeneracy. If convection is treated in the conventional way, i.e. if the boundary of the central convective regions is determined by the Schwarzschild criterion, the separation masses between the groups of stars are larger than when overshooting of convective cores is included. Table 6.1 shows the discriminating masses for the three groups as well for the Schwarzschild criterion as for overshooting. Since more and more observational evidence points to the presence of at least a moderate amount of overshooting, we adopt for the division in low mass, intermediate mass and massive stars the second column, i.e. the values based on overshooting.

6.2 Nuclear evolution - core and shell burning - motion of stellar layers

During the evolution nuclear fusion reactions in the stellar interior produce atomic nuclei of increasing charge and atomic weight, leading to an inhomogeneous chemical structure. The net result is that the number of particles per gram of matter decreases (e.g. 4 ^{1}H nuclei are converted into only one ^{4}He nucleus). Since the pressure is proportional to the *number* of particles (irrespective of their weight) the internal pressure decreases as nuclear burning proceeds. This implies that the central layers of the star will slightly contract and reach a new equilibrium at a slightly higher temperature, density and pressure.

As the nuclear evolution proceeds the star develops a high-density core of material processed by thermonuclear reactions, surrounded by a thin envelope. For massive stars

Table 6.1. The successive evolutionary stages for various initial masses. HB : hydrogen burning; GB : giant branch; DEG : degenerate; AGB : asymptotic giant branch; Plan. neb: planetary nebula; WD : white dwarf; CHB : core hydrogen burning; CONTR : contraction H shell : hydrogen shell burning; RG : red giant; WR : Wolf-Rayets stage SN : supernova; NS : neutron star; LBV : luminous blue variable.

	LOW-MASS	INTERMEDIATE STARS			MASSIVE STARS
MASSES: $M_{OVERSHOOTING}$	1.5	1.8	6	20	130
MASSES: $M_{SCHWARZSCHILD}$	1.3	2.3	9	40	130
CORES	RADIATIVE CORE	CONVECTIVE CORE			
SUCCESSIVE BURNING STAGES (TIME ↓)	HB	CHB	CHB	CHB	CHB CHB
	GB	CONTR	CONTR	CONTR	CONTR
	DEG	H-SHELL	H-SHELL	H-SHELL	LBV
	HE-FLASH			RG	WR
	THERMAL PULSES	GB	CEPHEID	WR	SN
	AGB		AGB		
	PLAN.NEB	TH.PU	WD	SN	NS
	WD	AGB		NS	
		WD			

the increasing temperature in the core allows the ignition of successive nuclear fuels (He, C, O, Si) and the star acquires an onion structure. The details of the evolution become more complicated: the nuclear burning processes proceed through more and more channels, the evolution proceeds extremely fast and dynamical effects occur. These different aspects lead to uncertainties in the structure of evolved stars, and to results that are more sensitive to the computational details.

At the end of a particular burning phase, the central parts of the star consist of the 'ashes' of the nuclear burning. Here, all nuclear reactions have ceased because the fuel is exhausted. However, just outside this inert core the temperature may become high enough to initiate nuclear burning in a shell around the inert core, consuming the last remainders of the nuclear fuel in a thin shell. This process is called *nuclear shell burning*. It is clear that nuclear shell burning generally occurs at composition discontinuities. For instance, during core hydrogen burning ^{4}He nuclei are formed in the central regions of stars at temperatures of a few tens of million degrees, below the threshold temperature for He burning. These helium layers are surrounded by regions still containing a considerable amount of hydrogen.

In thin shells of matter thermonuclear fusion processes may be active, producing energy; such shells are called *active shell sources*. If the temperature in such a shell drops below the threshold for nuclear reactions (e.g. by expansion) the nuclear energy production in the shell vanishes and the shell becomes inactive. Inactive shell sources may be reactivated if the temperature becomes sufficiently high, so that the nuclear fuel may be reignited.

The position of an active shell source tends to remain fixed as the star evolves: if matter at one side of an active shell source contracts, the matter at the other side tends to expand. This means that active shell sources are most likely to be found where nodes in the mass motions are present. This can be demonstrated in a simple way: Let us consider an active shell source s, situated at a distance r_s from the center, separating the inert core from the envelope. Let us further assume that we have an ideal gas with radiation:

$$P = N_0 k \frac{\rho}{\mu} T + \frac{1}{3} a T^4 \tag{6.1}$$

and that the envelope may be characterized by the polytropic relation $\rho = \kappa T^n$.

We can now write the pressure gradient as:

$$\frac{dP}{dr} = \frac{dP}{dT} \cdot \frac{dT}{dr} = \frac{N_0 k}{\beta \mu} \kappa (n+1) \rho \frac{dT}{dr} \tag{6.2}$$

where β is the ratio of the gas pressure to the total pressure. Using the condition of hydrostatic equilibrium we have

$$\frac{dP_s}{dr} = -\frac{GM_c}{r_s^2} \rho_s = \frac{N_0 k}{\beta \mu} \kappa (n+1) \rho_s \frac{dT_s}{dr_s} \tag{6.3}$$

where the index s refers to the parameters inside the shell source and M_c is the mass of the inert core. Straightforward integration of (6.3) yields

$$T_s = \frac{\beta \mu}{N_0 k} \frac{GM_c}{\kappa (n+1)} \frac{1}{r_s} + \text{constant} \tag{6.4}$$

This means that the temperature of an active shell source is inversely proportional to its distance and does not depend on other local parameters. If the shell source should be moved towards the center, the temperature inside the shell would rise. This would lead to an enhanced energy release, increasing the pressure and forcing the shell to return to its original position. If the shell should be moved outward, the temperature and hence the energy release would decrease, the pressure would decrease, and gravitation would pull the shell source to its original position. Thus, active shell sources remain at the same distance from the center of the star. It is now possible to determine the schematic motion of mass shells in the presence of one or more active shell sources (Figure 6.1).

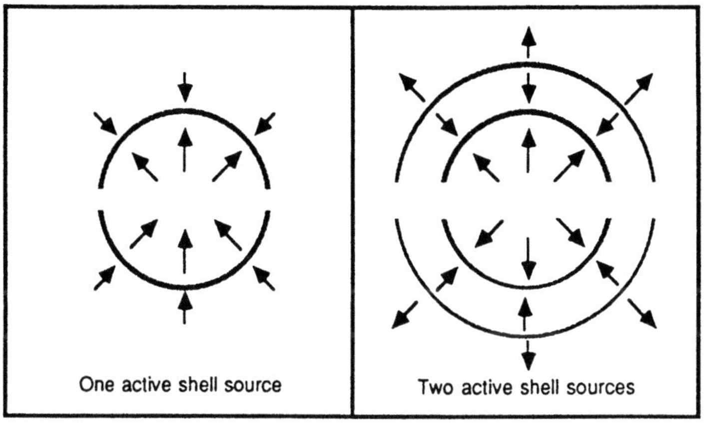

Fig. 6.1. The node theorem. In the case of a single active shell, core contraction leads to expansion of the envelope. When two active shells are present, core contraction leads to contraction of the envelope, and core expansion to expansion of the envelope.

6.3 The Virial Theorem

For a star or gas cloud in hydrostatic equilibrium a specific relation exists between the kinetic energy or gas energy and the potential energy. We derive this relation for a spherically symmetric gas cloud where the atoms satisfy the conditions of an ideal gas (i.e. no degeneracy or particle-particle interactions). The cloud is characterized by the temperature distribution as a function of the distance from the center, T(r) and the density distribution $\rho(r)$. Temperature and density distributions are related to the pressure distribution by the ideal gas law:

$$P(r) = N_0 k \frac{\rho(r)}{\mu} T(r) \tag{6.5}$$

The thermal energy distribution per unit volume of the gas is

$$E_{\text{th}}(r) = \frac{3}{2} N_0 k \frac{\rho(r)}{\mu} T(r) = \frac{3}{2} P(r) \tag{6.6}$$

The total thermal kinetic energy:

$$E_{\text{th}} = \int_{\text{volume}} E_{\text{th}} dV = \int_0^R E_{\text{th}}(r) 4\pi r^2 dr \tag{6.7}$$

of a uniform gas sphere (constant density and temperature) of mass M is

$$\frac{3}{2} N_0 k T \frac{M}{\mu}$$

The total thermal kinetic energy of a gas sphere of radius R in hydrostatic equilibrium is

$$E_{\rm th} = \frac{1}{2}\int_0^R \frac{GM(r)}{r}\rho(r)4\pi r^2 dr \qquad (6.8)$$

We can compare this result to the total potential energy of a gas cloud. The potential energy of a gas sphere of thickness dr at a distance r from the center is

$$dE_{\rm pot} = -\frac{GM(r)}{r}\rho(r)4\pi r^2 dr \qquad (6.9)$$

and the total potential energy of the gas cloud is then

$$E_{\rm pot} = -\int_0^R \frac{GM(r)}{r}\rho(r)4\pi r^2 dr \qquad (6.10)$$

The total potential energy of a uniform (constant density) gas sphere of mass M and radius R is

$$\frac{3}{5}\frac{GM^2}{R} \qquad (6.11)$$

From (6.9) to (6.11), one obtains *the virial theorem for gas spheres in hydrostatic equilibrium*:

$$E_{\rm pot} = -2E_{\rm th}$$

or: the potential energy is (minus) twice the thermal kinetic energy. We used only the equation for hydrostatic equilibrium and the ideal gas law. Hence the virial theorem is valid for all gas systems in hydrostatic equilibrium, independent of the energy source or the internal temperature distribution.

6.4 Hydrogen burning

Computations of stellar evolution start from an initial chemically homogeneous model, i.e. the X- and Z- values (hence also the Y- values) are the same at each point in the star. The changes

$$\Delta X = \frac{dX}{dt}\cdot\Delta t, \quad and \quad \Delta Y = \frac{dY}{dt}\cdot\Delta t$$

are calculated during a time interval Δt at each point of the star, and this new model is then used as input model for the next static model and so on. Also mass losses may be considered, e.g. mass transfer and mass loss for the evolution of close binaries, or mass losses by stellar winds for massive stars. Low-mass stars (mass below 1.6 to $2.3M_\odot$) where hydrogen burning occurs mainly via the pp-chain (mass below $1.3M_\odot$) develop radiative cores surrounded by convective envelopes, which extend to the base of the photosphere. Lower-mass stars have smaller cores. Initially the core is chemically homogeneous. The stellar energy losses are supplied by nuclear reactions, occurring at a nuclear time scale much larger than the dynamical time scale, so that the star is able to adapt its structure quasistatically to accommodate itself to the decreasing hydrogen fuel supply. Since the energy generation rate $\epsilon \sim \rho X^2 T^n$ the energy generation decreases with X, unless ρ and/or T decrease. This would cause a lower pressure and the star would contract while increasing X. According to the virial theorem a part of the released potential energy is used to increase the temperature. The remaining part goes into radiation. Hence ρ and T increase, leading to an enhanced energy output, and thus a larger L. On the other hand the conversion of hydrogen into helium in the core occurs at higher rates. The change in radius is rather small, as a result of which the change in T_{eff} is not too large either. The net effect is an increase in the stellar luminosity, and a small increase in T_{eff}.

The hydrogen content is further reduced, and helium is produced at the core; finally the core may become helium-dominated. Hydrogen burning is now mainly confined to a thick shell around a small, increasing helium core. The burning rate $\epsilon = dL_r/dM_r$ is larger than before, and the helium core is almost isothermal. The star evolves faster, the radius expands, and T_{eff} becomes lower. Hence in the HRD, the star continues to move upwards, but now to the right. For higher masses ($M > 1.3M_\odot$) the evolution occurs in a different way. For central temperatures larger than $2\ 10^7$, hydrogen burning occurs via the CNO tri-cycle in a convective core. Hence the composition throughout the core is homogeneous, even though the nuclear burning, proportional to the temperature, is more efficient in the central regions than in more peripheral layers. The energy generation ϵ is now determined by $\epsilon = \rho X Z_{CNO} T^n$ with $n \sim 18$, and Z_{CNO} the mass fractions of C, O and N. In this case the energy production is linear in X, instead of quadratic.

A decrease of X has now a smaller effect on the energy production rate than for smaller mass stars; the cores do not need to contract as much as for smaller masses. The luminosity increases and the star expands. The expansion rate is more pronounced than the luminosity increase, and the star moves slightly upward to the right. When nuclear burning has reduced the initial hydrogen abundance to about 0.05, the energy release rate is no longer able to support the core, and core contraction starts. The luminosity does not change very much, hence according to the Stefan- Boltzmann law the effective temperature increases: the star moves upwards to the left in the HRD (Figure 6.2). This marks the end of the MS stage.

The effect of main sequence evolution is the production of helium nuclei at temperatures below that of helium ignition, surrounded by a hydrogen-rich envelope. This has the following consequences:

1. Owing to the inhomogeneous composition the central mass concentration becomes larger.

2. As mentioned earlier shell burning sources have a tendency to remain at the same position as the star evolves.

3. According to the node theorem, volume changes (expansion or contraction) are reversed, at each active nuclear burning shell (node), but remain unaffected by inactive shells. The mean molecular weight of the ashes that are left generally exceeds that of the initial fuel (for helium, $\mu = 4/3$ and for hydrogen $\mu = 1/2$, according to $1/\mu = 2X + 3/4Y + 1/2Z$). This means that composition inhomogeneities lead to larger central condensations, i.e. mass concentration near the center. Shell burning sources have dramatic effects on the evolution.

Table 6.2. Separation values between stars of low, intermediate and high mass

	Schwarzschild	Overshooting
Low-mass stars	$M < 2.3\,M_\odot$	$M < 1.6 - 1.8\,M_\odot$
Intermediate-mass stars	$2.3\,M_\odot < M < 9\,M_\odot$	$1.6\text{-}1.8\,M_\odot < M < 6\,M_\odot$
Massive stars	$M > 9\,M_\odot$	$M > 6\,M_\odot$

6.5 Red giants

When a star has developed a shell source, the core becomes inert, although it may produce a small fraction of the luminosity by gravitational contraction. Owing to the shell source the envelope expands as the core contracts (node theorem). The shell luminosity L_s remains nearly constant, and consequently the surface temperature drops. The movement of the star in the HR diagram is nearly horizontal and to the right. The hydrogen shell is thin, and is not far from the stellar center. The star moves upward towards to the red giant branch; eventually the envelope may become convectively unstable. Due to convection the chemical profile of the envelope is uniform. The evolution is different for stars of different masses: In low-mass stars degeneracy sets in before core helium ignition is started. The degeneracy is only lifted by the helium flash, which causes the core to expand. Finally the core can readjust its structure so that helium burning can occur under normal, non-degenerate conditions.

In higher-mass stars a temperature gradient is developed by the increasing importance of gravitation. Consequently helium burning starts before degeneracy becomes dominant. After the core contraction the hydrogen burning shell supplies most of the energy. Helium ash is added to the core, which exceeds the Chandrasekhar-Schönberg limit, i.e.the maximum mass fraction that an isothermal, non-degenerate core can have in a star in complete (hydrostatic and thermal) equilibrium, and deriving its energy from a nuclear shell source surrounding the core, can have. Rapid core contraction occurs, the central temperature rises, the shell burns faster and the envelope expands rapidly. Most of the released energy is used to move matter to regions of lower potential energy by expansion. In the HRD the star moves to the right and slightly downwards. Finally these stars contract and evolve into white dwarfs. More massive stars end their evolution by explosive phases, and finally evolve into neutron stars or black holes.

The separation values between low-mass stars, intermediate-mass stars and massive stars are given in Table 6.1 for two distinct cases, the first one where the occurrence of central convection, and the extent of the convective cores, is calculated according to the Schwarzschild criterion, and the second one, where overshooting is taken into account. Evolving stars pass through consecutive nuclear burning stages. The nuclear ashes, produced by a given nuclear fusion stage can serve as nuclear fuel for a next burning phase if the temperature is high enough. The maximum temperature that can be attained in the stellar interior depends largely on the total mass of the star. It depends also on this temperature whether or not the produced "ashes" will be ignited for a subsequent nuclear fusion phase. For low mass stars, where the nature of the nuclear burning determines the core structure, the nuclear fusion processes will stop somewhere between hydrogen fu-

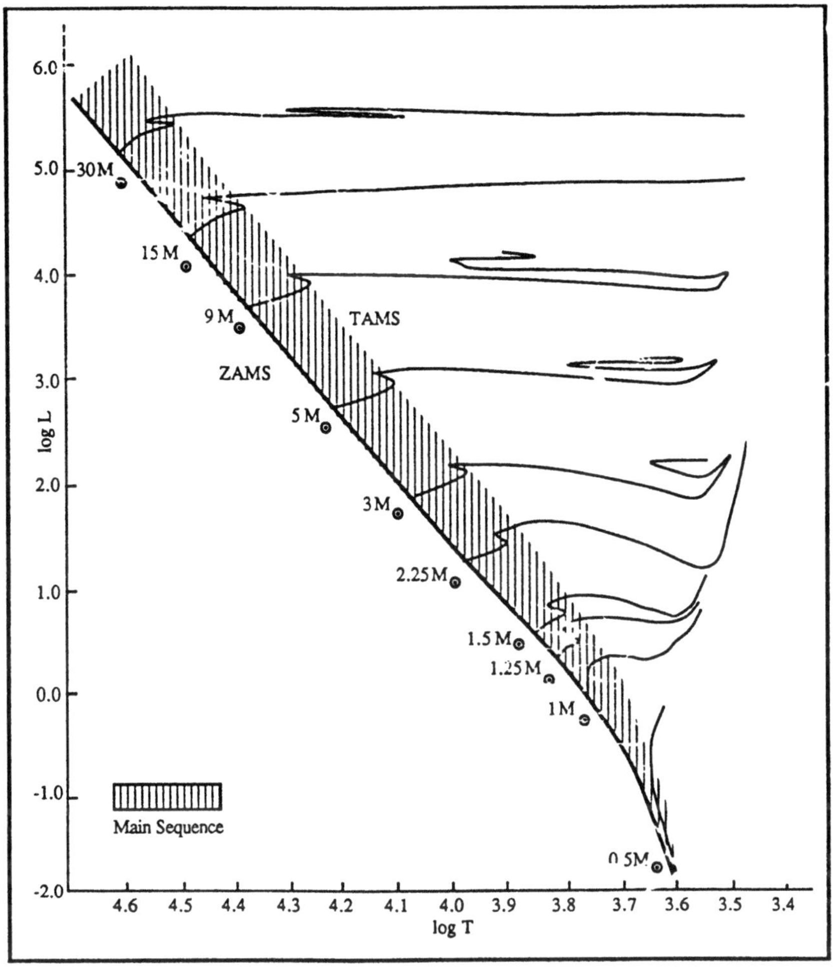

Fig. 6.2. The zero age main sequence, ZAMS, and the end points of hydrogen burning, the termination of the main sequence, the TAMS. The hatched area is the main sequence band.

sion and carbon burning. For high mass stars, neutrino processes become more and more important during the successive burning stages: they remove energy at extremely high rates and therefore play a very important role in the evolution. However, our knowledge of these phases is incomplete, and detailed computations of advanced evolutionary stages are not yet possible. Stellar evolution during hydrogen and helium burning is rather well established; the computation of carbon burning phases is more or less known, but for later phases many problems are only partially solved. Although the evolutionary scenario from ZAMS to the final stages is still under investigation some general ideas about the final configurations of stars of various masses are widely accepted. Configurations of cold matter with mass below the Chandrasekhar limit, $M \leq 1.44\ M_\odot$ and log ρ_c between 6 and 9, are stable against gravitational contraction and will become white dwarfs. Configurations with masses between 0.5 and $2.5M_\odot$ and with log ρ_c between 13 and 15 are identified with neutron stars.

According to Iben and Renzini (1983) the progenitor of the most massive white dwarf is near $8M_\odot$, depending on the He-abundance, metallicity and adopted overshooting. Therefore we will assume that stars with initial masses $M<8M_\odot$ end their life as white dwarfs and that higher mass stars evolve into neutron stars or black holes. Since the final masses of white dwarfs or neutron stars are maximum 1.4 to $3M_\odot$ these stars must have lost a considerable fraction of their initial mass. It is generally believed that stars with masses in the range 3 to $4M_\odot$ lose most of their matter during the red giant stage, leading to a.o. planetary nebulae. On the other hand for a star to become a supernova it has to be more massive than at least $5M_\odot$ as is found from observations in the Pleiades. Hence we may state that stars less massive than 4-$8M_\odot$ will evolve nonviolently and lose enough matter during their final stages to slim down below the Chandrasekhar limit. Stars with masses exceeding 6-$8M_\odot$ will evolve towards violent stages. If the explosion leaves a remnant smaller than $2.5M_\odot$ this will be a neutron star; higher mass remnants will collapse into black holes. The probability remains that stars are completely disrupted by supernova explosions, leaving no remnants at all.

6.6 Advanced stages-Evolution of progenitors of supernovae

6.6.1 Advanced stages of nuclear burning and nucleosynthesis

Nuclear fusion processes in stellar interiors lead to a general picture for the formation of most of the light elements, in agreement with the observations, i.e. with the relative cosmic abundances. By hydrogen burning ^{4}He is produced, and ^{3}He is formed by the incomplete pp-chain. The light elements, ^{2}H (D), Li, Be, B are not formed by these processes. The species ^{12}C, ^{16}O, ^{18}O, ^{22}Ne are formed by helium burning processes. ^{14}N is formed by the CNO tri-cycle during hydrogen burning, by the conversion of C, N and O. ^{20}Ne, Na, Mg, Al, Si are produced during carbon and oxygen burning. P and S are also formed during oxygen burning. Stable nuclei up to ^{32}S are formed by nuclei built up during fusion processes by ^{12}C and ^{16}O. In the core of a star that evolved beyond C and O-burning we find ^{20}Ne, ^{23}Na, ^{24}Mg, ^{25}Mg, ^{26}Mg, ^{27}Al, ^{28}Si, ^{29}Si, ^{30}Si, ^{31}P, ^{32}S, with Si and S as the most abundant species. Nuclear burning occurs in the hottest, densest regions of stars. Since the net effect is the conversion of lighter elements into heavier ones, nuclear burning leads to ashes of increasing atomic weight, and for advanced stages to highly concentrated matter in the stellar interior. As a consequence the dynamical timescale for the core becomes smaller. At the end of core helium burning the central density of a star of $5M_\odot$ is $\approx 10^6$ times larger than the ZAMS density, and the free fall time is reduced by a factor 1000 from 30 minutes to ≈ 2 seconds. On the other hand the density changes in the envelope are much less dramatic. The timescales for the core and the envelope, comparable during early evolutionary stages, become completely different for the later stages. Hence for advanced evolutionary stages, the timescales for core and envelope may differ by several orders of magnitude. The envelope is not able to adjust to the rapidly varying core's conditions. Expressed in the core's timescale and compared to the rapidly changing situation in the core, the envelope is static.

6.6.2 Stars that become type II supernovae

The set of stars becoming type II supernovae has on the lower end as boundary the most massive stars that can become white dwarfs, and on the upper end the most massive stars that still retain their hydrogen envelope at the time their core explodes. The precursors are massive, hydrogen rich population I stars. There exist stars with larger masses which also explode, but their light curves and spectra are such that they have to be disqualified as type II supernovae. As mentioned before, for single stars the progenitor of the most massive white dwarf is near $8M_{\odot}$. This value is also in agreement with statistics on the occurrence of SN (Tammann, 1982) the location of type II SN in spiral arms (Maza and Van den Bergh, 1976), observations of white dwarfs (Romanishin and Angel,1980; Weidemann and Koester, 1983), and theoretical models for white dwarfs (Iben, 1985). Stars with masses beyond $8M_{\odot}$ will ignite their carbon under non-degenerate conditions and avoid the formation of a thin helium burning shell that may eject the stellar envelope. Figure 6.3 shows evolutionary tracks in the T_c-ρ_c diagram for various masses. Before supernova 1987a it was generally believed that the progenitors of SNell were red supergiants (Woosley and Weaver, 1986a) and not blue supergiants as in the case of 1987a. We will consider two classes of stars more massive than $8M_{\odot}$, a first group in the range 8 - 12 $M_{\odot}$ and a second group containing all more massive stars.

Evolution of 8 – 10 $M_{\odot}$ *stars (He – cores of 2.2 to* $2.5M_{\odot}$*)*

At the lower end the stars ignite C degenerately, at the upper end the stars pass through all burning stages in their center, -,H, He, C, Ne, O, Si-, under non-degenerate conditions. Stars in this range may have off-center ignition, a high degree of degeneracy, and electron capture may occur. For helium cores between $\approx$2.2 and $\approx 2.5M_{\odot}$ (this corresponds with ZAMS masses of $\approx$9 to $10M_{\odot}$) a degenerate Ne-O-core develops after C-burning (Woosley, Weaver and Taam, 1985a,b; Miyaji et al. 1980; Nomoto, 1982, 1984a,b; Woosley and Weaver, 1985; Habets, 1985). Depending on details of the models, Ne and O- burning may (Woosley and Weaver, 1985) or may not (Nomoto, 1984ab, 1985ab) ignite off-center. In both cases the Ne-O composition in the center is maintained until ρ=2.5 10^{10} gcm^{-3}. Then O is ignited under extremely degenerate conditions in the core which is contracting rapidly because of electron capture on ^{20}Ne and ^{24}Mg. Nuclear burning proceeds to statistical equilibrium. Electron capture on Fe-group nuclei leads to to pressure decrement accelerating the collapse of the core to nearly free fall (Miyaji et al. 1980; Bruenn, 1972). Stable Si-burning and Fe-core formation are thus bypassed and the core with a mass of 1.35 to $1.4M_{\odot}$ collapses to nuclear density, while non processed fuel is still present in the outer layers of the core.

Evolution of 10 – 12 $M_{\odot}$ *stars (He – cores of 2.5 to* $3M_{\odot}$ *)*

After complicated stages of off-center burning and violent flashes an iron-core in hydrostatic equilibrium is eventually formed. In some calculated models (Woosley, Weaver and Taam, 1980) this leads to envelope ejection, several years before the core collapse. The

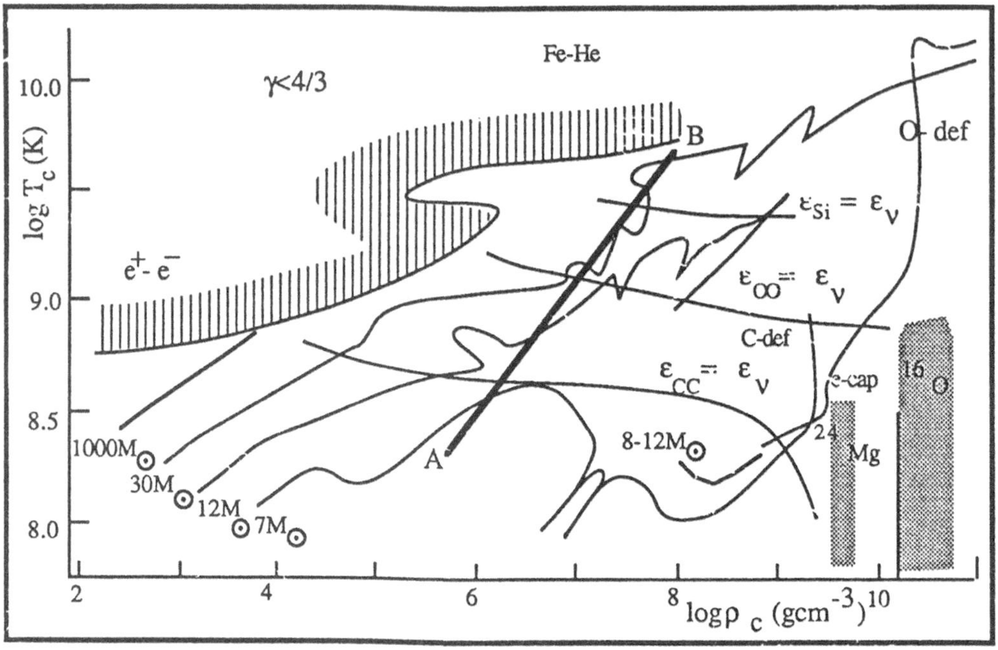

Fig. 6.3. Evolution of stars of different mass in the T_c-ρ_c plane. The electrons become significantly degenerate to the right of the thick line AB. The regions of ignition of C,O and Si are indicated, i.e. the regions where the energy liberated by the nuclear reactions equals the neutrino losses. The shaded regions show domains of dynamical instabilities caused by electron captures on ^{16}O and ^{24}Mg, iron photodisintegration or by pair production (data from Sugimoto, 1980; Ober et al. 1981; Mazurek and Wheeler, 1979).

collapse is triggered by photodisintegration and electron capture, just like in more massive stars. As a result an extremely bright supernova is produced, as a consequence of the interaction of a shock and the ejected envelope. The Crab Nebula, containing a neutron star, could possibly be formed by the evolution of a star in this mass range (Arnett, 1975; Woosley, Weaver and Taam, 1980; Nomoto, 1982, 1984ab, 1985ab; Davidson et al. 1982, Nomoto et al. 1982) because such objects produce neutron stars.

Evolution of stars with mass > 12M$_\odot$

Advanced burning stages for stars more massive than 12M$_\odot$ have been investigated in many studies (Fowler and Hoyle, 1964; Sugimoto, 1970; Ikeuchi et al. 1971; Woosley and Weaver, 1985; Habets, 1985).

Following C-ignition energy losses occur mainly by neutrino emission, by thermal processes and weak interactions. Since for each of the successive fuels the burning is characterized by an increasing Coulomb barrier, higher and higher temperatures are required for each stage. On the other hand neutrino losses are also strongly temperature dependent and therefore, owing to the temperature rise, the neutrino losses become more and more important and the total power of the star increases very rapidly. The specific energy per gram available from nuclear burning decreases with heavier fuels and therefore the lifetimes of advanced burning phases become shorter and shorter. At the moment carbon burning has started, the outer stellar layers have assumed their final configuration, which is typically a red supergiant, if mass loss has not removed the hydrogen envelope and if the star has a population I composition. Within this context, it is essentially as the explosion of a point source of $\approx 10^{51}$ erg that a Type II supernova occurs.

Table 6.2 lists for ZAMS-masses between 10 and 100 M the ZAMS-mass, the He core mass, the Fe core mass, the explosion energy, the neutron star mass and the mass of the ejected species with $Z \geq 6$.

Table 6.3. Presupernova models and explosions

ZAMS mass	He core mass	Fe core mass	Expl.energy 10^{50} erg	residual baryon mass	neutron star mass	ejected species($Z \geq 6$)
10	2.6	1.4	1.8	1.47	1.33	≈ 0
11	2.4	-	3.0	1.42	1.31	≈ 0
12	3.1	1.31	3.8	1.35	1.26	0.96
15	4.2	1.33	2.0	1.42	1.31	1.24
20	6.2	1.70	-	-	-	2.53
25	8.5	2.05	4.0	2.44	1.96	4.31
35	14	1.8	-	-	-	9.88
50	23	2.45	-	-	-	17.7
75	36	-	-	-	BH?	30?
100	45	≈ 2.3	≈ 4	-	BH?	39?

6.6.4 Type I supernova models

Many mechanisms have been proposed to explain type I supernovae: the hydrodynamical explosion of stellar cores (Lasher, 1975; Amett, 1979), pulsar driven explosion (Ostriker and Gunn, 1969; Bodenheimer and Ostriker, 1974; Shlovskii, 1975), the thermonuclear disruption of white dawrfs (Hoyle and Fowler, 1960; Schatzman, 1963; Amett, 1968,

1969; Paczynski, 1970). During the last decade there has been a convergence upon the thermonuclear model to explain the majority of Type I events.

Type Ia Sne are found in all kinds of galaxies, show no preference for spiral arms when they occur in spiral galaxies and do not show evidence for hydrogen in their spectra. The conclusion is that the progenitors are low mass stars, belonging to an old population (young stars are associated with spiral arms). For single stars it is hard to understand how they could lose their hydrogen rich envelopes. For binaries this poses no problem.

Hence it is suggested that the progenitors of Type Ia supernovae are mass accreting white dwarfs in binary systems. The white dwarf accretes matter expelled by its companion and grows to a critical mass (Whelan and Iben, 1973), ignites carbon or helium under extremely degenerate conditions; a substantial part of its matter is burned to nuclear statistical equilibrium. The burning energy disrupts the star at high velocity; in some cases a part of the white dwarf is left if the ignition occurs off-center. Generally no neutron star or black hole is expected in a Type I supernova. The abundance peak in nuclear statistical equilibrium lies around ^{56}Ni (Truran, Arnett, Cameron, 1967). Just as for Type II SN, almost all of the energy is kinetic.

Type Ib Sne occur in spiral arms, close to, or even in HII regions (Porter and Filippenko, 1987). This suggests massive, hydrogen deficient stars as progenitors, hence Wolf-Rayet stars (Wheeler and Levreault, 1985; Begelman ans Sarrazin, 1986; Schaeffer et al., 1987). The mass range for the progenitors is $\approx$20 to 30$M_\odot$.

A detailed analysis of supernova is given in the next chapter, "Supernovae: Observations, Theory, Models and Nucleosynthesis", by Ewald Muller.

REFERENCES

Allen, C.W., 1963, Astrophysical Quantities, The Athlone, University of London Press.

Aller, L.H., 1963, Astrophysics: The Atmospheres of the Sun and Stars, New York, Ronald Press.

Arnett, W.D., 1968, Nature, 219, 1344.

Arnett, W.D., 1979, Astrophys. J, 230, L37.

Begelman, M.C., Sarazin, C.L., 1986, Astrophys. J. Lett. 302, L59.

Bodenheimer, P., Ostriker, J.P., 1974, Astrophys. J., 191 , 465.

Bruenn, S.W., 1972, Astrophys. J. Suppl., 24, 283.

Cox, J.P., Giuli, R.T., 1968, Principles of Stellar Structure, Gordon and Breach, Science Publishers.

Crawford, J.A., 1955, Astrophys. J, 121, 71.

Davidson, K., Gull, T.R., Maran, S.P., Stecher, T.P., Fesen, R.A., Parise, R.A., Hawel, C.A., Kafatos, M., Trimble, V.L., 1982, Astrophys. J, 253, 696.

De Loore, C., Doom, C., 1990, Structure and Evolution of Single and Binary Stars, Reidel Publ. Cy, Dordrecht, the Netherlands.

Eddington, A.S., 1926, Internal Constitution of the Stars, Cambridge University Press (also Dover Publication, 1959).

Fowler, W.A., Caughlan, G.R, Zimmerman, B.A., 1975, Ann. Rev. Astron. Astrophys., 13.

Fowler, W.A., Hoyle, F., 1964, Astrophys. J. Suppl., 9, 201.

Habets, G., 1985, Ph. D. Thesis, Univ. Amsterdam.

Henyey, L.G., Wilets, L., Böhm, K.H., Lelevier, R., Lev, R.D., 1959, Astrophys. J., 129, 628.
Henyey, L.G., Forbes, J.E., Gould, N.L., 1964, Astrophys. J., 139, 306.
Hoyle, F., Fowler, W.A., 1960, Astrophys. J., 132, 565.
Iben, I., (Jr), 1985, Quart. J. Roy. Astron. Soc., 26, 1.
Iben, I., (Jr), Renzini, A., 1983, Ann. Rev. Astr. Astrophys., 21, 271.
Ikeuchi, S., Nakazawa, K., Murai, T., Hoshi, R., Hayashi, C., 1971, Prog. Theor. Phys. Kyoto, 46, 1713.
Johnson, H.L. and Sandage, A.R., 1956, Astrophys. J., 124, 379.
Kippenhahn R., Weigert, A., Hofmeister, E., 1967, in "Computational Methods in Physics", Vol. 7., eds. B. Alder et al. New York Academic Press 1967.
Kippenhahn R., Weigert, A., 1967, Zeitschrift für Astrophysik, 65, 221.
Kippenhahn R., Weigert, A., 1990, Stellar Structure and Evolution, Springer-Verlag.
Landau, L.D., Lifschitz, E.M. 1958, Statistical Physics, Pergamon Press.
Lasher, G., 1975, Astrophys. J., 201, 194.
Maza, J., van den Bergh, S., 1976, Astrophys. J., 204, 519.
Mazurek, T., Wheeler, J.C., 1979, Fund. Cosmic Phys., 5, 193.
Miyaji, S., Nomoto, K., Sugimoto, D., 1980, Publ. Astron. Soc. Japan, 32, 303.
Moffat, A.F.J., Seggewiss, W., 1978, Astron. Astrophys., 70.69.
Moore, C.E., 1970, Nat. Bur. Standards (Wash.), rpt. NSRDS-NBS 34.
Nomoto, K., 1982, Astrophys. J., 153, 798.
Nomoto, K., 1984a , Stellar Nucleosynthesis, eds, Chiosi and Renzini (D. Reidel,Dordrecht), p. 239.
Nomoto, K., 1984b, Astrophys. J., 277, 791.
Nomoto, K., 1985a , Proceedings of the 12th Texas Symposium on Relativistic Astrophysics, Ann. N.Y. Acad. Sci.
Nomoto, K., 1985b, Challenges and New Developments in Nucleosynthesis, ed. W.D. Arnett and J. Truran, (Univ. Chicago Press, Chicago), p. 202.
Ober, W.W., El Eid, M.F., Fricke, K.J., 1981, ed. C. de Loore in "Galactic Evolution", Study Series, Tijdschrift VUB, No.4.
Ostriker, J.P., Gunn, J.E., 1969, Astrophys. J., 157, 1395.
Paczynski, B., 1970, Acta Astronomica, 20, 47.
Porter, A., Filippenko, A., 1987, Astron. J. 93, 1372.
Press, W.H., Flannery, B.P., Teukolsky, S.A., Vetterling, W.T., 1986, Numerical Recipes, The Art of Scientific Computing, Cambridge University Press, Cambridge.
Romanishin, W., Angel, J.R.P., 1980, Astrophys. J., 235, 992.
Sandage, 1957, Astrophys. J., 125, 435.
Schaeffer, R., Cass, M., Cahen, S., 1987, Astrophys. J. Lett., 316, L31.
Schatzman, E., 1963, Star Evolution, Course 28, Proc. Int. School of Physics, "Enrico Fermi", ed. L. Gratton (Academic Press, New York).
Shlovskii, I.S., 1975, Astr. Zh. 52, 911 (see also Sov. Astron. Ap.19, 554).
Sugimoto, D., 1970, Prog. Theor. Phys. Kyoto, 44, 599.
Sugimoto, D., Nomoto, K., 1980, Space Sci. Rev. 25, 155.
Tammann, G.A., 1982, in " Supernovae: A Survey of Current Research", ed. M.J. Rees and R.J. Stoneham (Cambridge Univ. Press: Cambridge), p 371.
Truran, J.W. , Arnett, W.D., Cameron, A.G.W., 1967, Canadian J. Phys. 1979, 45, 2315.
Vardya, M.S., 1961, Astrophys. J., 133, 107.

Vardya, M.S., 1964, Astrophys. J. Suppl. 8, 80, 277.
Weidemann, V., Koester, D ., 1983, Astron. Astrophys., 121, 77.
Wheeler, J.C., Levreault, R.M., 1985, Astrophys. J. Letters 294, L17.
Whelan, J.C., Iben, I. Jr., 1973, Astrophys. J., 186, 1007.
Wilson, R.E., 1981, Astron. Astrophys. 99, 43.
Woosley, S.E., Weaver, T.A., Taam, R.E., 1980, Type I Supernovae, ed. J.C. Wheeler (Univ. of Texas, Austin), p 96.
Woosley, S.E., Weaver, T.A., 1985, Nucleosynthesis and its Implications on Nuclear and Particle Physics, Proc. Fifth Moriond Astrophysics Conference, ed. J. Audouze and N.Mathieu (Reidel, Dordrecht), p.145.
Woosley, S.E., Weaver, T.A., 1986a, Ann. Rev. Astron. Astrophys., 24, 205.
Woosley, S.E., Weaver, T.A., 1986b, Radiation Transport and Hydrodynamics, Proc. IAU Coll. 89, ed. D. Mihalas and K.H. Winkler, D. Reidel, Dordrecht.
Yamashita, Y., 1962, Publ. Astron. Soc. Japan, 1962, 14, 390.

General references

Clayton, D.D., 1968, Principles of Stellar Evolution and Nucleosynthesis, Mc Graw Hill Book Company, New York.
Cox, J.P., Giuli, R.T., 1968, Principles of Stellar Structure, Gordon and Breach, Science Publishers.
de Jager, C., 1980, The Brightest Stars, Reidel Publishing Company, Dordrecht, Holland.
Maeder, A., Renzini, A., eds., 1984, Observational Tests of the Stellar Evolution Theory, proc. IAU Symposium 105, Reidel, Dordrecht, the Netherlands.
Mihalas, D., 1978, Stellar Atmospheres, Freeman and Company, San Francisco.
Schwarzschild, 1958, Structure and Evolution of the Stars, Princeton, New Jersey, Princeton.
Unsöld, A., 1955, Physik der Sternatmosphren, Springer Verlag, Berlin, Göttingen, Heidelberg.

Literature per section

Section 1 - Introduction
Shlovskii, I.S., 1978, Stars: Their Birth, Life and Death, W.H. Freeman, San Francisco.
Abell, G.O., 1975, Exploration of the Universe, Rinehart and Winston.
Goldberg, H.S., Scadron, M., 1981, Physics of Stellar Evolution and Cosmology, Gordon and Breach Science Publishers.
Salpeter, E.E., 1955, The Luminosity Function and Stellar Evolution, Astrophys. J., 121, 161.

Section 2 - Equations of stellar structure.
Schwarzschild, M., 1958, The Structure and Evolution of the Stars,. Princeton University Press, Princeton, N.Y.
Cox, J.P., Giuli, R.T., 1968, Stellar Structure, Vol. I, Gordon and Breach.
Chiu, H.Y., 1968, Stellar physics, Waltham, Massachusetts, Blaisdell.
Novotny, E., 1973, Introduction to Stellar Atmospheres and Interiors, Oxford University Press, Oxford, England.

Section 3 - The equation of state
Chandrasekhar, S., 1939, An Introduction to the Study of Stellar Structure, University of Chicago Press, Chicago.

Section 4 - Nuclear reaction rates-nuclear burning
Arnett, W.D., Truran, J.W., eds, 1985, Nucleosynthesis, Challenges and New Developments, The University of Chicago Press.
Chandrasekhar, S., 1938, Stellar Structure, University of Chicago Press, Chicago, 254-261 (non-degenerate state), chapter X (degenerate state).
Chiosi, C., 1986, in "Nucleosynthesis and Chemical Evolution", 16th Advanced Course, Swiss Society of Astrophysics and Astronomy, Saas-Fee, eds. B. Hauck, A. Maeder and G. Meynet.
Clayton, D.D., 1968, Principles of Stellar Evolution and Nucleosynthesis, Mc Graw Hill, New York.
Fowler, W.A., Caughlan, G.R, Zimmerman, B.A., 1975, Ann. Rev. Astron. Astrophys., 13.

Section 5 - Numerical techniques
Kippenhahn R., Weigert, A., Hofmeister, E., 1967, Methods for Calculating Stellar Evolution in 'Computational Methods in Physics', Vol. 7, eds. B. Alder et al. New York Academic Press 1967.
Paczynski, B., 1970, Evolution of Single Stars, I, II, III, Acta Astronomica, Vol. 20, 2, 1970; 3, 1970; 4, 1970.
Press, W.H., Flannery, B.P., Teukolsky, S.A., Vetterling, 1986, W.T., Numerical Recipes, Cambridge University Press.

Chapter 6 - Successive burning stages and final evolution
Gursky, H., van den Heuvel, E.P.J., "X-ray Emitting Double Stars", Scientific American, 223, 3, 24.
Hewish, A., 1970, "Pulsars", Ann. Rev. Astr. Astrophys., 8, 265.
Iben, I. Jr, 1972, Normal Stellar Evolution, in Stellar Evolution, H.Y. Chiu and A. Murriel, eds. MIT Press, Cambridge, Mass.
Salpeter, E.E., "Stellar Evolution Leading up to White Dwarfs and Neutron Stars", in Relativity Theory and Stellar Structure, Lectures in Applied Mathematics, Vol. 10, American Mathematical Society, 1967.
Shapiro, S.L., Teukolsky, S.A., 1983, Black Holes, White Dwarfs and Neutron Stars, Wiley, New York.
Taylor, R.J., 1972, The stars: Their Structure and Evolution, Springer Verlag, New York.
Trümper, J., Lewin, W.H.G., Brinkmann, W., eds. " The Evolution of Galactic X-Ray Binaries", NATO ASI Series, published in Mathematical and Physical Sciences, vol. 167, Library, Reidel, Dordreacht, the Netherlands.
Woosley, S.E., 1986, in "Nucleosynthesis and Chemical Evolution", 16th Advanced Course, Swiss Society of Astrophysics and Astronomy, Saas-Fee, eds. B. Hauck, A. Maeder and G.Meynet.

Supernovae: Observations, Theory, Models and Nucleosynthesis

Ewald Müller

Max-Planck-Institut für Physik und Astrophysik Institut für Astrophysik
Karl-Schwarzschild-Str. 1 D-8046 Garching b. München, FRG

> Earlier, during the 5th month in the 1st year of the Chih-ho reign period (AD 1054, July 4th) the guest star appeared in the morning in the east guarding T'ien-kuan (Zeta Tauri). It was visible in the daytime, like Venus. It had pointed rays on all sides and its colour was reddish-white. Altogether it was visible for 23 days
>
> *Historical Chinese record about SN 1054 from the Sung-hui-yao ("Essentials of Sung History")*

Abstract: Observations, theory and models of supernovae are reviewed together with the explosive nucleosynthesis occuring during the events. In the observational part the classification, frequency, light curves, spectra and progenitors of supernovae in general and of SN1987A in particular are discussed. Theory and models of type II and type I supernovae are reviewed in detail including recent results from nonspherical core collapse calculations and from numerical simulations of the instabilities and mixing processes occuring during the explosion of type II supernovae. A general discussion of the explosive nucleosynthesis in both types of supernovae is presented together with recent numerical results obtained for SN1987A.

1 Introduction

Supernovae are one of the most exciting astrophysical phenomena, in particular if they occur in our own galaxy as the historical supernovae did or if they explode in our immediate galactic neighborhood as the spectacular SN1987A in the Large Magellanic Cloud (LMC). Besides being observationally exceptional events supernovae also pose one of the most challenging theoretical problems in astrophysics. Although the two basic energy sources of supernovae explosions, namely either the gravitational binding energy released during the collapse of a stellar core to a neutron star (or black hole) or the thermonuclear energy released during the explosive burning of about a Chandrasekhar mass of degenerate thermonuclear fuel (helium and/or carbon and oxygen), are already known for several decades, theoretical models of supernovae are still far from being satisfactory as will be discussed below. Other reviews of the subject can be found in Trimble (1982, 1983), Woosley (1986), Woosley and Weaver (1986a), Hillebrandt (1987), Khokhlov (1989a), Cooperstein and Baron (1990), Hillebrandt et al. (1990), Wheeler and Harkness (1990), and Woosley (1990).

As this review was presented at a school for students just before or at the beginning of their doctoral thesis in all fields of observational and theoretical astrophysics, I have concentrated on the basic physical mechanisms of supernovae explosions. The discussion of specific and sometimes still controversial numerical models is kept relatively brief. It mainly serves the purpose to illustrate and in some cases also to confirm the theoretical concepts presented earlier, which of course does not mean to imply that the theoretical concepts have been developed in advance of the respective numerical simulations. Actually, in the field of supernova research progress sometimes has been achieved just in the opposite way, namely by some unexpected numerical result, as e.g. in the case of the delayed explosion mechanism, or by a model independent numerical result, as e.g. the homologous collapse. As most of the numerical simulations and partially even the underlying theoretical concepts assume spherical symmetry, a restriction which in my opinion could be one of the causes of their failure, I have included a discussion of the influence of rotation on the dynamics of core collapse together with some recent numerical results. In addition a section on the results of multi-dimensional hydrodynamic simulations of the mixing and instabilities in the Type II supernova explosions is included in this review.

The outline of the review is as follows. In section 2 a detailed discussion of the present observational knowledge of supernovae including SN1987A is presented, which is divided into seven subsections about the history (2.1), the classification (2.2), the frequencies (2.3), the light curves (2.4), the spectra (2.5), and the progenitors (2.6) of supernovae followed by a subsection (2.7) on SN1987A.

Section 3 is devoted to Type II supernovae including a subsection (3.1) on theoretical models, where the topics weak interactions and neutrino transport, equation of state, collapse and bounce, prompt explosions and shock propagation, delayed explosions, and effectes due to rotation are discussed. In a further subsection (3.2) spherically symmetric and axially symmetric (i.e. rotating) numerical models are reviewed together with simulations of instabilities and mixing during the explosion. In the final subsection (3.3) the explosive nucleosynthesis in Type II supernovae in general and in SN1987A in particular is discussed.

Section 4 is concerned with Type I supernovae and is divided into four subsections. Whereas the first subsection (4.1) contains a discussion of the presupernova evolution, subsection (4.2) deals with the underlying physics of the various modes of burning in degenerate matter (i.e. detonations and deflagrations). Finally in subsections (4.3) and (4.4) the numerical simulations of and the explosive nucleosynthesis in Type I supernovae are reviewed.

2 Observations

2.1 History

Already many centuries ago supernovae (SNe) have been discovered by astronomers in the Far East (China, Japan and Korea), in Arabia and in Europe. According to Clark and Stephenson (1977) various historical records show that at least eight SNe have been observed in our galaxy in the past: AD 185, 386, 393, 1006, 1054, 1181, 1572 and 1604. These are the so-called historical supernovae. The two Renaissance supernovae were discovered by Tycho Brahe (1572) and by Johannes Kepler (1604), the latter one five years before Galileo Galilei for the first time used a telescope for astronomical purposes. In the last 30 years the remnants of the explosions have been identified by radio, optical and x-ray observations for most of the historical supernovae (see e.g. Weiler and Sramek, 1988) including the famous Crab nebula produced by the supernova of 1054. Very likely the (dim?) light of an additional galactic supernova reached the earth about 300 years ago, because the supernova giving rise to the Cas A supernova remnant occured around the year 1670, which can be inferred from the expansion velocity of the remnant. Cas A also is the prototype of the (up to now 7) so-called oxygen-rich supernova remnants (for a recent review on supernova remnants see Weiler and Sramek 1988).

Obviously most data about the historical supernovae are only available in indirect form. Except for Tycho's and Kepler's supernova this lack of exact data even holds for the position of the supernovae in the sky. In case of Kepler's supernova, for example, descriptions like "the supernova was much brighter than Jupiter", (around maximum), or "it was fainter than Xi Ophiuchi" (333 days after maximum) can be found in historical records. Using this kind of information it is possible to reconstruct the light curve of the historical supernovae with certain error bars. Additional information (e.g. about the composition of the exploded star) can be inferred from observations of the remnant today.

The history of extragalactic supernovae began on August 20th, 1885 when E. Hartwig of Dorpat Observatory (in Russia) discovered "a bright nova in M31", which is commonly known as S Andromedae. At this time, the extragalactic nature of M31 was not yet revealed and therefore SN 1885a, as it is called nowadays, was thought to be nova only. The long sought after remnant of SN 1885a was discovered recently by Fesen et al. (1989).

Based on 12 SNe accidentally discovered between 1900 – 1930, on S And, Z Cen (SN 1895b) and on Tycho's supernova Baade and Zwicky (1934a, 1934b, 1934c) first drew the definite distinction between "common novae" and "super-novae" and outlined the fundamental properties of SNe. It was also Zwicky (1940), who coined the word supernovae (first with a hyphen) in a CalTech course in 1931. The following properties of SNe were proposed by Baade and Zwicky based on a few points of the light curve and without any spectral information: (i) the total energy released in the event is $3\ 10^{51}$ – 10^{55} erg (assuming a photosphere of 10^{13} cm and a temperature of 10^5 – 10^6 K); (ii) SNe are transitions of ordinary stars into neutron stars; (iii) SNe emit cosmic rays; and (iv) SNe expel ionized gas shells at great speeds, which contain nuclei of heavy elements. As will be shown below these points still stand today except for some minor modifications.

Zwicky then began an organized supernova search program first with a small telescope at CalTech in 1934 and then with the 18 inch Schmidt telescope at Mt. Palomar in 1936 acquired for this purpose (Zwicky, 1965). Until 1975, 281 supernovae had been discovered almost half of them (122) by Zwicky himself (Trimble, 1982). A more recent supernova catalogue was published by Barbon et al. (1984), which contains 568 SNe discovered up to the end of 1983. The productivity of supernova discoveries has remained low ($\lesssim 20$ per year) in spite of automated SNe search programs, which are operative since a few years. Most of the discovered SNe are so faint ($m_V \geq 14$) that only a small fraction ($2-3\%$) is observed in more detail.

The naming convention of SNe used today was proposed by Zwicky et al. (1963). The events are assigned a year corresponding to the time of their maximum light output and a small letter reflecting the order of their discovery (not occurrence!) for that year. Some years ago the IAU has modified the naming system: Instead of small letters now capital letters are used. If more than 26 supernovae occur in one year (up to now only in 1954, namely SN 1954aa and SN 1954ab) the additional events should be numbered using small letters, i.e. SN 2001a would be the 27th supernova in the year 2001.

2.2 Classification

Supernovae are usually divided into two distinct types, called *Type I* and *Type II*, which are distinguished by the *absence* or *presence* of hydrogen lines in their spectra (Minkowski 1941; see Table 2). This immediately implies that if no spectrum is obtained for a supernova, its type cannot be determined. Due to a late discovery, incomplete coverage or a too small apparent brightness there is a lack of spectral data for many SNe. According to Barbon et al. (1984) only 2/3 of all SNe have been classified. In many cases even the light curve data, which are a much less reliable classification tool, are crude or incomplete. It is especially problematic to classify the historical SNe. The classification of SN 1054, e.g., has changed from Type I (Minkowski 1964) to Type II (Chevalier 1977) within 15 years.

Zwicky (1965) proposed three additional, physically distinct, but rare types of SNe: Type III, IV and V. His Type III SNe are similar to Type II SNe as far as the spectrum is concerned, but their light curves have a much broader maximum and they decline more slowly. The prototype is SN 1961i in NGC 4303. The spectrum of Type IV SNe (prototype is SN 1961f in NGC 3003) is like that of a Type I, but they show additional faint H_γ emission lines and exhibit a stepped decline in their light curve. Finally SN 1961v in NGC 1058 is the prototype (and only example) of Type V SNe. SN 1961v is extraordinary in another aspect, because its progenitor star was observed for more than 20 years before the explosion. The only other case where the progenitor star is known, is SN 1987A in the LMC. Utrobin (1984) has suggested that the progenitor star of SN 1961v was a $2000 M_\odot$ star. Nowadays Zwicky's classification is no longer used. Because all of his additional types show hydrogen lines in their spectra they are classified as Type II SNe.

In the last decade astronomers have accumulated increasing evidence that the both types do not form a homogeneous set, but should be divided (at least) into two distinct subtypes (see Table 1). Accordingly, SNe are nowadays classified as Type Ia and Ib (Porter and Filippenko 1987) and as Type II-L and II-P, where L and P denote a light curve of linear type and plateau type, respectively (Barbon et al. 1979; see also section 2.4). This classification scheme, which was initially suggested by observational data only, may reflect differences in the explosion mechanism and the progenitor star (Wheeler et al. 1987; Wheeler and Harkness 1990; Woosley 1990). Contrary to the "classical" Type I SNe, which are called Type Ia in the new classification scheme, Type Ib SNe do not show the Si II $\lambda 6355$ Å blueshifted absorption feature at $\lambda 6150$ Å (Wheeler and Levreault 1985; Uomoto and Kirshner 1985; see also Table 1). Type Ib are also intrinsically less luminous than Type Ia by about a factor 4 at maximum light (Panagia 1985; Uomoto and Kirshner 1985; Wheeler and Levreault 1985; Branch 1986) and their infrared spectra do not exhibit the absorption feature at ~ 20 days, which characterizes the J, H and K light curves of Type Ia (Elias et al. 1985). Type Ib are also, unlike Type Ia, strong radio emiters (Sramek et al. 1984; Panagia et al. 1986; Weiler et al. 1986; Weiler and Sramek 1988). As some SN Ib do only show very little evidence for He Wheeler and Harkness (1986, 1990) have proposed an additional class of Type I SNe, which they call Type Ic. However, whether SNe Ic are really a distinct additional class or whether they are just extremely helium deficient SNe Ib is still unclear.

Table 1. Identifying features of supernovae; adopted from Weiler and Sramek (1988)

Characteristic	Type Ia	Type Ib	Type II-L	Type II-P
Definition	No hydrogen	No hydrogen	Hydrogen	Hydrogen
Location	All galactic types, interarm and halo	Late-type galaxies, near H II regions	Late-type galaxies, near H II regions	Late-type galaxies, near H II regions
Optical spectrum	P Cygni lines, deep λ6150 Å absorption (Si II λ6355 Å)	P Cygni lines, no λ6150 Å absorption (Si II λ6355 Å), "prematurely old", distinct He lines	P Cygni lines, complex spectrum	P Cygni lines, complex spectrum
Ejection velocity	$\geq 10^4$ km/s	$\geq 10^4$ km/s	$\leq 10^4$ km/s	$\leq 10^4$ km/s
Absolute magnitude*	-20^m homogeneous	$-18^m \ldots -18^m.5$ homogeneous?	$\leq -18^m$ heterogeneous	$\leq -18^m$ heterogeneous
Infrared	Second H-band maximum at ~ 30 days, fast decay $\sim 0^m.05\ d^{-1}$	Peaks near optical maximum, no second maximum, slow decay $\sim 0^m.03\ d^{-1}$	Thermal, heated dust at late times	Thermal, heated dust at late times
Radio	No detections	Strong, rapid decay $t^{-1.6}$	Strong, slow decay $t^{-0.7}$	No well-studied cases

*$H_0 = 50$ km^{-1} s^{-1} Mpc^{-1}

2.3 Frequencies

Although the frequency of SNe has already been determined by Zwicky (1938, 1942) more than fifty years ago, the rates still remain rather uncertain. Qualitatively, however, the various studies have shown that the SNe frequency is strongly dependent on the intrinsic properties of the parent galaxy, in particular on its morphological type and luminosity, and that the SNe rates are also dependent on the supernova type (see e.g. Tammann 1982). The reasons for the quantitative uncertainties are (van den Bergh et al. 1987): (i) a significant fraction of all SNe are discovered by chance, rather than during systematic search programs; (ii) there is still not enough spectroscopic and photometric coverage of SN discoveries so that $V(\max)$, $E(B-V)$, and SN type remain unknown for a significant fraction of all discoveries; and (iii) inclination corrections to the SN discovery frequency in spirals are both large and uncertain. An additional complication is provided by the recently established existence of Type Ib SNe, which show Type I light curves, but are intrinsically a factor of four less luminous than Type Ia's (see Table 1 and section 2.2).

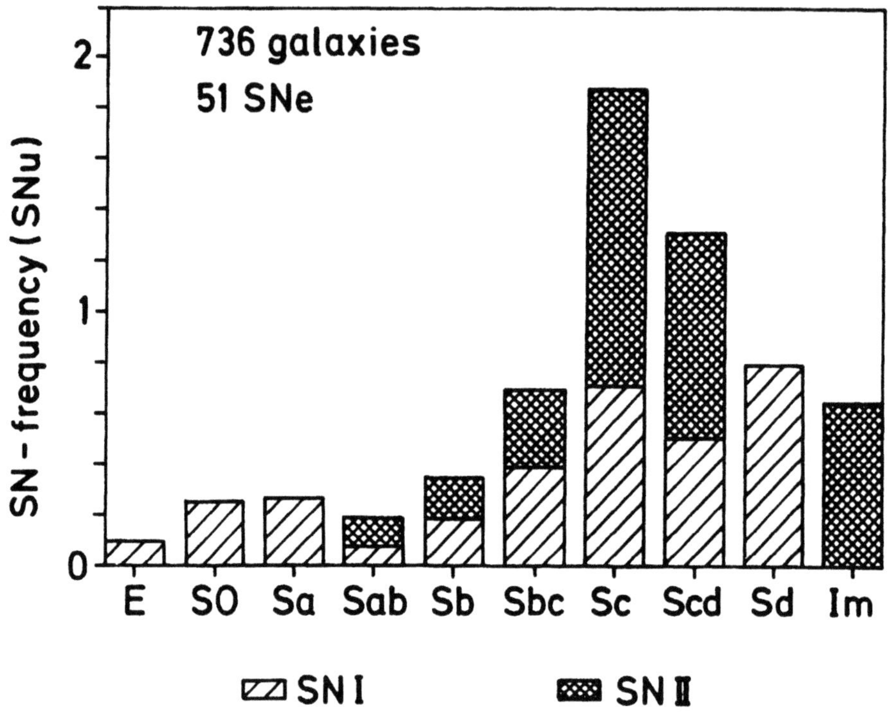

Fig. 1. SN frequency (SNe I + SNe II) in SNu (1 SNu = 1 SN per 100 yrs per 10^{10} $L_\odot$) as a function of galaxy type (from Cappellaro and Turatto 1988).

One of the latest determination of the frequency of both Type I and Type II SNe has been published by Cappellaro and Turatto (1988). Their data sample consists of 736 galaxies and 51 SNe taken from an updated version of the catalogue of Barbon et al. (1984). The distribution of SN frequency in different types of galaxies is shown in Fig. 1. The frequency is measured in so-called supernova frequency units or SNu, where 1 SNu = 1 supernova per 100 yr per 10^{10} $L_\odot$. Obviously the frequency of SNe (in SNu) increases from early to late type galaxies, and even more importantly Type II SNe only occur in spiral and irregular galaxies. Further results from the work of Cappellaro and Turatto (1988) are (i) the SN frequency is proportional to the luminosity of the parent galaxy, (ii) the frequency of Type I SNe increases from early- to late-type galaxies, which may be due to the occurrence of Type Ib SNe in late-type spirals, if they have indeed massive progenitors, as e.g. suggested by Wheeler et al. (1987).

According to Cappellaro and Turatto (1988) the total frequency of SNe I and SNe II is 0.27 h^2 and 0.22 h^2, respectively, where h is the value of the Hubble constant in units of 50 km^{-1} s^{-1} Mpc^{-1}. These rates are quite similar to those obtained by van den Bergh et al. (1987), who claim frequencies of 0.08 h^2, 0.10 h^2 and 0.28 h^2 for SNe Ia, SNe Ib and SNe II, respectively. Although there are still discrepancies between these two recent determinations (the frequency of SNe I differs by 50%, but the total frequency of SNe I and SNe II is practically identical: 0.49 h^2 vs. 0.46 h^2), both of them show that the previously used frequencies of Tammann (1982) are probably too high by a factor of $\propto 3$. Consequently for our own galaxy the interval between two SN explosions is more likely 50 – 100 years instead of 20 – 40 years as predicted by Tammann (1982).

2.4 Light Curves

Following Barbon et al. (1979) the light curves of SNe II can be separated into two distinct and fairly homogeneous subclasses, called "plateau" (SNe II-P) and "linear" (SNe II-L), while the light curves of SNe I (at least photometrically) seem to indicate a single homogeneous class. Note, however that e.g. Barbon et al. (1973), Pskovskii (1977,1978) and Branch (1982) have advocated a finer classification scheme, although the physical basis of the additional subclasses remains unclear (Dogget and Branch 1985). In Fig. 2 the mean blue light curves for SNe I, SNe II-L and SNe II-P are plotted on a common scale and can directly be compared. The light curves of SNe I and SNe II-L exhibit a fast ($\approx 1 - 2$ days) rise to a pronounced maximum, then a relative fast decay over a period of a few weeks followed by an exponential decay of 0.017 mag/day (SNe I) and 0.012 mag/day (SNe II-L), respectively (Dogget and Branch 1985). The shape of the mean light curve for SNe II-P is qualitatively different: Members of this class show a distinct phase of almost constant luminosity, the plateau, between days ≈ 40 to ≈ 80. The late light curve of SNe II-P also declines exponentially with a mean rate of 0.0075 mag/day (Dogget and Branch 1985).

Because of its distinctive plateau it seems to be safe to classify a SN as Type II-P on its light curve shape alone. However, the similarity of SNe I and SNe II-L (see Fig. 2) implies that it may be dangerous to distinguish between these two classes on the basis of their light curves, in particular if one keeps the scatter of the individual SN light curves around the mean light curve in mind. In case of incomplete light curve coverage a classification without spectral information will be almost certainly misleading.

A discussion of the theoretical modelling of SN light curves is beyond the scope of the present work (see e.g. Woosley (1986), Arnett et al. (1989a), and Lucy, this volume). However, it should be mentioned here, that very useful analytical solutions for SN light curves have been developed (Falk and Arnett 1973; Arnett 1980, 1982, 1990; Schaeffer et al. 1987) and successfully applied to the light curve of SN 1987A (see section 2.7).

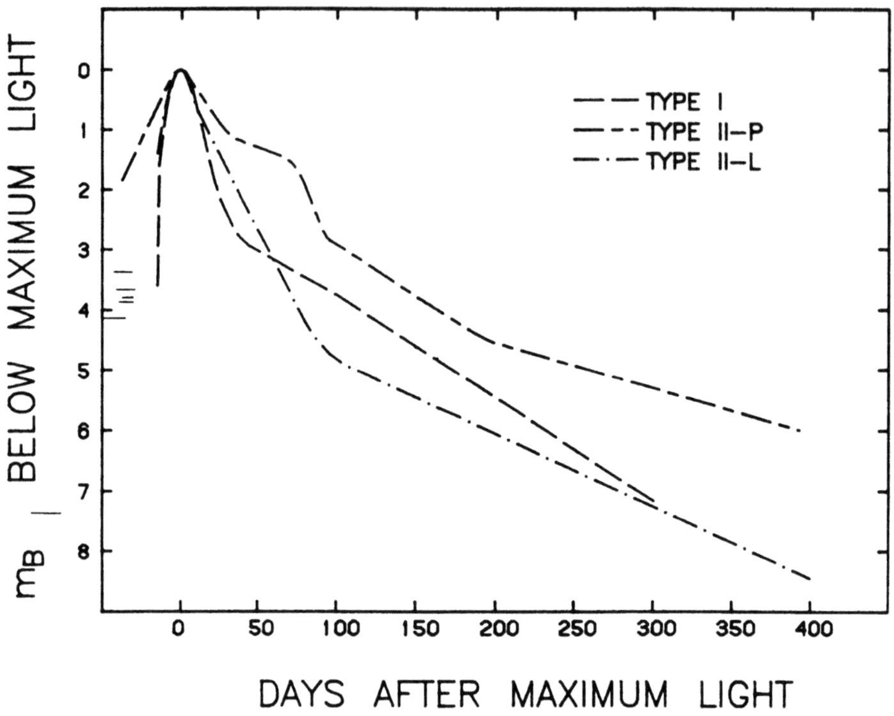

Fig. 2. Comparison of the mean blue light curves for Type I, II-P and II-L supernovae (from Dogget and Branch 1985).

2.5 Spectra

Typical spectra of Type Ia, Ib, II-P and II-L SNe are shown together in Fig. 3; note that the spectra have been shifted to fit onto one plot. Whereas the spectrum of the Type Ia SN (1981b) exhibits the characteristic $\lambda 6355$ Å absorption feature (blueshifted to $\lambda 6150$ Å), which is absent in the spectrum of the Type Ib SN (1984l). On the other side at maximum light the spectrum of a Type Ib SN contains a distinctive absorption dip at $\approx \lambda$ 5700 Å, which is due to a Doppler-shifted He I line at λ 5876 Å which is not present in the spectrum of a Type Ia SN (Wheeler and Harkness, 1990). The He I absorption dip also exists in in spectra of Type Ia SNe three weeks after maximum light (see Fig. 3 in Harkness et al. (1987). As the spectra of Type Ib SNe at maximum light resemble those of Type Ia SNe some weeks after maximum light at all wavelength less than 6000 Å, Type Ib SNe have been called "born-old" (Harkness et al., 1987).

Late time spectra of Type Ib SNe like that of SN 1983n 8 months (Gaskell et al. 1986) or that of SN 1985f 8 months after maximum light (Filippenko and Sargent, 1985, 1986) are very different from spectra near maximum light. Their late time spectra, which show strong lines of O I, Mg I and Ca II in emission (Gaskell et al 1986; Axelrod 1988; Fransson 1988), are also different from the late time spectra of Type Ia SNe, which are dominated by strong emission lines of [Fe II], [Fe III] and [Co III] with Ca H and K in absorption but no evidence for O (Axelrod 1980a, b). A peculiar case is SN 1987K, which first had a spectrum like that of a Type II SN, but at late times showed no hydrogen lines and resembled that of a Type Ib (Filippenko 1988).

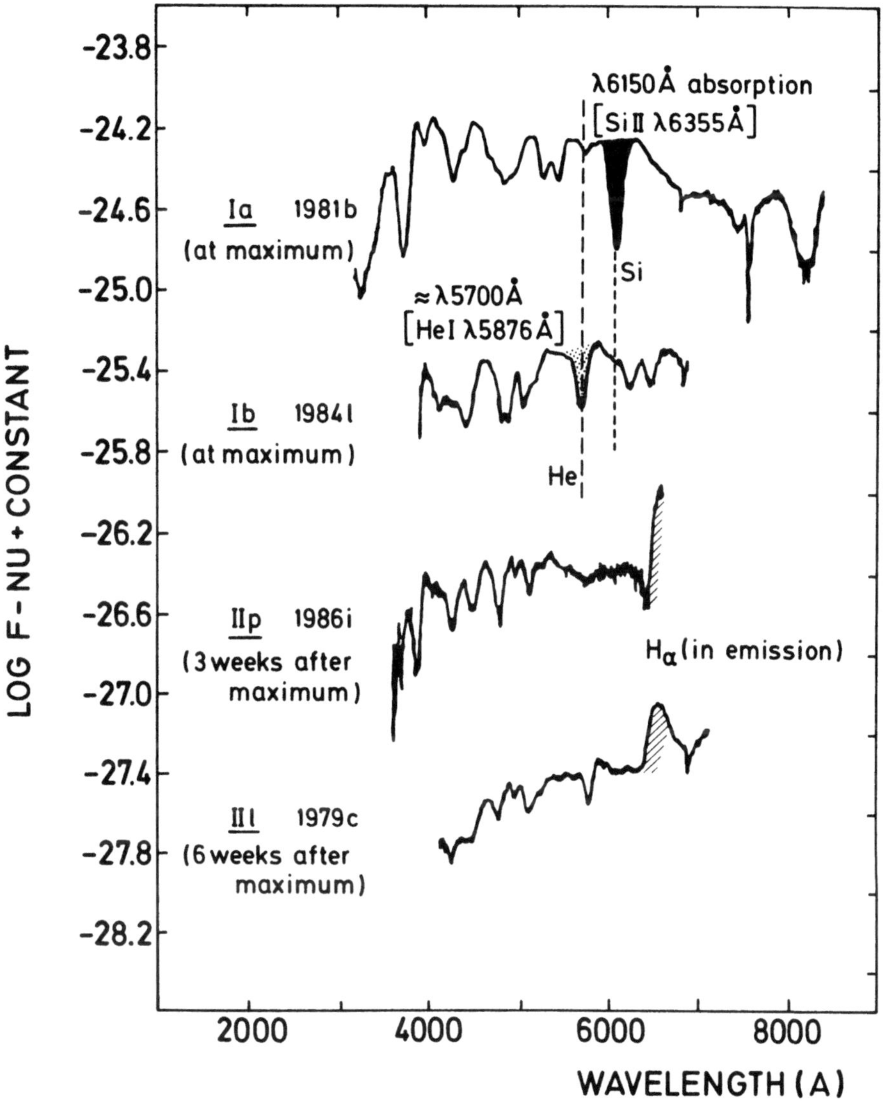

Fig. 3. Spectra of a Type Ia, Ib, II P and II L supernova (from Wheeler et al. 1987).

The spectra of SNe II are dominated by strong P-Cygni Balmer lines of hydrogen near maximum light and in the nebular phase (Zwicky 1965). A current estimate of the minimum hydrogen mass necessary to produce a Type II spectrum is $M_H \approx 0.1 M_\odot$, if the SN is detected near maximum light (Wheeler 1990). The spectra further show [O I] lines, which are strong in some events but only weak in others, and permitted and forbidden Ca II lines (see e.g. Panagia 1985; Hillebrandt and Höflich 1989). Whether Type II-L and Type II-P SNe have distinctively different spectra and whether the observed variations in Type II spectra reflect some intrinsic differences in their physical properties is unclear.

2.6 Progenitors

Type Ia SNe occur in all kind of galaxies and in particular in elliptical galaxies (see section 2.3), show no preference for spiral arms when they occur in spiral galaxies (Maza and van den Bergh 1976), and further show no evidence for hydrogen in their spectra (see section 2.5). This suggests that their progenitors are low mass stars and belong to an old population of stars, because massive (young) stars ($M \gtrsim 10 M_{\odot}$) are associated with spiral arms. As single stars are unlikely candidates (because it is difficult to understand, how they can get rid of their H-envelope and still explode; see e.g. Woosley 1990) the favourite model of Type Ia SNe involves an accreting white dwarf in a binary system, which when approaching the Chandrasekhar mass ignites its thermonuclear fuel (He or a mixture of C/O) under degenerate conditions. The resulting explosion completely disrupts the star leaving no compact remnant. The light curve is powered by the decay of the radioactive ^{56}Ni , which was synthesized during the explosion, into ^{56}Co and eventually into ^{56}Fe (Hoyle and Fowler 1960; Arnett 1969; Nomoto and Sugimoto 1977; see also section 4).

Type Ib SNe occur in spiral arms and close to or in H II regions (Porter and Filippenko 1987), which suggests massive ($\approx 20 - 30 M_{\odot}$) hydrogen deficient progenitors, i.e. Wolf-Rayet stars (Wheeler and Levreault 1985; Begelman and Sarazin 1986; Gaskell et al. 1986; Schaeffer et al. 1987). More massive stars as the only progenitors of Type Ib SNe are in disagreement with the Type Ib SN rate, which could be as large as those of Type Ia and Type II-L (Branch 1986; van den Bergh et al. 1987). Massive progenitors are further supported by the fact that Type Ib SNe, contrary to Type Ia SNe, are strong radio emiters (Weiler and Sramek 1988). This is most easily explained, if the progenitor star lost its hydrogen envelope by a stellar wind: the interaction of the supernova blast wave with the wind then can produce the observed radio emission. The small amount of ^{56}Ni synthesized in the explosion of a massive star is also consistent with the observation that Type Ib SNe are intrinsically dimmer than Type Ia SNe (Porter and Filippenko 1987; Table 1). However, in spite of these agreements the massive progenitor hypothesis is not generally accepted and several alternative models have been proposed (Branch and Nomoto 1986; Uomoto 1986; Woosley 1986; Iben et al. 1987; Ensmann and Woosley 1988).

Observational data strongly indicate that Type II SNe originate from massive hydrogen rich Population I stars, because they only occur in the arms of spiral galaxies (Tammann 1982; Maza and van den Bergh 1976), show strong hydrogen lines and are (at least the Type II-L) radio emiters (Weiler and Sramek 1988; see also Table 1). Before the discovery of SN 1987A it was generally believed that the progenitors of Type II SNe are red supergiants (see e.g. Woosley and Weaver 1986a) and not blue supergiants as in the case of SN 1987A (see section 2.7). Stellar evolution calculations show that only stars with a mass $M \gtrsim 8 M_{\odot}$ eventually form an iron core and explode (if their H-envelope has not been lost during the evolution in which case they may produce a Type Ib SN) as a Type II SN (see e.g. Woosley and Weaver 1986a).

A highly tentative classification scheme for SN remnants produced by massive progenitors let van den Bergh (1988) to suggest that the progenitors of SN Ib are O-stars or Wolf-Rayet stars, whereas early B-type stars produce Type II SNe. For a Hubble constant of 75 km s^{-1} Mpc^{-1} and the supernova rate of van den Bergh et al. (1987) the progenitors of Type II SNe have $8M_{\odot} \lesssim M \lesssim 18M_{\odot}$, whereas stars with $M \gtrsim 18M_{\odot}$ become Type Ib SNe (van den Bergh 1988).

The progenitors of Type II-L and Type II-P SNe probably only differ in the amount of mass of the H–envelope (Chevalier 1984). While the plateau in the light curve of Type II-P SNe, which is caused by the recombination of hydrogen, is indicative of a large hydrogen mass ($\gtrsim 10M_{\odot}$), its absence in Type II-L SNe implies, that their progenitors have lost most but not all of their H–envelope. In this picture both kinds of Type II SNe are caused by the same explosion mechanism (i.e. by the core collapse mechanism). A different picture was suggested by Dogget and Branch (1985) on the basis of the similarity of the light curves of Type Ia and Type II-L SNe. They argued that Type II-L might be caused by a thermonuclear explosion (see also Wheeler et al. 1987).

2.7 SN1987A

A detailed discussion of the observations and theoretical implications of SN1987A is far beyond the scope of this work. The subject was recently reviewed by Arnett et al. (1989a) and by Hillebrandt and Höflich (1989). In the following only the most important points about SN1987A will be presented in brief.

The progenitor star of SN1987A could be identified through IUE observations as the main component of the multiple system Sanduleak -69°202 (Gilmozzi et al. 1987; Wallborn et al. 1987). This is the only supernova up to now besides SN 1961v (see section 2.2) where the progenitor star is known. Instead of a red supergiant the progenitor star was a blue supergiant of spectral class B3 and luminosity class Ia (Rousseau et al. 1978). This surprising observation had important consequences for the light curve (see below) and questioned our knowledge of the evolution of massive stars (see e.g. Hillebrandt and Höflich, 1989).

The light curve of SN1987A was measured in the UV, optical and IR wavelength range, which for the first time made it possible to construct a bolometric light curve of a supernova (Fig. 4). Obviously the bolometric light curve of SN1987A is very different from that of other supernovae (see Fig. 2). Despite its peculiarities the bolometric light curve can be well understood within the general framework of Type II light curves (Falk and Arnett 1973; Arnett 1980, 1982, 1990; Nomoto and Shigeyama 1987; Woosley 1988a, b, c).

In particular the 10 to 100 times lower bolometric luminosity of SN1987A relative to average Type II SNe can be understood from the more compact structure of the progenitor and its lower metallicity ($Z_{LMC} \approx 1/3 Z_{\odot}$) (Arnett 1987; Hillebrandt et al. 1987). The relatively small luminosity of SN1987A has important consequences for the SN rate in general and in irregular galaxies in particular, because these dim, metal poor SNe may be quite common, but observationally rare (see e.g. van den Bergh and McClure 1989). The radiative release of the energy deposited by the shock wave in the envelope in form of thermal and ionization energy ($\approx 1/2$ of the total energy at shock break out) occurs on the diffusion timescale. In a more compact progenitor the diffusion timescale

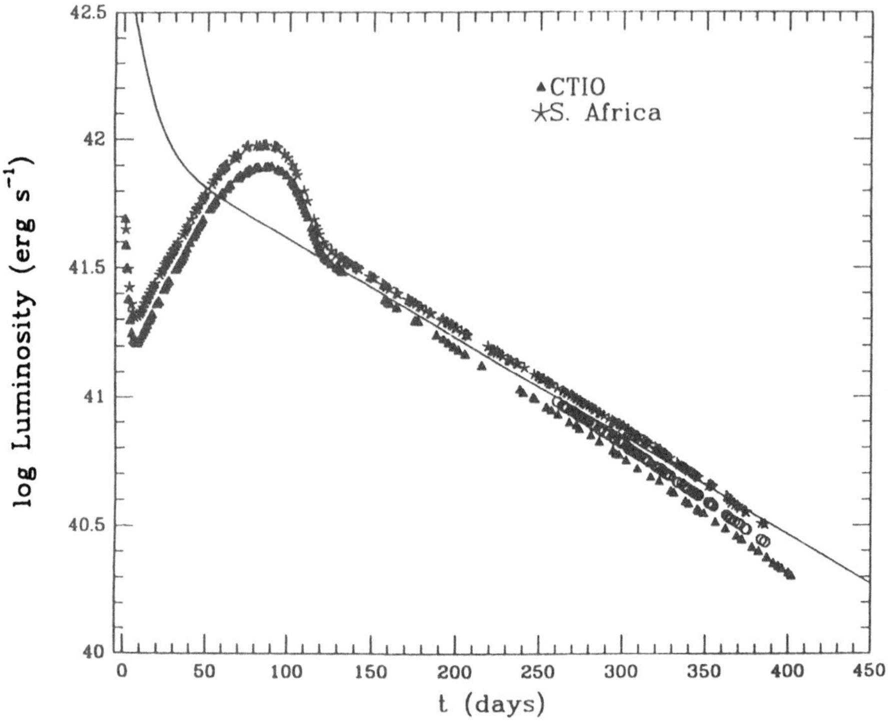

Fig. 4. Bolometric light curve of SN1987A. The solid line corresponds to the luminosity produced by a 100% conversion of the decay energy of 0.075 $M\odot$ of ^{56}Ni (from Arnett et al. 1989a).

is longer and a larger fraction of the internal energy is converted into expansion energy and a smaller fraction into radiation.

Because throughout the hydrogen envelope the energy (per mass) deposited by the shock wave is approximately constant, initially the bolometric light curve does not change strongly with time. However, when the diffusion timescale decreases during the expansion, recombination and energy input from radioactive decay start to contribute to the luminosity and cause it to rise until maximum light. Consequently the shape of the light curve in this epoch is governed by the diffusion timescale, which in turn depends on the mass of the hydrogen rich envelope and on the opacity, i.e. the chemical composition of the envelope. Woosley (1988a) showed in particular that the time at which the maximum luminosity is reached is a sensitive measure of the total hydrogen mass of the envelope. He found a value of order $10M_\odot$, a result which was confirmed by others (Arnett 1987, 1988; Nomoto et al. 1987; Utrobin 1988) and also by the spectral analysis (Höflich 1987, 1988a, b).

The agreement between the observed light curve and the ones obtained from numerical models could be improved significantly, when (*ad hoc*) mixing of heavy elements from the He-core into the H-envelope was introduced, because mixing increases the opacity and thereby the diffusion timescale. Mixing in SN1987A is further needed to explain the early discovery of x-rays and γ-rays, the almost constant flux of the highly energetic radiation up to day 570, the observed optical and IR spectra. Probably mixing is also required to understand the evolution of the progenitor star and its chemical composition as a function of depth as inferred from the spectral analysis (see Höflich 1988a; Arnett et al. 1989a; Hillebrandt and Höflich 1989; and section 3.2.3).

The light curve, which shows an exponential decline after day $\approx$ 110 with a mean half-life of 77 days (Fig. 4), is powered by the decay energy of ^{56}Co , which has the same half-life and which is the decay product of ^{56}Ni synthesized in the explosion (see section 3.3.2). From the light curve data a mass of $0.07 M_{\odot}$ of ^{56}Ni can be inferred (Arnett 1988; Nomoto and Shigeyama 1987; Woosley 1988a, b). Beyond day 300 the envelope starts to become transparent to the γ-rays produced in the ^{56}Co decay, i.e. the light curve then decreases faster than expected from the ^{56}Co half-life (Fig. 4). Up to now (January 1990) no doubtless evidence for a pulsar has been found in the light curve of SN1987A, although the almost constant bolometric light curve since day $\approx$ 900 could indicate the presence of a pulsar (Bouchet et al. 1989). The deviation of the bolometric light curve from a linear extrapolation from earlier epochs implies an upper limit for the total luminosity of such a possible pulsar of 10^{38} erg/sec (Bouchet et al. 1989), i.e. a luminosity at least five times less than that of the Crab pulsar.

During the first few months the observed spectra of SN1987A showed, like those of other Type II SNe, a flux distribution dominated by continuum radiation, but modified by lines, whereas at later times (after December 1987, in the so-called nebular phase) again in accordance with other Type II SNe the flux was mainly due to emission lines in the optical and IR wavelength range. However, the spectra of SN1987A also showed a number of peculiarities at ultraviolet and optical wavelengths: (i) A very rapid evolution during the first few days as a consequence of the rapid increase of the photospheric radius; (ii) a strong UV line blanketing due to the relatively steep density gradient of the compact progenitor and due to its relatively small luminosity; and (iii) an early occurrence of strong hydrogen lines again due to the steep density gradient and the early recombination phase of hydrogen (see Hillebrandt and Höflich 1989). The optical spectra including the hydrogen lines of the first seven months can be well reproduced by synthetic spectra calculated with NLTE models, provided a metallicity of $Z = 1/3 Z_{\odot}$ is assumed, which is consistent with the average metallicity of the LMC (Höflich 1987, 1988a, b; Fig. 5).

The observation of a neutrino pulse detected prior to the optical outburst of SN1987A, at February 23.32 (UT) (Hirata et al. 1987; Bionta et al. 1987), convincingly confirmed the theoretical prediction that the collapse of a stellar core to neutron star densities is the basic energy source of Type II SNe (Arnett et al. 1989a; Hillebrandt and Höflich 1989; and references therein). Using a maximum likelihood method Janka and Hillebrandt (1989a, b) deduced a total emitted energy in form of anti-electron neutrinos of $E_{\bar{\nu}_e}^{IMB} = 3.4^{+11.6}_{-2.4} \cdot 10^{52}$ erg (from data of Bionta et al. 1987) and $E_{\bar{\nu}_e}^{K2} = 6.4^{+8.1}_{-3.9} \cdot 10^{52}$ erg (from data of Hirata et al. 1987), respectively. The uncertainty bounds of the emitted energies correspond to a 95% confidence level. Note that both detections are only weakly consistent. The emitted energy of the anti-electron neutrinos translates into a total energy emitted in form of neutrinos and anti-neutrinos (i.e. including μ- and τ-neutrinos and anti-neutrinos) of several times 10^{53} erg, which is of the order of the gravitational binding energy of a neutron star.

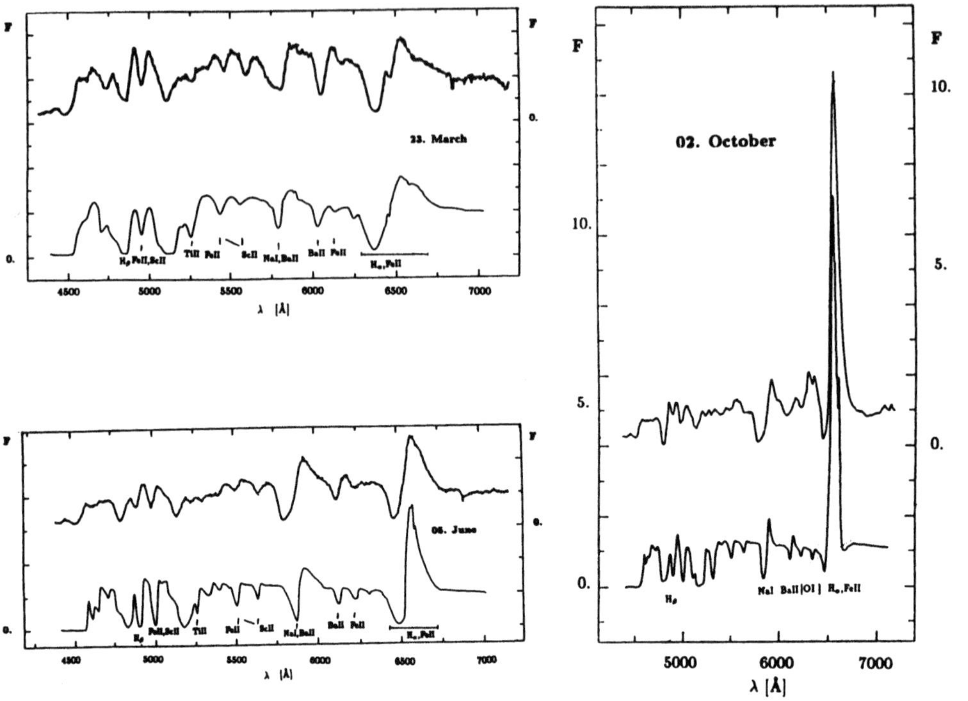

Fig. 5. Observed spectra of SN1987A about 1 month (top left), 2 months (bottom left), and 7 months (right) in comparison with reddened, synthetic spectra (lower spectrum, shifted by one unit; $E_{B-V} = 0^m.15$) (from Höflich 1988b).

3 Type II Supernovae

3.1 Theoretical Models

Theoretical models of Type II SNe are based on the idea, that the implosion of the iron core of a massive star ($M \gtrsim 9\ M_\odot$) at the end of its thermonuclear evolution, causes a Type II supernova explosion and the formation of a compact remnant, i.e. a neutron star (or a black hole). The energy source for the explosion, therefore, is the gravitational binding energy of the forming neutron star, which is of the order

$$E_{bind} \approx 3\ 10^{53} \left(\frac{M}{M_\odot}\right)^2 \left(\frac{R}{10\ \mathrm{km}}\right)^{-1} \mathrm{erg} \quad , \tag{3.1}$$

where M and R are the mass and the radius of the neutron star, and $M_\odot$ is the solar mass. This binding energy is much larger than the the kinetic energy of the expelled matter $E_{kin} \approx 10^{51}$ erg, which in turn is much larger than the total energy emitted in form of electromagnetic radiation $E_{em} \approx 10^{49}$ erg. Most of the binding energy (more than 99%), however, is carried away in form of neutrinos. Note that this was only a theoretical prediction, until the measurement of the neutrinos from SN1987A (see section 2.7).

All theoretical models of Type II SN explosions have to deal with the same basic problem, namely to find a mechanism by which a small fraction of the binding energy released during core collapse is transfered to the outer layers of the star resulting in a violent mass ejection with velocities $v_{eject} \approx 10^4$ km/sec and the formation of a neutron star with a mass $M \approx 1.4 M_{\odot}$. Obviously, if the explosion fails, matter outside the core eventually will be accreted onto the proto-neutron star until its critical mass is exceeded and a black hole forms.

The input physics and the dynamics of the collapse and the subsequent explosion will be reviewed in the following sections. Other recent reviews of the same topic have been presented e.g. by Woosley (1986), Woosley and Weaver (1986a), Hillebrandt (1987), Cooperstein and Baron (1990), and Hillebrandt et al. (1990).

3.1.1 Basic Considerations

At the onset of collapse the central density and temperature of the degenerate core of a massive star are $\rho_c \approx 10^{10}$ gcm^{-3} and $T_c \approx (8-10)\ 10^9$ K, respectively. The chemical composition of the core consists mainly of iron group nuclei, the reason why it is usually called the *iron core* (see e.g. Woosley and Weaver 1986a). The collapse timescale of the iron core $\tau_{collapse} \approx 0.1$ sec, while its dynamical (or free fall) timescale $\tau_{dyn} \propto \sqrt{\bar{\rho}}^{-1}$ ($\bar{\rho}$ the average core density) is of the order 1 msec.

Because of the high densities and temperatures in the iron core the strong and electromagnetic interactions are in equilibrium. In this (local) nuclear statistical equilibrium (NSE) the composition of matter is determined by the Saha equation and is, therefore, independent of the reaction rates (see e.g. Woosley 1986; and section 3.1.3). Furthermore the timescale of the strong and electromagnetic reactions $\tau_{rate} \ll \tau_{dyn}$, i.e. hydrodynamic processes can be considered to be adiabatic changes for these reactions.

In core matter the energy transport by diffusion of photons and electrons can be neglected, because the corresponding diffusion length λ_{diff} is much smaller than the typical density scale height $(d \ln \rho / dr)^{-1}$. Energy transport by heat conduction of degenerate electrons or ions is also unimportant, and the viscosity of core matter is negligible, too (Flowers and Itoh 1976). Consequently except for shock waves and energy transport by neutrinos *no* dissipative process or transport phenomenon is of importance for the dynamics of the core, i.e. core matter can be treated as an ideal fluid.

3.1.2 Weak Interactions and Neutrino Transport

For densities $\rho \lesssim 3\ 10^{12}$ gcm^{-3} weak interactions are not in equilibrium, because $\tau_{weak} \approx \tau_{collapse}$. Important weak processes, which have to be considered in core collapse simulations, are electron captures (EC) on *free* protons

$$e^- + p \longrightarrow n + \nu_e \tag{3.2}$$

and the following neutrino interactions

$$\begin{aligned} \nu_e + n &\longrightarrow p + e^- \ , \\ \nu_e + e^- &\longrightarrow \nu_e + e^- \ , \\ \nu_e + A &\longrightarrow \nu_e + A \ , \end{aligned} \tag{3.3}$$

i.e. neutrino absorption on neutrons, neutrino-electron scattering and coherent scattering of neutrinos off nuclei (Tubbs and Schramm 1975; Bruenn 1985). Note that electron captures on nuclei are blocked due to shell effects, because in a neutron-rich NSE composition the most abundant nuclei posses a closed neutron shell (Fuller 1982; Fuller et al. 1982).

In principle neutrinos and anti-neutrinos of all flavours

$$\nu_e, \bar{\nu}_e, \quad \nu_\mu, \bar{\nu}_\mu, \quad \nu_\tau, \bar{\nu}_\tau \tag{3.4}$$

can participate in the weak processes. However, due to the strong electron degeneracy ($\eta_e \gtrsim 20$) for $\rho \gtrsim 10^{13}$ gcm^{-3} the number density of all other kind of neutrinos is much smaller than the number density of electron neutrinos, i.e.

$$n_{\{\bar{\nu}_e, \nu_\mu, \bar{\nu}_\mu \dots\}} \ll n_{\nu_e} \quad , \tag{3.5}$$

because these neutrinos can only be created via pair and plasmon processes (see e.g. Bruenn 1985). In shock heated matter ($\rho < 10^{12}$ gcm^{-3}) on the other side, the rates of the pair and plasmon processes for μ- and τ-neutrinos are very low, and for $\bar{\nu}_e$ positron capture on neutrons is much more important than pair creation of electron anti-neutrinos.

Initially neutrinos can leak out of the core, because the density is relative low and consequently the opacity of core matter is low, too. The leakage of neutrinos results in a deleptonization of the core, i.e. the electron fraction (per nucleon) Y_e decreases. The neutrino opacity during collapse is dominated by coherent scattering off nuclei (due to neutral currents) in the collapse phase (Freedman 1974; Wilson 1974; Schramm and Arnett 1975). The mean free path of neutrinos, therefore, is given by (Hillebrandt 1987)

$$\lambda \approx 10^7 \text{cm} \left(\frac{\rho}{10^{12}\ \text{gcm}^{-3}}\right)^{-1} \frac{A}{N^2} \left(\frac{E_\nu}{10\ \text{MeV}}\right)^{-2} \quad . \tag{3.6}$$

In shocked matter the main contributions to the neutrino opacity are due to scattering off free nucleons and absorption on neutrons (see e.g. Bruenn 1985).

For densities $\rho \gtrsim 3\ 10^{11}$ gcm^{-3} the neutrino diffusion timescale $\tau_{diff} > \tau_{collapse}$, which implies that the total lepton number is conserved during further collapse (Mazurek 1974, 1975; Sato 1975a, b):

$$Y_L = Y_e + Y_\nu = constant \quad . \tag{3.7}$$

This epoch of the collapse is called ***neutrino trapping***. In analogy to photons a neutrino-sphere can be defined as the surface, where the optical depth is in the range

$$2/3 \lesssim \tau(r) = \int_\infty^r \frac{dr}{\lambda_\nu} \leq 1 \quad . \tag{3.8}$$

Note that the location of the neutrino-sphere depends on the energy of the neutrinos, i.e. the neutrino-sphere has a certain spatial extent. The typical extension is of the order $\Delta r_\nu \approx (10 - 30)$ km (see e.g. Janka and Hillebrandt 1989b). Whereas near the neutrino-sphere transport processes are of importance, the neutrinos can freely stream further outside.

Inside the neutrino-sphere the trapped neutrinos thermalize by neutrino-electron scattering. The chemical potential μ_ν of the Fermi distribution of the trapped neutrinos grows with increasing density until *β-equilibrium* is established at densities $\rho \gtrsim 3\ 10^{12}$ gcm^{-3}, i.e. the weak reactions and in particular

$$e^- \quad + \quad p \quad \rightleftharpoons \quad n \quad + \quad \nu_e \tag{3.9}$$

come into equilibrium and the chemical potentials (including the rest mass) of the electrons, protons, neutrons and neutrinos fulfill the relation:

$$\mu_e + \mu_p = \mu_n + \mu_{\nu_e} \quad . \tag{3.10}$$

An important consequence is that the entropy remains constant after β-equilibrium is reached, i.e. the further collapse proceeds adiabatically unless shocks are created.

3.1.3 Equation of State

Before discussing how to derive the equation of state (EOS), let us first estimate the various contributions to the pressure at the onset of collapse. For a density $\rho = 10^{10}$ gcm^{-3}, a temperature $T = 10^{10}$ K and matter mainly composed of iron the ion pressure is of the order

$$P_{ion} = \sum_i Y_i\, \rho k_b T \approx 2\ 10^{26}\ \mathrm{erg/cm^3} \quad , \tag{3.11a}$$

where Y_i are the number fractions of the ions and k_B is Boltzmann's constant. The radiation pressure is of the order

$$P_{rad} = \frac{a}{3} T^4 \approx 3\ 10^{25}\ \mathrm{erg/cm^3} \tag{3.11b}$$

and the pressure of the relativistic and degenerate electron gas ($n_e \approx 3\ 10^{33}$ cm^{-3}; $\mu_e \approx$ 10 MeV; $\mu_e \gg m_e c^2$ and $\mu_e \gg k_B T$) is of the order

$$P_e \approx \frac{2\pi}{3} \frac{1}{c^3 h^3} \mu_e^4 \approx 10^{28}\ \mathrm{erg/cm^3} \tag{3.11c}$$

with $P_e \propto (Y_e \rho)^{4/3}$. Equations (3.11a) to (3.11c) obviously show that

$$P_e \gg P_{ion} \gg P_{rad} \quad , \tag{3.12}$$

i.e. the iron core is stabilized against gravity by relativistic electrons. This further implies that the adiabatic index of the matter in the iron core at the onset of collapse is close to the critical value of 4/3 (see section 3.1.4):

$$\gamma = \left.\frac{\partial \ln P}{\partial \ln \rho}\right|_{Y_e, S} \approx \frac{4}{3} \quad . \tag{3.13}$$

A further small but not negligible contribution to the total pressure is the (negative!) pressure due to the Coulomb interaction of the ions (for its effect on the initial core mass see Nomoto and Hashimoto 1988; Woosley and Weaver 1988; Cooperstein and Baron 1990; Baron and Cooperstein 1990).

Detailed computations of the EOS are greatly simplified by the fact that strong and electromagnetic interactions are in equilibrium at temperatures above roughly 5 10^9 K (see section 3.1.1). It is therefore an excellent approximation to assume an equilibrium composition of nuclei and free nucleons for a given total number of neutrons and protons (NSE). Then the method of computation is (in principle) straightforward (see e.g. Hillebrandt 1987): One chooses a functional for the free energy $F = F(T, V, N_i)$ of the system, which depends on temperature, volume and composition. Next F is minimized with respect to the abundances at fixed temperature, volume and neutron and proton number

$$\left(\frac{\partial F}{\partial N_i}\right)_{T,V,Y_e} = 0 \quad . \tag{3.14}$$

Instead of keeping the neutron and proton number fixed in practice one minimizes the free energy at fixed electron number fraction Y_e or, if neutrinos are to be included, at fixed lepton number fraction Y_l or assuming β-equilibrium. Note that the lepton contribution to the EOS poses no severe problem, because the leptons form an almost homogeneous background of free fermions. After minimization all thermodynamic variables can be obtained by the usual thermodynamic relations, e.g. $P = -\partial F/\partial V$. Different approaches to obtain the EOS only differ in the models that are used in order to compute the free energy.

At "low" densities ($\rho \lesssim 10^{12}$ gcm^{-3}) free nucleons are rare and nuclei are far apart. Therefore, both can be treated as Boltzmann particles. The EOS is then determined, provided the ground state binding energies and the nuclear partition functions are known (see e.g. El Eid and Hillebrandt 1980). At "high" but subnuclear densities (10^{12} gcm$^{-3} \lesssim \rho \lesssim 2\ 10^{14}$ gcm^{-3}) the free nucleon density increases and starts to influence the properties of the nuclei. The nuclei become more closely packed and begin to interact. In this "high" density regime a microscopic model is required to compute the EOS, which incorporates nucleon-nucleon and nucleus-nucleus interactions, effects due to the finite size of the nuclei and lattice effects due to Coulomb correlations (see e.g. Barranco et al. 1989, Cooperstein and Baron 1990; Hillebrandt 1990). At nuclear saturation density a phase transition occurs and nuclei and free nucleons merge into homogeneous, asymmetric ($Y_e \approx 0.3$) nuclear matter (see e.g. Lattimer et al 1985a). The EOS gets very complicated, as many-body forces, relativistic effects, and extra hadronic and mesonic degrees of freedom must be taken into account (see e.g. Baldo et al. 1989; Glendenning 1990)

Feasible approaches are the usage of the temperature-dependent Hartree-Fock method (Bonche and Vautherin 1981, 1982; Hillebrandt and Wolff 1985; Vinas et al. 1987), the temperature-dependent Thomas-Fermi method (Barranco and Buchler 1980; Marcos et al. 1982; Ogasawara and Sato 1983; Suraud and Vautherin 1984; Suraud 1985), or the compressible liquid drop model (Lamb et al. 1978, 1981; El Eid and Hillebrandt 1980 Lattimer et al. 1985a). All these approaches are based on the idea to put a fixed number of protons and neutrons into a spherical box of radius R_c (Wigner-Seitz cell), which has to be minimized too in order to get the optimum radius of the cell. In more elaborate models Ravenhall et al. (1983; liquid drop model) and Lassaut et al. (1987; Thomas-Fermi method) the spherical Wigner-Seitz cell is replaced by a Wigner-Seitz cell reflecting the lattice structures, which result from Coulomb lattice correlations. For a given temperature the thermodynamic properties are then computed from a model Hamiltonian with an effective interaction, which is one of the major uncertainties of all approaches.

Published EOS show that the adiabatic index lies in the range $1.27 \lesssim \gamma \lesssim 1.325$ for densities $\rho \lesssim 2\ 10^{14}$ gcm^{-3} and entropies $s \approx 1$ k$_B$/nucleon (El Eid and Hillebrandt 1980; Hillebrandt and Wolff 1985 (Fig. 6); Lattimer et al 1985a; Barranco et al. 1989). If the density exceeds nuclear saturation density the EOS stiffens considerably due to the

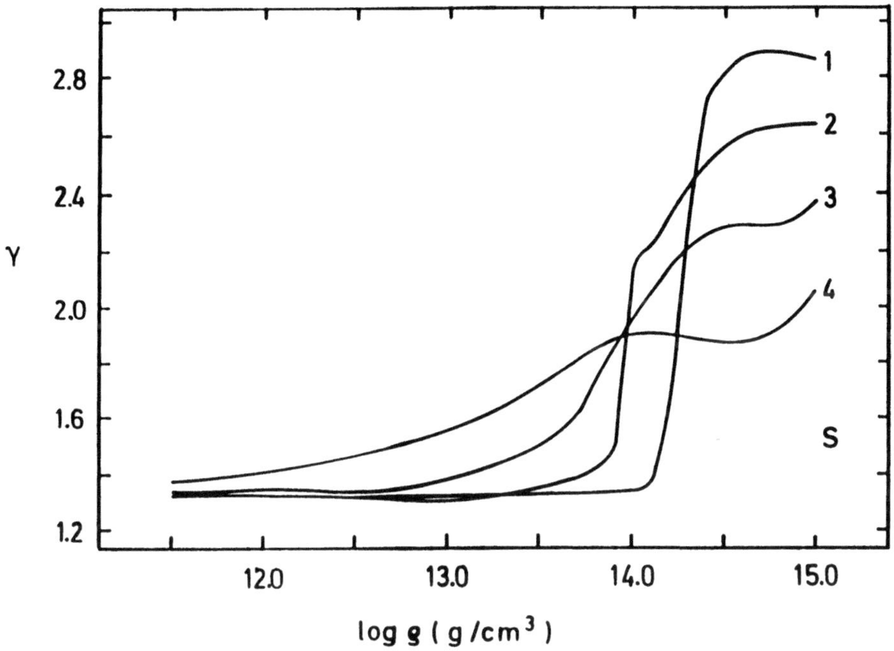

Fig. 6. Adiabatic index $\gamma = (\partial \ln p / \partial \ln \rho)_{S,Y_e}$ versus density for some selected isentropes (in units of k_B/nucleon). For low entropies ($S \approx 1$) γ is only slightly below 4/3 up to almost 10^{14} gcm^{-3}. Note the abrupt stiffening of the EOS at near nuclear matter density (from Hillebrandt and Wolff 1985).

repulsive part of the nuclear forces ($\gamma > 2.5$). Although it is somewhat problematic, it is quite common to characterize the nuclear equation of state (NEOS) only by means of three parameters, namely the high density adiabatic index, the nuclear incompressiblity K_0 and the bulk symmetry coefficient W_s (see e.g. Cooperstein and Baron 1990). The values of all three parameters and in particular the value of K_0 are the matter of a controversial debate experimentally (Blaizot 1980; Sharma et al. 1988, 1989) as well as theoretically (Brown and Osnes 1985; Vinas et al. 1987; Glendenning 1990).

For a qualitative discussion of the influence of the EOS on the collapse dynamics it has been proven useful to approximate the EOS by a simple polytropic law $P = K(Y_e, S)\rho^\gamma$, where the electron fraction Y_e, the entropy S and the adiabatic index are in general density-dependent (van Riper 1978, 1979; Müller et al. 1980; Tohline 1984; Cooperstein and Baron 1990; Mönchmeyer 1990). Obviously the quality of the polytropic approximation crucially depends on how well the behaviour of the realistic EOS is modelled by the density dependence of γ, Y_e and S. The same approach was used by Baron et al. (1985a, b) and van Riper (1988) to model the NEOS. They have introduced an additional zero temperature polytropic component of the nuclear matter pressure of the form

$$P_{nm} = \frac{K_0(Y_e)\,\rho_0(Y_e)}{9\gamma}\left[\left(\frac{\rho}{\rho_0(Y_e)}\right)^\gamma - 1\right] \qquad (3.15)$$

where

$$\rho_0(Y_e) = 0.16\,[1 - 3(0.5 - Y_e)^2] \tag{3.16}$$

is the saturation density for asymmetric nuclear matter (in units of fm^{-3}) and

$$K_0(Y_e) \equiv 9 \left(\frac{\partial P_{nm}}{\partial \rho}\right)_{Y_e}\Bigg|_{\rho=\rho_0(Y_e)} = K_0^{sym}\,[1 - 2(1 - 2Y_e)^2] \tag{3.17}$$

is the incompressibility for asymmetric matter. K_0^{sym} is the value of K_0 for symmetric matter.

3.1.4 Collapse and Bounce

According to Chandrasekhar (1939) a star, which is supported against gravity by the (zero temperature) pressure of its relativistic electrons, can only be stabilized, if its mass is less than a critical mass

$$M_{Ch} = 1.457\,(2Y_e)^2\,M_\odot \quad , \tag{3.18}$$

which is the famous Chandrasekhar mass. For a mixture of different relativistic leptons with average number fractions $< Y_i >$ and statistical weights g_i the generalized Chandrasekhar mass is given by the expression

$$M_{Ch} = 5.828 \left\{\sum_i < Y_i >\right\}^2 \left\{\left(\frac{2}{g_i}\right)^{1/3} \sum_i \left[\frac{< Y_i >}{\sum_i < Y_i >}\right]^{4/3}\right\}^{3/2} M_\odot \quad . \tag{3.19}$$

At finite temperature the critical mass of a stable iron core should increase, but this increase is almost perfectly canceled by a decrease of the critical core mass due to Coulomb lattice correlations (Baron and Cooperstein 1990).

At the end of Si-core burning the electron fraction in the iron core of a massive star has a value $0.42 \lesssim Y_e \lesssim 0.44$, because of electron captures which occured during oxygen and (core) silicon burning. The central entropy of the core is of the order 1 k_B/nucleon (Nomoto and Hashimoto 1988; Woosley and Weaver 1986a, 1988). The mass of the iron eventually exceeds the Chandrasekhar mass, because (i) the mass of the core grows as a result of Si-shell burning, (ii) EC's further reduce the lepton number in the core as neutrinos are lost from the star, and (iii) the pressure in the core is reduced due to (endothermic) photodisintegration of nuclei by energetic photons. The relative importance of the last two processes depends on the stellar mass. For very massive stars ($M \gtrsim 20\,M_\odot$) photodisintegrations are more important than EC's, but the opposite is true for less massive stars, because of their lower central entropies (Woosley 1986).

The crucial quantity for the dynamical behaviour of the core is the "effective" adiabatic index (van Riper and Lattimer 1982)

$$\Gamma \equiv \frac{\partial \ln P}{\partial \ln \rho}\Bigg|_M \quad , \tag{3.20}$$

which describes the change of the pressure along a collapse trajectory of a given mass element. If the collapse is neither adiabatic nor occurs at fixed Y_e the "effective" adiabatic index is given by

$$\Gamma = \left.\frac{\partial \ln P}{\partial \ln \rho}\right|_{S,Y_e} + \left.\frac{\partial \ln P}{\partial \ln Y_e}\right|_{\rho,S} \left.\frac{\delta \ln Y_e}{\delta \ln \rho}\right|_M + \left.\frac{\partial \ln P}{\partial \ln Y_\nu}\right|_{\rho,S} \left.\frac{\delta \ln Y_\nu}{\delta \ln \rho}\right|_M + \left.\frac{\partial \ln P}{\partial S}\right|_{\rho,Y_e} \left.\frac{\delta S}{\delta \ln \rho}\right|_M . \tag{3.21}$$

The first term ($\equiv \gamma$) is purely a function of the EOS. The next two terms in Eq. (3.21), which describe the change of Γ due to EC, are large (the third term only after neutrino trapping) compared to the fourth term, which gives the change of the "effective" adiabatic index due to the non-adiabaticity of the collapse (which is small; see below). If the EOS is approximated by a polytropic form $P = C(S) \cdot \exp(4/3 \ln Y_e) \cdot \exp(\gamma \ln \rho)$ the "effective" adiabatic index before neutrino trapping ($Y_l = Y_e$) can be written for $\delta S = 0$ as

$$\Gamma = \gamma + \frac{4}{3} \frac{1}{Y_e} \frac{\delta Y_e}{\delta \ln \rho} , \tag{3.22}$$

i.e. that $\Gamma < \gamma$ if $\delta Y_e < 0$. This demonstrates that the density dependent EC reduce the adiabatic index, and therefore destabilize the core.

The initial entropy of the core is small ($S \approx 1$ k$_B$/nucleon), a crucial fact first noticed by Bethe et al. (1979). Due to the smallness of the initial entropy the abundance of free protons is small, too: $Y_p \lesssim 10^{-4}$. However, EC on free protons are still more important than EC on nuclei. The latter are strongly suppressed due to nuclear shell effects because the most abundant nuclei all have closed neutron shells (Fuller 1982; Fuller et al. 1982). As weak processes are the only non-adiabatic processes during core collapse (see section 3.1.1), the entropy change before trapping is roughly given by

$$k_B T \frac{dS}{dt} = (\mu_e + \mu_p - \mu_n - < E_\nu >_{emit}) \cdot \left(-\frac{dY_e}{dt}\right) , \tag{3.23}$$

if small contributions due to neutrino thermalization are neglected. For EC on free protons the average energy of the emitted neutrinos is $< E_\nu >_{emit} \lesssim \frac{5}{6}\mu_e$ (since $\mu_e >> k_B T$), which gives rise to a small net entropy reduction of the order $-0.2 \lesssim \Delta S \lesssim -0.04$ (Bethe et al. 1979; Bruenn 1985, Hillebrandt 1987). After neutrino trapping the entropy change is approximately given by

$$k_B T \frac{dS}{dt} = (\mu_e + \mu_p - \mu_n - \mu_\nu) \cdot \left(-\frac{dY_e}{dt}\right) , \tag{3.24}$$

which is always positive. Detailed numerical calculations show that the total change of $Y_e \lesssim 0.1$ and that the difference of the chemical potentials $\Delta\mu \lesssim 5 k_B T$, which translates in a total entropy increase during collapse of $\Delta S \lesssim 0.5$ including the small contribution due to neutrino thermalization. This means the low initial entropy remains low and therefore the collapse proceeds almost adiabatically (Bethe et al 1979).

The near adiabaticity of the collapse has important consequences. Firstly most of the nucleons remain bound in nuclei, because the entropy of a nucleus of mass number A is much less (about a factor A) than the entropy of A free nucleons. Therefore, the pressure is dominated by relativistic electrons (and neutrinos after trapping) until nuclear matter density is reached, i.e. $\gamma \lesssim 4/3$. This in turn implies that the collapse of (non-rotating) iron cores cannot be stopped at densities less than nuclear matter density (see however section 3.1.7).

Analytic calculations (Goldreich and Weber 1981; Yahil and Lattimer 1982; Yahil 1983) and numerical simulations (see section 3.2) show that the dynamics of core collapse can be understood qualitatively by means of self-similar solutions, i.e. solutions which have the same structure at all times.

According to Goldreich and Weber (1981; henceforth GW) there exists an exact stable homologous (i.e. $v(r) \propto r$) solution for marginally stable polytropes with $P = \kappa_0 \rho^\gamma$ and $\gamma = 4/3$. Obviously, the size of the homologous region cannot extend out to arbitrarily large radii, because the velocity cannot exceed the local free-fall velocity. GW find, therefore, that their solution is in only applicable up to a maximum mass

$$M_{max} = 1.0449 \left(\frac{\kappa}{\kappa_0}\right)^{3/2} M_{Ch} \quad . \tag{3.25}$$

In the derivation of Eq. (3.25) they assumed that the collapse is initiated by a sudden radius-dependent pressure reduction, which results when κ_0 is decreased to κ. Note that (i) for a small pressure reduction ($\kappa \geq 0.9711\kappa_0$) the whole core can collapse homologously, and that (ii) (for $Y_e < 0.5$) the behaviour of the outer part of the core is not described by the GW solution. Applying the GW solution to realistic iron cores shows that the mass of the homologous part of the core shrinks (in mass) during collapse, because $\kappa \propto Y_e^4/3$, which decreases through EC's.

Yahil and Lattimer (1982) and Yahil (1983) (henceforth YL) have extended the work of GW to polytropes, where $\gamma \equiv \Gamma$ is constant in space and time and $6/5 \leq \gamma \leq 4/3$. They found self-similar solutions for the whole core, which contain the GW solution in the limit $\gamma = 4/3$, but which are not exactly homologous for smaller values of the adiabatic index. In the YL solutions the core splits into a subsonically and approximately homologously collapsing inner part and into an outer core, which is falling supersonically, the Mach number being in the range $2 \lesssim M \lesssim 3$ and $0.5 \lesssim v/v_{ff} \lesssim 0.8$, where v_{ff} is the local free-fall velocity. The mass of the homologous inner part M_h depends on the value of γ with $M_h(\gamma_1) > M_h(\gamma_2)$ if $\gamma_1 > \gamma_2$.

Near the edge of the homologous core exists a sonic point, where the infall velocity exceeds the local sound speed (Fig. 7). Somewhat further out in radius, at $r = r_{v_{max}}$, the (absolute value of the) infall velocity reaches a maximum (Fig. 7). The mass inside the (time-dependent) radius $r_{v_{max}}(t)$ is called the mass of the inner core (henceforth IC). If one defines the (time-dependent) Chandrasekhar mass for $\gamma < 4/3$ as the mass of a hydrostatic configuration with the same central density as the collapse model, and with zero total energy, the mass of the IC is given by

$$M_{IC} = (1 + f) M_{Ch} \quad , \tag{3.26}$$

where f depends on γ and $1.0 \lesssim f \leq 1.2$ (Yahil 1983). From the typical values of Y_l reached during collapse, one obtains

$$0.6\, M_\odot \lesssim M_{Ch} < M_{IC} \lesssim 0.9\, M_\odot \quad . \tag{3.27}$$

When the central density of the core exceeds nuclear matter density the EOS stiffens, i.e. γ increases from a value somewhat below 4/3 to a value 2.5 – 3.0 (see section 3.1.3). According to the virial theorem the core can reach a new equilibrium position, if the (pressure weighted) average $\bar{\gamma} > 4/3$ in the core. Starting at the center of the core successive mass shells are stopped. Pressure waves move outwards in radius and accumulate within a millisecond near the sonic point, where they steepen into a shock wave (Fig. 8). Due to its inertia the homologous core overshoots its equilibrium position and

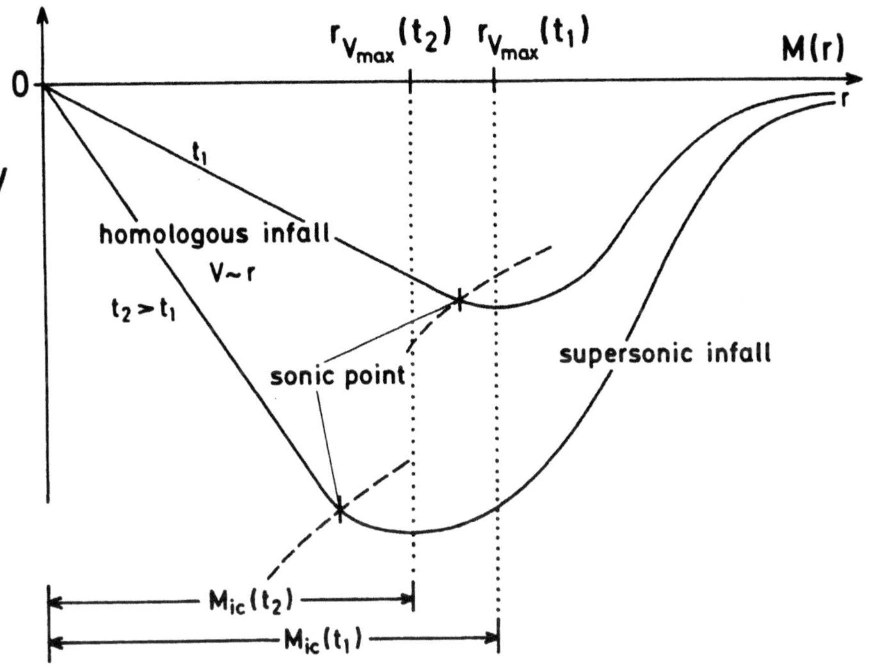

Fig. 7. Schematic profiles of the infall velocity versus radius at time t_1 and $t_2 > t_1$ showing the decrease (in radius and mass) of both the homologous core and the inner core (IC) due to electron captures occuring during core collapse. The edge of the homologous core coincides with the position of the sonic point, and the edge of the IC is defined by the minimum of the infall velocity.

rebounds behind the shock wave, which has already started propagating outward. The shock eventually detaches from the rebounding core at $r = r_{v_{max}}$, i.e. at the outer edge of the IC (Fig. 8).

This so-called bounce of the core occurs on the timescale of a radial pulsation, which is of the order 1 millisecond. For a stiff NEOS the maximum density reached during bounce is less than two times nuclear saturation density (see e.g. Hillebrandt and Müller 1981, 1984; Hillebrandt et al. 1984), whereas for a soft NEOS the collapse can continue up to $3-4$ times, and for a very soft NEOS (with $\gamma = 2$) even up to 10 times nuclear saturation density, if effects of general relativity are included in the calculation (see e.g. Baron et al. 1985a,b).

3.1.5 Prompt Explosions and Shock Propagation

The initial energy of the shock wave is approximately equal to the kinetic energy of the IC at the moment of last good homology (Brown et al. 1982; see also Lattimer et al. 1985b, Mönchmeyer 1990, and section 3.2.1), i.e.

$$E^{(i)}_{shock} \approx E^{IC}_{kin}(t = t_h) \quad . \tag{3.28}$$

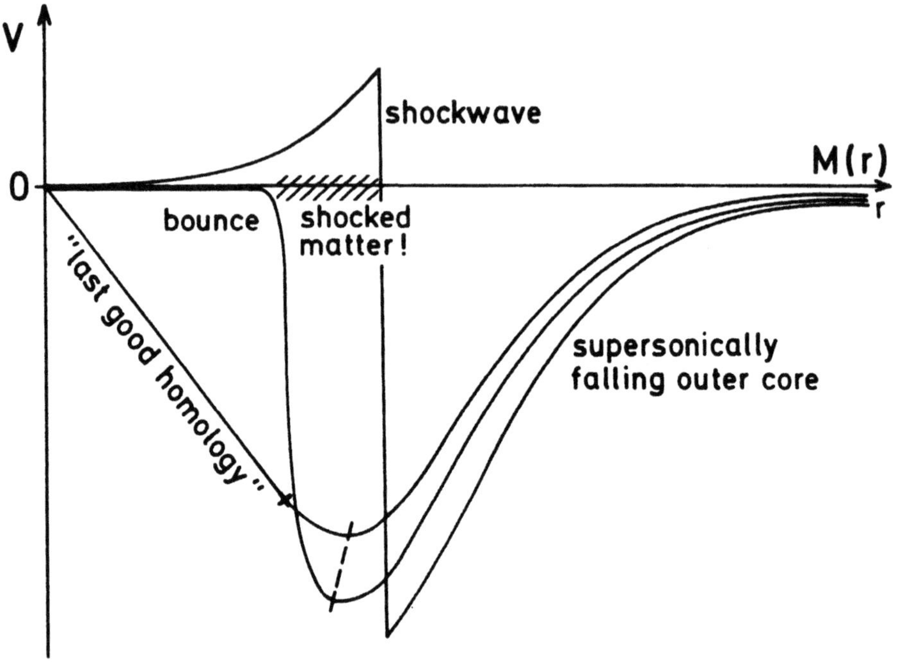

Fig. 8. Schematic velocity profiles versus radius at the point of "last good homology", at bounce and at the time, when the shock wave has just detached from the inner core.

During bounce the kinetic energy of the inner core is quickly transferred to the shock wave, which is an efficient absorber of sound waves. Note that the kinetic energy of matter between the sonic point (i.e. the edge of the homologous core) and $r_{v_{max}}$ (i.e. the edge of the IC) is dissipated before the shock detaches from the IC and, therefore, has to be included into the initial energy of the shock wave (Fig. 8). The moment of last good homology is defined as the time, when the center of the core begins to be decelerated by the stiffening NEOS.

Numerical simulations and analytical estimates (Mönchmeyer 1990) show that the initial energy of the shock wave is of the order of

$$E^{(i)}_{shock} \approx (4 - 10) \cdot 10^{51} \text{ erg} \tag{3.29}$$

(see section 3.2.1), much more than is required to explain the observed energy output of a supernova. However, while propagating outward the shock wave is severly damped by photodisintegration of heavy nuclei into free nucleons (if $S \gtrsim 5$ k$_B$/nucleon) or alpha particles (if 3 k$_B$/nucleon $\lesssim S \lesssim 5$ k$_B$/nucleon), the energy loss being of the order

$$E_{loss} = (1.6 - 1.8) \cdot 10^{51} \; erg \,/\, 0.1 \; M_{\odot} \tag{3.30}$$

(Hillebrandt and Müller, 1981; Mazurek 1982). After the shock wave has crossed the neutrino-sphere, it is further weakened by neutrino losses.

The shock wave also gains energy from the dissipated kinetic energy of matter passing through it. However, for spherically symmetric models it can be shown (Mönchmeyer 1990) that the kinetic energy is not sufficient to guarantee an expansion of the shock heated matter if $\gamma \lesssim 1.4$. Even in the layers, where nuclei are completely disintegrated ($\gamma \gtrsim 1.5$), most of the dissipated kinetic infall energy is needed to bring this matter into hydrostatic equilibrium while it is added to the quasi-static proto-neutron star. When most of the initial shock energy is lost due to photodistintegration, the value of the post-shock γ drops below 1.4 and the shock must turn into an accretion shock, if no further energy source is available. Such an additional source of energy is the heating of the shocked matter immediately behind the shock by energetic neutrinos diffusing out of the opaque central regions of the collapsed core (see next section). Applying the virial theorem to the shocked region further shows that the ram pressure at the shock front increases the need of thermal energy necessary to cause a continuous expansion of the shock heated matter.

The aim of the numerical simulations is to investigate, whether the above mentioned energy losses are so severe that the shock stalls and turns into an accretion shock, or whether the shock wave is able to reach the outer layers of the star and cause a SN explosion. As the (crucial) energy losses due to photodisintegration are proportional to the difference between the mass of the iron core and that of its IC (at bounce), it is obvious that the best candidates for Type II SNe are stars with small iron cores ($M \lesssim 1.2\ M_\odot$) and low initial entropy the latter resulting in a large mass of the IC (Hillebrandt 1987; see also section 3.2.1). In particular, the energy loss given in Eq. (3.30) implies a limiting mass, which can be photodisintegrated by the shock, of the order

$$M_{core} - M_{IC} \lesssim 0.45\ M_\odot \quad . \tag{3.31}$$

For an IC of mass $0.8\ M_\odot \lesssim M_{IC} \lesssim 0.9\ M_\odot$ as found in numerical simulations a so-called *prompt* explosion can only occur, if $M_{core} < 1.35\ M_\odot$ (see section 3.2.1).

3.1.6 Delayed Explosions

Besides the just discussed *prompt* SN explosion mechanism, there may exist another so-called *delayed* explosion mechanism by which the more massive stars (with larger iron cores) can become SN (Wilson 1985; Bethe and Wilson 1985; Wilson et al. 1986; Wilson 1990). Wilson (1985) has found that a few hundred milliseconds after core bounce the stalled shock wave is revived due to energy transport by neutrinos as the shocked matter is irradiated by a neutrino flux originating from the neutrino-sphere. Although only about 0.1% of the neutrino energy is absorbed in the outer layers, this neutrino flux leads to a slow accumulation of energy behind the stalled shock, and consequently to a gradual heating and expansion. However, this mechanism will lead to relatively weak explosions of a few times 10^{50} erg only.

According to Bethe and Wilson (1985) the net rate at which energy is absorbed by a gram of matter at large distance R_m from the center is

$$\dot{E} = \frac{7}{16} a c \kappa_{abs}(T_\nu) T_\nu^4 \left[\left(\frac{R_\nu}{2R_m} \right)^2 - \left(\frac{T_m}{T_\nu} \right)^6 \right] \mathrm{erg\,g^{-1}\,s^{-1}} \quad . \tag{3.32}$$

Here the subscript m refers to the matter element and ν to the neutrino-sphere. Generally, R denotes the radius and T the temperature in MeV. The neutrino absorption coefficient

κ_{abs} is in cm^2 g^{-1}, $a = 1.37\ 10^{26}$ erg cm^{-3} MeV^{-4}, c is the speed of light, and it has been assumed that the absorption coefficient for neutrinos is proportional to T^2.

The maximum matter temperature that can be obtained from this heating mechanism is determined from the equilibrium condition $\dot{E} = 0$. At 10 times the radius of the neutrino-sphere ($R_\nu \approx 20 - 70$ km) T_{max} is about $0.37\ T_\nu$ ($T_\nu \approx 2 - 3$ MeV) and thus about 1 MeV (Hillebrandt 1987). The heating time scale can be estimated to be a few hundred msec. Obviously the crucial quantity for the delayed explosion mechanism is the neutrino temperature at the neutrino-sphere, which in turn sensitively depends on the neutrino transport scheme used in the simulations. Further note that R_ν is determined both by the neutrino transport scheme and the EOS.

3.1.7 Effects due to Rotation

The most important difference between the collapse of rotating and non-rotating cores is that in rotating cores centrifugal forces may stop the collapse before nuclear matter densities are reached. This possibility was pointed out by Shapiro and Lightman (1976) and discussed in more detail by Tohline (1984) and Eriguchi and Müller (1984). Numerical examples of such a low-density bounce have been given by Müller et al. (1980), Symbalisty (1984) and Mönchmeyer and Müller (1989).

A theoretical argument for the stabilizing influence of rotation on radial modes of stars was derived by Ledoux (1945). To guarantee the stability of a configuration in rotational equilibrium the "effective" adiabatic index should fulfil the condition

$$\Gamma > \Gamma_{crit} = \frac{2}{3}\frac{(2-5\beta)}{(1-2\beta)} \tag{3.33}$$

with

$$\beta = \frac{E_{rot}}{|E_{pot}|} \quad . \tag{3.34}$$

Here E_{rot} and E_{pot} are the rotational and potential energy, respectively. Although Eq. (3.33) was only derived for slow rigid rotators with γ independent of density, numerical calculations have shown (see e.g. Tassoul, 1978) that Eq. (3.33) also holds under more general conditions.

For a given (average) Γ the critical value of β which a rotating core in equilibrium must exceed to be stable against pseudo-radial modes is

$$\beta > \beta_{crit} = \frac{1}{2}\frac{(4-3\Gamma)}{(5-3\Gamma)} \quad . \tag{3.35}$$

Note that this is only a necessary condition for a core bounce at subnuclear densities (Tohline, 1984; Mönchmeyer, 1990). However, Eq. (3.33) shows qualitatively that even a small amount of initial rotational energy can be sufficient to stabilize a core at densities less than nuclear matter density, provided that angular momentum is conserved during the collapse and that Γ has a value close to 4/3.

Due to conservation of angular momentum and the resulting increase of the centrifugal forces matter will not fall in on radial trajectories. In addition matter in the equatorial plane will not fall towards the center as fast as matter at the polar axis. Especially this last effect leads to a progressively flattening of the core. In comparison to a spherically symmetric configuration the collapse time scale is longer in rotating models.

In contrast to spherically symmetric cores, which come to rest soon after bounce, the kinetic infall energy of a rotating core bouncing due to centrifugal forces will be converted into oscillations, which are damped by non-spherical pressure waves. Therefore after bounce a rotating core will oscillate with a superposition of various axisymmetric radial and surface modes. The frequency of these modes is determined by the average density of the inner core. Simulations and analytical considerations show that the radial oscillations will become large and nonlinear, if the collapse is stopped just before nuclear matter densities are reached (Mönchmeyer 1990; see also Mönchmeyer et al. 1990).

It is also well known that rotation can drive meridional circulations in baroclinic regions (see e.g. Tassoul 1978). For type II supernova models the interesting question arises, whether there exist large scale circulation patterns behind the shock, that extend inwards to regions where neutrino energies are high. In the (neutrino) opaque matter those vortices could transport neutrinos towards the shock front much faster than diffusion (Müller and Hillebrandt, 1981).

From 1d–calculations one knows in addition that a negative entropy gradient is established behind the shock after photodisintegration losses have weakened it significantly. This region of decreasing entropy is unstable against convection provided that the stabilizing lepton gradient is not too large (Epstein 1979). Arnett (1985) has first pointed out the possible importance of convection for the type-II supernova mechanism. However, whether convection indeed helps or even harms the propagation of the shock is debated (Burrows 1987; Bethe, Brown and Cooperstein, 1987; Mönchmeyer 1990). Note that for rotating cores mixing of high and low entropy matter may be enhanced, if convectional currents are supported by vortices resulting from rotation in regions, where the deformed surfaces of constant pressure do not coincide with isopycnic surfaces. On the other hand side it has been argued that rotation may have a stabilizing effect on certain types of convective instability modes (see Tassoul 1978). However, especially for centrally condensed and differentially rotating objects the interaction of rotation and convection is not yet understood.

The propagation of the shock wave is influenced by rotation because of several effects (Mönchmeyer 1989, 1990; Mönchmeyer and Müller 1989):

1. Rotation can lead to a bounce at lower densities than in the non-rotating case, which implies that the kinetic infall energy of the inner core and consequently the initial shock energy of rotating cores is reduced.
2. Rotation on the other side tends to enlarge the mass of the inner core, because it changes the velocity profile and stops the collapse before (due to EC) the mass of the inner core can shrink to values found for spherical models at bounce. Numerical results show that the mass of the inner core increases by roughly 10% – 20% for initial values of $0.01 \lesssim \beta \lesssim 0.02$.
3. In the outer core the binding energy gained during collapse is transfered both into rotational *and* kinetic infall energy. According to the virial theorem rotation acts like a $\gamma = 5/3$ gas. It therefore helps to stabilize the shock heated matter in the gravitational potential, i.e. a larger part of the dissipated kinetic infall energy can support the expansion of the heated matter behind the shock front until equilibrium is achieved at larger radii and at lower potential energy. The resulting compressional work adds up to the work of the expanding inner core and strengthens the shock.

4. Centrifugal forces considerably reduce the ram pressure of the supersonic flow in comparison to 1-d models.
5. The asymmetry of the supersonic flow gives rise to an angular dependent propagation speed of the shock wave. This effect will be evident in the dissipation rates of kinetic energy and in the maximum entropy values obtained behind the shock front.

Whether the combined action of these effects will strengthen or weaken the propagation of the shock wave depends on the amount and distribution of the angular momentum of the core.

Due to conservation of angular momentum it cannot be excluded that configurations may form during core collapse, which are unstable against tri-axial deformations on secular or even dynamical time scales (Tohline 1984; Eriguchi and Müller, 1984), if $\beta \geq 0.14$ and $\beta \geq 0.27$, respectively (see e.g. Tassoul 1978). Whether these instabilities indeed do occur is a non-trivial question: If the equation of state at sub-nuclear densities is stiff, i.e. if the adiabatic index is very close to 4/3, the core may be stabilized before its rotational energy exceeds the critical value. If on the other hand the initial amount of rotation is small enough for the collapse to proceed to nuclear densities, $\beta > 0.14$ may not be reached before bounce.

Finally two effects of rotation should be mentioned, which are of importance for observations of SNe.

1. During the collapse of rotating cores gravitational waves will be emitted. Based on two dimensional hydrodynamic simulations Müller (1982, 1984) and Mönchmeyer et al. (1990; see also references therein) have calculated the resulting gravitational wave signal.
2. The non-spherical density stratification of a rotating core before and after bounce may modify the neutrino signal in a characteristic way and lead to a directional dependence of the ν-signal (Janka and Mönchmeyer 1989a, b).

3.2 Numerical Simulations

As it is far beyond the scope of the present work to review the now almost 30 year long history of SN simulations, which began with the pioneering calculations of Colgate and Johnson (1960), Colgate and White (1966), Arnett (1966, 1967, 1968), Le Blanc and Wilson (1970) and Wilson (1971), only some aspects of recent simulations will be briefly discussed below. Other more extensive reviews of the subject can be found e.g. in Trimble (1982, 1983), Woosley and Weaver (1986a) and Woosley (1986).

3.2.1 Spherically Symmetric Models

As discussed in section 3.1 there exist currently two (one dimensional) theoretical models of the Type II SN explosion mechanism: The prompt mechanism, which only seems to work for stars in the mass range $\approx (8-12)M_{\odot}$, and the delayed mechanism by which all and in particular more massive progenitors ($\gtrsim 20M_{\odot}$) may explode.

The simulations of the prompt mechanism have shown that the single most important factor determining the success or failure of the shock wave is the difference between the mass of the iron core and the mass of the inner core. If this difference gets too large ($\gtrsim 0.45M_{\odot}$) successful shock propagation cannot be expected, because energy losses due to photodisintegration of heavy nuclei then exceed the energy which is necessary for a continuous expansion of the shocked matter (Hillebrandt and Müller 1981; Mazurek 1982; Hillebrandt and Müller 1984; Cooperstein et al. 1984). The mass of the iron core at the onset of collapse depends on the main sequence mass of the stellar model, with the tendency that lower main sequence masses give also smaller iron cores. The actual value, however, depends on both, the treatment of nuclear reactions and energy transport (convection) during the late stages of evolution and ranges from (the up to now record low) $1.18 M_\odot$ for a $13 M_\odot$ stellar model (Nomoto and Hashimoto 1988) to $1.85 M_\odot$ for a $25 M_\odot$ star (Woosley and Weaver 1988).

The mass of the IC sensitively depends on the initial entropy and the neutrino transport, both of which determine the value of the final trapped lepton fraction Y_l of the inner core and thereby its mass (see section 3.1.4). In models with a low initial entropy less EC's occur due to the lower free proton fraction. This gives rise to a larger final lepton fraction and therefore a larger mass of the IC.

The importance of the neutrino transport for the final lepton fraction has been examined in detail by Bruenn (1985, 1986, 1989a), Myra et al. (1987), Myra and Bludman (1989) and Baron and Cooperstein (1990). It plays a crucial role for both the prompt and the delayed explosion mechanism. State of the art collapse simulations include non-LTE neutrino transport together with a large variety of neutrino absorption, emission and scattering processes (see e.g. Bruenn 1985; Cooperstein 1988). Contrary to simple trapping or leakage schemes non-LTE neutrino transport methods allow for the treatment of a non-equilibrium neutrino distribution which is not of Fermi-Dirac form. The simplest version of a non-LTE transport scheme is the two fluid (frequency integrated) flux limited diffusion method (Arnett 1981; Hillebrandt et al. 1984; van den Horn and Cooperstein 1986; Cooperstein et al. 1986, 1987). In this method one assumes that the distribution function is of Fermi-Dirac form, parameterized by a "temperature" and "chemical potential". A more elaborate approach, which can handle arbitrary distribution functions, is the multi-group flux limited diffusion method, where the neutrino distribution function is discretized in a set of energy groups and solves Boltzmann-like equations to determine the occupation probabilities for each energy group in space and time (Arnett 1977; Bowers and Wilson 1982; Mayle 1985; Bruenn 1985; Myra et al. 1987). Recently the multi-group method has been formulated within a fully general relativistic framework by Baron et al. (1989). For a static stellar background model the two fluid and the multi-group method can be compared with and calibrated by Monte Carlo transport calculations (see e.g. Janka and Hillebrandt 1989a, b). Finally note that both methods rely on a flux-limiter to describe the transition regime between complete trapping and free streaming, a purely technical tool, which can be the source of non-negligible errors (Janka and Hillebrandt 1989a, b).

A very important process for thermalizing the neutrino distribution is neutrino-electron scattering (NES) especially at low temperatures and densities ($T \lesssim 1$ MeV, $\rho \lesssim 10^{12}$gcm^{-3}). NES significantly contributes to the reduction of Y_l, because energetic neutrinos which have a short mean free path ($\lambda \propto E_\nu^{-2}$) can downscatter off the highly degenerate electrons, thereby increasing their mean free path and therefore their escape probability from the collapsing core (Bruenn 1985, 1986, 1989a; Myra et al. 1987; Myra and Bludman 1989). Note that the importance of NES is limited to the infall epoch and has only minor influence on the propagation of the shock wave.

Let us now turn to the influence of the EOS for the prompt mechanism. A soft EOS at densities at and beyond nuclear saturation has the effect that the collapse proceeds to higher densities. Therefore, the binding energy of the inner core is getting larger, i.e. the initial shock energy and the chance of a prompt explosion increase (see section 3.1.5). This effect becomes significant if general relativity is included into the model calculations (Baron et al. 1985a, b). In particular Baron et al. (1985a, b) have argued that for successful shock propagation the incompressibility of nuclear matter has to be significantly below the values obtained from fits to the monopole oscillations of lead and other nuclei and from empirically obtained effective nucleon-nucleon interactions (see section 3.1.3). The simulations show that for a soft EOS ($\gamma < 2.5$) the maximum density reached during core bounce is $\approx$ (3–12) times nuclear matter density (at $Y_e = 0.33$; Baron et al 1985a, b), whereas with a stiff EOS ($K_0(Y_e = 0.5) = 220$ MeV and $\gamma \approx 2.5 - 3.0$) the maximum density stays below two times nuclear matter density (Hillebrandt and Müller 1981, 1984; Hillebrandt et al. 1984; Baron et al. 1985a, b). The importance of the NEOS for the prompt mechanism has also been studied in great detail by Bruenn (1986, 1989b), and also by Takahara and Sato (1988) and van Riper (1988).

In addition the stiffness of the EOS below nuclear saturation density, i.e. the deviation of the adiabatic index from the critical value 4/3, is of importance because it also influences the size of the homologous core and its kinetic energy (Yahil and Lattimer, 1982). Note that the adiabatic index is the second derivative of the energy density with respect to density at constant entropy. Therefore small uncertainties in the energy density may cause significant effects on the dynamics of core collapse.

Among researchers in the field there is now general agreement that when the detailed effects of neutrino transport are taken into account currently available initial models and in particular models with iron cores more massive than about 1.35 $M_\odot$ cannot explode by the prompt mechanism, no matter what the right equation of state is (Bruenn 1989a, b; Myra and Bludman 1989; Cooperstein and Baron 1990; Hillebrandt et al, 1990). Conventional models of stellar evolution, therefore, exclude main sequence masses above about 12 $M_\odot$. Figure 9 shows one example of a successful simulation, where a 9 $M_\odot$ star exploded by the prompt mechanism the shock energy being enhanced by nuclear energy from oxygen burning (Hillebrandt et al. 1984; see however Baron et al. 1987). Figure 10 on the other side shows a typical example of a simulation where the shock stalled at about 1.2 $M_\odot$ (Hillebrandt 1987).

Because some supernova remnants (e.g., Cas A and Puppis A) show large oxygen overabundances indicating main sequence masses of at least 20 $M_\odot$ alternative explosion mechanisms had been searched for. A possibility which has been discussed extensively is the delayed explosion mechanism (see section 3.1.5). Wilson (1985), Wilson et al. (1986), Mayle and Wilson (1988), and Wilson and Mayle (1988) have computed the hydrodynamic evolution of several stellar models for approximately 1s after core bounce and found explosions in all cases considered. In most cases, however, the explosion energy ($\lesssim 4\ 10^{50}$erg) was too low to account for typical type II supernova light curves and in particular for the rather high kinetic energy seen in SN 1987A (see however Wilson 1990). Only for rather massive stars, $M \gtrsim 25 M_\odot$, did explosive oxygen burning add enough energy to the explosion to explain a typical outburst. This problem can be understood from a simple argument. Neutrino heating proceeds on a time scale much longer than the hydrodynamical time scale. Once, due to neutrino heating, a sufficient overpressure has been built up behind the accretion shock the heated zones will start to expand

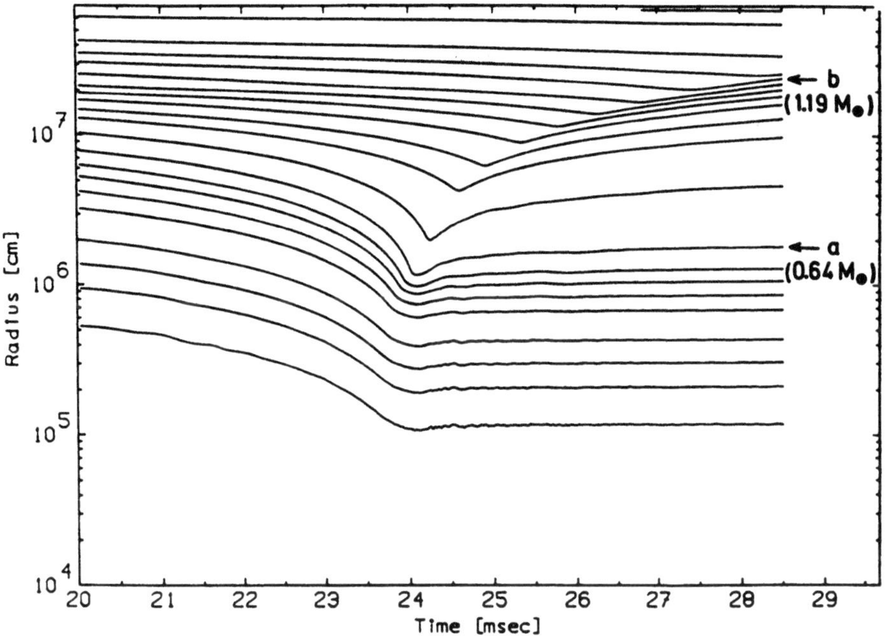

Fig. 9. Radius versus time for selected mass zones of a collapsing and exploding stellar model of 9 $M_\odot$. The unshocked inner core is labelled a, the first zone that reaches escape velocity, b. This computation has been performed with neutrino transport in non-equilibrium flux-limited diffusion approximation, a Hartree-Fock equation of state and general relativistic hydrodynamics. Oxygen burning was included by solving an appropriate nuclear reaction network (from Hillebrandt et al. 1984).

and further heating will be turned off. The thermal energy needed to build up such an overpressure will be of the order of the binding energy of the overlaying material, i.e., a few times 10^{50}erg. Consequently, one expects that the explosion energy is of the same order (Hillebrandt 1987).

From numerical experiments Wilson (private communication) has found that small modifications of his neutrino transport scheme can change successful explosions into failures. This, as well as results obtained by Hillebrandt (1985), indicate that the delayed explosion mechanism may not work at all. So we are left with the problem that also (and maybe in particular) for massive stars, $M \gtrsim 20 M_\odot$, the explosion mechanism is not understood. It may well be that one has to invent more complicated scenarios in order to be able to solve these problems. For example, the neglect of rotation, magnetic fields and/or nuclear energy generation may be an oversimplification (see, e.g., Müller and Hillebrandt 1981; Le Blanc and Wilson 1970; Symbalisty 1984; Bodenheimer and Woosley 1983; and next section).

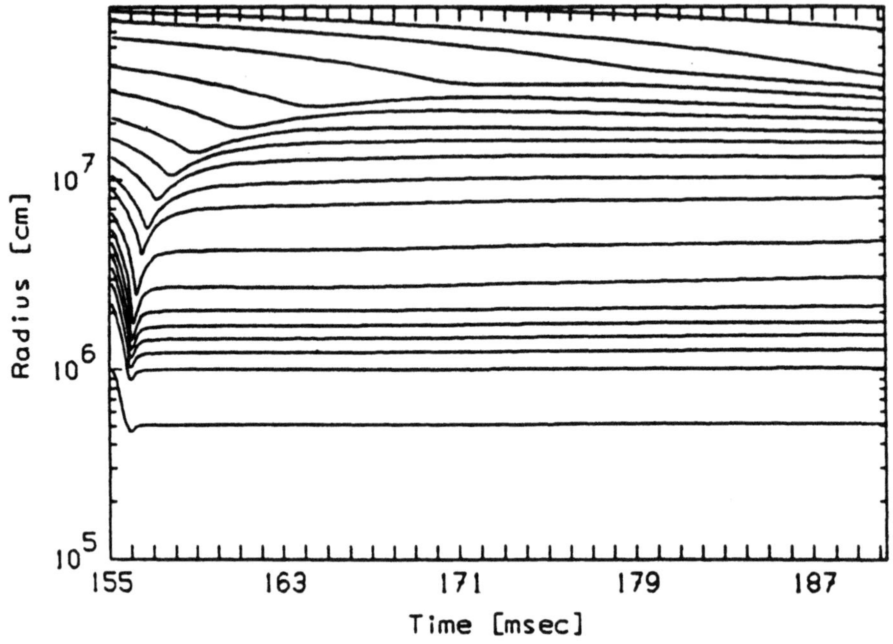

Fig. 10. Same as figure 9, but for a stellar model of 20 $M_\odot$. It is obvious that this time, because of the larger core mass, the shock stalls at about 300 km (from Hillebrandt 1987).

3.2.2 Axially Symmetric Rotating Models

Up to now only a few attempts have been made to relax the assumption of spherical symmetry and to perform axisymmetric, i.e. two-dimensional supernova simulations, which allow to study effects due to rotation (LeBlanc and Wilson 1970; Müller et al. 1980; Tohline et al. 1980; Müller and Hillebrandt 1981; Symbalisty 1984; Mönchmeyer and Müller 1989; Mönchmeyer 1989, 1990; see also Hillebrandt et al. 1990)

It is obvious that the simulation of rotational core collapse is computationally more difficult and more expensive than a spherical collapse calculation. However, it is still surprising that only a few simulations have been performed in the past, because (i) stars and especially massive stars rotate in general (see e.g. Tassoul, 1978), because (ii) at least some of the compact remnants of type II supernova explosions, the millisecond pulsars, do rotate with significant speed (see e.g. Taylor, 1987), and because (iii) of angular momentum conservation and the increasing centrifugal forces already small initial rotational energies may change the standard collapse picture completely. Therefore the inclusion of rotation may be crucial for the correct modelling of a type II supernova explosion. This argument is supported by the fact that up to now no spherically symmetric collapse simulation for stars with $M \gtrsim 15 M_\odot$, i.e. with an iron core mass $M_{core} \gtrsim 1.35\ M_\odot$ could convincingly produce a supernova explosion (see previous section).

The most recent axisymmetric simulations have been performed by Mönchmeyer (1990; see also Janka and Mönchmeyer 1989a, b; Mönchmeyer 1989; Mönchmeyer and Müller 1989). He has calculated four model sequences with different amounts and distributions of angular momentum. The choice of the initial angular momentum distribution is one of the most problematic aspects of any 2-d simulation, because there exist no evolutionary calculations of rotating massive stars up to the iron core collapse. Therefore the amount as well as the distribution of the initial angular momentum can be chosen arbitrarily within certain limits imposed by stability considerations (see e.g. Tassoul, 1978).

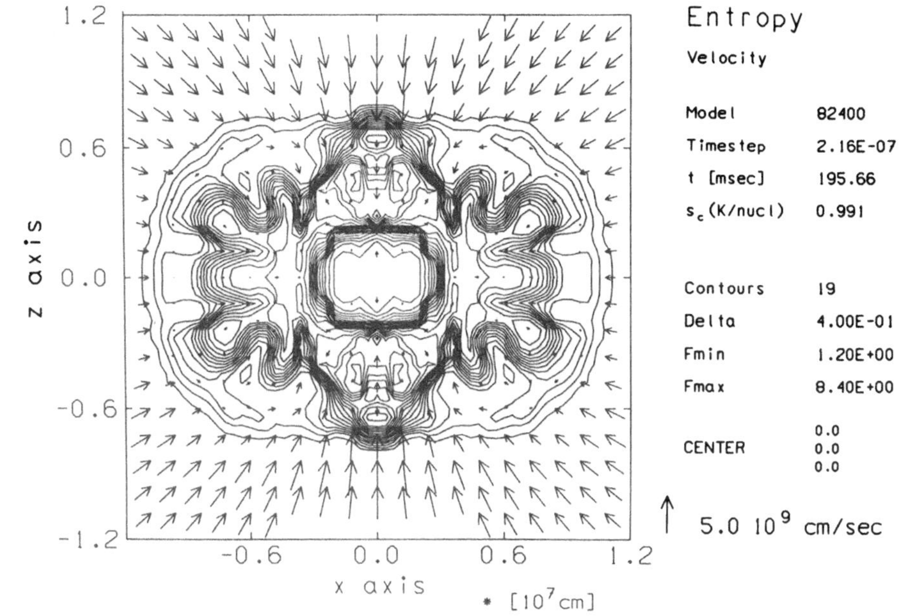

Fig. 11. Profiles of the specific entropy and flow pattern 6.5 msec after bounce showing the instability at the edges of the high entropy region. The contours cover a range from Fmin to FMax with a spacing of Delta. The time, the central entropy and the velocity scale are given in the legend of the figure (from Mönchmeyer and Müller 1989).

Before discussing the properties of the rotating models of Mönchmeyer (1990) one has to point out that the concept of homology cannot be used for rotating cores. Yet there exists a surface, which separates subsonically from supersonically falling matter, and the inner core (IC) can be defined as the matter inside a surface of constant density for which the (absolute) value of the angular averaged radial infall velocity has a maximum.

During the collapse a subsonically falling IC forms in each model. About $10-20$ msec before bounce the IC begins contracting rapidly, whereby it significantly flattens due to angular momentum conservation. At bounce the ratio of its polar to equatorial diameter is roughly $1/2$ in each model. The profile of the infall velocity is asymmetric during collapse, especially in the supersonic flow region. At bounce the ratio of the maximum polar infall velocity to the maximum equatorial infall velocity typically ranges from 1.8 to 2.5, and the maximum compression of the IC in polar direction occurs before the equatorial contraction is stopped completely.

In one model with an initial value of $\beta_i = 0.02$ centrifugal forces stop the collapse already at a central density of $2\ 10^{13}$ gcm^{-3}. Behind a rather weak shock front a rotating equilibrium configuration is formed. The further evolution of this configuration, which occurs due to neutrino cooling on a timescale of 0.1 sec, has not yet been investigated.

In a second more slowly rotating model ($\beta_i = 0.005$) core bounce is caused by the stiffness of the equation of state beyond nuclear saturation density (similar to non-rotating models). As a consequence most of the kinetic infall energy of the IC is transfered to the shock within 0.3 msec after bounce. Therefore and also due to the stiffness of the NEOS the amplitudes of the oscillations of the IC are only small. Despite a rather small initial shock energy of only $3.6\ 10^{51}$ erg the shock reaches a mass coordinate of $\approx 1.3 M_\odot$ before it stalls, because of the stabilizing effect of rotation on the shocked matter. In contrast to one dimensional models a Rayleigh-Taylor instability develops behind the shock front (Fig. 11). However, as discussed by Bethe et al. (1987) and as found in the simulation the decrease of the entropy gradient caused by the instability weakens the shock.

In a third model with $\beta_i = 0.01$, i.e. intermediate to the two previously discussed models, and with an initially more differential rotation law the bounce occurs due to centrifugal forces at a sub-nuclear central density of $1.5\ 10^{14}$ gcm^{-3}. Therefore the kinetic infall energy of the IC is only $1.6\ 10^{51}$ erg. But due to the small adiabatic index of matter at sub-nuclear densities the IC can be much easier compressed. In this model the kinetic infall energy of the IC is not directly transferred to the shock, but powers a large scale post-bounce expansion of the IC, which pushes the shock like an expanding piston. In addition centrifugal forces support the expansion of the shocked matter (see section 3.1.7), and by the resulting adiabatic cooling of the matter part of the disintegration energy can be regained. Due to these favourable effects the shock reaches and even penetrates the silicon shell with a positive (radial) velocity indicating a weak explosion. When the calculation was stopped the mass surrounded by the shock surface is $1.42 M_\odot$ (Fig. 12).

3.2.3 Instabilities and Mixing During the Explosion

There appears to be increasing evidence that the observations of Supernova 1987A can not be explained unless substantial nonradial motion and mixing occurred during the explosion (see section 2.7). The idea that nonradial motion would occur in Type II supernovae was first discussed by Falk and Arnett (1973). Chevalier (1976) used the stability analysis of Chandrasekhar (1961) to show that in the idealized case of a blast wave propagating down a power law density gradient, Rayleigh-Taylor (henceforth RT) instabilities could develop for a range of power law indices. The criterion for the gas behind the shock front to become unstable was that the density and pressure gradients should be in opposite directions. Note that, as pointed out by Bandiera (1984) and Benz and Thielemann (1990), the RT growth rate, which is derived for an incompressible fluid, only gives a lower limit to the actual "convective" growth rate of a compressible fluid.

The observations of mixing in SN1987A has caused renewed theoretical interest in this problem. A number of numerical simulations have been performed recently to investigate whether RT instabilities do occur in supernovae and, if they do, how much mixing would result. The first of these calculations was performed by Nagasawa et al. (1989) using a three dimensional smooth particle hydrodynamics (SPH) code for the idealized case of a shock propagating through a polytropic density distribution. They found that a small amount of mixing occurred, but probably not enough to explain the observations.

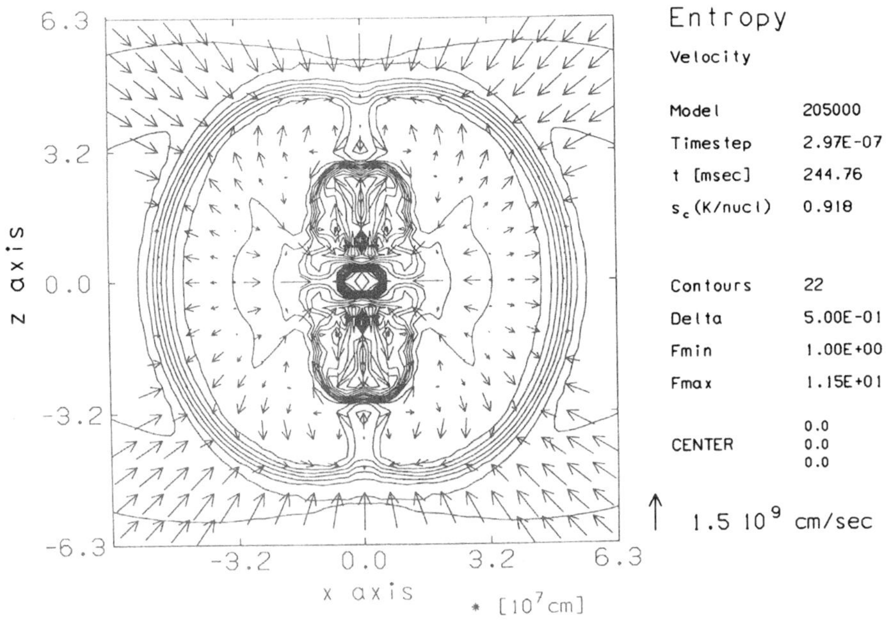

Fig. 12. Profiles of the specific entropy and flow pattern for a model bouncing due to centrifugal forces about 45 msec after bounce. Note the shock front, which has already penetrated the silicon shell, and the 'jet', which rapidly expands into polar direction (from Mönchmeyer 1989).

However, Benz and Thielemann (1990) repeated their calculations also using SPH and found that the development of the instability depended very sensitively on how the explosion was initiated. They concluded that the instability found by Nagasawa et al. (1989) was probably a numerical artifact.

The same conclusion was reached by Müller et al. (1989) who repeated the same calculations using several different finite difference codes in both two and three dimensions. The results for a polytropic density distribution showed no instability. However, a power law density distribution with a power law index in the range which was predicted to be unstable by Chevalier (1976), did show signs of a RT instability.

High resolution two dimensional (Arnett et al. 1989b, 1990) and three dimensional (Müller et al. 1990) calculations using a realistic density distribution for a Type II supernova progenitor found a very strong RT instability, which causes substantial mixing on a time scale of a few hours after the explosion (Fig. 13). The calculations were performed with a state of the art difference scheme (the PPM method of Collela and Woodward 1984) and kept track of 10 different elements (treated as separate fluids) to determine the amount of mixing of various nuclear species and included a non gamma-law equation of state (radiation + 10 Boltzmann gases). The calculations show that: (i) hydrogen is mixed inward into the helium shell and even into the C/O core where the mixed in hydrogen mass fraction reaches values of up to 10%; (ii) significant amounts of helium are mixed outward into the hydrogen envelope and inward into the C/O core where the increase in helium mass fraction can exceed 20%; (iii) heavy elements are mixed outward from the C/O core into the helium shell and to a less extent even into the hydrogen envelope.

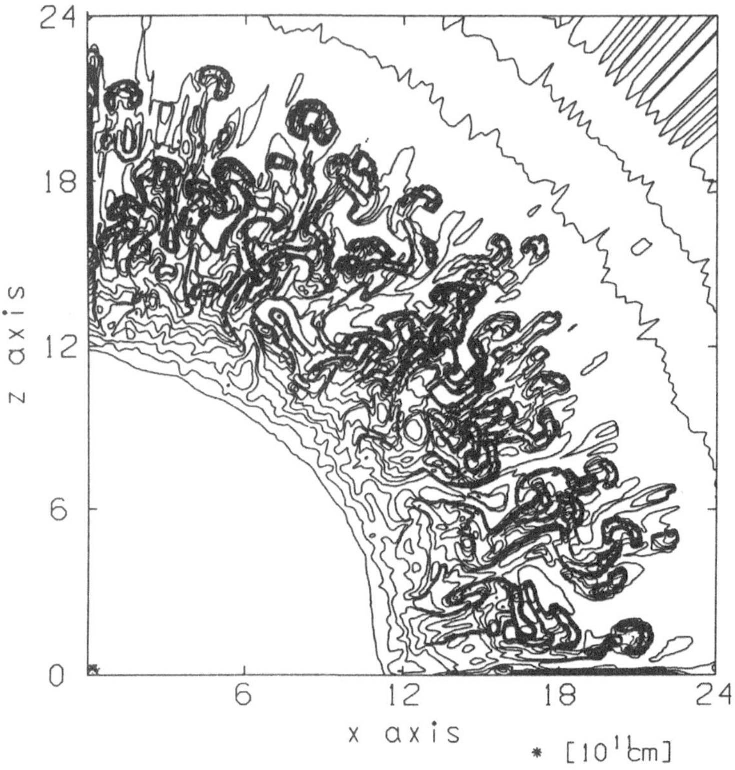

Fig. 13. Density contours showing the instability 2 hrs after the begin of the explosion ($E_{ex} = 2\ 10^{51}$ erg). A random perturbation of 10% amplitude was imposed at 282 s. The results are obtained on a 500 × 200 grid in spherical coordinates. There is rotational symmetry about the vertical axis and equatorial symmetry about the horizontal axis. The 49 contour lines are equally spaced on a linear scale: $\rho_{max} = 1.9\ 10^{-3}$ gcm^{-3}, $\rho_{min} = 1.6\ 10^{-6}$ gcm^{-3} (from Müller et al. 1990).

3.3 Explosive Nucleosynthesis

3.3.1 General Remarks

The modification of the initial chemical and isotopic composition of a star occuring on long time scales during its various hydrostatic burning phases is called *hydrostatic nucleosynthesis*. The nuclear transmutations taking place on dynamical time scales during explosive events (e.g. in novae and supernovae explosions) are called *explosive nucleosynthesis* (Woosley et al. 1973). Comprehensive recent reviews of hydrostatic and explosive nucleosynthesis can be found in Truran (1985), Woosley (1986), Woosley and Weaver (1986a) and Thielemann (1989).

The explosive nucleosynthesis in Type II SNe comprises explosive helium, carbon, neon, oxygen and silicon burning, caused by the passage of the shock wave through the respective mass layers of the stellar envelope. The shock wave heats and compresses matter in the envelope. The resulting temperature increase accelerates nuclear reactions, which have already been operative on long (i.e. evolutionary) time scales, by orders of magnitude and triggers additional reactions. Note that the matter outside the carbon core is ejected essentially unaltered by explosive processesing (see e.g. Thielemann et al. 1990). The mass range for the various explosive burning conditions can be estimated, if one assumes a homogeneous density and temperature distribution behind the shock and that the post-shock pressure is radiation dominated (Thielemann et al. 1990). Numerical simulations show that both assumptions are approximately correct and that therefore the explosion energy can be equated with the energy in the radiation field inside the radius R of the shock wave, i.e. $E_{ex} \approx \frac{4\pi}{3} R^3 aT^4$, or $R \sim E_{ex}^{1/3} T^{-4/3}$. For a given explosion energy and the minimum temperature for a specific explosive burning stage (e.g. $T \gtrsim 5\ 10^9$ K for explosive Si-burning) the (outer) radius of the respective burning zone can then be estimated and the stellar model provides the corresponding mass coordinate (see next section).

A consistent numerical treatment of the explosive nucleosynthesis and the hydrodynamics of a Type II (and Type I; see section 4.4) supernova explosion is unfortunately not possible. Firstly there still exist open problems with the explosion mechanism (see section 3), and secondly, even if those would be solved, the numerical problems involved in computing core collapse, bounce and shock propagation through the whole star together with all the necessary nuclear reactions are far from being trackable. Therefore, the commonly used approach consists of the following steps (see e.g. Thielemann 1989): First one chooses a stellar model at the verge of collapse, which was evolved computing the hydrostatic nucleosynthesis with a small ($\lesssim$ 50 species) optimized nuclear reaction network. In a second step one deposites in a more or less sophisticated way an energy of the order 10^{51} erg in the center of the model and follows the propagation of the resulting shock wave with a hydrodynamic code. The second step implicitly assumes that the composition of the matter which is finally ejected did not experience significant changes between the onset of collapse and the arrival of the shock wave. This assumption is justified outside the Si-shell but is problematic for the inner parts of the Si-shell close to the mass-cut dividing the ejecta from the neutron star remnant. The temperature $T(m,t)$ and density $\rho(m,t)$ evolution of each mass shell computed in step 2 is then used in step 3 to calculate the abundance changes by ***post-processing*** each mass zone with an extended ($\lesssim$ 300 species) reaction network. Post-processing is justified, because the shock energy is much larger than the energy released during explosive nucleosynthesis.

In order to compare abundances of explosive nucleosynthesis in Type II SNe with solar abundances, which is the commonly used reference point, the computed nucleosynthesis results have to be integrated over the whole mass range of Type II SNe and also over the galactic age. However, when weighting the ejected mass of heavy elements in Type II SNe with the (present) initial mass function, the "average supernova" of importance for heavy element nucleosynthesis lies in the range $20-30M_\odot$ (Woosley and Weaver 1986a), a result which considerably simplifies the comparison of theory and observation.

3.3.2 SN1987A

From optical observations of the progenitor of SN1987A a radius of $R = (3 \pm 1)\ 10^{12}$ cm and a main-sequence mass of $M = 19 \pm 3 M_{\odot}$ can be inferred (Shigeyama et al. 1988; Woosley 1988a). This corresponds to a star with a He-core mass of $6 \pm 1 M_{\odot}$ after core hydrogen burning (Nomoto and Hashimoto 1988; Woosley and Weaver 1988). Observations and several arguments concerning the observed light curve, the H-envelope structure and the synthetic spectra further suggest that the progenitor star lost about 3 to $7 M_{\odot}$ prior to its explosion in form of a stellar wind (see e.g. Arnett et al. 1989a). The explosive nucleosynthesis in such a star has been calculated by Woosley et al. (1988) by post-processing the phases of hydrostatic and explosive nucleosynthesis with an extended network including 150 nuclei. Thielemann et al. (1990) have only post-processed the phases of explosive nucleosynthesis inside the He-core with a large reaction network containing 300 nuclei using, however, an updated data base of nuclear reaction rates. In the latter calculation the hydrostatic phases of the presupernova evolution were covered by Nomoto and Hashimoto (1988) with a smaller network, containing 30 nuclei, for the phases from He-burning through O-burning and a quasi-equilibrium network containing 250 nuclei for Si-burning.

When comparing the results of both calculations, large deviations show up for all nuclei which are produced in small amounts during hydrostatic burning stages and not included in the 30 nuclei network. This shortcoming does not affect the mass zones inside $2M_{\odot}$, where the final abundances are dominated by explosive processing, but the outer layers, which are only slightly changed by the passage of the shock wave, i.e. their abundances are determined by hydrostatic burning. Mostly affected are light nuclei with small abundances (^{13}C, ^{15}N, ^{17}O, and ^{19}F) and (neutron-rich) s-process nuclei between Ne and the Fe-group (Thielemann et al. 1990). Other differences occuring are either due to the neglect of the composition of the H-envelope in the work of Thielemann et al. (1990) or are due to the structure and composition of the presupernova models, which give rise to small (factor of 2) but systematic deviations for nuclei from Ne to Sc. Due to the overall agreement of both calculations I will only review some of the results obtained by Thielemann et al. (1990) in the following discussion. A more detailed review and comparison is beyond the scope of this work.

Explosive Si-burning requires temperatures above 5 10^9 K, which according to the simple analytical estimate proposed by Thielemann et al. (1990; see section 3.3.1) can be obtained for an explosion energy of 10^{51} erg inside a radius $R \approx 3500$ km. This radius corresponds to $1.7 M_{\odot}$ for the presupernova model of Nomoto and Hashimoto (1988), in excellent agreement with the hydrodynamic calculation. Depending on the peak temperatures and densities reached during passage of the shock explosive Si-burning can be divided into three sub-regimes called incomplete Si-burning and complete Si-burning with Si-exhaustion undergoing either normal or alpha-rich freeze-out (Woosley et al. 1973). In complete Si-burning the composition is determined by NSE (see section 3.1.1) and normal (alpha-rich) freeze-out occurs with decreasing temperature at high (moderate) densities favouring few (many) free alpha particles due to the strongly density-dependent triple-alpha reaction. In the $6M_{\odot}$ He-core studied by Thielemann et al. (1990) only alpha-rich freeze-out and incomplete Si-burning (for $m(r) > 1.69 M_{\odot}$) are encountered, because densities in excess of $10^8 \mathrm{gcm}^{-3}$ required for normal freeze-out (and encountered in Type I SNe; see section 4.4) are not attained in the ejecta. The most abundant nucleus in alpha-rich freeze-out is ^{56}Ni provided that $Y_e \geq 0.49$, which is fulfilled for the ejecta in case of Thielemann et al. (1990; see also Fig. 14).

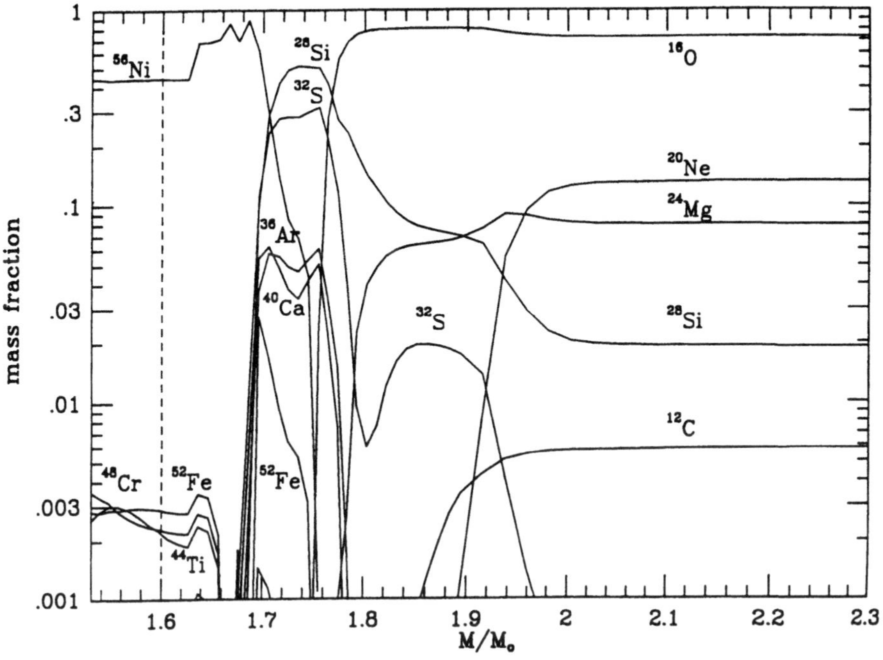

Fig. 14. Mass fractions of a few major nuclei resulting from explosive nucleosynthesis in a $20M_\odot$ stellar model of SN1987A for an explosion energy of 10^{51} erg. Matter outside $2M_\odot$ is essentially unaltered, whereas zones further in experience explosive Ne, O and Si-burning. Inside $1.7M_\odot$ Fe-group nuclei dominate. The drop in the ^{56}Ni abundance at $1.63M_\odot$ coincides with a change in Y_e, which favours more neutron-rich nuclei in the innermost part of the ejecta. The dashed line indicates the position of the mass-cut, if only $0.07M_\odot$ of ^{56}Ni should be ejected (from Thielemann et al. 1990).

Incomplete Si-burning extends out to $1.74M_\odot$. The corresponding peak temperatures of $4-5\ 10^9$ K are not high enough to establish a complete NSE, but only allow for two quasi-equilibrium (QSE) groups of nuclei above and below the proton magic number $Z = 20$, while the abundance ratio between both QSE-groups is out of equilibrium (Woosley et al. 1973; Thielemann and Arnett 1985). The most abundant nuclei in the ejecta of this mass range are the fuel nuclei ^{28}Si and ^{32}S, the alpha-nuclei ^{36}Ar and ^{40}Ca from the lower QSE-group, and furthermore ^{56}Ni , ^{54}Fe and to some extent ^{52}Fe, ^{58}Ni, ^{55}Co and ^{57}Ni from the upper QSE-group (see Fig. 14; and Fig. 7 in Thielemann et al. 1990).

Explosive oxygen burning occurs in the mass range $1.74 < m(r)/M_\odot < 1.8$, where temperatures in excess of roughly $3.3\ 10^9$ K lead to a QSE for nuclei with mass numbers in the range $28 < A < 45$. The main burning products are ^{28}Si, ^{32}S, ^{36}Ar, ^{40}Ca, ^{38}Ar and ^{34}S (see Fig. 14). Mass zones between $1.8M_\odot$ and $2.0M_\odot$, which reach peak temperatures in excess of 2.1 10^9 K, undergo a combined version of explosive Ne and O-burning. Explosive (similar to hydrostatic) Ne-burning mainly produces ^{16}O, ^{24}Mg and ^{28}Si, and explosive C-burning contributes ^{20}Ne, ^{23}Na, ^{24}Mg, ^{25}Mg and ^{26}Mg (see Fig. 14).

Thielemann et al. (1990) have also examined whether in SN1987A r-process nuclei have been produced in the neutron-rich ejecta close to the mass cut and/or in explosive He-burning. They find that this is not the case, i.e. that a $20M_\odot$ star does not eject r-process nuclei consistent with recent galactic chemical evolution studies, which suggest that only Type II SNe with $M \lesssim 11M_\odot$ contribute to the observed r-process abundances (Matthews and Cowan 1989).

The exponential decline of the light curve, which is powered by the radioactive decay of ^{56}Co (produced by the decay of ^{56}Ni) into ^{56}Fe , is consistent with an amount of $0.07 \pm 0.01 M_\odot$ of ^{56}Ni synthesized during the explosion (Shigeyama et al. 1988; Woosley 1988a; Arnett and Fu 1989). Infrared observations of SN1987A indicate further a total amount of stable Ni isotopes for SN1987A of $\approx 0.005M_\odot$ (see e.g. Arnett et al. 1989a). When this amount is compared with the $0.0197M_\odot$ of (stable) Ni predicted in the model of Thielemann et al. (1990), the observations, assuming that all of the Ni is seen, put a serious constraint on the stellar model. A smaller Ni-abundance requires either a mass cut further out in the star but still consistent with the ^{56}Ni -mass inferred from the light curve, or a modified presupernova model. Note, however, that the Ni-abundance also depends on the questionable assumption that the composition of the ejecta close to the mass-cut is not changed between the onset of the collapse and the arrival of the shock wave.

The first solution would suggest a mass-cut at $1.63M_\odot$ resulting in a ^{56}Ni -mass of $0.059M_\odot$, which is marginally consistent with the observations. For an explosion energy larger than 10^{51} erg the agreement can be improved, because due to the stronger shock ^{56}Ni is also produced at larger radii. Including the uncertainty of the explosion energy Thielemann et al. (1990) derive for their stellar model a mass-cut at $1.6 \pm 0.045M_\odot$, which corresponds to a (gravitational) neutron star mass of $1.43 \pm 0.05M_\odot$ consistent within the limits inferred from the neutrino observations (see e.g. Arnett et al. 1989). The second solution requires a larger Y_e just outside the mass-cut, which can be obtained according to Thielemann et al. (1989) either by a smaller size of the convective O-burning shell and/or less mixing of Si shell-burning products into the outer layers.

The yields computed by Thielemann et al. (1990) are shown in Fig. 15 relative to solar abundances (see remark at the end of the previous section) for a mass-cut at $1.6M_\odot$ corresponding to $0.07M_\odot$ of ^{56}Ni . The abundance ratios are normalized to ^{28}Si, which has an over-production factor of 25.2 if only the He-core is considered and a factor of 7.6 for the entire $20M_\odot$ star. Nuclei heavier than Si and P are on average underproduced by a factor 2 to 4 as compared to ^{28}Si. This mainly reflects the size of hydrostatic zones to the explosively processed layers (which is a function of stellar mass), because all elements heavier than Si and P are only produced during the explosion. The C and Ne-burning products, i.e. C, O, Ne, Na, Mg and Al, originate from hydrostatic burning phases only, and have large abundances due to an extended shell of combined C and Ne-burning in the progenitor star. This nicely illustrates the dependence of the results on the stellar model, because the $18M_\odot$ model of Woosley et al. (1988) does not have this extended shell and, therefore, does not show such large abundances of the C and Ne-burning products. Further note that the $^{12}C/^{16}O$ ratio reflects the "effective" $^{12}C(\alpha,\gamma)^{16}O$ rate during core He-burning (see Thielemann et al. 1990). Finally the Ni isotopes with mass number 58, 61

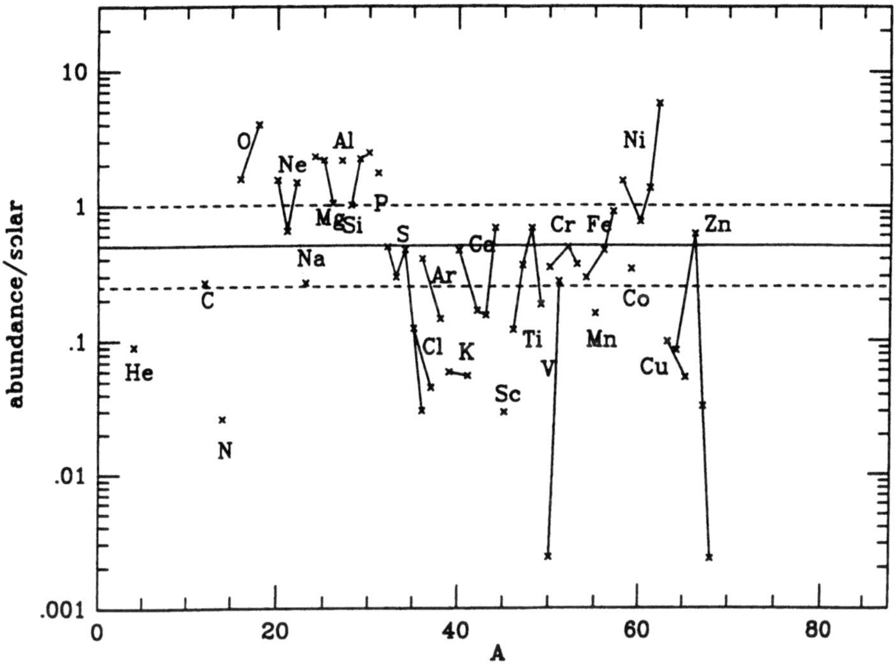

Fig. 15. Composition of the supernova ejecta in comparison to solar abundances (normalized to ^{28}Si) obtained for the $6M_{\odot}$ He-core of a $20M_{\odot}$ stellar model of SN1987A for an explosion energy of 10^{51} erg and a mass-cut at $1.60M_{\odot}$. Isotopes of each element are connected by lines (from Thielemann et al. 1990).

and 62 show a large overabundance, whereas all other nuclei between S and Cu fall well along a line of roughly constant overproduction (Fig 16). The problematic Ni isotopes are produced in form of the neutron-rich species ^{58}Ni and ^{61,62}Zn. Thielemann et al. (1990) have demonstrated that their production is very sensitive to Y_e and, therefore, depends on the location of the mass-cut.

4 Type I Supernovae

4.1 Presupernova Evolution

The progenitors of Type I (or more precise Type Ia) SNe are, according to the commonly accepted scenario, mass-accreting white dwarfs in a binary system, which when growing to the Chandrasekhar mass explode and are completely incinerated and disrupted (for a review see e.g. Iben and Tutukov 1984, 1985; Woosley and Weaver 1986a; Nomoto and Hashimoto 1987; Canal et al. 1990; Woosley 1990). All chemical compositions of the white dwarf are possible in principle: helium, carbon-oxygen or oxygen-neon-magnesium. Most studies, however, have concentrated on C+O white dwarfs, because He white dwarfs almost certainly detonate (which is inconsistent with observations; see section 4.4) and

O-Ne-Mg white dwarfs are thought to collapse rather than explode (Miyaji et al . 1980; Miyaji and Nomoto 1987; see however Canal et al. 1990). Scenarios leading to the formation of C+O white dwarfs, which eventually explode as Type Ia SNe have been studied in great detail by Iben and Tutukov (1984,1985). These scenarios and further complications resulting from an eventually occuring crystallization of the white dwarf and from a phase separation of carbon and oxygen (see Canal et al. 1990) will not be addressed further in this review.

Because of mass accretion white dwarfs in binary systems evolve quite differently from single white dwarfs, which slowly cool down and end up as dark matter. Accreting white dwarfs, however, have a steady supply of energy in form of gravitational binding energy of the accreted matter. The gravitational energy gained during the transfer of matter from the companion to the white dwarf is radiated away in the accretion shock near the white dwarf's surface. However, additional gravitational energy is released when the accreted matter settles on the surface of the white dwarf and thereby compresses the whole star. This energy, which is liberated inside but close to the surface of the star is partially transported to its surface and radiated away. The remaining part heats the interior of the white dwarf. The temperature inside the white dwarf is thus determined by the balance between cooling and heating, i.e. it is a sensitive function of the mass accretion rate $\dot{M}$ (Nomoto 1982, 1986).

The evolution of the temperature profile of the white dwarf shows that the temperature first rises near the surface and then heat begins to diffuse inward (Fig. 16). According to Nomoto (1982) the diffusion time scale decreases with increasing $\dot{M}$, because a larger mass accretion rate leads to a larger heat flux and steeper temperature gradient. For $\dot{M} = 10^{-6} M_\odot/\mathrm{yr}$ the heat wave needs about 2 10^5 yr to reach the center of the star, whereas it takes 5 10^6 yr for $M = 4\ 10^{-8} M_\odot/\mathrm{yr}$ (Nomoto 1986). Therefore, in both cases the white dwarf mass has grown by $0.2 M_\odot$ before the central temperature changes due to the inward heat flow. If in addition the initial white dwarf is massive (and cold) enough, i.e. $M \gtrsim 1.2 M_\odot$, it reaches the Chandrasekhar mass still being cold and degenerate in the center, and explosive carbon burning will be ignited in the center at relatively high densities ($\rho \lesssim 10^{10}\mathrm{gcm}^{-3}$; Isern et al. 1983).

The previous considerations obviously show that the ultimate fate of an accreting C+O white dwarf depends sensitively both on the mass accretion rate $\dot{M}$ and on the initial white dwarf mass. The fate of the white dwarf further depends on whether the accreted matter is composed of hydrogen, helium or carbon and oxygen. A detailed discussion of the various evolutionary paths is beyond the scope of this review (see e.g. Nomoto 1986; Nomoto and Hashimoto 1987; Canal et al. 1990; Woosley 1990). A schematic overview of possible evolutionary results is shown in Fig. 17 taken from Nomoto (1986). Note that Nomoto (1986) following the suggestion of Branch and Nomoto (1986) favours an off-center He-detonation in a C+O white dwarf as a model for Type Ib SNe, which is debated by other workers in the field (see e.g. Woosley 1990; and section 2.6). The relatively high accretion rates ($\dot{M} \gtrsim 10^{-8} - 10^{-10} M_\odot/\mathrm{yr}$) required for the above models are, however, not observed. Furthermore theoretical evolutionary paths leading to such systems have only be discussed on the level of plausible scenarios (common envelope evolution; merging white dwarfs; etc), which may produce the binary systems with frequencies consistent with the Type Ia SN rate (see e.g. Iben 1986).

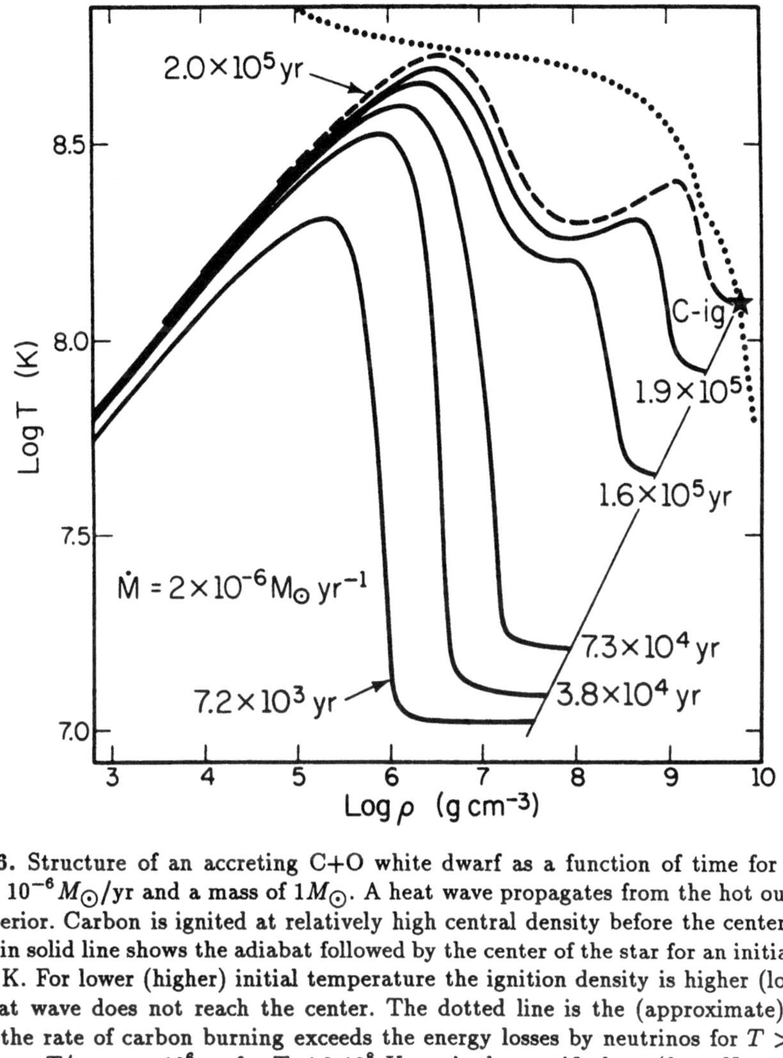

Fig. 16. Structure of an accreting C+O white dwarf as a function of time for a model with $\dot{M} = 2\ 10^{-6} M_\odot/\mathrm{yr}$ and a mass of $1 M_\odot$. A heat wave propagates from the hot outer layers into the interior. Carbon is ignited at relatively high central density before the center is heated up. The thin solid line shows the adiabat followed by the center of the star for an initial temperature of 10^7 K. For lower (higher) initial temperature the ignition density is higher (lower) provided the heat wave does not reach the center. The dotted line is the (approximate) ignition line, where the rate of carbon burning exceeds the energy losses by neutrinos for $T > 2\ 10^8$ K and $\tau_{C+C} = c_P T/\epsilon_{C+C} = 10^6$ yr for $T \leq 2\ 10^8$ K; c_P is the specific heat (from Nomoto 1986).

4.2 Thermonuclear burning in degenerate matter

There are two separate time scales for burning which must be considered. The first is the *ignition time scale* of the fuel which is defined to be the temperature e-folding time

$$\tau_T = T/\dot{T} \approx C_V T/\dot{\epsilon}_{nuc} \quad , \tag{4.1}$$

where $\dot{\epsilon}_{nuc}$ is the energy release rate of the nuclear processes, and C_V is the specific heat. Because charged-particle reactions are heavily modified by Coulomb barrier penetration, this time scale strongly decreases with increasing temperature. The second important timescale is the *burning time*, i.e. the time to significantly reduce the abundance of fuel, which is defined as

$$\tau_i = X_i/\dot{X}_i = Y_i/\dot{Y}_i \quad , \tag{4.2}$$

where X_i is the mass faction of species i, and Y_i is the mole number, obtained by dividing the mass fraction by the atomic weight of the species. In simple cases, this differs from τ_T by the ratio of the thermal energy content to the Q-value for the reaction (per unit mass burned). For example, for $^{12}C + ^{12}C$ this ratio is 0.25 T_9, where T_9 is temperature in units of 10^9 K; this is of order unity at explosive temperatures. For comparison, the ratio is about 0.005 for hydrostatic hydrogen burning by the CNO cycle. This ratio can also approach zero in explosive situations, if the matter is degenerate. In this case the specific heat approaches zero, so that consumption of a small amount of fuel gives a large change in temperature.

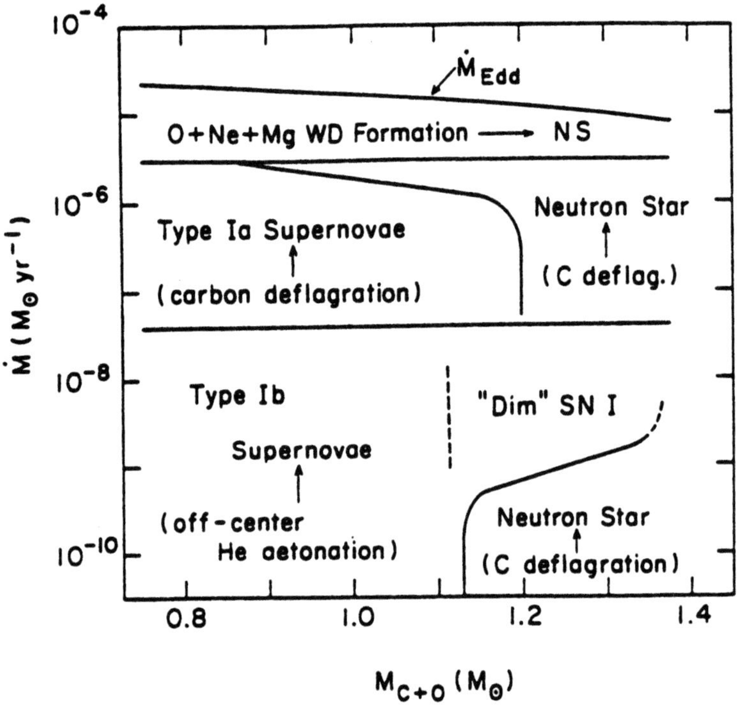

Fig. 17. The final fate of accreting C+O white dwarfs expected for their initial mass M_{CO} and accretion rate $\dot{M}$. In two regions of the $\dot{M} - M_{CO}$ plane indicated by *Neutron Star*, a carbon deflagration is ignited in the center when the density is as high as $\rho_c \approx 10^{10}$gcm^{-3}, which induces a collapse of the white dwarf. "Dim" SN occur, when a small amount of accumulated helium ($\sim 0.01 - 0.1 M_\odot$) detonates on a very massive white dwarf (from Nomoto 1986).

Finally, two timescales associated with hydrodynamic motion must be considered. The first is the time for a region to react to a pressure imbalance, which is taken to be the *sound travel time*

$$\tau_{hyd} = \delta r / c_s \quad , \tag{4.3}$$

where c_s is the local sound speed. Finally, there is the time for a convective element to move through the region in which convection occurs (obviously an oversimplification of a complex process). This *convective time scale* is defined by

$$\tau_{conv} = \delta r_{conv} / v_{conv} \quad , \tag{4.4}$$

where δr_{conv} is the width of the convective zone and v_{conv} is the typical velocity of a convective blob.

Depending on the relative size of the various time scales, very different requirements for the numerical treatment arise. If the nuclear time scales τ_i are all large compared to τ_{conv}, one might approximate the convective zone as uniform in abundances, which slowly evolve on the nuclear timescales. If *some* of the τ_i are shorter than τ_{conv}, it is not correct to ignore the abundance gradients that this implies. In particular, these gradients may interact with the burning to modify the convective flow itself. If the nuclear time scales are all small compared to τ_{conv}, the problem simplifies again, and each region is loosely coupled to its neighbors. For more dynamic problems, such as pulses or explosions, the convenient fiction of steady state convection is untenable, and the hydrodynamics must be treated as an equally important aspect of the problem.

4.2.1 Detonations

Astrophysical detonations generally occur only under degenerate conditions. During the initial stages of a thermonuclear runaway, before the temperature rises significantly, the reaction rate is still relatively small. In nondegenerate matter, the pressure increase produced by the reactions will cause the burning region to expand and cool, preventing the runaway from proceeding. In other words, $\tau_T > \tau_{hyd}$. On the other hand, if the matter is degenerate, the temperature increase created by the burning will not create a significant increase in pressure. Thus the temperature will continue to increase until the matter becomes nondegenerate. At this point, the energy generation rate will be too large for hydrodynamic motion to stop it, and an explosion will result. If the resulting shock is sufficiently strong to raise the fuel above the ignition temperature, a detonation wave will propagate outwards from the point of ignition.

In its simplest form (i.e. for planar steady one dimensional detonations) detonation wave theory is easily understood (see e.g. Courant and Friedrich 1948; Fickett and Davis 1979). As the shock propagates into the unburned fuel, it compresses and heats the material beyond the ignition point. Immediately behind the shock is the reaction zone, in which the fuel burns. In detonation theory the width of the reaction zone is frequently neglected, so that the detonation front is treated as a sharp discontinuity. In this case jump conditions can be derived for the change in the hydrodynamic variables across the front in much the same way as is done for a simple shock. From mass conservation one obtains the condition

$$\rho_1 \, D = \rho_2 \, (D - u_2) \ , \tag{4.5}$$

where D is the velocity of the detonation front, ρ is the mass density, u is the fluid velocity, and the subscripts 1 and 2 denote the pre-detonation and post-detonation states, respectively. The velocity of the material ahead of the front, u_1, is taken to be 0. A second jump condition, which expresses momentum conservation across the front, can be written as

$$P_2 - P_1 = \rho_1 \, u_2 \, D \, . \tag{4.6}$$

where P is the pressure. The third jump condition, based on energy conservation, is

$$e_2 - e_1 = P_1 \, V_1 + \frac{1}{2} \, D^2 - P_2 \, V_2 - \frac{1}{2} \, (D - u_2)^2 + q \, , \tag{4.7}$$

where e is the specific internal energy, V is the specific volume and q is the amount of energy generated by the complete burning of the fuel. By eliminating the velocity from the first two jump conditions, one obtains the equation for the "Rayleigh line"

$$\mathcal{R} = \rho_1^2 \, D^2 - \frac{P_2 - P_1}{v_1 - v_2} = 0 \, . \tag{4.8}$$

The equation for the "Hugoniot curve" is obtained by eliminating the fluid velocity and detonation speed from the energy jump condition, using the jump conditions for mass and momentum. The resulting equation can be written as

$$\mathcal{H} = e_2 - e_1 - \frac{1}{2} \, (P_2 + P_1) \, (v_1 - v_2) - q = 0 \, . \tag{4.9}$$

For a given equation of state $p = p(\rho, e)$, the intersection of the Rayleigh line and the Hugoniot curve determines the post-detonation state. Note, however, that in order to obtain the post-detonation state, one must first choose a detonation velocity. Unlike the case for simple shocks, the front velocity is not determined from the jump conditions. Depending on the value chosen for the detonation velocity, the Rayleigh line intersects the Hugoniot curve at 0, 1, or 2 points. If there is no intersection, no detonation wave is possible for that detonation velocity. If there are two points of intersection, there are two possible solutions. These two solutions correspond to "strong" and "weak" detonations. A strong detonation propagates at a speed slower than the post-detonation sound velocity with respect to the fluid behind the shock, i.e., $D < u_2 + c_2$, so that disturbances generated behind the front will eventually catch it. Thus this solution is unstable. Weak detonations, which propagate faster than the post-detonation sound velocity with respect to the fluid behind the shock, i.e., $D > u_2 + c_2$, are generally considered to be unphysical except under certain special conditions. The detonation which usually occurs in nature is the one corresponding to the speed at which the Rayleigh line and Hugoniot curve have only one point of intersection. This detonation speed, called the Chapman-Jouguet velocity, is equal to the sum of the post-detonation fluid velocity and sound speed, i.e.

$$D_{CJ} = u_2 + c_2 \, . \tag{4.10}$$

By assuming that the front propagates with this velocity, the post-detonation state is completely determined. Both the pressure and the density increase across the front, while in the frame in which the shock is stationary, the fluid velocity decreases.

Although the above theory provides a satisfactory explanation for simple detonations, it has significant limitations because of the assumptions that the reaction rate is infinite and that the reaction zone has zero width. A slightly more complex treatment is used in the Zeldovich-von Neumann-Doering (ZND) model (Zeldovich 1940; von Neumann 1942; Doering 1943), which, however, is still restricted to planar steady one dimensional detonations. This theory assumes that the shock, which is taken to be infinitely thin, is followed by a reaction zone of finite width. The primary difference from the equations given above is that the term in the Hugoniot curve involving the energy generated by the reaction must now be multiplied by the extent of the reaction. Thus, each state within the reaction zone can be determined by the intersection of the Hugoniot curve for the appropriate extent of reaction and the Rayleigh line. The final state obtained after the fuel is completely burned is exactly the same as for the simpler theory described above. Additional complications in the theory arise due to multi-dimensional effects, such as cellular detonations and spinning detonations, and due to the instability of steady planar ZND detonation waves (see e.g. Fickett and Davis 1979; Khokhlov 1989a).

4.2.2 Deflagrations

Deflagrations represent a much less violent form of burning than detonations, but in many ways are more complex (see e.g., Chap. 5 of Williams 1985; Zeldovich et al. 1985). They result when the burning is unable to produce sufficient overpressure to create a shock which is strong enough to ignite the fuel. The motion of the front is usually very subsonic. Burning is initiated by the diffusive transfer of heat from the hot ashes behind the front into the cold fuel. Although, for the case of a thin front, deflagrations must obey the same jump conditions as detonations, the propagation velocity now depends on the rate of heat transfer. Another major difference is that the pressure and density decrease behind the deflagration front, and in the reference frame in which the front is stationary, the velocity increases. In the case of a deflagration which begins at the origin in spherical symmetry, the velocity behind the front must eventually become zero to satisfy the boundary condition at the origin. The only way in which this can happen is if the deflagration is preceded by a compression wave which accelerates the material away from the front. The passage of the deflagration will then provide exactly the correct jump in velocity so that the material behind the front will come to rest. This can happen since the deflagration velocity is subsonic, and therefore, there is communication between the origin and the rest of the flow by sound waves.

Unlike the case of Chapman-Jouguet detonations, where it is possible to compute the exact propagation velocity, the propagation speed of deflagrations can only be crudely estimated (Landau and Lifshitz 1959; Fryxell and Woosley 1982; Nozakura et al. 1984). For the simplest case of a laminar front which propagates as a result of radiative diffusion or conduction, it is fairly easy to obtain an order of magnitude estimate for the velocity of the wave. The width of the deflagration can be approximated by setting the diffusion time scale τ_{diff} equal to the burning timescale τ_i. Thus the width of the front is given by

$$\delta \sim \sqrt{\lambda c \tau_i} \ , \tag{4.11}$$

where λ and c are the mean free path of photons or electrons and the speed of light, respectively. The velocity D of the deflagration can then be estimated as

$$D \sim \frac{\delta}{\tau_i} \sim \sqrt{\lambda c / \tau_i} \ . \tag{4.12}$$

If the deflagration is propagated by turbulent convection rather than by diffusion, two situations can arise. If the scale of the turbulence is small compared to the width of the front, the deflagration will remain laminar, and the above equations still apply if the mean free path of the photons or electrons is replaced by the typical length scale for the turbulence, and the velocity of the diffusing objects is replaced by the convective velocity. If, on the other hand, the thickness of the front is small compared to the length scale of the turbulence, the front will be wrinkled. In this case, the velocity of the front is increased compared to that of the diffusively propagated deflagration both by the increase in the heat transfer rate and by the increase in the surface area of the front. The upper limit to the speed of the burning front which can be obtained in an extremely wrinkled front is approximately the convective velocity.

Another complication which arises is that many astrophysical deflagrations fronts are Rayleigh-Taylor unstable (Müller and Arnett 1982, 1986; Woosley and Weaver 1986b; Woosley 1990). Since the density decreases behind the front, when a deflagration propagates outwards from the center of a star against the force of gravity, the flow will be unstable. The situation which results is a hot low density bubble trying to expand outward into a denser medium. Under these conditions, the flame front will almost certainly be wrinkled, and the average rate of propagation of the front will be determined by the growth rate of the instability (see Müller and Arnett 1986). It is also possible that the distortions could become so extreme, that the front might not remain simply-connected, with blobs of burning material being scattered throughout the star ahead of the main deflagration wave.

The fact that the propagation velocity of a deflagration depends upon the width of the front can cause severe computational problems. In order to calculate the correct structure, it is necessary to resolve the width of the front on the computational grid. In many cases of interest, the width of the front will be much smaller than one zone. In this case, the only way to obtain the correct deflagration structure is to use an adaptive grid to obtain very high resolution near the front or a front tracking method. In one dimensional calculations, this procedure is difficult but probably feasible. For two or three dimensional calculations, moving the grid to obtain the required resolution at the front, especially if the front is severely wrinkled, is virtually impossible with current technology.

4.3 Detonations and Deflagrations in Astrophysical Models

Detonations and deflagrations play a significant role in the understanding of Type I supernova explosions, i.e. in the thermonuclear explosion of accreting white dwarfs (for recent reviews see Woosley and Weaver 1986a and Khokhlov 1989a). Therefore, the numerical simulation of detonations and deflagrations has been a major effort in this research area over the last two decades. It is far beyond the scope of this review to cover all the results and astrophysical implications of these simulations. Instead, we will focus on the numerical techniques and the physical approximations which have been used. Some implications of the models for the nucleosynthesis will be discussed in the next section.

For detonations, astrophysicists have tried to answer two major questions: (i) under what conditions does a detonation wave form, and (ii) if it forms, will it propagate through the star? Let us first address the second question. Starting with the calculation of Arnett (1969), it has been demonstrated in many numerical simulations, that once a detonation is formed, it will propagate self-consistently in a white dwarf composed of either helium or a mixture of carbon and oxygen (Buchler et al. 1972; Bruenn 1972; Mazurek et al. 1974; Ivanova et al. 1974; Mazurek et al. 1977; Woosley et al. 1986; Müller and Arnett 1986). In these calculations the detonation was initiated artificially, and in all except three calculations (Mahaffy and Hansen 1975; Denisov et al. 1986; Müller and Arnett 1986) spherical symmetry was assumed.

This symmetry restriction is a major shortcoming of the so-called "off-center detonation" models. In those models the runaway starts at the bottom of the helium layer surrounding the central carbon-oxygen region of the white dwarf. Depending on the mass accretion rate (see section 4.1) a single outward propagating helium detonation develops or a so-called *double detonation* forms, where a carbon detonation wave propagates into the C-O core and at the same time a helium detonation incinerates the accreted helium layer (Nomoto 1982; Woosley et al 1986). Due to the strong temperature dependence of the nuclear reaction rates, it is very unlikely that the fuel starts to detonate simultaneously in a spherical shell. It is more likely that the detonation will start from a point, and therefore, at least two-dimensional calculations are necessary to determine the outcome of such an off-center detonation. The major error associated with the one dimensional calculations is that a detonation which ignites simultaneously in a spherical shell will be amplified by geometrical convergence as it approaches the center of the star. For multi-dimensional detonations which initiate at a point, the detonation will perhaps even weaken as it propagates toward the center. Thus, the one dimensional calculations overestimate the amount of burning which occurs in the core (see however Dgani and Livio 1989). For central detonations geometrical effects seem to be less important. Using a two-dimensional axisymmetric hydrodynamics code, Müller and Arnett (1986) showed that a detonation initiated in the center of a C-O white dwarf propagates through the star as a spherical wave.

Another question concerning the propagation of the detonation wave is whether it will die out when it reaches the increasing density and temperature gradients near the surface of the star (Mazurek et al. 1977; Woosley and Weaver 1986a). Due to the lower pre-shock temperatures in the outer layers, the shock compression might be insufficient to heat the matter beyond the runaway temperature, i.e. the detonation dies. If one tries to attack that problem numerically, one has to make absolutely sure that the numerical method used is capable of accurately describing the propagation of the detonation wave down a steep gradient in density and temperature. All methods used up to now relied on artificial viscosity to handle the shock. The resulting spreading of the shock front over a few zones will produce incorrect shock jumps if there is a steep gradient ahead of the shock. Thus, the question of whether the detonation continues to propagate becomes very difficult. It is clear that methods which produce very thin shock fronts are to be preferred (see e.g. Fryxell et al. 1990)

Furthermore, as Imshennik and Khokhlov (1984) have shown, the common claim (see e.g. Nomoto 1982) that lighter ($A \leq 40$) elements are unable to form during detonation propagation is incorrect. This claim is due to the neglect of the finite width of the burning zone in a detonation and the assumption that nuclear statistical equilibrium (NSE) will instantly be established behind the front. The calculations of Imshennik and Khokhlov (1984), which resolved the burning zone, convincingly show that for densities below 10^7 g cm^{-3}, the above claim is untenable. This result is confirmed by recent calculations of Khokhlov (1989b), who has studied the structure of carbon and helium detonation waves in great detail.

The problem of detonation formation is much more difficult (for a recent review see Khokhlov 1989a). Shock tube analysis in planar geometry (Mazurek et al. 1977) indicates that the formation of a detonation in carbon-oxygen white dwarfs is unlikely, unless the white dwarf has an almost isothermal core (see also discussion below). However, these authors did not really address the question of whether a detonation does or does not form in the conditions actually encountered in the evolutionary calculations, namely in a strongly driven, convective, turbulent core (Arnett 1969). Furthermore, because of the high degeneracy pressure in the center of carbon-oxygen white dwarfs, the resulting overpressure due to nuclear burning is only of the order 20%. This energy argument led Nomoto et al. (1976) to conclude that the formation of a detonation in such white dwarfs is impossible, and that instead, a subsonic deflagration incinerates the star. The same argument can be used to show that in the case of a helium white dwarf, a detonation will form, since helium ignition occurs at a significantly lower density ($\approx 10^7$ g cm^{-3} instead of $\approx 10^9$ g cm^{-3} for carbon and oxygen) and hence the resulting overpressure is larger in this case. However, the validity of their conclusions for the carbon detonation question is doubtful (see below). But even if a deflagration initially forms a sudden abrupt transition into a detonation can occur, an effect often observed in laboratory experiments (see e.g. Lee and Moen 1980: Khokhlov 1989a).

In addition, due to convection, the temperature gradient at the time of runaway will be adiabatic (Arnett, 1969). Under those conditions a carbon detonation is more likely to form (Mazurek et al. 1977), since it is easier to start a detonation in matter that is already on the verge of exploding than in cold fuel (Mazurek et al. 1974). Calculations by Woosley and Weaver (1986b) suggest that the formation of either a detonation or a deflagration is possible. The outcome is critically dependent upon both the mass zoning and especially the temperature at which one relinquishes a mixing-length approximation to the energy transport preceding the runaway. These calculations are also confirmed by the results of Müller and Arnett (1986), who showed through their numerical experiments that a detonation will form if more than a critical mass of material is exploded as a unit.

The critical mass aspect for detonation formation has been studied in more detail by Blinnikov and Khokhlov (1986a, b) and independently by Woosley and Weaver (1986b). Consider two points in the core of a white dwarf having a finite temperature gradient. As the burning time is shorter for higher temperatures, the fuel is first incinerated at the point of higher temperature. This difference in burning times leads to the propagation of the boundary separating burnt from unburnt matter. If the corresponding phase velocity is larger than the local speed of sound, the burning at one point does not influence the burning at the other point. The phase velocity of this so-called spontaneous burning (Zeldovich 1980) is obviously given by

$$D_{sp} = \left(\frac{d\tau_{nuc}}{dr}\right)^{-1} , \tag{4.13}$$

where τ_{nuc} is the nuclear burning time at an initial temperature T and r is the radial coordinate. For the $^{12}\mathrm{C} + ^{12}\mathrm{C}$ rate, Woosley and Weaver (1986b) have given a simple analytical approximation for the burning time, which leads to

$$D_{sp} = -3.26 \times 10^8 \, T_9^{21.2} \rho_9^{3.05} (X_{12}/0.5)^2 \left(\frac{dT}{dr}\right)^{-1} \ \mathrm{cm\ s^{-1}} , \tag{4.14}$$

in the temperature range $0.6 \le T_9 \le 1.2$, where T_9 and ρ_9 are the temperature and density in units of 10^9 K and 10^9 g cm^{-3}, respectively. Note that the phase velocity is very sensitive to the initial temperature, and that it approaches infinity for an isothermal temperature distribution (Mazurek et al. 1977). If the initial temperature gradient is small enough, the phase velocity will exceed the sound speed, and hence the corresponding region of the white dwarf will run away on less than a sound crossing time. According to Blinnikov and Khokhlov (1986a, b) this will give rise to a detonation, if the ignition temperature (i.e. the temperature where nuclear energy generation begins to exceed the energy losses by neutrinos) $T_{ign} \lesssim (6-7)10^8$ K. The appearance of a deflagration in such a case would undoubtedly be an artefact of an insufficient spatial grid resolution (Khokhlov 1989a).

The simulation of deflagrations in C-O white dwarfs still poses one of the most challenging numerical problems in astrophysics, although many attempts have been made to attack the problem (Nomoto et al. 1976; Nomoto et al. 1984; Sutherland and Wheeler 1984; Woosley and Weaver 1986b; Müller and Arnett 1986; Woosley 1990). As discussed in section 4.2.2, the deflagration is either propagated by heat diffusion or (turbulent) convection. Talking of a convective deflagration is, however, a misleading and oversimplified description. From a microscopic point of view, a deflagration wave in a C-O white dwarf always propagates by (electron) heat conduction. According to Zeldovich et al. (1985), the corresponding conductive velocity may be estimated from that of the normal conductive speed of a laminar flame. Numerical studies involving very fine zoning ($\sim 10^{-5}$ cm) of the burning front show that for a (conductive) carbon deflagration

$$v_{cond} \approx 50 \left(\frac{\rho}{2\ 10^9 \mathrm{gcm}^{-3}}\right)^{0.8} \left(\frac{X_{12}}{0.5}\right) \mathrm{km\ s}^{-1} \quad , \tag{4.15}$$

where X_{12} is the carbon mass fraction and that the width of the burning front is $\sim 10^{-3}$ cm, only (Woosley 1990).

From a macroscopic viewpoint, i.e. when considering the propagation through the white dwarf, the deflagration speed is, as pointed out in section 4.2.2, much more difficult to calculate. Since the pressure increases and the density decreases behind the deflagration front, it is Rayleigh-Taylor unstable when propagating outward in the star's gravitational field, and therefore, it will become wrinkled, as demonstrated by Müller and Arnett (1986; see Fig. 18). In such a situation, the fuel consumption rate is determined by the laminar conductive velocity (which is normal to the front everywhere) multiplied by the surface area of the front. Since the surface area depends on the amount of wrinkling of the front, i.e. on the growth rate of the Rayleigh-Taylor instability, the complicated wrinkled structure has to be resolved numerically in order to calculate the correct deflagration speed. Up to now, this has been attempted only in the work of Müller and Arnett (1982, 1986) using an explicit axisymmetric Eulerian hydrodynamic code. Although severely hampered by grid resolution issues and, to a lesser extent, by the symmetry restrictions imposed, they were able to show that the front is Rayleigh-Taylor unstable and that the overall propagation speed indeed depends on the amount of wrinkling of the burning front.

All other workers in the field have used a more or less sophisticated ad hoc parameterization of the burning velocity within their one dimensional Lagrangian stellar evolution codes. Nomoto et al. (1976; see also Nomoto et al. 1984) have modelled the deflagration front by employing the time-dependent mixing length theory of convection developed by Unno (1967). This approach is based on the idea that the velocity of the subsonic deflagration front depends on the timescale of convective energy transport, and therefore can

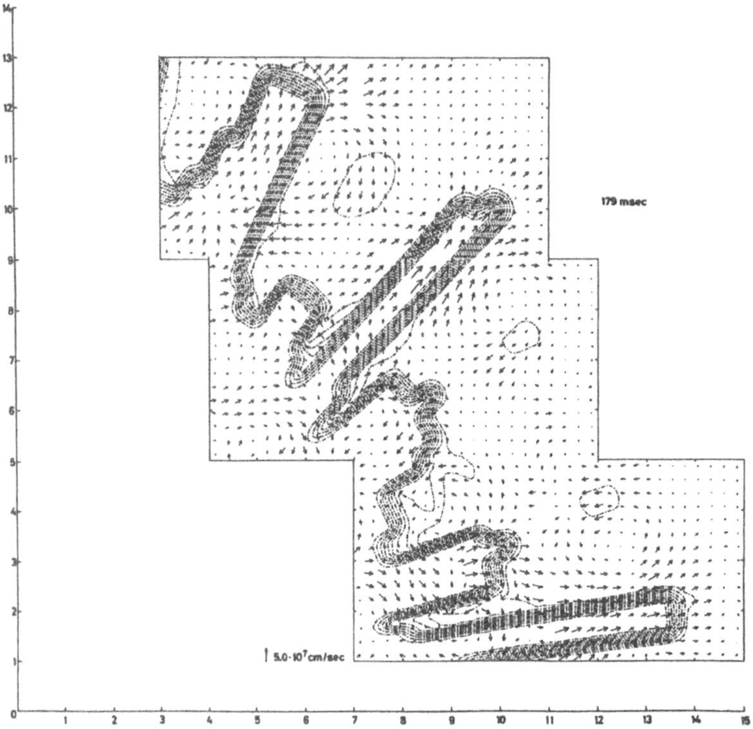

Fig. 18. Section of a Rayleigh-Taylor unstable burning front in an exploding C-O white dwarf. The snapshot is taken 0.179 sec after the (artificial) initiation of the burning. Contours are surfaces of constant carbon abundances, the spacing being 01. on a linear scale (from Müller and Arnett 1986).

be described within the mixing length theory. The deflagration velocity is determined by a free parameter $\alpha = l/H_p$, which is the ratio of the mixing length l and the pressure scale height H_p. Actually the mixing length l is the free parameter of the theory, and α only measures it in units of the pressure scale height.

Although the parameter α is time-independent, due to variations of the pressure scale height through the white dwarf, the deflagration velocity is time-dependent, and varies from $v_{def} \approx 0.08c_s$ at early times to $v_{def} \approx 0.3c_s$ at late times of the evolution, if $\alpha = 0.7$ is used, where c_s is the local sound speed. Nomoto et al. (1984) find that for $\alpha = 0.7$, the deflagration takes about six times longer to incinerate the star than a detonation, and that this choice of the parameter provides the best fit with observed spectra. However, the latter agreement is only achieved if an additional ad hoc complete mixing of all material with velocity greater than 8000 km s^{-1} (approximately the outer half of the white dwarf by mass) during the explosion is assumed (Branch et al. 1985).

In spite of the relative success of their simulation, the approach of Nomoto et al. (1984) remains flawed from a theoretical point of view. Besides the ad hoc parameterization of the deflagration velocity, they also rely on a theory of convection which was not devised for the regime in which they apply it. The mixing length theory of convection requires that the variation of quantities be small over length scales shorter than the mixing length. Here, the opposite is true, as finite mass zones flash individually, giving rise to enormously steep temperature variations (see also Woosley 1990).

Woosley and Weaver (1986b) used a parameterization obtained from dimensional analysis of the Rayleigh-Taylor instability and from nucleosynthesis constraints imposed by observations of Type-I supernova. They suggest that the deflagration velocity is given by

$$v_{def} = F c_s (1 - \exp(-r/R_0)) \ . \tag{4.16}$$

For small radii r this results in $v_{def} \propto r$ and approaches a fraction F of the sound speed at large radii. The dependence for small radii reflects the fact that the fastest propagating deformation of the front, i.e. the maximum wavelength, can be no larger than the radius of curvature of the burning front. The actual values of the two free parameters F and R_0 are to be determined ideally by observational constraints on the light curve, the explosion energy, and nucleosynthesis (see next section). In their model calculations they used $F = 0.5$ and $R_0 = 2 \times 10^7$ cm.

Although this is a more physically based parameterization, the burning velocity is determined in an artificial fashion by limiting the rate of increase of the convective luminosity which couples the zone currently burning to the one just ahead of it. While the physical justification for this treatment is phenomenological, it provides a convenient means for numerically generating a variable flame speed.

Recently Woosley (1990) has suggested a novel determination of the effective flame speed based upon concepts of fractal geometry. Similar to Woosley and Weaver (1986b; see above) he finds that the amount of burned mass increases quite slowly initially but more rapidly towards the end. Even though Woosley's (1990) approach is one of the most promising quasi-analytical attempts up to now to determine the flame speed, it suffers, like all previous attempts, from the existence of a free parameter (here: the fractal dimension) in the model.

4.4 Explosive Nucleosynthesis

Type Ia SNe are thought to be a major source of iron-peak elements and of intermediate mass elements. The first point is supported by observations of Type Ia late time spectra and light curves, both of which can be well explained, if a large amount of iron was produced during the explosion in form of ^{56}Ni , which then decays via ^{56}Co to stable ^{56}Fe . The observations require that a total amount of $0.4 - 1.0 M_\odot$ of ^{56}Ni should be ejected in each Type Ia event (see e.g. Thielemann 1989 and references therein). The evidence for intermediate mass elements stems from early time (close to maximum light) spectra, where lines of Ca, Si, S, Mg and O could be identified (Branch et al. 1981). Taken together these observations suggest that the outer layers, visible at maximum light, are composed of intermediate mass elements, whereas the central layers, visible only at late times, contain mainly ^{56}Ni .

Detailed nucleosynthesis calculations of carbon deflagration models have been performed by Nomoto et al. (1984), Woosley et al. (1984), and Thielemann et al. (1986). Similar studies for carbon detonation models have been carried out, e.g., by Arnett (1969)

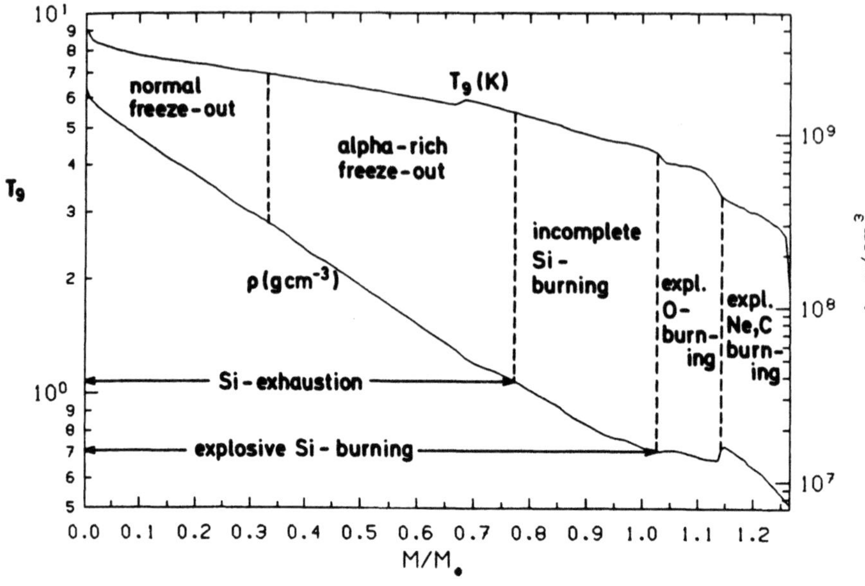

Fig. 19. Maximum temperatures and densities attained during the propagation of a deflagration front as a function of the radial mass coordinate. Zones of different explosive burning conditions are indicated (from Thielemann et al. 1986).

and Woosley et al. (1986; see also Woosley 1990), but these calculations had severe difficulties to reproduce the observations, because the whole star was incinerated into ^{56}Ni (see, however, Imshennik and Khokhlov 1984; Khokhlov 1989b). A further discussion of the nucleosynthesis of detonation models can be found in Woosley and Weaver (1986a) and in Khokhlov (1989a).

The maximum temperatures and densities attained in the carbon deflagration model of Nomoto et al. (1984) are shown in Fig. 19, and the resulting nucleosynthesis (Thielemann et al. 1986) is displayed in Fig. 20. The various burning regimes seen in Fig. 20 have already been discussed in section 3.3. Contrary to the explosive nucleosynthesis in Type II SNe mass layers with normal and alpha-rich freeze-out are encountered here in explosive Si-burning with Si-exhaustion. In this model the inner $0.8M_\odot$ are composed only of iron-group nuclei, where except for the very center ^{56}Ni , which has the largest binding energy per nucleon of all N=Z nuclei, is the dominant nucleus (Fig. 20). Near the center the densities are high enough that electron captures on free protons and to a minor extent on heavy nuclei (20%) can change the total proton to neutron ratio. In this more neutron-rich environment the most abundant nuclei become first ^{54}Fe and ^{58}Ni and finally ^{56}Fe . The composition of this model was used by Branch et al. (1985) to compute synthetic spectra. Assuming an *ad hoc* homogeneous mixing (caused by a Rayleigh-Taylor instability; see previous section) of the mass layers containing the intermediate mass elements and the unburned surface layer they could reproduce observed spectra quite well.

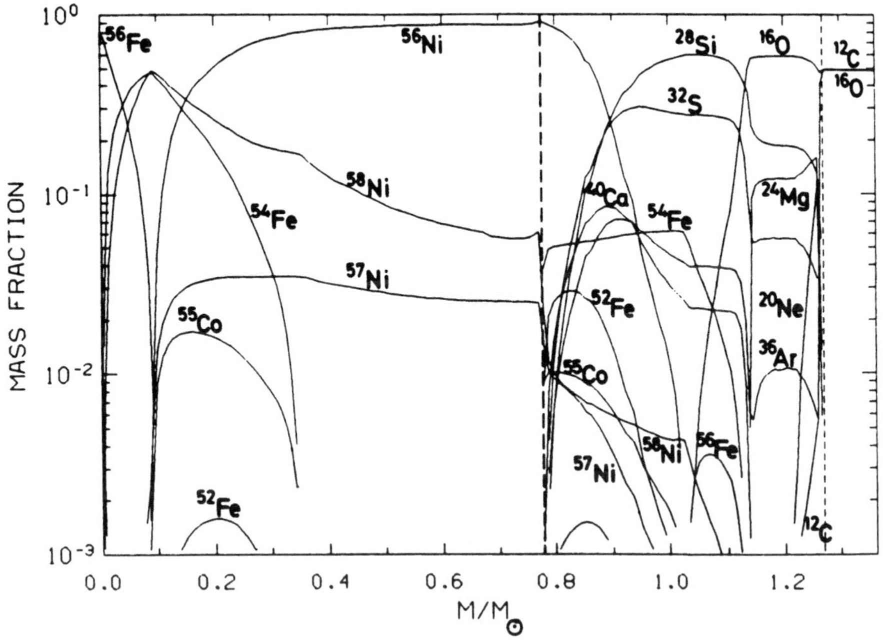

Fig. 20. Resulting nucleosynthesis (major abundances only) in a Type Ia SN deflagration model. In the inner $0.8 M_{\odot}$ only iron-group nuclei are produced. The inner dashed line shows the transition to incomplete Si-burning, followed further out by products of explosive O, Ne and C-burning. The outer dashed line marks the quenching of the burning front beyond which unburned C and O are left (from Thielemann et al. 1986).

Fig. 21 shows the yields (normalized to ^{56}Fe) obtained with the model of Thielemann et al. (1986) after the decay of unstable nuclei. Assuming that all Type Ia SN progenitors are white dwarfs with a mass equal to the Chandrasekhar mass, all events produce the same nucleosynthesis, i.e. a meaningful comparison with solar abundances is possible without averaging over a complete sample of events. Obviously the production of iron-group elements is a factor of two larger than that of intermediate mass elements from Si to Ca (Fig. 21). This implies that Type Ia SNe are the dominant production sites of iron-group nuclei and that Type II SNe have to produce the missing intermediate mass elements.

The yields displayed in Fig. 21 show another disturbing feature, namely large variations of the isotopic abundance ratios within the iron-group elements, which should not occur, if Type Ia SNe are indeed the main contributors of iron-group nuclei to the interstellar medium. In particular, the overproduction obtained for the neutron-rich isotopes ^{54}Fe (slightly), ^{58}Ni and ^{62}Ni is problematic. This overproduction is caused by a too large neutron excess of the material, i.e. a more constant overproduction within the isotopes of the iron-group elements can be achieved when reducing the total neutron to proton ratio (Thielemann et al. 1986; Woosley and Weaver 1986b).

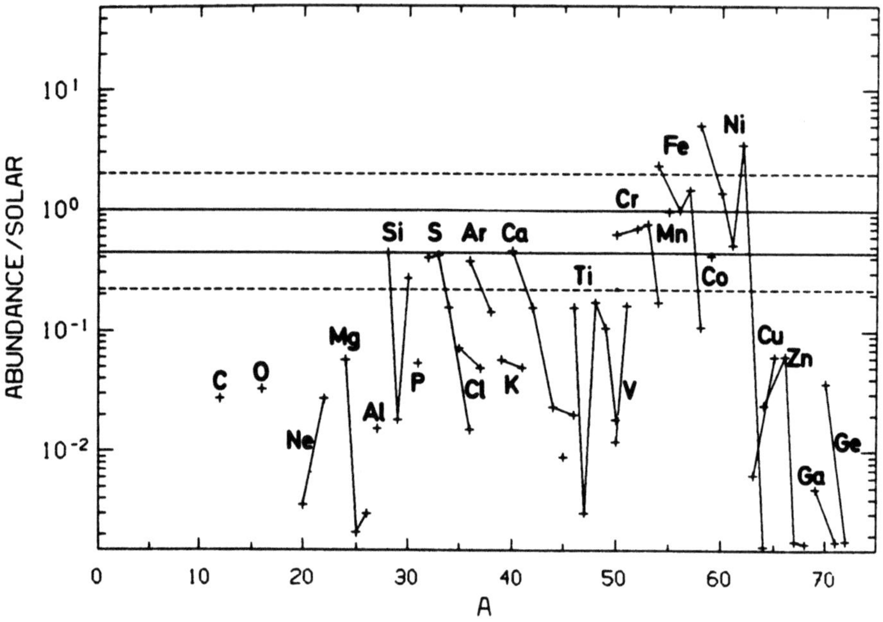

Fig. 21. Nucleosynthesis of a carbon deflagration model. The abundances of stable isotopes are shown relative to their solar values. The ratio is normalized to ^{56}Fe. Note the large overabundances of ^{58,62}Ni and ^{54}Fe (from Thielemann et al 1986).

Several possibilities connected with the initial white dwarf model (e.g. variations in the ignition density, metallicity, C/O ratio) have been suggested by Thielemann et al. (1986) to avoid an overproduction of neutron-rich species. However, the most likely solution may be connected with the speed of the deflagration front. If the deflagration speed is small initially, the fuel ahead of the front can expand. If one further assumes that the flame then accelerates most of the iron will be synthesized at lower density, which implies less electron captures, and hopefully cures the neutronization problem (see e.g. Woosley and Weaver 1986b; Woosley 1990).

Acknowledgement

This work has benefitted from helpful comments and valuable suggestions by my colleagues W. Hillebrandt, P. Höflich and R. Mönchmeyer.

References

Arnett, W.D.: 1966, *Canad. J. Phys.* **44**, 2553

Arnett, W.D.: 1967, *Canad. J. Phys.* **45**, 1621

Arnett, W.D.: 1968, *Astrophys. J.* **153**, 341

Arnett, W.D.: 1969, *Astrophys. Spac. Sci.* **5**, 180

Arnett, W.D.: 1977, *Astrophys. J.* **218**, 215

Arnett, W.D.: 1980, *Astrophys. J. Lett.* **230**, L37

Arnett, W.D.: 1981, unpublished

Arnett, W.D.: 1982, *Astrophys. J.* **253**, 785

Arnett, W.D.: 1985, unpublished

Arnett, W.D.: 1986, *Astrophys. Spac. Sci.* **5**, 180

Arnett, W.D.: 1987, *Astrophys. J.* **319**, 136

Arnett, W.D.: 1988, *Astrophys. J.* **331**, 377

Arnett, W.D.: 1990, *Astrophys. J.* , in press

Arnett, W.D., Bahcall, J.N., Kirshner, R.P., Woosley, S.E.: 1989a, *Ann. Rev. Astronom. Astrophys.* **27**, 629

Arnett, W.D., Fryxell, B.A., Müller, E.: 1989b, *Astrophys. J. Lett.* **341**, L63

Arnett, W.D., Fryxell, B.A., Müller, E.: 1990 in Proc. of the Santa Cruz Workshop on *Supernovae*, ed. S.E. Woosley, Springer-Verlag, Berlin, in press

Arnett, W.D., Fu, A.: 1989, *Astrophys. J.* **340**, 396

Axelrod, T.S.: 1980a, Ph.D. Thesis, University of California, Santa Cruz

Axelrod, T.S.: 1980b, in *Type I Supernovae*, ed. J.C. Wheeler, University of Texas, Austin, p. 80

Axelrod, T.S.: 1988, in *IAU Coll. 108 on Atmospheric Diagnostics of Stellar Evolution: Chemical Peculiarity, Mass Loss, and Explosion*, ed. K. Nomoto, Springer-Verlag, Berlin, p.375

Baade, W. Zwicky, F.: 1934a, *Proc. Natl. Acad. Sci. USA* **20**, 259

Baade, W. Zwicky, F.: 1934b, *Proc. Natl. Acad. Sci. USA* **20**, 254

Baade, W. Zwicky, F.: 1934c, *Phys. Rev.* **45**, 138

Baldo, M., Ferreira, L.S., Lombardo, U.: 1989, in Research Rep. Phys. *Nuclear Astrophysics*, eds. M. Lozano, M.I. Gallardo and J.M. Arias, Springer-Verlag, Berlin, p.279

Bandiera, R.: 1984, *Astron. Astrophys.* **139**, 368

Barbon, R., Ciatti, F., Rosino, L.: 1973, *Astron. Astrophys.* **25**, 241

Barbon, R., Ciatti, F., Rosino, L.: 1979, *Astron. Astrophys.* **72**, 287

Barbon, R., Cappellaro, E., Turatto, M.: 1984, *Astron. Astrophys.* **135**, 27

Baron, E., Cooperstein, J.: 1990, *Astrophys. J.* , in press

Baron, E., Cooperstein, J., Kahana, S.: 1985a, *Phys. Rev. Lett.* **155**, 126

Baron, E., Cooperstein, J., Kahana, S.: 1985b, *Nucl. Phys.* **A440**, 744

Baron, E., Cooperstein, J., Kahana, S.: 1987, *Astrophys. J.* **320**, 300

Baron, E., Myra, E.S., Cooperstein, J., van den Horn, L.J.: 1989, *Astrophys. J.* **339**, 978

Barranco, M., Buchler, J.R.: 1980, *Phys. Rev.* **C22**, 1729

Barranco, M., Garcias, F., Pi, M., Suraud, E.: 1989, n Research Rep. Phys. *Nuclear Astrophysics*, eds. M. Lozano, M.I. Gallardo and J.M. Arias, Springer-Verlag, Berlin, p.279

Begelman, M.C., Sarazin, C.L.: 1986, *Astrophys. J. Lett.* **302**, L59

Benz, W., Thielemann, F.-K.: 1990, *Astrophys. J. Lett.* **348**, L17

Bethe, H.A., Brown, G.E., Applegate, J.H., Lattimer, J.M.: 1979, *Nucl. Phys.* **A324**, 487

Bethe, H.A., Brown, G.E., Cooperstein, J.: 1987, *Astrophys. J.* **322**, 201

Bethe, H.A., Wilson, J.R.: 1985, *Astrophys. J.* **295**, 14

Bionta, R.M. et al.: 1987, *Phys. Rev. Lett.* **58**, 1494

Blaizot, J.P.: 1980, *Phys. Rep.* **64**, 171

Blinnikov, S.I., Khokhlov, A.M.: 1986a, *Inst. Theor. Exp. Phys.*, Report No. 29

Blinnikov, S.I., Khokhlov, A.M.: 1986b, *Inst. Theor. Exp. Phys.*, Report No. 190
Bodenheimer, P., Woosley, S.E.: 1983, *Astrophys. J.* **269**, 381
Bonche, P., Vautherin, D.: 1981, *Nucl. Phys.* **A372**, 496
Bonche, P., Vautherin, D.: 1982, *Astron. Astrophys.* **12**, 268
Bouchet, P., Danziger, I.J., Lucy, L.B.: 1989, *IAU circular* 4933
Bowers, R.B., Wilson, J.R.: 1982, *Astrophys. J.* **263**, 366
Branch, D.: 1982, *Astrophys. J.* **258**, 35
Branch, D.: 1986, *Astrophys. J. Lett.* **300**, L51
Branch, D., Dogget, J.B., Nomoto, K., Thielemann, F.-K.: 1985, *Astrophys. J.* **294**, 619
Branch, D., Falk, S.W., McCall, M.L., Rybski, P., Uomoto, A.K., Wills, B.J.: 1981, *Astrophys. J.* **244**, 780
Branch, D., Nomoto, K.: 1986, *Astron. Astrophys.* **164**, L13
Brown, G.E., Bethe, H.A., Baym, G. 1982, *Nucl. Phys.* **A375**, 481
Brown, G.E., Osnes, E.: 1985, *Phys. Lett.* **B154**, 223
Bruenn, S.W.: 1972, *Astrophys. J. Suppl.* **24**, 283
Bruenn, S.W.: 1985, *Astrophys. J. Suppl.* **58**, 771
Bruenn, S.W.: 1986, *Astrophys. J. Suppl.* **62**, 331
Bruenn, S.W.: 1989a, *Astrophys. J.* **340**, 955
Bruenn, S.W.: 1989b, *Astrophys. J.* **341**, 385
Buchler, J.-R., Wheeler, J.C., Barkat, Z.: 1972, *Astrophys. J.* **172**, 713
Burrows, A.: 1987, *Astrophys. J. Lett.* **318**, L57
Canal, R., Isern, J., Labay, J., Lopez, R.: 1990, in Proc. of the Santa Cruz Workshop on *Supernovae*, ed. S.E. Woosley, Springer-Verlag, Berlin, in press
Cappellaro, E., Turatto, M.: 1988, *Astron. Astrophys.* **190**, 10
Chandrasekhar, S.: 1939, *An Introduction to the Study of Stellar Structure*, Dover Publ., New York
Chandrasekhar, S.: 1961, *Hydrodynamic and Hydromagnetic Stability*, Clarendon Press, Oxford
Chevalier, R.A.: 1976, *Astrophys. J.* **207**, 872
Chevalier, R.: 1977, in *Supernovae*, ed. D.N. Schramm, Reidel, Dordrecht, p.53
Chevalier, R.: 1984, in *Ann. N.Y. Acad. Sci.* **422**, 215
Clark, D.H., Stephenson, F.R.: 1977, *The Historical Supernovae*, Pergamon, Oxford
Colella, P., Woodward, P.R.: 1984, *J. Comp. Phys.* **54**, 174
Colgate, S.A., Johnson, H.J.: 1960, *Phys. Rev. Lett.* **5**, 235
Colgate, S.A., White, R.H.: 1966, *Astrophys. J.* **143**, 626
Cooperstein, J.: 1988, *Phys. Rep.* **163**, 95
Cooperstein, J., Baron, E.: 1990, in *Supernovae*, ed. A. Petschek, Springer-Verlag, Berlin
Cooperstein, J., Bethe, H.A., Brown, G.E.: 1984, *Nucl. Phys.* **A429**, 527
Cooperstein, J., van den Horn, L.J., Baron, E.: 1986, *Astrophys. J.* **309**, 653
Cooperstein, J., van den Horn, L.J., Baron, E.: 1987, *Astrophys. J. Lett.* **321**, L119
Courant, R., Friedrich, K. O.: 1948, *Supersonic Flow and Shock Waves*, Interscience, New York
Denisov, A.A., Koldoba, A.V., Poveshenko, Yu.A., Popov, Yu.P., Chechetkin, V.M.: 1986, preprint IPM AN SSSR-99
Dgani, R., Livio, M.: 1989, preprint TECHNION-PH-89-22, *Astrophys. J.* , in press
Doering, W.: 1943, *Ann. Phys.* **43**, 421
Doggett, J.B., Branch, D.: 1985, *Astron. J.* **90**, 2303
El Eid, M.F., Hillebrandt, W.: 1980, *Astron. Astrophys. Suppl.* **42**, 215

Elias, J.H., Mathews, K., Neugebauer, G., Persson, S.E.: 1985, *Astrophys. J.* **296**,37
Ensman, L.M., Woosley, S.E.: 1988, *Astrophys. J.* **333**, 754
Epstein, R.I.: 1979, *Mon. Not. Roy. Astronom. Soc.* **188**, 305
Eriguchi, Y., Müller, E.: 1984, *Astron. Astrophys.* **147**, 161
Falk, S.W., Arnett, W.D.: 1973, *Astrophys. J. Suppl.* **33**, 515
Fesen, R.A., Hamilton, A.J.S., Saken, J.M.: 1989, *Astrophys. J. Lett.* **341**, L55
Fickett, W., Davis, W.C.: 1979, *Detonation*, Univ. California Press, Berkeley
Filippenko, A.V.: 1988, *Astron. J.* **96**, 1941
Filippenko, A.V., Sargent, W.L.W.: 1985, *Nature* **316**, 407
Filippenko, A.V., Sargent, W.L.W.: 1986, *Astrophys. J.* **91**, 691
Flowers, E., Itoh, N.: 1976, *Astrophys. J.* **206**, 218
Fransson, C.: 1988, in *IAU Coll. 108 on Atmospheric Diagnostics of Stellar Evolution: Chemical Peculiarity, Mass Loss, and Explosion*, ed. K. Nomoto, Springer-Verlag, Berlin, p.383
Freedman, D.Z.: 1974, *Phys. Rev.* **D9**, 1389
Fryxell, B.A., Müller, E., Arnett, W.D.: 1990, in *Numerical Methods in Astrophysics*, ed. P.R. Woodward, Academic Press, New York, in press
Fryxell, B.A., Woosley, S.E.: 1982, *Astrophys. J.* **261**, 332
Fuller, G.M: 1982, *Astrophys. J.* **252**, 741
Fuller, G.M, Fowler, W.A., Newman, M.J.: 1982, *Astrophys. J.* **252**, 715
Gilmozzi, R., Cassatella, A., Clavel, J., Fransson, C., Gonzalez, R., Gray, C., Panagia, N., Talavera, A., Wamsteker, W.: 1987, *Nature* **328**, 318
Gaskell, C.M., Cappellaro, E., Dinerstein, H.L., Garnett, D., Harkness, R.P., Wheeler, J.C.: 1986, *Astrophys. J. Lett.* **306**, L77
Glendenning, N.K.: 1990, in *NATO ASI on The Nuclear Equation of State*, eds. W. Greiner and H. Stöcker, Plenum, New York, in press
Goldreich, P., Weber, S.V.: 1980, *Astrophys. J.* **238**, 991
Harkness, R.P., Wheeler, J.C., Margon, B., Downes, R.A., Kirshner, R.P., Uomoto, A., Barker, E.S., Cochran, A.L., Dinerstein, H.L., Garnett, D.R., Levreault, R.M.: 1987, *Astrophys. J.* **317**, 355
Hillebrandt, W.: 1985, in *Proc. 3rd Workshop on Nuclear Astrophysics*, eds. W. Hillebrandt, E. Müller and F.-K. Thielemann, MPA-report 199, Garching, p.59
Hillebrandt, W.: 1987, in *NATO ASI on High Energy Phenomena Around Collapsed Stars*, ed. F. Pacini, Reidel, Dordrecht, p.73
Hillebrandt, W.: 1990, in *Int. School of Physics Enrico Fermi on High-Presssure Equation of State: Theory and Applications*, ed. S. Eliezer and R.A. Ricci, North Holland, Amsterdam, in press
Hillebrandt, W., Höflich, P.: 1989, *Rep. Progress Phys.* **52**, 1421
Hillebrandt, W., Höflich, P., Truran, J.W., Weiss, A.: 1987, *Nature* **327**, 597
Hillebrandt, W., Müller, E.: 1981, *Astron. Astrophys.* **103**, 147
Hillebrandt, W., Müller, E.: 1984, in *NATO ARW on Problems of Collapse and Numerical Relativity*, eds. D. Bancel and M. Signore, Reidel, Dordrecht, p.183
Hillebrandt, W., Müller, E., Mönchmeyer, R.: 1990, in *NATO ASI on The Nuclear Equation of State*, eds. W. Greiner and H. Stöcker, Plenum, New York, in press
Hillebrandt, W., Nomoto, K., Wolff, R.G.: 1984, *Astron. Astrophys.* **133**, 175
Hillebrandt, W., Wolff, R.G.: 1985, in *Nucleosynthesis: Challenges and New Developments*, eds. W.D. Arnett and J.W. Truran, Univ. Chicago Press, p.131

Hirata, K., et al.: 1987, *Phys. Rev. Lett.* **58**, 1490
Höflich, P.: 1987, in *Proc. 4th Workshop on Nuclear Astrophysics, Lect. Notes Phys.* **287**, eds. W. Hillebrandt, R. Kuhfuß, E. Müller and J.W. Truran, Springer-Verlag, Berlin, p.307
Höflich, P.: 1988a, *Proc. Astron. Soc. Austr.* **7**, 434
Höflich, P.: 1988b, in *IAU Coll. 108 on Atmospheric Diagnostics of Stellar Evolution: Chemical Peculiarities, Mass Loss, and Explosion*, ed. K. Nomoto, Springer-Verlag, Berlin, p.388
Hoyle, F., Fowler, W.A.: 1960, *Astrophys. J.* **132**, 565
Iben, I.Jr.: 1986, in *Cosmogonical Processes*, eds. W.D. Arnett, C.J. Hansen, J.W. Truran, S. Tsuruta, VNU Science Press, Utrecht, p.155
Iben, I.Jr., Nomoto, K., Tornambe, A., Tutukov, A.V.: 1987, *Astrophys. J.* **317**, 717
Iben, I.Jr., Tutukov, A.V.: 1984, *Astrophys. J. Suppl.* **54**, 335
Iben, I.Jr., Tutukov, A.V.: 1985, *Astrophys. J. Suppl.* **58**, 661
Imshennik, V.S., Khokhlov, A.M.: 1984, *Sov. Astron. Lett.* **10**, 262
Isern, J., Labay, J., Hernanz, M., Canal, R.: 1983, *Astrophys. J.* **273**, 320
Ivanova, L.N., Imshennik, V.S., Chechotkin, V.M.: 1974, *Astrophys. Spac. Sci.* **31**, 497
Janka, H.-T., Hillebrandt, W.: 1989a, *Astron. Astrophys.* **224**, 49
Janka, H.-T., Hillebrandt, W.: 1989b, *Astron. Astrophys. Suppl.* **78**, 375
Janka, H.-T., Mönchmeyer, R.: 1989a, *Astron. Astrophys.* **209**, L5
Janka, H.-T., Mönchmeyer, R.: 1989b, *Astron. Astrophys.* **226**, 69
Khokhlov, A.M.: 1989a, *Astrophysics and Space Physics Reviews* **8**, 1
Khokhlov, A.M.: 1989b, *Mon. Not. Roy. Astronom. Soc.* **239**, 785
Lamb, D.Q., Lattimer, J.M., Pethick, C.J., Ravenhall, D.G.: 1978, *Phys. Rev. Lett.* **41**, 1623
Lamb, D.Q., Lattimer, J.M., Pethick, C.J., Ravenhall, D.G.: 1981, *Nucl. Phys.* **A360**, 459
Landau, L.D., Lifshitz, E.M.: 1959, *Fluid Mechanics.*, Pergamon, Oxford, Vol. 6 of Course in Theoretical Physics
Lassault, M., Flocard, H., Bonche, P., Heenen, P.H., Suraud, E.: 1987, *Astron. Astrophys.* **183**, L3
Lattimer, J.M., Burrows, A., Yahil, A.: 1985b, *Astrophys. J.* **288**, 644
Lattimer, J.M., Pethick, C.J., Ravenhall, D.G., Lamb, D.Q.: 1985a, *Nucl. Phys.* **A432**, 646
LeBlanc, J.M., Wilson, J.R.: 1970, *Astrophys. J.* **161**, 541
Ledoux, P.: 1945, *Astrophys. J.* **102**, 143
Lee, J.H.S., Moen, I.O.: 1980, *Progress in Energy and Combustion Sciences* **6**, 359
Mahaffy, J.H., Hansen, C.J.: 1975, *Astrophys. J.* **201**, 695
Marcos, S., Barranco, M., Buchler, J.R.: 1982, *Nucl. Phys.* **A381**, 507
Matthews, G.J., Cowan, J.J.: 1988, in *Heavy Ion Physics and Nuclear Astrophysical Problems*, eds. S. Kubono, M. Ishihara and T. Nomura, World Scientific, Singapore, p.143
Mayle, R.W.: 1985, *PH.D. thesis*, LLL-report, UCRL-53713
Mayle, R., Wilson, J.R.: 1988, *Astrophys. J.* **334**, 909
Maza,J., van den Bergh, S.: 1976, *Astrophys. J.* **204**, 519
Mazurek, T.J.: 1974, *Nature* **252**, 287
Mazurek, T.J.: 1975, *Astrophys. Spac. Sci.* **35**, 117

Mazurek, T.J.: 1982, in *Supernovae: A Survey of Current Research*, eds. M.J. Rees and R.J. Stoneham, Reidel, Dordrecht, p.371
Mazurek, T.J., Meier, D.L., Wheeler, J.C.: 1977, *Astrophys. J.* **213**, 518
Mazurek, T.J., Truran, J.W., Cameron, A.G.W.: 1974, *Astrophys. Spac. Sci.* **27**, 261
Minkowski, R.: 1941, *Publ. Astron. Soc. Pac.* **53**, 224
Minkowski, R.: 1964, *Ann. Rev. Astronom. Astrophys.* **2**, 247
Miyaji, S., Nomoto, K.: 1987, *Astrophys. J.* **318**, 307
Miyaji, S., Nomoto, K., Yokoi, K., Sugimoto, D.: 1980, *Publ. Astron. Soc. Japan* **32**, 303
Mönchmeyer, R.: 1989, in *Proc. 5th Workshop on Nuclear Astrophysics*, eds. W. Hillebrandt and E. Müller, MPA-P1, Garching, p.92
Mönchmeyer, R.: 1990, *Ph.D. thesis*, Tech. Univ. München, unpublished
Mönchmeyer, R., Müller, E.: 1989, in *NATO ASI on Timing Neutron Stars*, eds. H. Ögelman, and E. van den Heuvel, Kluwer, Dordrecht, p.549
Mönchmeyer, R., Schäfer, G., Müller, E., Kates, R.: 1990, *Astron. Astrophys.* , in press
Müller, E.: 1982, *Astron. Astrophys.* **114**, 53
Müller, E.: 1984, in *NATO ARW on Problems of Collapse and Numerical Relativity*, eds. D. Bancel and M. Signore, Reidel, Dordrecht, p.271
Müller, E., Arnett, W.D.: 1982, *Astrophys. J. Lett.* **261**, L109
Müller, E., Arnett, W.D.: 1986, *Astrophys. J.* **307**, 619
Müller, E., Fryxell, B.A., Arnett, W.D.: 1990, in *Proc. Elba Workshop on Chemical and Dynamical Evolution of Galaxies*, eds. F. Ferrini, F. Matteucci and J. Franco, Casa Editrice Giardini, Pisa, in press
Müller, E., Hillebrandt, W.: 1981, *Astron. Astrophys.* **103**, 358
Müller, E., Hillebrandt, W., Orio, M., Höflich, P. Mönchmeyer, R., Fryxell, B.A.: 1989, *Astron. Astrophys.* **220**, 167
Müller, E., Różyczka, M., Hillebrandt, M.: 1980, *Astron. Astrophys.* **81**, 288
Myra, E.S., Bludman, S.A.: 1989, *Astrophys. J.* **340**, 384
Myra, E.S., Bludman, S.A., Hoffman, Y., Lichtenstadt, I., Sack, N., van Riper, K.A.: 1987, *Astrophys. J.* **318**, 744
Nagasawa, M., Nakamura, T., Miyama, S.: 1989, *Publ. Astron. Soc. Japan* **40**, 691
Nomoto, K.: 1982, *Astrophys. J.* **253**, 798
Nomoto, K.: 1986, in *Accretion Processes in Astrophysics*, eds. J. Audouze and J. Tran Thanh Van, Edition Frontieres, Gif sur Yvette, p.137
Nomoto, K., Hashimoto, M.: 1987, *Astrophys. Spac. Sci.* **131**, 395
Nomoto, K., Hashimoto, M.: 1988, *Phys. Rep.* **163**, 13
Nomoto, K., Shigeyama, T.: 1987, in *Proc. 4th Georg Mason Astrophysics Workshop: SN1987A in the LMC*, eds. M. Kafatos and A.G. Michalitsianos, Cambridge Univ. Press, Cambridge, p.273
Nomoto, K., Shigeyama, T., Hashimoto, M.: 1987, in *ESO Workshop on SN1987A*, ed. I.J. Danziger, ESO, Garching, p.325
Nomoto, K., Sugimoto, D.: 1977, *Publ. Astron. Soc. Japan* **29**, 165
Nomoto, K., Sugimoto, D., Neo, S.: 1976, *Astrophys. Spac. Sci.* **39**, 37
Nomoto, K., Thielemann, F.-K., Yokoi, K.: 1984, *Astrophys. J.* **286**, 644
Nozakura, T., Ikeuchi, S., Fujimoto, M.Y.: 1984, *Astrophys. J.* **286**, 221
Ogasawara, R., Sato, K.: 1983, *Prog. Theor. Phys.* **68**, 222
Panagia, N.: 1985, in *Supernovae as Distance Indicators, Lec. Notes Phys.*, Vol. 224, Springer-Verlag, Berlin, p.14

Panagia, N., Sramek, R.A., Weiler, K.W.: 1986, *Astrophys. J. Lett.* **300**, L 55
Porter, A., Filippenko, A.: 1987, *Astron. J.* **93**, 1372
Pskovskii, Yu.P.: 1977, *Sov. Astron.* **21**, 675
Pskovskii, Yu.P.: 1978, *Sov. Astron.* **22**, 200
Ravenhall, D.G., Pethick, C.J., Wilson, J.R.: 1983, *Phys. Rev. Lett.* **50**, 2066
Rousseau, J., Martin, N., Prevout, L., Reheirot, E., Robin, A., Brunet, J.P.: 1978, *Astron. Astrophys. Suppl.* **31**, 243
Sato, K.: 1975a, *Prog. Theor. Phys.* **53**, 595
Sato, K.: 1975b, *Prog. Theor. Phys.* **54**, 1325
Schaeffer, R., Casse, M., Cahen, S.: 1987, *Astrophys. J. Lett.* **316**, L31
Schramm, D.N., Arnett, W.D.: 1975, *Phys. Rev. Lett.* **34**, 113
Shapiro, S.L., Lightman, A.P.: 1976, *Astrophys. J.* **207**, 263
Sharma, M.M., Borghols, W.T.A., Brandenburg, S., Crona, S., van der Woude, A., Harakeh, M.N.: 1988, *Phys. Rev.* **C38**, 2562
Sharma, M.M., Stocker, W., Gleissl, P., Brack, M.: 1989, *Nucl. Phys.* **A504**, 33
Shigeyama, T., Nomoto, K., Hashimoto, M.: 1988, *Astron. Astrophys.* **196**, 141
Sramek, R.A., Panagia, N. Weiler, K.W.: 1984, *Astrophys. J. Lett.* **285**, L59
Suraud, E.: 1985, *Astron. Astrophys.* **143**, 844
Suraud, E., Vautherin, D.: 1984, *Phys. Lett.* **B138**, 325
Sutherland, P.G., Wheeler, J.C.: 1984, *Astrophys. J.* **280**, 282
Symbalisty, E.: 1984, *Astrophys. J.* **285**, 729
Takahara, M., Sato, K.: 1988, *Astrophys. J.* **335**, 301
Tammann, G.: 1982, in *NATO ASI on Supernovae: A Survey of Current Research*, eds. M.J. Rees and R.J. Stoneham, Reidel, Dordrecht, p.371
Tassoul, J.-L.: 1978, *Theory of Rotating Stars*, Princeton Univ. Press, New Jersey
Taylor, J.H.: 1987, in *Proc. 13th Texas Symposium on Relativistic Astrophysics*, ed. P.M. Ulmer, World Scientific, Singapore, p.467
Thielemann, F.-K.: 1989, in Research Rep. Phys. *Nuclear Astrophysics*, eds. M. Lozano, M.I. Gallardo and J.M. Arias, Springer-Verlag, Berlin, p.106
Thielemann, F.-K., Arnett, W.D.: 1985, *Astrophys. J.* **295**, 604
Thielemann, F.-K., Hashimoto, M., Nomoto, K.: 1990, *Astrophys. J.* **349**, 222
Thielemann, F.-K., Nomoto, K., Yokoi, K.: 1986, *Astron. Astrophys.* **158**, 17
Tohline, J.E.: 1984, *Astrophys. J.* **285**, 721
Tohline, J.E., Schombert, J.M., Boss A.P.: 1980, *Space Sci. Rev.* **27**, 555
Trimble, V.: 1982, *Rev. Mod. Phys.* **54**, 1183
Trimble, V.: 1983, *Rev. Mod. Phys.* **55**, 511
Truran, J.W.: 1985, *Ann. Rev. Nucl. Part. Sci.* **34**, 53
Tubbs, D.L., Schramm, D.N.: 1975, *Astrophys. J.* **201**, 467
Unno, W.: 1967, *Publ. Astron. Soc. Japan* **19**, 140
Uomoto, A.: 1986, *Astrophys. J. Lett.* **310**, L35
Uomoto, A., Kirshner, R.P.: 1985, *Astron. Astrophys.* **149**, L7
Utrobin, V.: 1984, *Astrophys. Spac. Sci.* **98**, 115
Utrobin, V.: 1988, *Inst. Theor. Exp. Phys.*, Report No. 177
van den Bergh, S.: 1988, *Astrophys. J.* **327**, 156
van den Bergh, S., McClure, R.D.: 1989, *Astrophys. J. Lett.* **347**, L29
van den Bergh, S., McClure, R.D., Evans, R.: 1987, *Astrophys. J.* **323**, 44
van den Horn, L.J., Cooperstein, J.: 1986, *Astrophys. J.* **300**, 142

van Riper, K.A.: 1978, *Astrophys. J.* **221**, 304
van Riper, K.A.: 1979, *Astrophys. J.* **232**, 558
van Riper, K.A.: 1988, *Astrophys. J.* **326**, 235
van Riper, K.A., Lattimer, J.M.: 1982, *Astrophys. J.* **249**, 270
Vinas, X., Barranco, M., Treiner, J., Stringari, S.: 1987, *Astron. Astrophys.* **182**, L34
von Neumann, J.: 1942, in *John von Neumann, collected works*, ed. A.J. Taub, Vol.6, p.203
Walborn, N.R., Lasker, B.M., Laidler, V.G., Chu, Y.-H.: 1987, *Astrophys. J. Lett.* **321**, L41
Weiler, K.W., Sramek, R.A.: 1988, *Ann. Rev. Astronom. Astrophys.* **26**, 295
Weiler, K.W., Sramek, R.A., Panagia, N., van der Hulst, J.M., Salvati, M.: 1986, *Astrophys. J.* **301**, 790
Wheeler, J.C.: 1990, in Proc. of the Santa Cruz Workshop on *Supernovae*, ed. S.E. Woosley, Springer-Verlag, Berlin, in press
Wheeler, J.C., Levreault, R.M.: 1985, *Astrophys. J. Lett.* **294**, L17
Wheeler, J.C., Harkness, R.P.: 1986, in *NATO ARW on Galaxy Distances and Deviations from Universal Expansion*, eds. B.F. Madore and R.B. Tully, Reidel, Dordrecht, p.45
Wheeler, J.C., Harkness, R.P., Cappellaro, E.: 1987, in *Proc. 13th Texas Symposium on Relativistic Astrophysics*, ed. M.P. Ulmer, World Scientific Publ., Singapore, p.402
Wheeler, J.C., Harkness, R.P.: 1990, *Rep. Progress Phys.* , in press
Williams, F.A.: 1985, *Combustion Theory*, Benjamin/Cummings, Menlo Park, California
Wilson, J.R.: 1971, *Astrophys. J.* **163**, 209
Wilson, J.R.: 1974, *Phys. Rev. Lett.* **32**, 849
Wilson, J.R.: 1985, in *Numerical Astrophysics*, eds. J.M. Centrella J.M. LeBlanc and R.L. Bowers, Jones and Bartlett, Boston, p.422
Wilson, J.R.: 1990, in *NATO ASI on The Nuclear Equation of State*, eds. W. Greiner and H. Stöcker, Plenum, New York, in press
Wilson, J.R., Mayle, R.: 1988, *Phys. Rep.* **163**, 63
Wilson, J.R., Mayle, R., Woosley, S.E., Weaver, T.A.: 1986, in *Proc. of 12th Texas Symp. on Relativ. Astrophys., Ann. N.Y. Acad. Sci.* **479**, 267
Woosley, S.E.: 1986, in *Nucleosynthesis and Chemical Evolution*, 16th Advanced Course, Swiss Society of Astrophysics and Astronomy, eds. B. Hauck, A. Maeder and G. Meynet, Geneva Observatory, Geneva, p. 1
Woosley, S.E.: 1988a, *Astrophys. J.* **330**, 218
Woosley, S.E.: 1988b, in *Proc. 4th Georg Mason Astrophysics Workshop: SN1987A in the LMC*, eds. M. Kafatos and A.G. Michalitsianos, Cambridge Univ. Press, Cambridge, p.289
Woosley, S.E.: 1988c, in *IAU Coll. 108 on Atmospheric Diagnostics of Stellar Evolution: Chemical Peculiarities, Mass Loss, and Explosion*, ed. K. Nomoto, Springer-Verlag, Berlin, p.361
Woosley, S.E.: 1990, in *Supernovae*, ed. A.G. Petschek, Springer-Verlag, Berlin, in press
Woosley, S.E., Arnett, W.D., Clayton, D.D.: 1973, *Astrophys. J. Suppl.* **26**, 231
Woosley, S.E., Axelrod, T.S., Weaver, T.A.: 1984, in *Stellar Nucleosynthesis*, eds. C. Chiosi and A. Renzini, Reidel, Dordrecht, p.263
Woosley, S.E., Pinto, P.A., Weaver, T.A.: 1988, *Proc. Astron. Soc. Austr.* **7**, 355
Woosley, S.E., Taam, R.E., Weaver, T.A.: 1986, *Astrophys. J.* **301**, 601
Woosley, S.E., Weaver, T.A.: 1986a, *Ann. Rev. Astronom. Astrophys.* **24**, 205

Woosley, S.E., Weaver, T.A.: 1986b, in *Radiation Hydrodynamics in Stars and Compact Objects.*, eds. D. Mihalas and K.H. Winkler, IAU Colloqium No.89, *Lecture Notes in Physics* **255**, Springer, Berlin, p.91

Woosley, S.E., Weaver, T.A.: 1988, *Phys. Rep.* **163**, 79

Yahil, A.: 1983, *Astrophys. J.* **265**, 1047

Yahil, A., Lattimer, J.M.: 1982, in *Supernovae: A Survey of Current Research* eds. M.J. Rees and R.J. Stoneham, Reidel, Dordrecht, p.53

Zeldovich, Ya.B.: 1940, *Zh. Eksp. Teor. Fiz.* **10**, 542, (English translation: NACA TM 1261 (1960))

Zeldovich, Ya.B.: 1980, *Arch. Combust.* **1**, 1

Zeldovich, Ya.B., Barenblatt, G.I., Librovich, V.B., Makhviladze, G.M.: 1985, *The Mathematical Theory of Combustion and Explosions*, Plenum, New York, p. 487f

Zwicky, F.: 1938, *Astrophys. J.* **88**, 529

Zwicky, F.: 1940, *Rev. Mod. Phys.* **12**, 71

Zwicky, F.: 1942, *Astrophys. J.* **96**, 8

Zwicky, F.: 1965, in *Stellar Structure*, ed. L.H. Aller and D.B. McLaughlin, Univ. Chicago Press, Chicago, p. 367

Zwicky, F., Berger, J., Gates, H.S., Rudnicki, K.: 1963, *Publ. Astron. Soc. Pac.* **75**, 236

Circumstellar Envelopes of Late Type Stars

Alain Omont

Observatoire de Grenoble, Université Joseph Fourier et Institut d'Astrophysique de Paris, CNRS, 98bis, Boulevard Arago, F-75014 Paris

1 General Features of Circumstellar Envelopes (CSE)

1.1 Introduction

Planetary nebulae and red giants are conspicuous objects in the sky which have been known for centuries. As early as 1605 Mira Ceti was discovered. It is now known to be a typical representative of the very late and brief stages that red giants live through before becoming a planetary nebula. However, it was only in 1935 (20) that the discovery of blue shifted spectral lines in red giants indicated the presence of circumstellar matter, which was definitely confirmed in 1956 by such absorption on the line of sight to the companion of α Her (19). The same year 1956, Schlovskii (18) gave convincing arguments that the progenitors of planetary nebulae should be the red giants. Indeed, because the matter of CSEs is cold and mostly molecular, it is hardly directly detectable by optical astronomy. Detailed studies of CSEs had thus to wait for the advent of infrared and millimeter astronomy, and even the discovery of most of them relied on infrared surveys. The year 1970 $\pm$ 1 marked several crucial steps in the understanding of CSEs : a clear pattern of the late stages of evolution of red giants then emerged with the helium flash at the end of the red giant branch, followed by the asymptotic giant branch (AGB), terminated by double shell burning with thermal pulses. Simultaneously, the 2μm Sky Survey (TMSS, IRC) (21) revealed a new class of CSEs, emitting practically all their energy in the infrared. Infrared astronomy thus confirmed the presence of large amounts of dust and its formation in the cooling ejected gas. Infrared spectroscopy proved the presence of silicates (22), while molecular spectroscopy rapidly developed with Fourier transform spectroscopy of CO (24), discovery of OH masers (25) and the advent of millimeter radioastronomy (23).

All these studies have been vigorously developed in the last 20 years with the discovery of several tens of, often quite exotic, circumstellar molecules, the understanding of the basic features of rotational molecular excitation and of millimetre line formation, as well as maser pumping and circumstellar chemistry. However, the most important progress was certainly brought about by the IRAS mission in 1983, which detected probably more than 10^5 CSEs, and the follow-up studies still quite active. In the meantime the theory of stellar evolution had a parallel development, with in particular the emergence of the notion of "superwind" and of its importance for the termination of the red giant stage and the transition to planetary nebulae.

This course is aimed at giving a general view of these newly discovered astronomical objects. This first chapter presents their broad features, the related classification, the properties of the stellar wind which determines the structure and the physical conditions of CSEs, as well as the problems of stellar evolution involved.

Chapters 2 and 3 describe in some detail the observational data, mainly from millimeter radioastronomy and infrared observations. Millimeter lines give rich information on molecular abundances, and their study provides the input for modelling circumstellar chemistry, as well as some information on element abundance and nucleosynthesis through the measurement of isotope ratios. Infrared radiation is emitted by dust; accordingly, IR photometry and broad band spectroscopy, such as provided by IRAS, inform us about the amount of dust, its spatial distribution, its composition, and thus on the complex processes of its formation.

This body of data is the basis of our knowledge of the properties of the stellar wind discussed in Chapter 4. It is often called a superwind because of the very large values of the mass-loss rates. The physics of the origin of the superwind is still debated and it remains a major problem of the field.

Chapter 5 comes back to our understanding of the evolution of these objects and the numerous open questions. A central point is the termination of this evolution and the transition to the planetary nebula and white dwarf stages which are mainly determined by the intensity of the mass-loss.

General references can be found in (1–16).

1.2 Internal Structure and Evolution

1.2.1 General Features of Red Giants

The effective surface temperature of red giants spans a relatively broad range from ~ 2500 K to ~ 4500 K. It has a general tendency to decrease with time, so that most of the CSEs which correspond to the very late stages of evolution, are situated to the lower end of this range.

There is an extended convective zone in most of the star. However, the energy flow in fact controlled by the opacity near the surface, which is dominated by H^-. Evolved red giants harbor a generally degenerate central core either composed of helium, or of carbon and oxygen on the AGB where the core indeed approaches the properties of a white dwarf. In nearly all cases the thermonuclear energy is generated in a thin shell, and the luminosities of red giants span quite a large range, although the luminosities of most CSEs of AGB red giants are restricted to a much narrower range of about 10^4 $L_\odot$.

We summarize below the main features of the evolution and the internal structure of stars during the last stages of red giants (see e.g. 1–8 and Figure 1.1). We essentially restrict the discussion to the evolution of isolated stars, but it should not be forgotten that the evolution of close binaries, with mass-transfer, can be quite different, while being relatively frequent (Section 5.6).

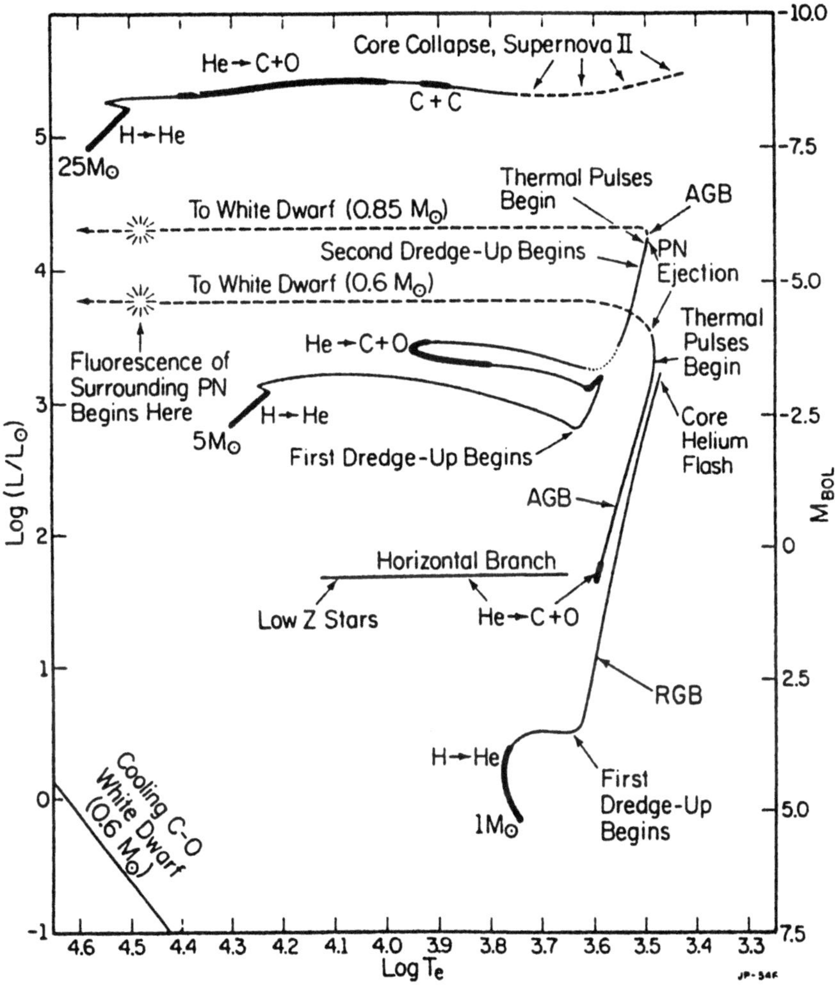

Figure 1.1 – Representative evolutionary tracks from I.Iben in ref . 4.

1.2.2 First Ascent of the Red Giant Branch (RGB)

After the exhaustion of the hydrogen in its central regions, the star moves to the red giant region, with a central helium core. The energy is then provided by hydrogen burning in a thin shell. The temperature of the helium core progressively increases with its mass and luminosity, until the temperature is large enough ($\sim 8\ 10^7$ K) to ignite the helium flash, which converts ^{4}He into ^{12}C by the triple α process.

Typical characteristics of the core at the helium flash are the following (Figure 1.1) : mass $M_c \sim 0.5\ M_\odot$, luminosity $\sim$ 2–3 $10^3\ L_\odot$, delay after leaving the main sequence $\sim 10^9$yr for an initial mass $M_i = 1\ M_\odot$, $\sim 10^7$yr for $M_i \sim 2.25\ M_\odot$.

The first dredge up occurs at the beginning of the ascent of the RGB. Complex convective motions bring to the outer layers of the star material from the central layers, leading in particular to a strong enrichment in ^{14}N formed from ^{12}C during the main sequence.

After the helium flash, the expansion of the core removes the electron degeneracy, leading to quiet helium burning with a very strong drop of the luminosity for low mass stars ($L \sim 50\ L_\odot$ for $M_i \sim 1\ M_\odot$).

On the other hand, for $M_i \gtrsim 2.25\ M_\odot$, helium burning begins before the electrons become degenerate and proceeds without a strong helium flash. The luminosity remains nearly constant, but the star return to conditions when it is bluer than red giants, describing an horizontal branch with different kinds of instabilities (Cepheids, etc..).

1.2.3 Second Ascent: The Asymptotic Giant Branch (AGB)

We will not discuss the case of massive stars ($M_i \gtrsim 8\ M_\odot$), where the core temperature rises enough to ignite carbon burning and subsequent processes, with the formation of a supergiant and the final core collapse and explosion of a type II supernova.

For the other stars, helium combustion products a carbon and oxygen degenerate core, which in the final stage is quite similar to a white dwarf with $T \sim 10^8$ K and $M_c \sim 0.5\ M_\odot$ for $M_i \sim 2\ M_\odot$, and $M_c \sim 1.1\ M_\odot$ for $M_i \sim 8\ M_\odot$. While this core contracts, its temperature increases and the luminosity strongly rises up to 0.5–5 $10^4\ L_\odot$ (ascent of the AGB). While the early AGB phase is characterized by a quiet He burning shell, the region in the last phases of the AGB is a double shell burning with alternate burning of hydrogen and helium. The short helium burning phases reproduce periodically and begin by helium flashes, denoted "thermal pulses".

1.2.4 Thermal Pulses on the AGB

At the end of the phase of helium burning, which occurs because the helium shell becomes exhausted and is partially mixed with hydrogen by convection, hydrogen is reignited in a thin shell. Then H burning proceeds for most of the time, with a luminosity which was shown by Pascinski to depend only on the mass of the CO core M_{CO} (16)

$$L_H = 5.9\ 10^4 L_\odot\,(M_{CO}/M_\odot - 0.5) \tag{1.1}$$

Periodically, as the helium shell grows and the temperature rises, the triple alpha process is reignited giving a very short helium flash. Its luminosity is very large ($L_{He} \sim 10^7\ L_\odot$), but the energy generated is not immediately conveyed to the surface. It induces an expansion and a cooling of the core, which stops the hydrogen burning. A phase of quiescent helium burning then follows, which is much shorter than that of hydrogen burning. And finally hydrogen burning is reignited when helium burning stops, and a new cycle starts. Table 1.1 gives typical values of the duration of the hydrogen burning phase τ_H (which is not very different from the period of thermal pulses) and of the mass of hydrogen burnt during each cycle.

The effects of the different phases of the thermal pulses cycle on the observational properties of the star, and in particular on its mass–loss rate, are still debated (see Chapter 4). Indeed, the models (see e.g. 4) show that immediately after the helium flash of each thermal pulse both the radius and the luminosity of the star decrease. Convection is always active in the outer layers of the star. After each thermal pulse, it can mix the inner layers and bring down the products of hydrogen burning, while it dredges up to the surface both the products of H and He burning.

Table 1.1. Typical parameters between thermal pulses: M_{CO} is the mass of the CO core ; L_H, τ_H and ΔM_H are the luminosity, duration and mass accreted to the core, respectively, relative to a H–burning phase.

$M_{CO}/M_\odot$	0.6	0.8	1.0
$L_H/L_\odot$	$6\ 10^3$	$1.8\ 10^4$	$3\ 10^4$
τ_H (yr)	$2\ 10^5$	10^4	10^3
$\Delta M_H/M_\odot$	10^{-2}	10^{-3}	$2\ 10^{-3}$

1.2.5 Synthesis of s-Elements

Red giants are believed to be important contributors to the synthesis of s–elements (by successive "slow" capture of neutrons by the Fe group elements). Overabundances of s-elements are currently observed in some classes of red giants (7). A possible source of neutrons is the reaction ^{22}Ne + $\alpha \rightarrow$ ^{25}Mg + n, which takes place during the helium flash of massive stars. Another possibility which has been recently proposed to be quite important (7, 27), is the reaction ^{13}C + $\alpha \rightarrow$ ^{16}O + n. It should proceed during thermal pulses in all stars, when convection brings ^{13}C, produced in the incomplete CNO cycles of hydrogen burning, down to the bottom layers of the He shell, which are hot enough to induce this reaction. Detailed models of the synthesis of s-elements by the slow capture of these neutrons reproduce quite well their solar abundances (27). It is thus quite possible that AGB red giants are the major site of synthesis of the s–elements.

1.2.6 Termination of the AGB

The very last stages of the AGB are quite important to determine and understand the final products of stellar evolution; planetary nebulae and white dwarfs, and also the precise dividing point that determines whether a star becomes a supernova or a white dwarf. Accordingly the termination of the AGB and the transition to the planetary nebula stage will be discussed in detail in Chapter 5. Let us sketch here its main feature to prepare the discussion of the observational properties of transition objects (pre–planetary nebulae) in Chapters 2 and 3, and that of the importance of the properties of mass–loss in Chapter 4.

Mass–loss dominates the evolution of the late stages of the AGB, since the mass–loss rates of the "superwinds" can reach $10^{-5} - 10^{-4}$ $M_{\odot}$/yr and are then much larger than the rate of accretion into the CO core ($10^{-7}-10^{-6}$ $M_{\odot}$/yr, see Table 1.1). Accordingly, it is the superwind which exhausts the matter in the star outside its central core (in the "stellar envelope", not to be confused with the "circumstellar envelope", CSE). Then, the star will become a planetary nebula, at least in most cases. Mass–loss almost stops by exhaustion of the gas, and the previously ejected CSE moves away. The central star is then basically a white dwarf with a thin H/He "stellar envelope". Its effective temperature rapidly increases and its radiation eventually becomes hot enough to ionize the CSE, giving rise to a planetary nebula (PN).

Objects in the transition stages between the AGB and PN are obviously particularly interesting for understanding this brief and crucial phase of evolution. They are usually denoted as "post–AGB", or "pre–planetary nebulae" or "proto–planetary nebulae" (PPNe). They are characterized by a central star generally of intermediate (F or G) spectral type, with a semi–detached CSE. They should normally rapidly evolved into planetary nebulae, although other evolutionary schemes (return to the AGB, or direct evolution to a white dwarf) are not excluded.

1.3 CSE Structure and Physical Parameters

1.3.1 Overview and Scheme

A general view of the structure and of the physical parameters of CSEs is sketched in Figure 1.2. The radius of the photosphere of the AGB red giant is typically a few AU. Indeed, it varies following the pulsations of the star with periods of a few hundred days. A complex region lies above the photosphere, where pulsations periodically raise masses of gas which are believed to partly return to the photosphere, and to partly be ejected to give mass–loss; the gas thus levitated cools, allowing dust formation which is pushed away by radiation pressure, and carries away the gas by friction.

Then, the ejected gas and dust move away practically in free expansion with a velocity in the range 10–20 km/s. The temperature and density progressively decrease in these intermediate layers formed of stable molecules and dust. The conditions approach those of interstellar molecular clouds, and finally the molecules are photodissociated by the interstellar UV radiation. Before the gas is completely reduced to atomic form, photodissociation generates unstable radicals and ions which can further react to produce complex unstable molecules.

1.3.2 Internal Layers

As already mentioned, the structure and the conditions are very complex and uncertain in the layers above the photosphere until the wind is fully accelerated at a distance of $\sim 10^{+15}$ cm. We shall discuss in Chapter 4 the gas motions in these regions, which are related to pulsations and wind acceleration, probably with shocks. Grain formation which takes place in the range 1500–500 K, will be discussed in Chapter 3. The temperature law depends on different heating and cooling processes (see e.g. 29), and remains quite uncertain. The diagnostics of these regions rely mainly on infrared observations of molecular lines and of the continuum emission and scattering by dust, and in addition on maser emission. They are still difficult because of the lack of spatial resolution of infrared observations, and of the complexity of radiative transfer effects, dust formation and maser excitation. It should be noted that there is some evidence that this inner envelope could often be asymmetric (e.g. IRC+10216), which could be related to the presence of companions as identified in o Ceti (Mira).

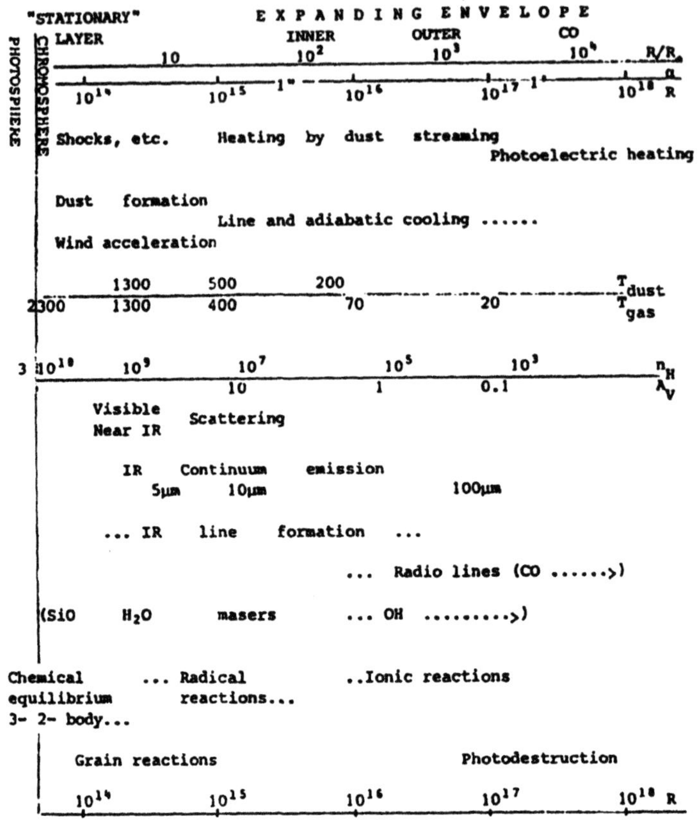

Figure 1.2 – Structure of a massive envelope (IRC+10216). Assumed parameters : d = 200 pc, $R_* = 1000\,R_\odot = 7\,10^{13}$ cm, $\dot{M} = 4 10^{-5}\,M_\odot/yr$. In the outer envelope : $n_H = 5\,10^{37}\,r^{-2}\,cm^{-3}$, $A_V = 2.5\,10^{-22}\,N_H$. The range of maser emission is indicated for similar O–rich envelopes, it is obviously irrelevant for C–rich shells such as IRC+10216. From ref. 28.

1.3.3 Density

When the mass–loss is isotropic, the density n at radius r is simply related to the local value $\dot{M}(r)$ of the mass–loss rate :

$$\dot{M} = n\,\mu\,4\pi\,r^2\,V_e \tag{1.2}$$

where μ is the mean molecular weight and V_e the expansion velocity. Thus, if the mass–loss rate is constant, the density obeys a simple r^{-2} law in the outer envelope where V_e has reached its constant asymptotic value.

Numerically the hydrogen molecule density is given by

$$n_{H_2}(cm^{-3}) \approx 2\,10^5\,\dot{M}_{H-5}\,r_{16}^{-2}\,V_{e15}^{-1} \qquad (1.3)$$

where $\dot{M}_{H-5}$, r_{16} and V_{e15} are the hydrogen mass–loss rate, the radius and the expansion velocity in units of 10^{-5} $M_{\odot}$/yr, 10^{16} cm and 15 km/s respectively. As the mass–loss rate lies in the range

$$\dot{M} \sim 10^{-7} - 10^{-4} M_{\odot}/yr \qquad (1.4)$$

one sees that the density is comparable to that of interstellar molecular clouds, $10^3 - 10^5$ cm^{-3}, for $r \sim 10^{16} - 10^{17}$ cm.

It is obvious that this simple r^{-2} density law is not valid in the inner envelope and in the region of dust acceleration, there being many cases known where the mass–loss is clearly asymmetric and/or has varied with time (see Sections 1.4.4 and 1.4.5).

1.3.4 Opacity

The opacity of circumstellar envelopes is important for the absorption and the reddening of the visible photospheric spectrum, and for the penetration of the interstellar UV which photodissociates the molecules in the outer layers. As little is actually known about the visible and UV absorption properties of circumstellar dust, one generally assumes that the ratio of the visible extinction A_V to the hydrogen column density $N(H_2)$ is the same as in the interstellar medium

$$A_V \sim N(H_2)\,/\,10^{21} cm^{-2}$$

Then in the region of constant expansion velocity and mass–loss rate, the visible extinction from a radius r to infinity is

$$A_V \simeq \frac{\dot{M}_{H-5}}{r_{16}\;V_{e15}}$$

1.3.5 Temperature

The temperature is more difficult to estimate in CSEs, even in the outer envelope with well defined velocity and density laws. There are few and limited observational diagnostics of the temperature. Because except for CO in massive envelopes, the rotational excitation is generally due to infrared absorption, the rotation temperatures (defined from the ratios of the different rotational populations) are not directly related to the kinetic temperature T_K. However, in massive envelopes, measurements of the rotational temperature of CO from infrared spectroscopy can provide a reasonable estimate of T_K in the relatively inner layers where the infrared CO lines are formed.

In the more outer layers, important for millimeter observations, one has generally to rely on theoretical modelling of the kinetic temperature (30,

31). In the regions shielded from interstellar UV radiation, the main heating process has been identified as friction due to grain streaming through the gas under the action of radiation pressure. Comparable amounts of cooling are produced by the adiabatic expansion and by the CO rotational lines. The temperature law is determined by the equation (see 31, 32)

$$\frac{dT}{dr} = -\frac{4}{3}\frac{T}{r} + \frac{8\pi r^2}{3k\dot{N}}\left\{\Gamma(r) - \Lambda(r)\right\}$$

where the first term represents adiabatic cooling, and $\Gamma(r)$ and $\Lambda(r)$ correspond to the heating and line cooling, respectively. $\dot{N}$ is the mass-loss rate expressed in the number of H_2 molecules per second. If collisions with IR excited H_2O molecules in inner layers, and the grain photoelectric effect can be important heating contributions, the dominant heating process is the friction due to the streaming of grains pushed by radiation pressure, with a drift velocity with respect to the gas of

$$v_d = (Q\, L\, V_e\, /\, \dot{M}c)^{1/2} \tag{1.5}$$

where $Q = \sigma_{abs}/\sigma_{geom}$ is the grain absorption efficiency (typically ~ 0.02 at $\lambda \sim 5\mu$m).

Then

$$\Gamma(r) = \left\{n_{grain}(r)\, \sigma_{geom} v_d\right\} \times \frac{1}{2}\, \rho(r)\, v_d^2 \;\propto\; v_d^3 \tag{1.6}$$

For a CSE with $\dot{M} \sim 10^{-5}\ M_\odot/yr$ such as IRC+10216, $v_d \sim 1$ km/s, and one has the typical temperature law

$$T \sim 100K\, (r/10^{16} cm)^{-0.7} \tag{1.7}$$

However, such a law is valid only for mass-loss rates in the range of $10^{-5}M_\odot$/yr. It has been shown (32) that for $\dot{M} \sim 10^{-7}M_\odot$/yr with $v_d \sim 10$ km/s, T_K can be an order of magnitude larger at $r \sim 10^{16}$cm. On the other hand, T_K should be much lower than given by Eq (1.7) for $\dot{M} \sim 10^{-4}M_\odot$/yr ($v_d \sim 0.3$ km/s), such as in extremely cold OH/IR stars.

1.4 Overview of Classification and of Wind Properties

1.4.1 Overview of Visible and Infrared Classification

a) Visible Spectra. (see e.g. 12) It is well known that the visible spectra of red giants are completely different depending on the value of the ratio C/O of carbon to oxygen abundances. Visible AGB stars thus belong to one of the following classes:

- M-Stars. They are oxygen rich with C/O < 1. Their visible spectrum is dominated by the molecular bands of TiO and of other oxides. They also display absorption in the infrared bands of CO, OH, SiO, H_2O, etc.

- Carbon stars. C/O > 1. They are characterized by optical bands of CN, C_2, C_3, SiC_2, CH, etc, and by infrared bands of CO, CN, C_2, HCN, C_2H_2, etc. The hotter ones are called of R-type, and the cooler ones of N-type. They contain several peculiar classes: J-stars which are ^{13}C-rich; H deficient stars such as R CrB stars, etc; etc.
- S-Stars, where C/O $\sim$ 1. Their spectra are characterized by bands of hydrides and of ZrO.

As already mentioned, post–AGB objects have generally an intermediate spectral type (F or G). Very often these spectra present the characteristics of supergiants, although most of these stars are quite different from regular supergiants (Chapter 5).

b) Infrared Objects. When the circumstellar envelope is massive and optically quite thick, the star is very faint in the optical and its spectral classification can be difficult from visual spectra. Its infrared properties are then more characteristic of the exact nature of the CSE. Both infrared colour–colour diagrams and spectra (such as the IRAS Low Resolution Spectra) are used to characterize with precision the effective temperature and the composition of CSEs (Chapter 3). More vaguely, one can distinguish among the infrared stars: IRC objects, which were discovered by the Two Micron Sky Survey (21), and similarly AFGL sources, which are generally cooler, revealed by the AFGL survey (33); oxygen rich OH/IR stars, displaying a strong OH maser emission, which can be quite cold; cold carbon stars, which are IRC objects or cooler; etc.

1.4.2 Overview of Pulsations of AGB Stars

Most of and possibly all AGB stars with CSE are variable. The pulsations are more or less regular. Their effects show up not only in the optical luminosity, colours and spectrum, but also in the infrared emission and at least in some radio lines such as OH masers. One generally distinguish the following classes:

- Miras. Their pulsations are regular with long periods $\sim 60 - 500$ days and large amplitudes (magnitude variation $\Delta m > 2$). Emission lines appear around maximum yielding Me spectral types. The mode of pulsation, fundamental or first overtone, is still debated (34). The majority of Mira stars are oxygen rich. However, there is a whole class of carbon rich Miras.
- Semi–Regular Variables. They display irregular light curves with smaller amplitudes ($\Delta m < 2$). SRb stars are more irregular than SRa ones. One finds a larger proportion of C–rich than O–rich semi–regular variables.
- Irregular Variables. Their variations appear quite random. AGB irregulars are denoted Lb.

- Infrared stars. Their period is larger than for Mira stars. It increases with the thickness of the envelope, up to 2000 days for extremely cold OH/IR stars (35). They probably pulsate in the fundamental mode.
- Post-AGB stars. It has been shown that pulsations have stopped in most post-AGB stars (35).

1.4.3 Expansion Velocities

The expansion velocity V_e is easily measured from the total extension, $\Delta V \sim 2\ V_e$, of the line profiles. Most expansion velocities lie in the range 10–25 km/s. However, V_e is in the range 5–10 km/s for the warmest Miras; and it exceeds 30 km/s for the few "OH-supergiants" known and a few carbon stars (36). There is a trend of V_e to increase for small galactic latitudes (36), which should mean large initial masses.

Because the dust grains are pushed by radiation pressure through the gas, their velocity probably exceeds that of the gas by 1–10 km/s (31, 32). This large velocity difference, which has not yet been directly observed, has some important consequences, in particular on the gas heating (Section 1.3.5).

Some very high ejection velocities have been observed in several post-AGB objects. They appear as "wings" in the profiles of radio lines, with velocities much larger than V_e, the expansion velocity of most of the envelopes. These objects have probably a bipolar structure (Section 1.4.5). The largest velocities observed approach 100–200 km/s, and are then comparable to the wind velocities of planetary nebulae.

1.4.4 Overview of Mass-Loss Rates

As already mentioned, mass-loss rates $\dot{M}$ will be discussed in detail in Chapter 4. Their values range from $10^{-7} M_\odot$/yr for Miras and bright carbon stars to $10^{-4} M_\odot$/yr for extremely cold OH/IR stars and OH supergiants. Such large values cannot last very long without exhausting the mass of the "stellar envelope" which has a mass $\lesssim 1\ M_\odot$. The evolution of $\dot{M}$ with time and its dependence on the star's parameters (initial mass, composition, etc..) are rather uncertain. There is a probable trend of $\dot{M}$ to increase with time, with possible interruptions or sharp decreases examplified in a number of stars presenting semi-detached shells, and/or excess of cold dust emitting in the 60μm IRAS band. Such interruptions have been proposed to be related to thermal pulses and to accompany O-rich to C-rich transitions. The semi-detached shells of post-AGB objects clearly point to the interruption, and probably the termination of strong mass-loss.

1.4.5 Bipolar Structure

As mentioned most AGB CSEs have an approximate spherical symmetry, although internal asymmetry can be present. On the other hand, many post-AGB objects display characteristic bipolar structures. Bipolar nebulae are reflection nebulae which scatter the light from the central star which easily propagates on the axis of the nebulae where the density is reduced. Some of them have quite spectacular shapes, such as the Egg Nebula (AFGL 2688), the Red Rectangle (HD 44179), the Calabash Nebula (OH 231.8 + 4.2) and Frosty Leo (IRAS 09371 + 1212). Such nebulae are suggestive of a general disk-flow structure. The central disk, which generally masks the star, is probably most often a thick toroid expanding at V_e. However, it could also be a thinner keplerian disk (37). The most likely explanation for such asymmetric disk structures is some effect of binarity on mass-loss (see e.g. 37). The origin of the high velocity wind observed (with clear evidence of shocks in a few objects) is very probably related to this disk structure, with possible similarities to other comparable structures such as for young stellar objects. However, a detailed modelling of such complex objects is still to be worked out.

1.5 Influence on Chemical Galactic Evolution

From a direct estimate of their total mass-loss in the solar vicinity, AGB stars are believed to be the main contribution to the return of matter from stars to the interstellar mediumm, with a total rate $\sim 3 - 6\ 10^{-4}$ $M_{\odot}/yr/kpc^2$ in the solar neighbourhood (38, 40). Most of this matter will then appear in planetary nebulae, but this gas has been ejected almost totaly during the AGB phase.

The gas returned, together with unprocessed material (H, O, heavy elements), is strongly enriched in products of nucleosynthesis. It is thus possible that most of the following elements of the interstellar medium could have been synthetized in red giants : ^{13}C and ^{14}N from the CNO incomplete cycle; ^{12}C from the triple α process; s-elements from neutrons of ^{13}C in hot bottom burning (Section 1.2.5).

Both O- and C-rich massive envelopes give an important contribution to the gas return, which is probably dominated by CSEs with $\dot{M} \sim 10^{-5}\ M_{\odot}/yr$ (40). It should be noted that such cold CSEs, which are quite evolved and close to the end of the AGB, have abundances of most of the processed elements completely different from that of visible red giants. Direct studies of abundances in massive CSEs is thus essential to understand and model the chemical evolution of galaxies.

2 Radio Observations and Molecular Abundances

2.1 Overview of Molecular Abundances

A general view of the molecules presently detected in CSEs is presented in Table 2.1. Their number is about twice smaller than the larger set of molecules detected in interstellar clouds. Generally speaking, CSEs are much less massive than prominent millimeter sources in interstellar molecular clouds, and the detectability of a given species requires as a rule that its abundance be several orders of magnitude larger in CSEs than its detection limit in interstellar clouds. This conclusion is examplified in 0–rich CSE (Section 2.1.2) where less than 20% of the known interstellar molecules have been detected. However, the presence of a single source, IRC+10216 (CW Leo), tempers this conclusion quite a lot; the overabundance of carbon makes the chemistry much richer, and this massive CSE ($\dot{M} \sim 10^{-5} M_{\odot}/yr$), is exceptionally near (100–150 pc). IRC+10216 is thus one of the most important sources of millimeter radioastronomy, where many new cosmic molecules have been discovered.

Table 2.1 Molecules detected at radio wavelengths in circumstellar envelopes (adapted from ref. 14)

Molecule	Approximate Number of Sources	
	C-rich	O-rich
CO	150	150
CN	15	
CS	15	3
SiO ("Thermal")	4	30
SiS	7	1
SiC_2	7	
SO		7
SO_2		15
H_2S	1	15
OCS		1
HCO^+	3	2
HCN	80	13
CH_3CN	3	
HNC	5	2
NH_3	2	1
HC_3N	6	
HC_5N	4	
HC_7N	2	
HC_9N	1	
$HC_{11}N$	1	
C_2H	7	
C_3H	2	
C_3H_2	3	

C_4H	4	
C_5H	1	
C_6H	1	
C_3N	5	
C_2S	1	
C_3S	1	
$HSiC_2$ (HSC_2) (?)	1	
NaCl	1	
AlCl	1	
KCl	1	
AlF	1	
SiC	2	
C_4Si	1	
CP	1	
OH (maser)		1500
H_2O (maser)	4	200
SiO (maser)		140

2.1.1 Molecular Abundances in IRC+10216 (CW Leo)

The molecules detected in IRC+10216 are displayed in Table 2.2 with an order of magnitude of their abundances. Because of different problems discussed in Sections 2.2 and 2.3, the accuracies of the abundances displayed are certainly not better to a factor 3, and possibly 10 in some cases.

Compared with the interstellar medium, there are very few oxygen bearing molecules (CO, SiO, possibly HCO^+) because $C/O > 1$ and most of the oxygen is locked in CO. The abundances of most of the other molecules are larger than in the interstellar medium (by about one order of magnitude). There are no deuterium compounds since deuterium has been destroyed in the star.

A few small stable molecules are most abundant and contain most the main elements (CO, C_2H_2, CS, SiS, SiO, tc). This is not surprising since a similar situation is found in the photosphere where thermodynamical chemical equilibrium is achieved, and it is believed that these stable molecules found in the outer shells have survived from the photosphere in the ejected envelope.

Some simple unstable radicals (CN, C_2H, SiC, etc) are very probably direct products of photodissociation.

There are a lot of complex carbon molecules, as expected in a C–rich environment. However, at least in the molecules detected, this complexity implies only acetylenic chains in most cases, and a few examples of small rings. Indeed, a large proportion of these unstable and exotic species were discovered in this source, even before laboratory detection.

Table 2.2. Observed abundances in IRC+10216 updated from table 1 of ref. 15, and in OH231.84.2 (210). All abundances are uncertain by factors > 3. $x(CO)/H_2 \sim 6\ 10^{-4}$.

	$x(i)/H_2$		IRC+10216				OH231.8+4.2
	C_2H_2						SO_2
10^{-5}	HCN						
							SO
					CN		NH_3, OCS
10^{-6}	NH_3				C_3		H_2S
					C_2H		HCN, CS
				HC_3N	C_4H		
	C_2H_4	SiS	CS	HC_5N	C_3N		HNC
10^{-7}	CH_4					ClAl	SiO, HCO^+
		SiC_2		HC_7N	C_6H	FAl	
	HNC	SiO					
			C_2S		C_3H_2	ClNa	
10^{-8}			C_3S	$HC_{11}N$	C_3H	ClK	
		SiH_4			C_5H		
	CH_3CN						
			H_2S				
10^{-9}							

2.1.2 Abundances in Oxygen-Rich CSEs

The chemistry of O-rich CSEs is much less rich, and in addition there is no massive O-rich CSE as close as IRC+10216. Accordingly, few molecules have been detected except for the major species: CO for carbon, H_2O and OH, together with CO for oxygen, SO, SO_2 and H_2S for sulfur, SiO for silicon, NH_3 for nitrogen (N_2 being undetectable). However, a few C-bearing molecules, besides CO, are relatively common, in particular HCN. Other unexpected species such as CS, SiS, HCO^+, prove the existence of none equilibrium processes (shocks, photochemistry, etc.). The PPN OH231.8+4.2 is particularly rich in various molecules which could result from an active shock chemistry.

2.2 Basic Features of Molecular Spectroscopy

Electron spectra (UV or visible) are not very important in CSEs, (outside the photosphere). We briefly recall below a few basic features of vibrational and rotational spectroscopy, relevant for understanding the problems of abundance determinations (Figure 2.1).

Vibrational spectra are mostly confined to the near infrared range (2-20 μm), especially for small molecules. They are composed of bands corresponding to transitions between a given pair of vibrational levels (v',v). Each band is composed of many lines connecting a pair of rotational levels (v'J'-v J), grouped in series ("branchs" P, Q, R) corresponding to transitions with J'-J = +1, 0 and -1 respectively.

Rotational lines correspond to transition betweeen rotation levels of a given vibration state (generally v = 0). The most important circumstellar lines are in the millimeter range. However, depending on the mass of the molecule and the value of J, they can spread to the centimeter range (e.g. HC_5N), or to the submillimeter range (in particular for the AH_x hydrids).

Linear molecules with closed shells (and zero electron angular momentum, such as CO, HCN, $HC_{2n+1}N$, CS, SiO, HCO^+) have a very simple structure of rotation level (Figure 2.1a) and a spectrum with harmonic frequencies

$$\nu = 2\,B\,J \qquad (2.1)$$

where J is the upper rotation level and B the rotation constant.

Others have more complex energy patterns, but the same order of magnitude for the frequencies, except for possible centimeter transitions inside doublets (OH, NH_3, H_2CO). One distinguishes symmetric tops (NH_3) with regular energy level patterns, asymmetric tops (H_2O, SO_2, etc. (Figure 2.1b)) with irregular patterns, molecules with non zero electron angular momentum which are mainly radicals with an unpaired electron giving simple harmonic transitions with doublets (CN, C_2H) or more complex patterns (OH) (Figure 2.1c).

2.3 Infrared Spectroscopy

The observation of near infrared molecular lines requires high resolution (more than 10^5 to get full information on the velocity profile), which is achieved with Fourier Transform Spectrocopy. The lines generally result from, simple absorption by cold molecules on the line of sight of the continuum source; i.e. warm dust close to the region of dust formation. However, complex effects of radiative transfer are not excluded.

Derivation of the molecular abundances from such infrared observations implies the following steps:

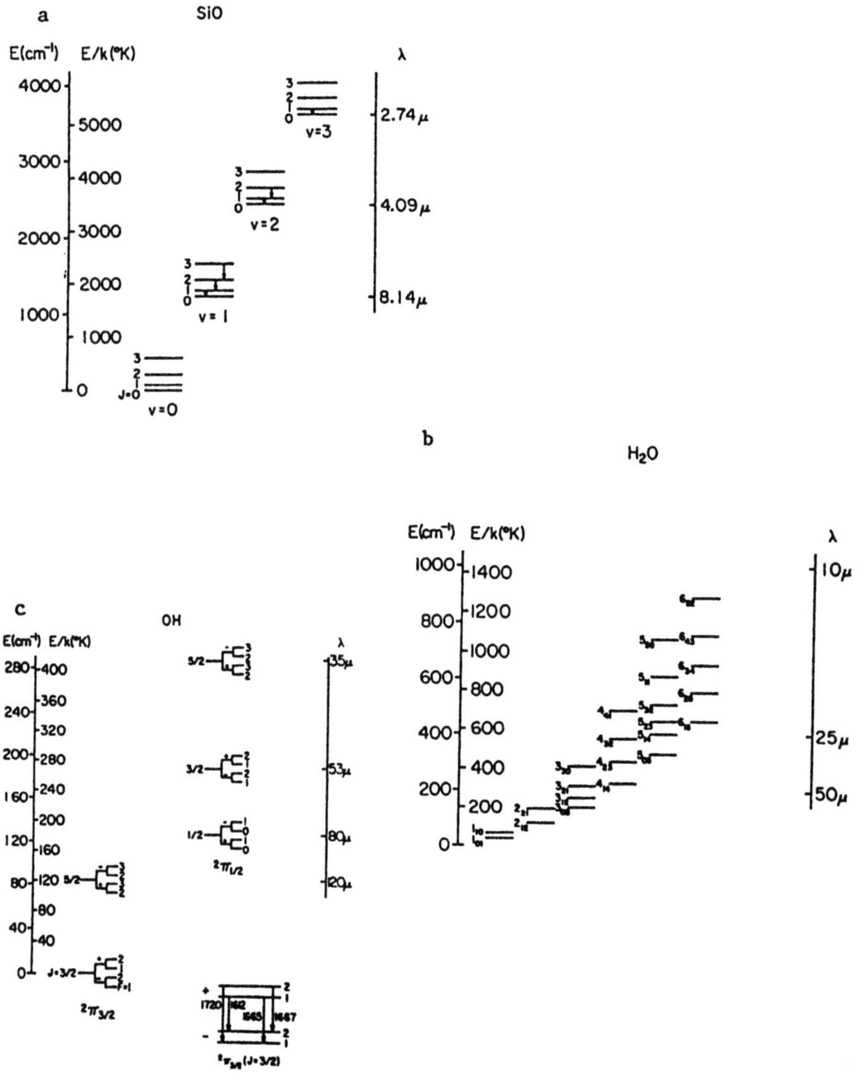

Figure 2.1- Diagrams of energy levels of rotation-vibration levels of SiO, and of rotation levels of ortho-H_2O and OH. From Elitzur in ref. 4, pp. 364–366.

- Identification of the infrared lines (from laboratory data which generally do not exist for exotic species), and determination of their equivalent width (integrated area) after correcting for possible telluric lines and blends, and with the usual problems of noise, baseline, etc.
- Determination of the column density $N_{i\ell}$ (cm^{-2}) in the absorbing level ℓ of molecule i from the equivalent width. This is very simple (if the line strength is known) for optically thin lines formed in regions well separated from that of continuum emission. But it can be relatively uncertain in other cases where complex radiative transfer effects can occur.
- Derivation of the total column density N_i of molecule i, assuming a value for the rotational temperature T_R, which can be determined from the ratios $N_{i\ell}/N_{i\ell'}$,

$$N_{i\ell} = g_\ell \, exp(-E_\ell \;/\; kT_R) \; N_i / \Sigma g'_\ell \, exp(-E'_\ell \;/\; kT_R) \tag{2.2}$$

- Then the abundance of molecule i is :

$$x_i = N_i / N_{H2} \tag{2.3}$$

where the column density of H_2, N_{H2}, is given by :

$$N_{H2} = \int n_{H2}(r)\, dr \;= C/r_o \tag{2.4}$$

For a constant mass-loss rate $\dot{M}$. The radius r_o is the inner radius of the absorption range, i.e. in practice the radius of the region of continuum emission.

It is seen that many of these steps can be uncertain. However, high resolution infrared spectroscopy provides rich information, mainly for inner layers, about the presence and often the abundances of many molecules (see Figure 2.1), the kinetic temperature from the few collisionally excited species, the expansion velocity (with possible evidence for stepwise acceleration, multishell ejection, etc), isotopic ratios ($^{13}CO/^{12}CO$), etc.

2.4 Radio Line Profiles

For spherically symmetric CSEs the basic profiles of emission in the radio lines are relatively simple (Figure 2.2). Their shapes depend on the optical depth in the line and on the relative size of the emitting envelope Φ_E and the telescope beam Φ_B. As already stated, the range of emission is $\nu_o\ [1 + (V_0 \pm V_e)/c]$ in frequency units, or $V_o \pm V_e$ in velocity units, where ν_o is the rest frequency of the line, V_o the central velocity of the star with respect to the Earth and V_e its expansion velocity.

If the line is optically thin and $\Phi_B > \Phi_E$, projection effects give an equal weight to each frequency interval and the profile is square (Figure 2.2a).

On the other hand, if $\Phi_B < \Phi_E$, the signal at $V_o \pm V_e$ is much enhanced with respect to that at V_o. The profile has then sharp spikes at both ends (Figure 2.2b). When the line is optically thick, the profile is round topped in both cases (Figure 2.2d). The optical thickness is larger along a radius than perpendicularly, the emission at $V_o \pm V_e$ then comes from outer layers with smaller excitation temperatures.

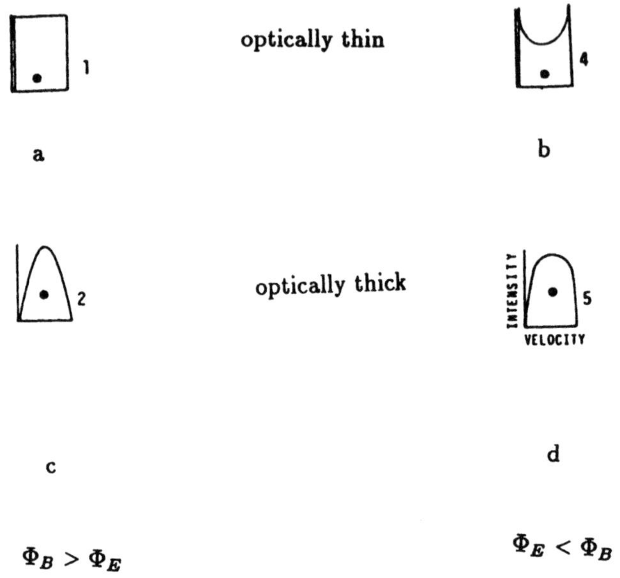

Figure 2.2 – Various typical lineshapes as a function of the optical depth and of the relative angular sizes of the source Φ_E and of the telescope beam Φ_B.

Indeed, several complications arise : self-absorption makes the intensity larger in the red wing than in the blue wing (41), (42). The emission by expanding thick disks is slightly different, and bipolar winds generate high velocity wings. Lines formed in the regions where the envelope is not fully accelerated, have a narrower profile, which is often irregular. Maser profiles have peculiar profiles because of non-linear amplification. Accidental blends of weak lines of different species are frequent because of the large number of detectable lines ; blends between hyperfine components of N-bearing molecules are also frequent. Contamination by interstellar emission of molecular clouds on the line of sight is frequent for sources near the galactic plane, especially for CO lines.

2.5 Abundance Determination from Radio (mm) Observations

For an optically thin line, the total power detected in the line (proportional to its integrated area W) is just

$$P = \mathcal{N}_{iu} \quad A_{u\ell} \quad h\nu \quad \frac{\pi R_T^2}{4\pi D^2} \tag{2.5}$$

where $\mathcal{N}_{iu}$ is the total number of molecules i in the upper level of the line in the telescope beam, $A_{u\ell}$ the Einstein coefficient, D the distance and R_T the radius of the telescope.

Knowing the rotational excitation temperature T_R one derives the total number $\mathcal{N}_i$ of molecules i in the beam :

$$\mathcal{N}_{iu} = \mathcal{N}_i g_u \; exp(-E_u / \; k \; T_R)/Z(T_R)$$

$$Z(T_R) = \Sigma \; g_i \; exp(-E_i \; / \; k \; T_R) \tag{2.6}$$

and dividing by the total number of hydrogen molecules in the beam, one obtains the abundance of molecule i (in fact the mean abundance averaged over the emitting region of radius r_i).

$$f_i = \frac{n_i}{n_{H2}} = \frac{W \; V_e \; \Phi_B^2 \; exp\,(E_u/k \; T_R) \; Z\,(T_R)}{\nu \; S_{u\ell} \; \mu_0^2 \; r_i \; \dot{M}/D^2} \tag{2.7}$$

The different requirements to derive abundances and the corresponding uncertainties are obvious from this formula :

- The integrated area W is affected by noise and the difficult calibration of the radiotelescope.
- The excitation temperature T_R is approximately determined from line ratios or from theoretical modelling. It varies with r and one has to use some average value.
- The permanent dipole momentum μ_o and to a less extent the line strength $S_{u\ell}$ may be unknown, and thus only guessed, for exotic molecules.
- The mass–loss rate $\dot{M}$ supposed to be constant, is always uncertain. However, the uncertainty is smaller for the ratio $\dot{M}/D^2$.
- The extension r_i of the emission is generally determined by photodissociation. In most of the cases it has to be somewhat guessed.
- The above formula is only valid for optically thin lines. The use of optically thick lines implies a modelling of always uncertain radiation transfer.

2.6 Goals and Results of Millimeter Observations

2.6.1 CO Observations

CO is by far the most abundant molecule with a non zero dipole momentum, except for H_2O and OH which have no millimeter transitions between low energy levels. It is also one of the most stable molecules against photodissociation. Accordingly, it is much more widely used as a general diagnostic of CSEs, than other molecules which are generally studied for themselves, their abundances, their chemistry and the peculiarities of their excitation.

CO has now been detected in more than 300 CSEs with a comparable number of O– and C–rich envelopes (43 and references therein). Its studies in CSEs present many interests: its typical circumstellar profile is a good way to identify CSEs , e.g. among IRAS sources with ambiguous IR colours. Its detection directly provides the values of the expansion velocity V_e, and also of the central velocity V_o which can give a kinematic distance. Its intensity is one of the best ways to determine the mass–loss rate $\dot{M}$ since in certain conditions $W(CO) \propto \dot{M}/D^2$ (see e.g. 38). CO mapping with a good angular resolution can enable the determination of the envelope structure and extension. It is also the best way to detect high velocity wings. Detailed observations of CO are required to test the elaborate modellings of line formation and of CO photodissociation, and more generally of the kinetic temperature in CSEs.

2.6.2 Other Molecules

Obvious goals of the observations of other molecules are :

- The identification of new molecules in CSEs, and more generally, in astrophysical media.
- The determination of molecular abundances in various CSEs (e.g. HCN, SiO, S–bearing molecules, HCO^+ in C– or O–rich CSEs), with their implications for element abundances, circumstellar chemistry and dust formation.
- Measurement of isotopic ratios (Section 2.9).
- Etc...

2.6.3 Studies of the Spatial Distribution of Molecules

Millimeter studies with a good angular resolution are particularly important to detect possible variations of $\dot{M}$ with time and to identify the important chemical processes through spatial variations of abundances. However, such effects are difficult to disentangle from the spatial variations of the excitation. One can distinguish:

- Single disk studies. The angular resolution $\Delta\alpha \sim \lambda/2R_T$ is thus practically limited to 10" (2–1 line of CO with $\lambda = 1.3$ mm and telescope diameter $2R_T = 30$ m). However, it should be possible to map many tens of CSEs with such a resolution with the advantage that this method is rapid and straightforward.

- Interferometers, with baselines ~ a few 10^2 m, allow to reach a resolution ~ 1". However, making a full map with such a resolution is long and laborious, especially for weak sources. Most of the programs are actually limited to a resolution of a few arc secondes. Important results have already be obtained in particular by Bieging and Rieu (44) who have proved that in IRC+10216 C_2H, HC_3N and HNC are distributed in a hollow shell, showing that they are formed by photochemistry in the outer layers. Many developments are expected in the coming years with the advent of large millimeter arrays.

A still better resolution can be achieved with millimeter very long baseline interferometry. However, only very strong lines are observable, and the main results to be expected from this method are the localization and the size of the regions of maser emission.

2.7 Masers

Strong circumstellar masers are known in several transitions of OH, H_2O, SiO and HCN (Figure 2.2). Maser action (Microwave Amplification by Stimulated Emission of Radiation) requires strong non LTE effects to achieve population inversion, which in practice involves some radiative process in CSEs. The large negative optical depths of strong masers are limited to abundant molecules. Exponential amplification gives rise to strong non linearities and to anisotropic emission. Small differences in optical depth, due to density or velocity inhomogeneities, are greatly enhanced, so that maser emission often appears irregular and clumpy. Circumstellar masers are often polarized, which gives some information on the magnetic field. Strong masers are also often saturated, with a power proportional to the pumping rate (e.g. 1612 MHz OH masers whose intensity follows the variable far IR flux of the OH/IR stars).

Basic pumping mechanisms of circumstellar masers are relatively well identified. However, their emission is difficult to model, because both the pumping mechanisms and the emitting medium (clumps, velocity and radiation fields, chemistry) are complex.

The most important masers in CSEs are OH masers at 1612 MHz. They are the rule with a few exceptions in massive O–rich AGB CSEs. They provide an easy identification of this type of CSEs in most of the Milky Way. The very large number of detections (> 1500 (45)) allows statistical studies of the distributions of these stars in the galactic disk and bulge. They provide much information on these CSEs (see e.g. 35): expansion and central velocity, empirical value of the mass–loss rate, structure of the emission by interferometric mapping, photodissociation of H_2O and OH, distance of the star by measuring the angular radius α_{OH} of the OH emission region and the phase lag Δt between the OH and near IR emissions due to the propagation of the IR pumping radiation along the OH radius r_{OH}

$$\Delta t = r_{OH}/c \tag{2.8}$$

$$D = r_{OH}/\alpha_{OH} = c\,\Delta t/\alpha_{OH} \tag{2.9}$$

"Main line" OH masers (at 1665 and 1667 MHz) are weaker. They are important in CSEs of Mira stars (46). The other strong masers (H_2O, SiO, HCN) imply radio transitions between levels lying at a large energy above the ground state ($E_i \sim$ 500–2000 K). Accordingly, their emission is confined to inner regions with warm gas and strong IR radiation. H_2O masers (22, 183, GHz) occur between nearby excited rotational levels of the ground vibrational state. Their interpretation is difficult. SiO masers are also common in O–rich CSEs. They bring useful insights on the structure of the complex regions just above the photosphere. HCN masers could play a comparable role in C–rich CSE (they are the only strong masers known there). But they are weaker and more difficult to detect.

2.8 Circumstellar Chemistry

(see e.g. 15, 28, 48, 49, 50, 51 for general references)

2.8.1 Comparison with and Relation to Photospheric Abundances

Because of the large temperatures and densities in photospheres and regions just above, the abundances are well determined there by thermodynamical chemical equilibrium. Their computation is then relatively easy (see e.g. 47), except for a few exotic species with uncertain energies of formation. It is thus found that a few stable abundant molecules are the main reservoirs of the different elements:

C–rich CSEs : H_2, CO, C_2H_2, N_2, HCN, CS, SiS, SiO, etc

O–rich CSEs : H_2, CO, H_2O, N_2, H_2S, SiO

The situation is quite similar in CSEs for these main reservoirs, except for grain formation which takes from the gas most silicon and probably sulfur, and a large proportion of oxygen in O–rich CSEs and carbon in C–rich CSEs.

However, it is believed that there is very little relation between the abundances of minor species in photospheres and in CSEs. The observation of C–bearing molecules other than CO (HCN, CS...)in O–rich envelopes proves the existence of non–LTE processes (shocks?) which take away the carbon from CO and synthetize other C–bearing molecules (as in the interstellar medium).

Radicals (such as OH, CN, CH) are present both in photosphere and in outer regions of CSEs. However, they are believed to be mostly destroyed in the intermediate layers, and to be regenerated by photodissociation in the outer shells (52).

Complex exotic species observed in CSEs (C_n, C_nH, C_nN, C_nS, $HC_{2n+1}N$, C_3H_2, HNC), and also SO_2 and SO in O-rich envelopes are probably formed by photochemistry in the outer layers, e.g.:

$$H_2O \xrightarrow{h\nu} OH \xrightarrow{S} SO \xrightarrow{OH} SO_2 \qquad (2.10)$$

2.8.2 General Models of Circumstellar Chemistry

(see e.g. 11, 48, 49, 50)

Chemical LTE is probably achieved in dense warm layers above the photosphere, at least approximately until T $\sim$ 1200 K. However, coupling of chemical, thermal and dynamical processes can generate instabilities, especially in the gas motions generated by pulsations (53.)

Then the abundances of stable (closed shell) species are essentially "frozen out", in the more outer layers, except for grain formation. The behaviour of atoms and radicals is probably dominated by atomic hydrogen (whose abundance is relatively uncertain), e.g.

$$H_2S + H \rightleftharpoons SH + H_2 \qquad (2.11)$$

$$SH + H \rightleftharpoons S + H_2 \qquad (2.12)$$

There are certainly shocks in these layers, but their chemical consequences are quite uncertain. Grain formation itself is far from obeying chemical LTE, and is quite complex (see Section 3.4.3) and incomplete : e.g. an important proportion of C_2H_2, H_2O, SiO, etc, is still present in the gas of the outer layers.

2.8.3 Photochemistry

Finally, photodissociation in the outer layers, penetrated by the interstellar UV radiation, destroys all the molecules. For H_2 and CO, photodissociation occurs through UV absorption in narrow spectral lines, so that relatively inner molecules are UV shielded by the outer ones, and the photodissociation radius is quite large (see e.g. 55). Other molecules are mostly destroyed by the continuum UV radiation, and their UV shielding is mainly performed by dust. Photodissociation also, directly generates most of the simple radicals (such as OH, CN, C_2H, SiC) observed in the CSEs as well as probably a few ions such as $C_2H_2^+$ and C^+.

Such abundant radicals, atoms or ions can then react with other abundant species to initiate a rich photochemistry. Examples of radical reactions are given in Eq. (2.10–12). Ionic reactions could be the initial step in the synthesis of complex carbon chains in C–rich CSEs :

$$C^+ + C_2H_2 \longrightarrow C_3H^+ + H \xrightarrow{H_2} C_3H_2^+ \xrightarrow{e^-} C_3H \tag{2.13}$$

$$C_2H_2^+ + HCN \begin{cases} \nearrow & H_2CN^+ + C_2H \xrightarrow{e^-} HCN \,,\ HNC \\ \searrow & C_3H_2N^+ + H \xrightarrow{e^-} HC_3N + H \end{cases} \tag{2.14}$$

Since most of the complex species detected in IRC+10216 seem to be distributed in hollow shells, they are formed in the outer layers and very probably by photochemistry through ionic and radical reactions of this type (11, 56, 57).

2.9 Isotope Ratios

Relative abundances of the various nuclei provide basic information about nucleosynthesis in stars. However, in CSEs the determination of the ratio of the abundances of two elements from molecular observations is not easy. Even if the molecular abundances are accurately known, the deduced element abundances generally suffer from the necessity to model the chemistry and the grain composition. Isotope ratios are relatively free from this difficulty, since most of the chemical processes are thought to be the same for the different isotopes, except for a few fractionation processes such as selective photodissociation of the isotopic varieties of CO (fractionation by chemical reactions at low temperature which is important in interstellar molecular clouds, is generally negligible in CSEs, because the temperatures are higher). Accordingly, measurement of isotope ratios is a fundamental diagnostic of nucleosynthesis and dredge up in AGB stars.

However, it is not always easy to derive isotope ratios from the measurement of the intensities of molecular lines of the different varieties of isotope. If the lines are optically thick, the abundances of the molecules are not proportional to their intensities, and their determination relies on complex and uncertain radiative transfer modelling. One must thus choose optically thin lines, but it is often impossible to find a transition where the lines of the two varieties of isotope are optically thin and detectable; when the line of the abundant isotope is optically thin, the other is very often undetectable. One has then to use indirect methods combining for instance several isotope ratios (see e.g. 58, 59).

Measurements have been performed on millimeter lines of many isotopes: ^{12,13}C, ^{14,15}N, ^{16,17,18}O, ^{28,29,30}Si, ^{32,33,34}S, ^{35,37}Cl. The most important work involves oxygen and especially carbon. For the latter, ^{12}C/^{13}C is also measurable from circumstellar infrared lines of CO and in visible stars from photospheric lines of CH and CN. Measurements in circumstellar lines are the only ones possible in massive CSEs, and hence for probing nucleosynthesis and dredge-up in the last stages of the AGB.

Millimeter lines of ^{13}CO have been observed in more than 40 envelopes. However, it is difficult to deduce precise values for ^{12}C/^{13}C because the lines of ^{12}CO are generally optically thick. Use of other molecules such as CS is preferable. For instance, Kahane et al. (59) have determined that in IRC+10216 ^{12}C/^{13}C $\sim$ 50. ^{12}C/^{13}C is large ($>$ 100) in the two young PN NGC 7027 and IRAS21282+5050. This ratio can reach $\sim$ 10 in some O-rich CSEs, and the extreme values 3-4 in a few peculiar bright carbon stars (J-Stars) (60, 32)

^{18}O and ^{17}O have comparable abundances (a few 10^{-3} that of ^{16}O). They are detected in a few massive CSEs, where it is clear that ^{17}O is enhanced with respect to the interstellar medium, but it does not seem to be correlated with ^{13}C.

3 Infrared Observations and Dust

3.1 The IRAS Point Source Catalog and Infrared Colour-Colour Diagrams

Red giants, and especially CSEs are the main sources of the different infrared sky surveys: The Two Micron Sky Survey (TMSS or IRC) (21) detected the strongest relatively warm northern sources (a few thousands). The discovery of a new class of objects, the infrared stars, has generated a lot of follow-up studies (see the general catalog of infrared observations by Gezari et al. 61).

The AFGL catalog (33) extended the survey to longer wavelengths (4 bands between 4 and 30 μm), and hence to colder sources.

The IRAS survey is devoted to the far infrared and is much more sensitive. Its Point Source Catalog (PSC, 62) gives mean values of flux densities (measured in Jansky, 1 Jy = 10^{-26} W/m^2/Hz) in 4 broad bands centered on 12, 25, 60 and 100 μm. It covers almost the whole sky, and most of the sky was observed 3 times, which enables us to obtain an idea of the variability of the sources on time scales $\sim$ 6 months. The angular resolution is $\sim$ 1 x 5 (arcmin)^2, with a position accuracy of 10-20". The PSC reports the detection of 282 000 sources in at least one band (more detections are reported in the recently released Faint Source Catalog). At least 56% of these sources are stars, and 82% of these stars show circumstellar emission. The PSC thus contains probably more than 10^5 CSEs. It is clear that, together with the LRS spectral catalog (Section 3.2), the IRAS data represent a tremendous increase of our knowledge on CSEs.

Infrared Colour-Colour Diagrams

Colour–colour diagrams are a good way to visualize the properties of the infrared sources and to distinguish between broad classes. The most useful one relates the 25/12 μm and 60/25 μm colours, i.e. PSC flux ratios (plot of $\log(F_{25}/F_{12})$) versus $\log(F_{60}/F_{25})$, (Figure 3.1). Other diagrams are also useful, involving the 100/60 μm, or combined with near IR photometry (63).

Two separate classes of red giants are clearly visible in Figure 3.1a. Regular red giants without circumstellar emission display colours quite close to those of a black body at photospheric temperatures. On the other hand, CSEs are much colder, and their colours can be quite different from a black body reflecting the peculiarities of dust emission and the distribution of temperatures in the envelope. Different classes of CSEs are located in various regions of this diagram. Van der Veen and Habing (64) have proposed a classification of these different regions (Figure 3.1b), with an analysis of the different sources found in each of them. When the mass–loss $\dot{M}$ increases, the representative point moves to colder temperatures (to the right and up), because near IR radiation is self–absorbed. Most of the O–rich CSEs are distributed on a relatively well defined "sequence" corresponding to increasing mass–loss rates (from Miras to cold OH/IR stars), which could be an evolutionary sequence (Section 5.1). C–rich CSEs have larger values of F_{60}/F_{25} than most O–rich ones. CSEs in regions VI and V correspond to spectacular time variations of $\dot{M}$: in region VI warm dust close to the star (or photospheric emission), emitting at 12 μm, coexists with cold dust, emitting at 60 μm, ejected in the past at a time where $\dot{M}$ was strong. Sources of region V display a maximum at 25 μm, with very little emission at 12 μm showing a nearly complete absence of dust close to the star. Such semi–detached envelopes are believed to be characteristic of pre–planetary nebulae (Section 5.3). Except for the close mixing with planetary nebulae in region V, it is remarkable how well CSEs are distinguishable from other classes of sources on this diagram. In particular, galaxies and young stellar objects are mostly located in region VIII (with a very few in the upper parts of regions V and VI, see also (65, 66)).

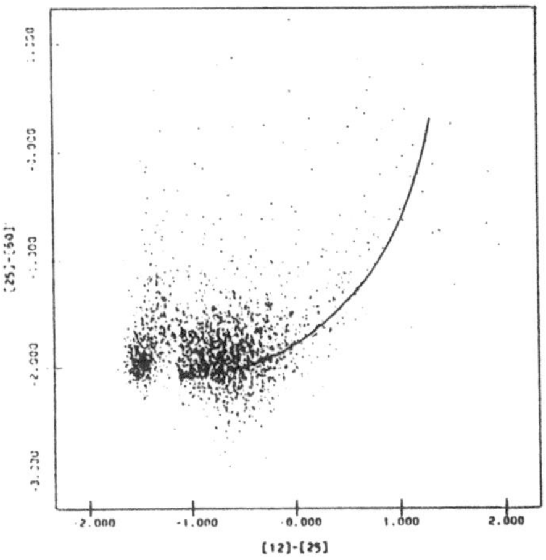

Figure 3.1a – IRAS 2–colour diagram. All IRAS point sources with a LRS spectrum: the abscissa is 2.5 $^{10}\log(F_{25}/F_{12})$ and the ordinate is 2.5 $^{10}\log(F_{60}/F_{25})$. F_{12} is the IRAS point source flux density at 12 μm. (Van der Veen and Habing) (64).

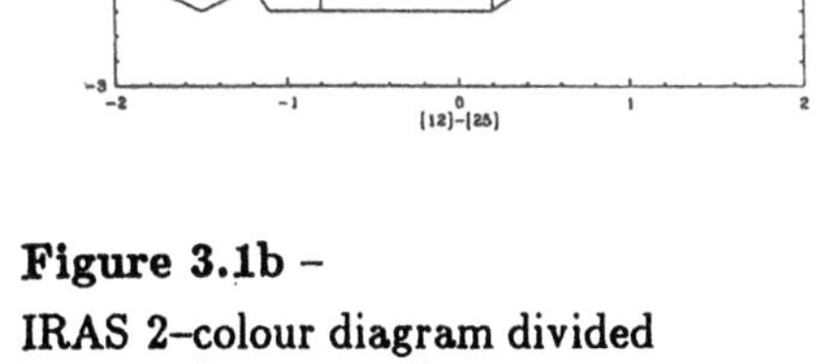

Figure 3.1b –
IRAS 2–colour diagram divided in 10 regions according to Van der Veen and Habing (64).

Star counts and averaged variability for different regions in the IRAS two-colour diagram as indicated

Region	#	<VAR>	#(LRS)	LRS class	characteristic type of object
I	791	12.7	76%	17-19 (89%)	oxygen-rich non-variable stars without circumstellar shells
II	869	35.0	65%	14-16 (64%) 21-29 (23%)	variable stars with "young" O-rich circumstellar shells
IIIa	2657	55.1	55%	21-29 (76%) 14-16 (14%)	variable stars with more evolved O-rich circumstellar shells
IIIb	617	64.7	20%	31-39 (66%) 21-24 (11%) 41-43 (9%)	variable stars with thick O-rich circumstellar shells
IV	204	29.6	8%	39 (50%) 71-79 (31%)	variable stars with very thick O-rich circumstellar shells
V	176	16.4	15%	91-96 (69%) 71-79 (27%)	planetary nebulae and non-variable stars with very cool circumstellar shells
VIa	501	19.9	20%	16-18 (64%) 42-45 (17%) 31-32 (8%)	non-variable stars with relatively cold dust at large distances; a significant part of the stars is carbon-rich
VIb	605	35.9	5%	22-29 (47%) 14-15 (37%)	variable stars with relatively hot dust close to the star and relatively cold dust at large distances; some of the objects have proven to be oxygen-rich
VII	2048	40.6	44%	41-47 (30%) 21-29 (22%) 14-16 (35%) 17-18 (11%)	variable stars with more evolved C-rich circumstellar shells
VIII	315	27.1	10%	71-79 (43%) 91-95 (30%) 33-36 (13%)	different sort of objects (see text)

3.2 IRAS Low Resolution Spectra

The published LRS catalog (67) contains 5425 sources. However, the LRS data base contains much more data on about 50000 sources, often of poor quality, but from which the useful information is still far from being extracted. About 70% of these published spectra belong to stars with CSEs (see e.g. Figure 3.2). A spectral classification has been proposed in the published LRS catalog, based on the color of the continuum emission and on the main features shown. The main classes correspond to the following categories of CSEs (subclasses are also defined within each class, corresponding to the feature intensities):

- Most of O-rich CSEs display the 10 μm feature of silicates. The most numerous (*class 2n*) show 10 μm emission on a blue background (decreasing with λ). The band goes into absorption when $\dot{M}$ increases, firstly on a blue background (*class 3n*), and then on a red background for a few very cold CSEs (*class 7n*). A very few peculiar sources display 10 μm emission superposed on a red background (*class 6n*).
- Most of the C–rich CSEs display the 11.4 μm emission band of SiC (on a blue background), which defines the *4n class.*
- Our definition of CSEs excludes ionised envelopes of planetary nebulae, and, hence, the sources of LRS *class 9n* and most of those of *class 8n*, which display ion emission lines. However, a few PPN (such as the Red Rectangle) display the 11.3 μm emission line attributed to Polycyclic Aromatic Hydrocarbons (PAH, Section 3.4.5), and are in LRS *class 80.*
- An important proportion of the sources of featureless classes are also CSEs: *0n*, noisy spectra; *1n*, close to black bodies, mostly stars without envelopes, but which includes also featureless dust emission of some CSEs (see e.g. 68); *50*, cold featureless CSEs (PPN).

The simple criteria used in this broad classification sometimes make the identification of the chemical composition ambiguous, when radiative transfer effects strongly modify the appearance of the 10 μm band of the silicates or the SiC 11.4 μm band: some O-rich CSEs are thus classified *4n* (69, 65), and some C-rich ones *21* or *22* (65).

Besides the main features used in this classification, the LRS spectra contain much more information, e.g. : the 18 μm band of silicates is generally clearly identifiable in the O-rich spectra; the shape of the 10 μm silicate band display considerable variations, probably related to various chemical and crystalline compositions of the "silicate" minerals implied (see e.g. Tielens (70), who also proposes that possible additional features in the range 12-22 μm could be attributed to various oxides present in this mineral). Many C–rich CSEs show absorption below 8.5 μm. Kwok et al. (71) have showed that a few probably C–rich PPN, display a still unidentified spectacular band at 21 μm.

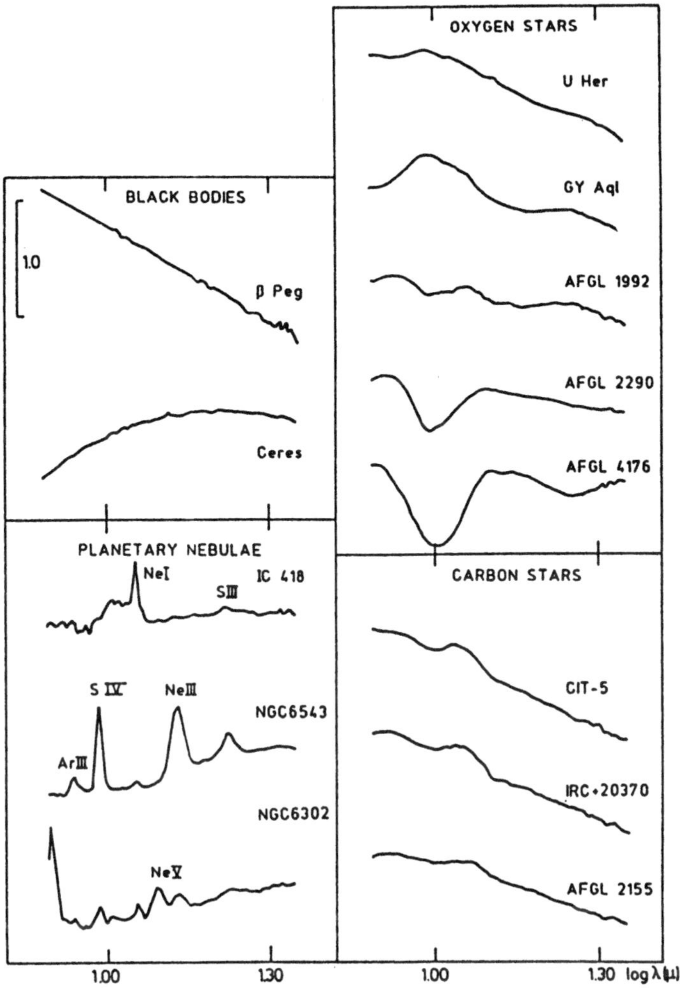

Figure 3.2 – A sample of low resolution spectra from the IRAS Atlas (67). From ref. 11, p. 204.

An attempt to fit most of these features in a new classification scheme of the LRS spectra, based on artificial intelligence methods has been recently proposed by (72).

3.3 Other Infrared Studies

3.3.1 Near Infrared Spectrophotometry

Atmospheric windows in the near infrared allow detailed and extensive work on photometry and spectrometry of CSEs. Photometry is generally performed with standard filters, J, H, K, L, M, N, corresponding to bands centered at 1.25, 1.75, 2.2, 3.6, 5, and 10 μm respectively. Combined with the IRAS data, it provides a general view of the spectral distribution of the continuum emission of the various classes of CSEs (Figure 3.4). Indeed, most CSEs have the maximum of their emission between 2 and 10μ m. The emission profile directly reflects the dust temperature, and it enables one to model dust emission, incorporating heating by photospheric radiation, IR radiative transfer and dust emissivity laws. Large bodies of data combining near IR and IRAS photometry provide new classification criteria (63) and permit statistical studies giving some insight on the progenitors and the evolution of CSEs (see e.g. 73).

Ground based spectrometry reveals most of the prominent bands, formed either in the photosphere or in the CSE of the various classes of sources (see e.g. 74, and Figure 3.3). Besides silicates and SiC, the strongest ones are those of C_2H_2 and HCN at 3 μm. CO bands at 4.6 and 2.3 μm are also detectable with some difficulty; but the strongest band of H_2O is blocked by the Earth atmosphere. The 3.1 μm band of ice is detected in the coldest OH/IR stars with very large mass-loss rates.

3.3.2 Far Infrared and Submillimeter Observations

While waiting for the ISO and SIRTF space missions, the knowledge of the far IR observations relies on airborne observations. Besides photometry with a good angular resolution, far IR spectrometry gives unique information on dust in very cold CSEs.A broad band at about 30 μm in C–rich CSEs has been tentatively attributed to MgS (75). It is extremely strong in some very young PN. Emission in the ice bands between 40 and 70 μm has been detected in a few CSEs, which are mainly cold OH/IR stars. It is quite strong in the unique CSE IRAS09371+1212 (77).

Submillimeter and millimeter continuum emission enables one to probe the very cold dust of extended envelopes (78, 79), and also the photospheric and chromospheric emission (80).

3.4 Grain Properties and Formation

3.4.1 Importance of Grains

Interstellar and circumstellar dust grains are the main infrared emitters. CSE are believed to be one of the most important location of refractory interstellar grains. In addition grains play a major role in many physical and chemical processes in CSEs : they are quite important for circumstellar chemistry by depleting many elements from the gas, and by the possibility of surface reactions. They are essential for the presence of molecules in CSEs by shielding them from the interstellar ultraviolet radiation. They drive the final wind acceleration by radiation pressure (Chapter 4). They are believed to be the main agent for gas heating, by the friction of their streaming through the gas (Section 1.3.5) (and by the photoelectric effect in the very outer layers). Etc.

3.4.2 Composition

Information on the composition of interstellar and circumstellar grains is generally provided by the features present in the infrared spectra. Some additional indications are given by the behavior of the continuum IR emission. IRAS LRS and PSC data have much enriched our knowledge on circumstellar grains in both areas.

Refractory materials condense first at high temperature. They are the essential constituents of circumstellar grains:

In O–rich CSEs the major constituant of dust is generally silicates as attested by the prominent strength of the 10 μm band (Figures 3.2, 3.3,3.4). However, the various shapes of this band reflect the many possible compositions and structures of silicates (70). There is also some spectral evidence for oxides (70), which are thought to condense first, such as Al_2O_3 for example, and they are thought to be the major compounds in some Mira stars without silicate bands (68). Other materials including S, Fe, carbonates, etc, are not excluded.

The bulk of dust of C–rich CSE is certainly some form of carbon material although there is no clear spectral signature. There is no direct evidence of graphite or diamond and the most widespread material is probably some kind of amorphous carbon possibly more or less hydrogenated. Indeed, the most characteristic spectral signature of C—rich CSEs is the 11.4 μm band of SiC which should condense before carbon.

In addition to these refractory materials, mantles of different more volatile compounds are possible. The only clear case is that of ice with its 3.1 μm band in massive O–rich CSEs. MgS has been proposed as the carrier of the broad 25–45 μm feature of C–rich cold CSEs (75). The emitter of the 21 μm band encountered in some PPN (71) is probably some transient compound not yet identified.

3.4.3 Formation of Grains in CSEs

Modelling the formation of dust in the cooling outflows of CSEs is an extremely difficult problem. Even the exact treatment of the ideal situation of thermodynamical equilibrium would be hard because of the complexity of the gas and solid compositions. But thermodynamical equilibrium is certainly never achieved in this range because of supersaturation effects and the rapidity of the envelope evolution. However, thermodynamical calculations, with simplified compositions, are a useful first step and a guide to foresee possible compounds and the temperature hierarchy of the condensation.

Different attempts have been made to describe dust formation in CSEs with the use of nucleation theory (see e.g. 12, 81 and references therein). Condensation of gas into solid particles must go through all the intermediate steps of clusters with N particles. The latter are less stable than the bulk solid since the average binding energy of a particle is reduced because of

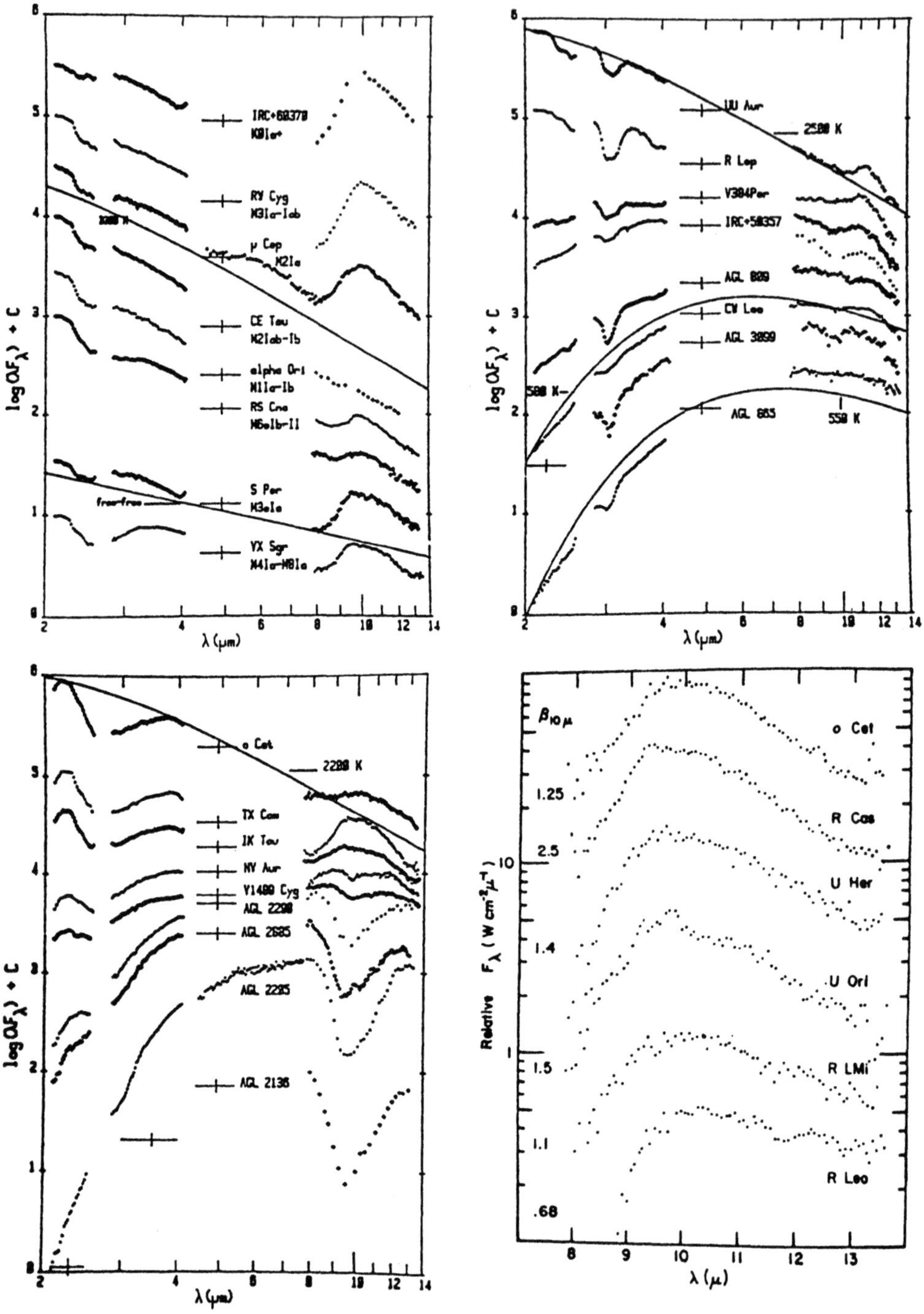

Figure 3.3 – Typical near infrared features of O-rich CSEs, and of C-rich CSEs (top and right). From ref. 12 p. 170.

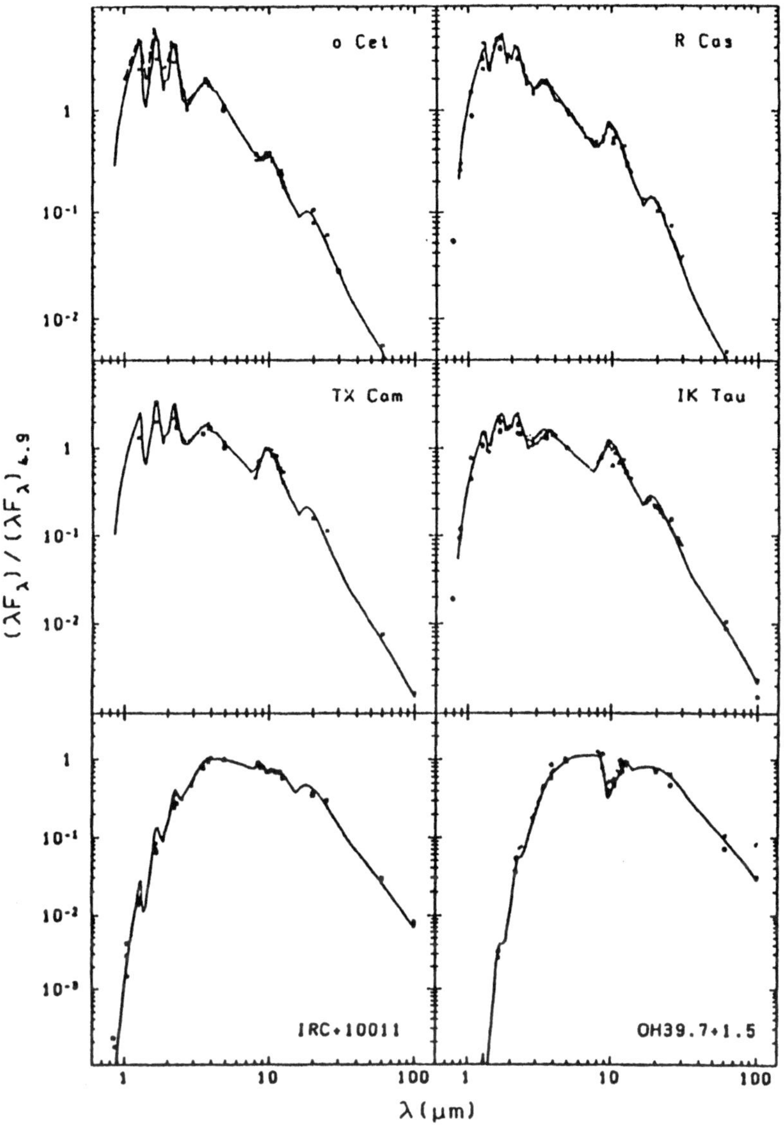

Figure 3.4 – Typical complete infrared spectra of O-rich CSE's. From ref. 90 p. 208.

surface effects. Only homogeneous nucleation theory is relatively simple and well developed ; it accounts for the case where a single chemical compound is present. Then, the sursaturation factor ($p/p_{sat}(T) \gg 1$), required to overcome the cluster instability for N$\leq$10, can be computed if one can estimate the cluster binding energies. But the required inhomogeneous nucleation theory is much more difficult because many chemical species are present, and one must incorporate the detailed complicated reactions between gas and clusters. Several additional complications arise; for instance because of the importance of radiation there is no unique temperature ($T_{kin} \neq T_{dust}$, etc). The inner temperature of a cluster or of a very small grain fluctuates when it absorbes a photon or accretes an atom or a molecule. There are great uncertainties in the gas parameters and their evolution : pulsations, shocks and clumpiness of the transition layers, instabilities related to the dust itself on the dynamics, temperature and chemistry of the gas. There is a complex time evolution of the size distribution of grains and clusters, with important coagulation processes. On the other hand, the hierarchy of condensation of the different elements depending on whether they are more or less refractory, simplifies the condensation of the most volatile products which can just condense on preexisting more refractory grains. It is thus possible that the nucleation problem is completely irrelevant for the most abundant compounds of both O– and C–rich CSEs, with silicates condensing on oxides, and carbon on SiC grains, both probably at T$\leq$1000 K. Laboratory simulations are also very difficult because of the time constants involved in CSEs. Accordingly it is very likely that most of progress in our knowledge of dust formation in CSEs will continue to come from observations. It is thus fundamental to obtain detailed information on the spatial distribution of dust, and of its temperature and composition, in the different types of CSEs.

3.4.4 Grain Growth and Destruction in CSEs

The problem of grain growth by accretion of gas molecules, outside of the main region of grain formation, is relatively simple in principle for smooth winds. The only detailed treatment is that of ice by Jura and Morris (82). They prove that ice condenses at T~110 K, quite efficiently for $\dot{M} \sim 10^{-4} M_{\odot}$/yr, as observed in OH/IR stars. But for smaller values of $\dot{M}$, the density is too small for efficient accretion when T reaches $\sim$ 110 K.

Grain collisions also are certainly frequent in dense layers because of their velocity dispersion. With the relatively large relative velocities expected in the fully accelerated wind, the most likely effect is shattering which could generate a size distribution similar to that in the interstellar medium (83).

It is also quite possible that such grain–grain collisions generate large molecules as suggested in (84) for HC_7N in the bipolar flow of AFGL2688, or PAH's (Section 3.4.5).

Other possible grain destruction processes in CSEs are sputtering by gas atoms or molecules in shocks or fast winds ($V_{gas/grain} \gtrsim 50$ km/s) or reevaporation if the cool medium in which dust has formed, is warmed up again, for instance in the pulsation cycle.

3.4.5 Polycyclic Aromatic Molecules (PAH) in CSEs ?

Over the last ten years, evidence has accumulated for the presence of very small grains, or rather very large molecules (N $\sim$ 20 – 100 atoms) in different astronomical objects (see e.g. (85) and references therein). Infrared emission with very high colour temperature independent of the UV intensity strongly suggests transient heating by a single UV photon, and, hence, such a very small size of gas particles (rather than IR fluorescence of fluffy solid material). In addition, strong bands in the emission, at 3.3, 6.2, 7.7, 8.6 and 11.3 μm, suggest that at least a large proportion of these large molecules are polycyclic aromatic molecules (PAH) (86, 85).

Such "PAH" emission is quite widespread in various sources with strong UV radiation : H_{II} regions, starburst galaxies, reflection nebulae, planetary nebulae, and some C–rich PPN such as the Red Rectangle. Large abundances of PAH in C–rich PN and PPN raise the question of their presence in AGB C–rich CSEs which are the progenitors of these objects.

The answer is still quite uncertain : PAH emission needs excitation by UV (or at least optical) radiation which is absent in AGB CSEs. IR absorption bands of PAH's would be too weak and practically undetectable. Accordingly, PAH's could quite well be present but undetectable in C–rich CSEs such as IRC + 10216. On the other hand, strong PAH emission is absent in the transition objects AFGL 2688 (bipolar PPN) and AFGL 618 (extremely young PN). It is thus also quite possible that the PAHs observed in the PN and some PPN are mostly generated in the transition PPN–PN, by some process related to the fast wind, grain acceleration, etc. ?

4 Properties and Origin of Winds of Circumstellar Envelopes

4.1 Overview of Moderate Winds of Red Giants and Supergiants

Evidence of circumstellar influence in the profile of optical stellar lines goes back to at least 1935 (20). The existence of extended circumstellar matter was definitely proved in the binary system α Her by Deutsch in 1956 (19). Indeed, optical (and even UV) circumstellar lines are a general feature of evolved red giants on the late RGB and early AGB, as well as red supergiants (but they are more difficult or impossible to observe in stars with massive CSEs because of optical and UV extinction). One can distinguish various optical indicators of such relatively modest winds ejected by moderately evolved red giants and supergiants:

P Cygni profiles are quite common and characteristic of circumstellar line formation (Figure 4.1). They generally appear superposed on broad photospheric absorption lines, and are essentially due to photon scattering by the atoms or ions of the circumstellar gas. There is thus a blue shifted circumstellar absorption line, corresponding to the photons removed by circumstellar scattering on the line of sight of the star. This scattered light is emitted over the whole frequency range $\nu_o(1 \pm V_e/c)$, and it appears as an excess redshifted emission. The effects of circumstellar line formation are more complex on the optically thick Hα line, yielding a characteristic asymmetry. In a few cases, one can also spatially resolve the scattered emission of a very strong resonanceline such as that of K_I in α Ori (87).

Binaries often provide even more conspicuous evidence of circumstellar matter of red giants by line absorption studies on the line of sight to a hot companion (e.g. α Her, 19).

Such optical studies have given a lot of information on these moderate winds of red giants and supergiants ; winds are often already present in chromospheres. Velocity gradients revealed by lines formed at various heights, reflect the wind acceleration. Many components of lines prove a multishell structure generated by a variable wind. The presence of the strong wind apparently inhibits the formation of a corona.

These optical lines enable various determinations of the mass–loss rates of such winds from analysis of P Cygni profiles, of resolved scattering intensities, of Hα profiles, of absorption lines in the spectrum of the hot companion, etc ...

Results of such determinations have lead to the popular empirical Reimers formula for $\dot{M}$ (88). It is partially justified by the dimensional argument stating that the power required for the wind escape, G $\dot{M}$ M_* / R_*, must be proportional to the stellar luminosity L :

$$\dot{M} = 4\ 10^{-13}\ \eta \frac{L}{L_\odot} \frac{g_\odot}{g} \frac{R_\odot}{R_*} M_\odot/yr \qquad (4.1)$$

or

$$\dot{M} = 2\ 10^{-6}\ \eta \frac{L}{10^4 L_\odot} \frac{R_*}{500 R_\odot} \frac{M_\odot}{M_*} M_\odot/yr \qquad (4.2)$$

where g is the gravity at the surface of the star,and η a numerical factor $\sim 0.3 - 1$.

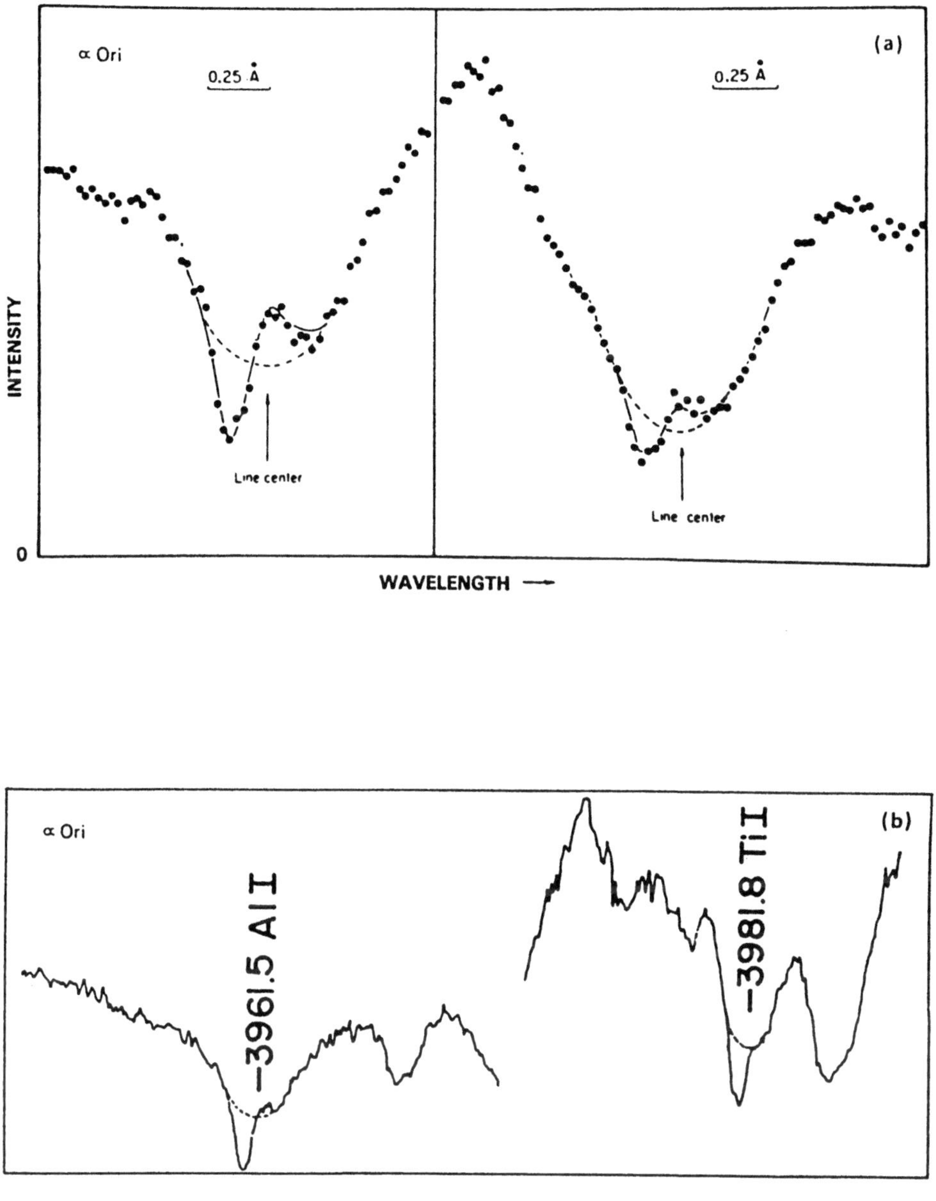

Figure 4.1 – Selected circumstellar lines in the M supergiant, α Ori: (a) Mn resonance lines at λ4030.8 (left) and λ 4033.1 (right), showing a P Cygni type profile. The interpolated core of the photospheric line is shown by the dashed line, while the predicted shell absorption core with redward emission is shown by the solid line; (b) λ3961.5 Al I and λ3981.8 Ti I CS lines. From ref. 12 p. 151.

4.2 Observational Derivation of Mass–Loss Rates of CSEs

(see 92 for general references)

4.2.1 Infrared Methods

The methods discussed in the previous sections, based on optical lines of atoms and ions, are not well suited for the determination of the mass–loss rates of molecular CSEs. However, millemeter and infrared observations provide many ways for estimating the mass–loss rates of these "superwinds".

a) Near IR Spectroscopy. The equivalent widths of optically thin molecular absorption lines yield the column density of molecule i (section 2.3)

$$N_i = \int_{r_o}^{\infty} x_i \frac{C}{r^2} dr = x_i \frac{C}{r_o} \tag{4.3}$$

with $C = \dot{M}/4\pi \ \mu \ V_e$

If one can assume reasonable values for x_i and r_o, one directly deduces $\dot{M}$ from N_i. In practice only CO has a well enough determined abundance.

b) 10 μm Feature of Silicates. Its quasi–universal presence in IRAS LRS spectra of O–rich CSEs makes it very attractive. However, the derivation of $\dot{M}$ from its strength and shape relies on relativelycomplex modelling of radiative transfer and silicate composition and abundance (89). This method has thus not yet reached the state of establishing a straightforward correspondence between observed silicate profiles and $\dot{M}$.

c) Infrared Colours. Modelling the full IR spectrum of CSE is obviously a powerful method for deducing $\dot{M}$ (90). However, it again implies radiative transfer and dust properties. A simple empirical formula has been proposed by van der Veen (91) with

$$\dot{M} \ = \ 7\ 10^{-6}\ L_4\ V_{e15}^{-1}\ (F_{25}/F_{12})^3\ \ M_\odot/yr \tag{4.4}$$

where $L_4 \ = \ L/10^4 L_\odot$ and $V_{e15} \ = \ V_e/15kms$.
It seems appropriate for a CSE with constant $\dot{M}$.

d) Far IR (or submm) Continuum Emission. Dust is approximately optically thin in the outer envelope emitting most of the IR radiation. Its temperature is thus just deduced from the balance between near IR heating and far IR emission. Assuming a certain value for the dust emissivity and abundance, Jura (96) has proposed to deduce $\dot{M}$ from the 60μm IRAS PSC flux :

$$\dot{M} = 1.3\ 10^{-7}\ V_{e15}\ F_{60}(Jy)\ D(kpc)^2\ L_4^{-0.5} \quad (4.5)$$

e) Conservation of Momentum. For outflows "driven" by radiation pressure, one can show (see e.g. 93, 76) that the momentum in the wind is approximately proportional to the radiation momentum multiplied by the mean number (β) of times radiation is (re)absorbed before escaping from the envelope.

$$\dot{M}\ V_e \sim \beta\ L/c \quad (4.6)$$

which yields

$$\dot{M} = 1.5\ 10^{-5}\ \beta\ L_4\ /\ V_{e15}\ M_\odot/yr \quad (4.7)$$

However, if for optically thin CSEs β is approximately equal to their optical depth, its evaluation is complex for thick CSEs, where $\beta \sim 2$ could be a first guess.

4.2.2 Radio Observations

a) CO Intensity. From the modelling of CO rotational excitation by collisions or IR pumping, and of mm line formation, Knapp and Morris (38, see also 92) have shown that one has roughly

$$\dot{M} \propto T_A\ V_e^2\ x_{CO}^{-1}\ D^2 \quad (4.8)$$

where T_A is the CO (1–0) (or 2–1) line intensity and x_{CO} the abundances of CO. However, this formula relies on an assumed kinetic temperature law which is not appropriate for very large $\dot{M}$ ($\sim 10^{-4}\ M_\odot/yr$) and for PPN (see section 1.3.5).

b) OH Maser Flux. Baud and Habing (94) have proposed the empirical formula

$$\dot{M} = 3\ 10^{-5}\ V_{e15}\ (F_{OH}/100Jy)^{0.5}\ D(pc)\ M_\odot/yr \quad (4.9)$$

where F_{OH} is the flux density in the 1612 MHz maser line.

c) Photodissociation Extension. The extension of CO or OH is governed by photodissociation and hence by τ_{UV} which is proportional to $\dot{M}$ (95, 97, 55).

4.2.3 Discussion

The most interesting methods are those which are:
- either independent of D, which is nearly always uncertain, i.e. near IR spectroscopy, 10μm feature;
- or straightforward, general and accurate, i.e. CO emission (for $\dot{M} < 5 10^{-5}\ M_\odot/yr$), infrared colours, 60$\mu$m emission.

However as obvious from the comparison of the values of $\dot{M}$ deduced from different methods, the accuracy is never better than a factor $\sim$ 2–3, plus the uncertainty in D. It should also be remembered that the various methods mostly weight different layers of the CSEs: inner shells and hence recent values of $\dot{M}$ for the infrared methods; outer layers ejected in the past for mm CO emission. They can therefore give different values for $\dot{M}$, if it has varied.

4.3 Possible Mechanisms for Mass–Loss of CSEs

The momentum equation governing the wind acceleration can be written

$$u\ \frac{du}{dr} + \frac{GM_*}{r^2} + g_T + g_R + g_W = 0 \tag{4.10}$$

u(r) is the wind velocity ($u(\infty) = V_e$), g_T, g_R and g_W are the different acceleration terms : thermal, radiative and by "waves" (acoustic shocks, Alfvén waves, etc), respectively.

Thermally driven winds need a high temperature (such as in the solar corona). They are irrelevant for cool stars. Generally $\dot{M}\ V_e\ \sim\ L/c$, so that acceleration by radiation pressure is possible. Radiative forces are also normally larger than gravity, justifying an outward net acceleration. Generally

$$\frac{GM_*}{r_*} > \frac{1}{2}\ V_e^2 \tag{4.11}$$

so that more energy is required in order to overcome the gravitational energy than to subsequently accelerate the gas to the velocity V_e.

Based on these comparisons, there is a certain consensus that the most likely driving force of the wind acceleration in CSEs is radiation pressure on dust which induces the acceleration of the gas by grain–molecule collisions (section 1.3.5). There is indeed a good correlation between the presence of dust and of superwinds in CSEs. However, probably most of the dust forms far from the photosphere, and a very efficient mechanism is required to move the gas up to such a height.

It has been suggested that radiation pressure itself could also achieve, or at least contribute to such a levitation of the gas, either through absorption and multiple scattering in H_2O lines in O–rich CSEs (98 and references therein), or by early dust formation close to the photosphere in cold clumps of gas, such as those which show up in SiO maser emission. However, there is good evidence that the mass-loss of CSEs is associated with their long period pulsations (99). It seems clear that this levitation is related to the motions in the upper atmosphere induced by the pulsation. The most natural way is just to assume that this levitated stationary layer is periodically filled by strong photospheric shock waves driven by pulsations (100). Then, this extracted gas cools enough to allow complete dust formation. Once the dust has formed, radiation pressure coupled by friction to the gas, drives the massive dusty superwind.

5 Evolution from Miras to Planetary Nebulae

5.1 General Considerations

CSEs correspond to a very brief phase in stellar evolution from the very late stages of the AGB to the beginning of ionization in PNe. This evolution results from a complex interplay between double shell burning, dredge-up and mass-loss. Observational data are thus essential to constrain the theoretical modelling, and in particular to establish relations, and to give indications on the duration of the different phases from counts of the corresponding objects. Boundary conditions are also provided by properties and statistics of their progenitors, main sequence stars and red giants, and of their progeny, PN and white dwarfs. Many questions and difficulties remain. Some important phases are very brief, with thus a few objects and poor statistics. Distances are quite uncertain. The influence of M_{in} on the evolution is difficult to estimate; in the absence of direct indications of M_{in} of CSEs, one has generally to rely on galactic distributions, or to some probable trends in other parameters: L, V_e, etc. The mass remaining of the star at each stage of evolution depends of the total mass-loss in the preceding phases, with thus many uncertainties. Finally, binaries deserve special considerations.

5.2 Evolution of O-rich AGB Stars

In the IRAS colour-colour diagram (Figure 3.1), most of the O-rich CSEs are located along a relatively well defined "sequence" stretching from warm Miras of region II to the coldest OH/IR stars of region IV. It is clear that $\dot{M}$ increases (from 10^{-7} to 10^{-4} $M_\odot$/yr) and the effective temperature of the emitting dust decreases along this sequence. The period of pulsations also regularly increases, and there are indications that the luminosity, the initial mass M_{in} and the age (at constant M_{in}) do the same thing. This sequence could thus be an evolutionary sequence. However, it is at present quite difficult to disentangle the effects of the age and the initial mass. If this general scheme seems likely, many points of such evolution phase remain obscure. The pulsation mode and hence the mass of Miras are still debated

(34). The number of Miras seems smaller than expected, which should imply that the duration of the Mira phase is shorter (34). A large proportion of O–rich CSEs lie relatively far from this sequence; this could result from a recent variation of $\dot{M}$, or from peculiar dust properties. The effect of the thermal pulses on the observed properties of the CSEs ($\dot{M}$, pulsations, dust properties, etc) is not yet clear, as well as the rate of change of $\dot{M}$ with time.

5.3 C–rich CSEs

A large proportion of CSEs (20–25%) are C–rich. The fact that about half of the PNe are C–rich implies that half of the stars pass through the C–rich stage of CSEs, which lasts a shorter time than the stage of O–rich CSEs. Most of the C–rich CSEs also lie on a kind of evolution (?) sequence, well separated from the O–rich one in most of the colour–colour diagram (Figure 3.1). However, the detailed understanding of their evolution raises the same problems as for O–rich ones. C–rich CSEs are numerous among those with variable $\dot{M}$; but they seem relatively rare among those with very large $\dot{M}$ (65). The proportion of C–rich PPNe is still uncertain, but it may approach 50% as inferred from the PNe. The S stars are easily identifiable through their visual spectrum; and they are clearly relatively numerous (1/3 of carbon stars). However, there is not yet any practical criterion to identify S stars with massive and cold CSEs. It is clear that C and S stars are the result of the final dredge–up at the end of the AGB. However, the scenario of the transition from O– to C–richness is still debated.

5.4 Transition from O– to C-Richness

Many questions are still open about this turning point in the evolution of AGB stars: When in the evolution ? For which stars ? What is their initial mass ? Which are the accompanying phenomena (on $\dot{M}$, pulsations, etc) ? However, a detailed model, albeit one still debated, has recently been proposed by Willems and de Jong from the analysis of IRAS data (102, 54, see also 101). The main suggestions of these authors are that the change of chemical composition is not gradual, with an intermediate stage of S star, but quite sharp following the dredge–up after a (last) thermal pulse which brings to the surface large quantities of ^{12}C ; and that it is accompanied by an interruption of the mass–loss. The latter could be induced by the temporary end of pulsations and/or absence of dust. The result should be a loop in the colour–colour diagram, with a semi–detached O–rich CSE moving away and cooling, surrounding a bright carbon star, which after some time should resume its own mass–loss, leading to the observed C–rich CSEs which become more and more massive.

Such a scenario appears attractive. It has been recently refined by de Jong (101). However it is still debated. Besides several theoretical arguments related to thermal pulses, dredge–up and mass–loss physics, it well accounts for many observational features, the proportion of carbon stars is large in the region of the colour–colour diagram corresponding to the predicted loop of semi–detached envelopes, although it is still impossible to say whether such envelopes are O– or C–rich. It can explain the existence of carbon stars with silicates and H_2O masers, which should correspond to the very beginning of such loops before the envelope is really semi–detached. The model predict reasonable proportions of C, S and O CSEs and PNe. It also well explains the IR colours of bright carbon stars.

However, many questions remain open (103, 17). Firstly concerning the proportion oftransition objects it seems that there are too many carbon stars with silicates, but too few objects with semi–detached envelopes. There is also a relatively large proportion of O–rich stars among the sources with semi–detached envelopes. The O–richness of semi–detached envelopes around carbon stars is unproved, and at least some of them are C–rich. The model (101) predicts that carbon stars have mainly large initial masses, and that most O–rich PN have small initial masses; there is no obvious observational evidence for any of these statements.

In any case, after the O–>C transition, C–rich CSEs probably follow their own evolutionary sequence of mass–loss rate with time, and the final transition to PPN and PN. However, there are uncertainties comparable with the case of O–rich CSEs concerning the rate of evolution, the role of the initial mass, the scatter of individual CSEs in the IR colour–colour diagram and the question of very large mass–loss rates.

5.5 Post–AGB CSEs (PPNe ?)

There are a variety of objects with non ionized circumstellar matter, which are believed to be in some post–AGB stage of evolution. We shall not discuss here those whose main characteristics seem to directly result from their binarity, such as RV Tau stars. Most of the others form a relatively well defined class with infrared colours corresponding to the region V of van der Veen and Habing (64) (Figure 3.1), where they are relatively well separated from young stellar objects but not from planetary nebulae. They have very weak emission at 12 μm, but relatively large 25μm intensities, characteristic of a detached envelope where the mass–loss has almost stopped, although not completely. The CO mm emission of the expanding CSE is relatively strong, and most of them are associated with a star of intermediate spectral type (mostly F or G), which is more or less hidden. All these features fit nicely with the evolutionary model of pre–planetary nebulae. As mentioned, a certain proportion display visible bipolar nebulae, and high velocity winds are relatively common. It is relatively difficult to distinguish whether they are O– or C–rich because most of their IRAS LRS spectra are featureless (which does not imply the absence of silicates or of SiC, because they emit very little in the 10μm range). However, indications from their optical or near IR spectra, and from radio HCN or OH emission prove that the proportions of both O– and C–rich objects are large.

Their visual spectral type is rather systematically of supergiant type, although they are clearly not massive stars like regular supergiants at least in most cases. They share this characteristic as well as a far IR excess with other post–AGB objects of earlier spectral type (A or B) and warmer IR colours such as HR 4049 (see e.g. 104 and references therein).

5.6 Transition from AGB to White Dwarf

There is no reason to believe that the majority of the AGB stars do not evolve to PNe and finally to white dwarfs along the evolutionary path sketched in Section 1.2 ; the mass–loss rate becomes strong ($\sim 10^{-5}$ $M_\odot$/yr) at the end of the AGB, and possibly very strong ($\sim 10^{-4}$ $M_\odot$/yr) at the very end. The star leaves the AGB when the gas of the "stellar envelope" is exhausted (M–*env* $\lesssim 10^{-3}$ $M_\odot$). Then it warms, while pulsations and the mass–loss stop, leading to the characteristic PPN structure with a detached CSE around a visible star. When the star is hot enough to ionize hydrogen, a planetary nebula forms. Then its central star cools and its luminosity strongly decreases while the nebula fades and moves away into the interstellar medium, leaving a white dwarf.

However, alternative routes to the PN stage are also possible:

i) "Lazy PNe". As the dispersion of the detached CSE is relatively fast (a few 10^4 yr), it has been suggested that the evolution of the star is not always rapid enough to reach the stage of emitting radiation in the meantime. Accordingly some low–mass stars, and more precisely some of the observed PPN, could go directly from the stage of PPN to a white dwarf, without ever displaying the ionized nebula of a PN (105). However, there is no clear cases of such "lazy PNe" with a really remote neutral shell ($r_{inner} \gtrsim$ 0.03–0.1 pc) about a central star of intermediate spectral type. It is quite possible that the mass–loss rate in the PPN stage remains large enough to accelerate the final evolution into a hot star ionizing the circumstellar gas into a planetary nebula.

ii) Post–AGB Helium Flash. Most of the PNe are probably formed with a H–burning central star, which will evolve into a DA white dwarf (displaying a hydrogen spectrum). However, a certain proportion ($\gtrsim$ 10–15%) of the white dwarfs have no hydrogen. They could correspond to stars which have suffered their last helium flash ("thermal pulse") at the very end of the AGB or during the PPN or PN phases, with the ejection of all the H rich shells. It has been proposed that the occurence of such a last He flash when the star already is a PPN or a PN, could induce a loop in the HR diagram, making the star return to the AGB stage for a brief time (106).

iii) Binary Systems. The possibility of mass-exchange in binary systems, especially by accretion onto the secondary of the gas ejected by the mass-loosing red giant, makes possible many new evolutionary schemes (107,2, 108, 109). We will not develop these topics here for lack of space, although they are certainly extremely important for a large proportion of post-AGB objects:probably many stellar systems are multiple rather than single stars ; and it has been estimated that as many as one AGB giant out of 4 or 5 may be significantly affected by the presence of a companion. The latter can deeply modify the mass-loss of the primary, its rate and time evolution, by tidal effects. It can either prevent the ejection of an envelop by swallowing it via Roche lobe overflow, or it could dramatically enhance the process by common envelop and spiral-in evolution. Each star can in turn go to the giant-PN evolution. One can expect in principle that following properties of the initial binary system, it could lead to different kinds of symbiotics or of peculiar red giants or post-AGB stars, and finally to PN with wide binary, or close binary, or even in some cases single nuclei.

ACKNOWLEDGEMENTS

I would like to thank M. Friedjung for his critical reading of the manuscript and V. Demailly, C. Douillet, M.C. Pantalacci, B. Raban for their editorial work.

REFERENCES

1. V. Weidemann: In The Second Conference on Faint Blue Stars, ed. by A.G. Davis Philip, D.S. Hayes and J.W. Liebert (Davis Press Inc., Schenectady, New York 1987) p.19
2. I. Iben: Quat. J. roy. Astr. Soc. 26, 1 (1986)
3. D. Schonberner: In The Second Conference on Faint Blue Stars, see [1] p.201
4. Physical Processes in Red Giants, ed. by I. Iben and A. Renzini (Reidel, Dordrecht 1981)
5. Mass Loss in Red Giants, ed. by M. Morris and B. Zuckerman (Reidel, Dordrecht 1985)
6. Late Stages of Stellar Evolution, ed. by S. Kwok and S.R. Pottasch (Reidel, Dordrecht 1987)
7. Evolution of Peculiar Red Giants, ed. by H.R. Johnson and B. Zuckerman Cambridge University Press (1989)
8. I. Iben and A. Renzini: Ann. Rev. Astron. Astrophys. 21, 271 (1983)
9. B. Zuckerman: Ann. Rev. Astron. Astrophys. 18, 263 (1980)

10. Effects of Mass-Loss on Stellar Evolution, ed. by C. Chiosi and R. Stalio (Reidel, Dordrecht 1981)
11. Circumstellar Matter, Symp. IAU No 122, ed. by I. Appenzeller and C. Jordan (Reidel, Dordrecht 1987)
12. M–Type stars, ed. by H.R. Johnson and F. Querci (NSF-CNRS 1988)
13. H. Olofsson: In (Sub)millimeter Astronomy, ed. by P.A. Schaver and K. Kjar (ESO Publications, Garching 1985)
14. H. Olofsson: Space Sci. Rev., 47, 145 (1988)
15. A.E. Glassgold and P.J. Huggins: In M, S and C Stars, ref 12 p.291
16. S.R. Pottasch: Planetary Nebulae (Reidel, Dordrecht 1984)
17. From Miras to Planetary Nebulae: Which Path for Stellar Evolution, ed by M.O. Menessier and A. Omont (Editions Frontieres, Gif-sur-Yvette 1990)
18. I.S. Shklovskii: Astron. J. U.S.S.R. 33, 315 (1956)
19. A.J. Deutsch: Astrophys. J. 236, 567 (1956)
20. W.S. Adams and E. MacCormack: Astrophys. J. 81, 119 (1935)
21. G. Neugebauer, R.B. Leighton: NASA SP 3047 (1969)
22. N.J. Woolf and E.P. Ney: Astrophys. J. (Letters) 155, L181 (1969)
23. P.M. Solomon, A. A. Penzias and K.B. Jefferts: Astrophys. J. (Letters) 169, L35 (1971)
24. T.R. Geballe, E.R. Wollmann and D.M. Rank: Astrophys. J. (Letters) 177, L27, (1972)
25. R.W. Wilson and A.H. Barrett: Astron. Astrophys. 17, 385 (1972)
26. B. Paczynski: Acta Astr. 20, 47 (1970); 20, 287 (1970)
27. R. Gallino, M. Busso, G. Picchio, C.M. Raiteri, A. Renzini: Astrophys. J. Lett. 334, L45 (1988)
28. A. Omont: In Mass Loss in Red Giants, ref 5 p.269
29. A.G.G.M. Tielens: Astrophys. J. 271, 702 (1983)
30. P. Goldreich, N.Z. Scoville: Astrophys. J. 205, 144 (1976)
31. J. Kwan, R.A. Linke: Astrophys. J. 254, 587 (1982)
32. M. Jura, C. Kahane, A. Omont: Astron. Astrophys. 201, 80 (1988)
33. S.D. Price, R.G. Walker: AFGL TR 76-0208 (1976) S.D. Price, T.L. Murdock: AFGL TR 83-0161 (1983)
34. P.D. Wood: In Physical Processes in Red Giants, ref 4 p.205 In From Miras to Planetary Nebulae: Which Path for Stellar Evolution, ref 17
35. J. Herman, H.J. Habing: Phys. Reports 124, 256 (1985)
36. B. Zuckerman, and H.M. Dyck: Astron. Astrophys. 209, 119 (1989)
37. M. Morris: In From Miras to Planetary Nebulae: Which Path for Stellar Evolution, ref 17
38. G.R. Knapp, M. Morris: Astrophys. J. 292, 640 (1985)
39. H. Olofsson, L.E.B. Johansson, A. Hjalmarson and Nguyen-Q-Rieu: Astron. Astrophys. 107, 128 (1982)
40. S.G. Kleinmann, M. Jura: Astrophys. J. 341, 359 (1989)
41. Nguyen-Q-Rieu, V. Bujarrabal, H. Olofsson, L.E.B. Johansson and B.E. Turner: Astrophys. J. 286, 276 (1984)
42. M. Morris, R. Lucas and A. Omont: Astron. Astrophys. 142, 107
43. C. Loup, T. Forveille, J.F. Paul and A. Omont: Astron. Astrophys. Supp. in preparation
44. J.H. Bieging, Nguyen-Q-Rieu: Astrophys. J. Lett. 329, L107 (1988)
45. P. te Likkel Hekkert, J. Caswell, R. Haynes, R. Norris, A.M. Le Squeren, P. Sivagnanam, J. Wouterloot: in preparation.

46. P. Sivagnanam, A.M. Le Squeren, F. Foy and F. Tran Minh: Astron. Astrophys. 211, 341 (1989)
47. T. Tsuji: Ann. Tokyo Astron. Obs. 2nd Ser. 9, 1 (1963); Astron. Astrophys. 23, 411 (1973)
48. A. Omont: In Astrochemistry, IAU Symp. N. 120, ed. by M.S. Vardya and S.P. Tarafdar (Reidel, Dordrecht 1987) p.357
49. T.J. Millar: In Rate Coefficients in Astrochemistry, ed. by T.J. Millar and D.A. Williams (Kluwer Academic Pub. 1988) p.287
50. A. Omont: in Cosmic Chemistry, AG Meeting Koln, ed. by G. Klare (Springer-Verlag, Heidelberg 1988)
51. A. Omont: in The Physics and Chemistry of Interstellar Molecular Clouds, ed. by G. Winnewisser and J.T. Armstrong. Lecture Notes in Physics (Springer–Verlag, Heidelberg, 1989)
52. S. Lafont, R. Lucas and A. Omont: Astron. Astrophys. 106, 201
53. R.E. Stencel: In Circumstellar Matter, Symp. IAU No 122, ref 11
54. F.J. Willems : Astron. Astrophys., 203, 51 and 65 (1988)
55. G.A. Mamon, A.E. Glassgold, P.J. Huggins: Astrophys. J. 328, 797 (1988)
56. A.E. Glassgold, R. Lucas, A. Omont: Astron. Astrophys. 157, 35 (1986)
57. D.A. Howe and T.J. Millar: M.N.R.A.S. 244, 444
58. P.G. Wannier, R.A. Linke: Astrophys. J. 225, 130 (1978)
59. C. Kahane, J. Gomez-Gonzalez, J. Cernicharo, M. Guelin: Astron. Astrophys. 190, 167 (1988)
60. D.L. Lambert, B. Gustafsson, K. Eriksson and K.H. Hinkle: Astrophys. J. Supp. 62, 273 (1986)
61. D.Y. Gezari, M. Schmitz and J.M. Mead: Catalog of Infrared Observations, NASA Reference Publication 1196 (1987)
62. IRAS, Point Source Catalogue: US Government Publication Office (1985)
63. N. Epchtein, T. Le Bertre and J.R.D. Lepine: Astron. Astrophys. 227, 82 (1990)
64. W.E.C.J. van der Veen and H.J. Habing: Astron. Astrophys. 194, 125 (1988)
65. A. Omont, C. Loup, T. Forveille, P. te Lintel Hekkert, J. Caswell, P. Sivagnanam and H.J. Habing: in preparation
66. L. Likkel, T. Forveille, A. Omont and M. Morris: Astron. Astrophys. submitted
67. IRAS, LRS Catalogue, eds F.M. Olnon and E. Raymond: Astron. Astrophys. Supp. 65, 607 (1986)
68. M.S. Vardya, T. de Jong and F.J. Willems: Astrophys. J. Lett. 304, L29 (1986)
69. P. Sivagnanam : Thesis (Paris 1989)
70. A.G.G.M. Tielens: From Miras to Planetary Nebulae: Which Path for Stellar Evolution, ref. 17, p. 186
71. S. Kwok, K.M. Volk and B.J. Hrivnak : Astroph. J. Lett. **345**, L51 (1989)
72. P. Cheeseman, J. Stutz, M. Self, W. Taylor, J. Goebel, K. Volk and H. Walker: Automatic Classification of Spectra from the Infrared Astronomical Satellite (IRAS), NASA Reference Publication 1217 (1989)
73. M.J. Claussen, S.G. Kleinmann, R.R. Joyce, M. Jura: Astrophys. J. Supp. 65, 385 (1987)
74. K.M. Merrill and S.T. Ridgway: Ann. Rev. Astron. Astrophys. 17, 9 (1979) P.G. Wannier: In Mass Loss in Red Giants, ref 5, p. 65

75. J.H. Goebel and S.H. Moseley: Astrophys. J. Lett. 290, L35 (1985)
76. M. Jura : In From Miras To Planetary Nebulae which Path for Stellar Evolution, ref. 17, p. 41
77. A. Omont, H.M. Moseley, T. Forveille, W.J. Glaccum, P.H. Harvey, L. Likkel, R.F. Loewenstein and C.M. Lisse: Astrophys. J. Lett. 355, L27 (1990)
78. R. Sopka, R. Hildebrand, D. Jaffee, I. Gatley, T. Roellig, M. Werner, M. Jura, B. Zuckerman: Astrophys. J. 294, 242 (1985)
79. M. Jura: Saint Andrew Summer School (1987)
80. C.M. Walmsley, R. Chini, E. Kreysa, H. Steppe, T. Forveille and A. Omont: in preparation
81. E. Seldmayr: In From Miras to Planetary Nebulae: Which Path for Stellar Evolution, ref. 17 p.179
82. M. Jura, M. Morris: Astroph. J. 292, 487 (1985)
83. P. Bierman and M. Harwit: Astrophys. J. Lett. 241, L105 (1980)
84. M. Jura and H. Kroto: Astrophys. J. 351, 222 (1990)
85. J.L. Puget and A. Leger: Ann. Rev. Astron. Astrophys. 27, 161 (1989) L.J. Allamandola, A.G.G.M. Tielens and J.R. Barker: Astrophys. J. Supp. 71, 733 (1989)
86. A. Leger and J.L. Puget: Astron. Astrophys. 137, L5 (1984)
87. L. Goldberg: In M, S and C Stars, ref (12), p. 245
88. D. Reimers: Mem. Soc. Roy. Sci. Liege 6th Ser. 8, 369 (1975)
89. W. Schutte, A.G.G.M. Tielens: Astron. Astrophys. in press (1988)
90. P.J. Bedijn: Astron. Astrophys. 186, 136 (1987)
91. W. van der Veen: Astron. Astrophys. 210, 127 (1989)
92. W. van der Veen and H. Olofsson: From Miras to Planetary Nebulae: Which Path for Stellar Evolution, ref. 17, p. 186
93. H.P. Gail, E. Sedlmayer: Astron. Astrophys. 161, 201 (1986)
94. B. Baud and H.J. Habing: Astron. Astrophys. 127, 73 (1983)
95. P.J. Huggins and A.E. Glassgold: Astron. J. 87, 1828 (1982)
96. M. Jura: Astrophys. J. 313, 743 (1987)
97. N. Netzer, G.R. Knapp: Astrophys. J. 323, 734 (1987)
98. M. Elitzur, J.A. Brown and H.R. Johnson : Astrophys. J. Lett. (1990) in press
99. M. Jura: Irish Astron. J. 17, 322 (1986)
100. A.G. Hearn: In From Miras to Planetary Nebulae: Which Path for Stellar Evolution, ref. 17 p.121
101. T. de Jong: Astron. Astrophys. Lett. 223, L23 (1989) In From Miras to Planetary Nebulae: Which Path for Stellar Evolution, ref. 17 p.289
102. F.J. Willems, T. de Jong: Astron. Astrophys. 309, L39 (1986), 196, 173 (1988)
103. B. Zuckerman and R.J. Madalena: Astron. Astrophys. Lett. 223, L20
104. L.B.F.M. Waters et al. : Astron. Astrophys. F.M. Waters, C. Waelkens, N.R. Trams: In From Miras to Planetary Nebulae: Which Path for Stellar Evolution, ref. 17 p.449
105. A. Renzini: In Physical Processes in Red Giants, ref 4 p.438
106. I. Iben, Jr: In Late Stages of Stellar Evolution, ref 6
107. C.de Loore: In Effects of Mass-Loss on Stellar Evolution, ref (10), p.405
108. P.P. Eggleton: In From Miras to Planetary Nebulae: Which Path for Stellar Evolution, ref. 17 p.513

109. I. Iben, Jr and A.V. Tutukov: In Planetary Nebulae, IAU Symp. N. 131, ed. S. Torres-Peimbert (Kluwer, Dordrecht, 1989), p. 505
110. M. Morris, S. Guilloteau, R. Lucas, A. Omont : Astrophys. J. 321, 88 (1987)
111. Nguyen Q.Rieu : in M-Type Stars

FIGURE CREDITS

Figure 1.1. Physical Processes in Red Giants, ed. by I. Iben and A. Renzini (Reidel, Dordrecht 1981).

Figure 1.2. Mass Loss in Red Giants, ed. by M. Morris and B. Zuckerman (Reidel, Dordrecht, 1981), p.269.

Figure 2.1. Physical Processes in Red Giants, ed. by I. Iben and A. Renzini (Reidel, Dordrecht 1981), p.354–356.

Figure 3.1. Astron. Astrophys. **194**, 125 (1988)

Figure 3.2. Circumstellar Matter, Symp. IAU N^0 122, ed. by I. Appenzeller and C. Jordan, (Reidel, Dordrecht 1987), p. 204.

Figure 3.3. M–Type Stars, ed. by H.R. Johnson and F. Querci (NSF–CNRS 1988), p.170.

Figure 3.4. Astron. Astrophys. **186**, 136, 1987

Figure 4.1. M–Type Stars,ed. by H.R.Johnson and F. Querci (NSF–CNRS 1988), p. 151.

Evolution of Close Binary Systems: Application to X-Ray Binaries

G.J. Savonije

Astronomical Institute 'Anton Pannekoek'
University of Amsterdam
The Netherlands

1 Tidal Effects in Close Binary Systems

1.1 Secular Evolution due to Dissipation in Tidal Flow

Binary systems are stellar configurations in which two stars are gravitationally bound together and move about a common centre of gravity. One speaks of Close Binary Systems (CBS) if the separation of the two stars is not much larger than the dimension of the biggest star in the system. In that case the star(s) are noticeably deformed by their mutual gravitational attraction (tides). Observations show that CBS tend to have orbits with small eccentricities. Koch and Hrivnak (1981) show a picture of the orbital eccentricity e as a function of the orbital period P for a sample of main sequence components with spectral type later than F8. It appears that binaries with orbital periods less than about 7 days have almost circular orbits, while wider binaries show a large scatter of orbital eccentricities. This is most naturally explained in terms of tidal interaction, which tends to circularize the orbits, and which is only effective in close binaries. The stars in KH's sample are low mass stars with convective envelopes. However, from observations of massive X-ray binary systems we know that close binaries with massive components can also be subject to efficient tidal interaction. The orbits of Cen X-3 and SMC X-1 have negligeable eccentricities, while the binary must have been violently disturbed by the supernova explosion that formed the neutron star in the system. The tidal circularization time must have been less than the lifetime of the massive OB-star in the system, which is less than roughly 10^7 years.

The gravitational forces in a CBS lead to a deformation of the stellar components. If the orbit is eccentric, or if the spin of the stars is not synchronized with the orbital revolution, or if the rotation axes are not perpendicular to the orbital plane each star will be subject to tidal forces that vary in time. Only if the orbit is circular and all the other conditions mentioned fulfilled will the tidal deformation be stationary in a frame rotating with the binary. In all other cases there exists a time dependent tidal flow in the stars in which inevitably dissipation takes place. The dissipation causes the orbital energy E to decrease, so that the binary parameters evolve in time: the binary system undergoes tidal evolution.

Equation 1 relates the orbital eccentricity e to the orbital angular momentum H and orbital energy $E = -(GM_pM_s)/2a$ of the binary system (e.g. Landau and Lifshitz 1959):

$$e = \sqrt{1 + \frac{2(M_p + M_s)}{G^2 M_p^3 M_s^3} E H^2} \tag{1}$$

where G is the constant of gravity and M_p and M_s are the masses of the primary and secondary star. For a fixed H and fixed stellar masses the orbital energy cannot decrease beyond the minimum value:

$$E_{\min} = -\frac{1}{2} \frac{G^2 M_p^3 M_s^3}{(M_p + M_s)} H^2$$

which corresponds to a circular orbit with $e = 0$. However, in reality the orbital angular momentum H is not exactly conserved, because the dissipation causes the tidal deformation to be phase-lagged with respect to the perturbing tidal forces. This gives rise to a **tidal torque** whereby angular momentum is exchanged between the orbit and the stellar spin. In general the dissipation causes the orbit to circularize and causes the stellar spins to align and synchronize with the orbital revolution. Once this equilibrium situation has been reached the tidal perturbations become stationary in the corotating binary frame and the dissipation ceases.

The semi-minor axis b of the elliptic orbit depends on both the orbital energy E and orbital angular momentum H, while the semi-major axis a and the orbital period P_b are independent of H:

$$a = \frac{GM_pM_s}{2|E|}$$

$$b = \frac{H}{\sqrt{2\mu|E|}}$$

$$P_b = \pi G M_p M_s \sqrt{\frac{\mu}{2|E|^3}}$$

We have introduced here the reduced mass $\mu = \frac{M_pM_s}{M_p+M_s}$ of the binary system.

For negligeable stellar rotational (spin) angular momenta and fixed stellar masses, a and P_b *always decrease* due to the dissipation of orbital energy E during the tidal evolution. However, if the stellar spin is large the orbit may in fact expand by absorbing spin energy and angular momentum (Earth-Moon system) or may even become more eccentric (see below) by tidal effects.

It is evident that the timescale of the tidal evolution is determined by the dissipation rate of the tidal flow. In general the dissipation rate is very difficult to determine. In calculations of tidal evolution one therefore often adopts a mathematically simple ad hoc relation between the phase lag of the tidal deformations and the relative orbital angular velocity of the companion star (e.g. Alexander 1973, Hut 1981). For a full treatment one has not only to determine the time dependent tidal potential in the binary, but also the response of the stars on this tidal perturbation, thereby taking into account the dissipative processes.

1.2 The Tidal Potential

Let us now derive the tidal potential of the perturbing primary star at the position of the secondary star, assuming for the moment that the binary has a circular orbit with radius a. The gravitational potential of the primary star (we neglect its own tidal deformation as a higher order effect), in a spherical coordinate system centred at the gravity centre of the secondary star, is given by:

$$\Psi_p(r,\theta,\phi) = \frac{GM_p}{|r-a|}$$

We can expand the $|r-a|^{-1}$ term in terms of spherical harmonics, using the addition theorem (e.g. Jackson 1962):

$$\frac{1}{|r-a|} = 4\pi \sum_{l=0}^{\infty} \sum_{m=-l}^{+l} (2l+1)^{-1} \frac{r^l}{a^{l+1}} Y(\frac{\pi}{2}, \Omega t) Y^*(\theta,\phi) \tag{2}$$

where

$$Y(\theta,\phi) = P_l^m(\theta) e^{im\phi}$$

and a superscript * denotes complex conjugation. P_l^m are the associated Legendre functions. For simplicity we assume here and in the following that the spin axes of the two stars are aligned. The first term in the above expansion which yields a tidal (deformation) force on the secondary star is the $l = 2$ term. Hence, for small perturbations we may approximate the tidal potential of the primary star, as the real part of:

$$\Psi_p(r,\theta,\phi) = \left(\frac{GM_p}{4a^3}\right) r^2 \Psi \tag{3}$$

For a circular orbit Ψ equals:

$$\Psi_0 = P_2^2(\cos\theta) e^{2i(\Omega t - \phi)}$$

For moderately eccentric orbits ($e^2 << 1$) Ψ may be approximated by a sum of four harmonics: $\Psi = \Psi_0 + \Psi_1 + \Psi_2 + \Psi_3$, with:

$$\Psi_1 = 6eP_2(\cos\theta) e^{i\Omega t}$$

$$\Psi_2 = \frac{e}{2} P_2^2(\cos\theta) e^{i(\Omega t - 2\phi)}$$

$$\Psi_3 = -\frac{7}{2} e P_2^2(\cos\theta) e^{i(3\Omega t - 2\phi)}$$

where $\Omega = 2\pi/P_b$ is the mean orbital angular velocity.

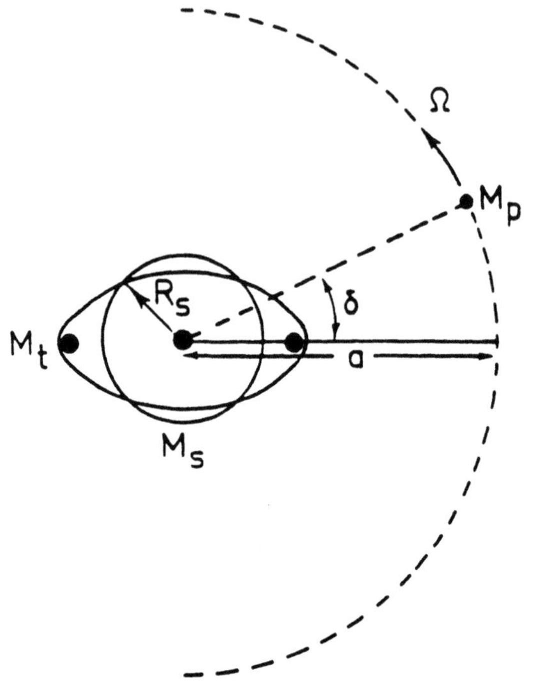

Fig. 1. Schematic picture of star which is tidally distorted and spun up by a pointmass companion with mass M_p. The mass distribution of the deformed star is approximated by three pointmasses, where $M_t << M_s$ is the mass in the two tidal bulges. The phase lag δ is greatly exaggerated.

1.3 Crude Estimate of Tidal Spin-Up Torque

Let us consider a slowly rotating binary component ($\omega < \Omega$) in a circular orbit. Fig. (1) sketches the situation schematically, in which the tidal phase lag δ between the perturbing (point mass) companion with mass M_p and the tidal bulge is greatly exaggerated for clearity. In order to get a rough idea how the tidal torque depends on the binary parameters, we will crudely approximate the secondary's mass distribution by three colinear point masses. The central pointmass has the mass M_s, whereas the two other pointmasses have much smaller masses M_t and represent the mass in the two tidal bulges. For the time being we will neglect dynamical effects and use a simple hydrostatic approximation to estimate the stellar reponse to the tidal perturbation. From (3) we can estimate the height h of the secondary's tidal bulges, by setting equal the potential energy of one gram of mass in the tidal bulge (raised over the height h) to the primary's tidal potential at the secondary's surface:

$$\left(\frac{GM_s}{R_s^2}\right) h \simeq \left(\frac{3GM_p}{4a^3}\right) R_s^2$$

This yields for h:

$$h \simeq \left(\frac{M_p}{M_s}\right)\left(\frac{R_s}{a}\right)^3 R_s$$

The mass in the tidal bulges is difficult to determine, but let us simply take (we are here only interested in a dimensional analysis) :

$$M_t = f\left(\frac{M_s}{R_s}\right) h \simeq f M_p \left(\frac{R_s}{a}\right)^3$$

where the dimensionless factor f depends on the stellar structure. As a result of the phase lag δ the secondary is subject to a spin-up torque T. The distances r_1 and r_2 between M_p and the two M_t's are given by: $r_1 \simeq a + R_s \cos(\delta)$ and $r_2 \simeq a - R_s \cos(\delta)$. The gravitational forces on the tidal bulges are: $F_1 = (GM_tM_p)/r_1^2$ and $F_2 = (GM_tM_p)/r_2^2$. The spin-up torque is then:

$$T = |\sum_i \mathbf{r}_i \times \mathbf{F}_i| = R_s F_2 \sin(\delta) - R_s F_1 \sin(\delta)$$

Substituting F_1 and F_2 and using $(R_s/a) << 1$ and $\delta << 1$ yields:

$$T \simeq \frac{GM_tM_p}{a^2}\left(\frac{R_s^2}{a}\right)\delta$$

Substituting the formal expression for M_t yields finally:

$$T \simeq f\left(\frac{GM_s^2}{R_s}\right)\left(\frac{M_p}{M_s}\right)^2\left(\frac{R_s}{a}\right)^6 \delta \qquad (4)$$

The tidal torque appears proportional to the phase lag δ which depends, as we have seen, on the dissipation rate in the tidal flow. Note that the tidal torque is very sensitive to the ratio (R_s/a), which is the reason why tidal effects are important only in CBS.

1.4 Equations for Tidal Evolution

We will now discuss briefly some work in which the tidal torque is calculated more accurately. We will express the tidal torque in terms of a characteristic spin up timescale for a non-rotating secondary in a binary with circular orbit (angular velocity Ω) as:

$$t_{sp} = -\frac{I\Omega}{\frac{dH}{dt}} = \left(\frac{M_p}{M_s}\right)^{-2}\left(\frac{M_p}{M_s}\right)^{-2}\left(\frac{R_s}{a}\right)^{-6} t_0(\Omega) \qquad (5)$$

where I is the secondary's moment of inertia about its rotation axis and $t_0(\Omega)$ is an 'intrinsic' tidal timescale for which we have factored out, according to (4), the dependence of the tidal torque $T = \frac{dH}{dt}$ on the orbital geometry and mass ratio. The timescale t_0 is a convenient measure for the secondary's (dissipative) response to the tidal perturbations. In terms of the crude (hydrostatic) approximation of Sect. 1.3 t_0 would be given by:

$$t_0(\Omega) = I\Omega\left(f\frac{GM_s^2}{R_s}\right)^{-1}\delta^{-1}$$

where the phase lag δ of the stellar response is undetermined and depends on Ω. In the so-called 'weak friction' approximation -applicable to the weak tides in solid bodies, but sometimes also used for stars- one assumes that the phase lag δ of the hydrostatic tide is simply linearly proportional to Ω. In that case t_0 would be independent of the tidal frequency. In (gaseous) stars, however, we have to take into account dynamical effects in calculating $t_0(\Omega)$. Since consistent models of (differentially) rotating stars are not available, $t_0(\Omega)$ has only been determined for non-rotating stars. And then only in the case were the deformed outer layers of the star are convectively stable, that is for massive stars. The problem of calculating the non-adiabatic stellar response to the tidal perturbations was first addressed by Zahn (1977) in an asymptotic approximation for low tidal frequencies and more generally, with a numerical scheme, by Savonije and Papaloizou (1984). Both are linear calculations in which the tidally forced (small amplitude) non-radial oscillations of a massive star around a spherical equilibrium state are solved. The dissipation occurs through radiative damping of the forced g(ravity) mode oscillations. The damping takes place near the stellar surface, where the (local) thermal timescale of the star becomes small. The intrinsic timescale $t_0(\Omega)$ varies considerably with the tidal (forcing) frequency Ω and, to a smaller extent, as a function of the stellar evolution phase

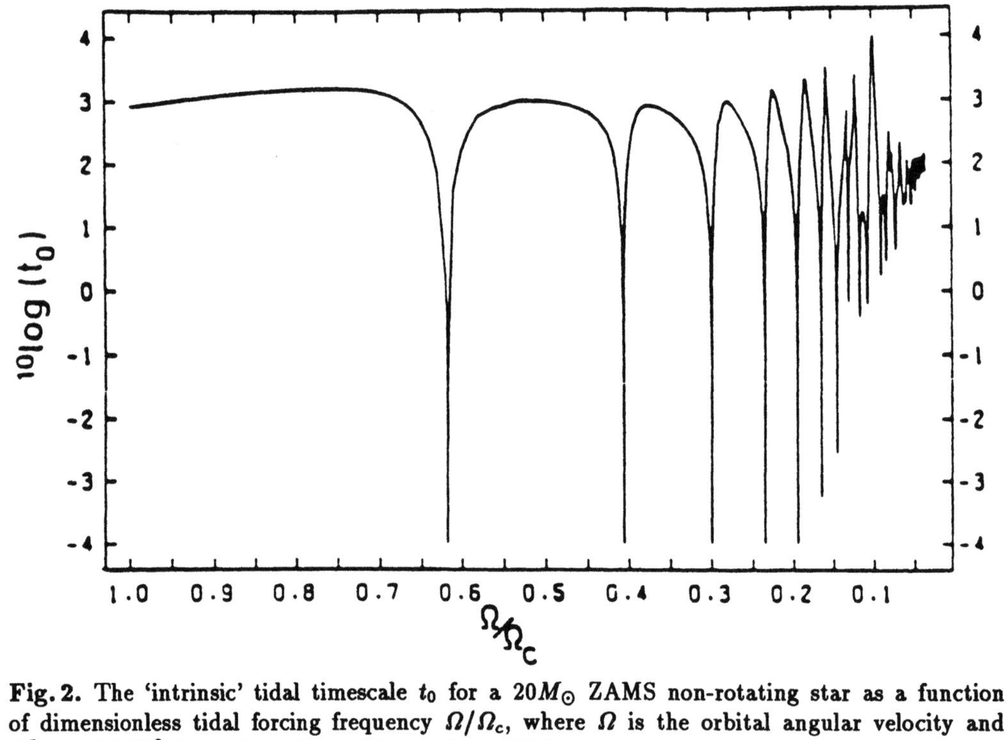

Fig. 2. The 'intrinsic' tidal timescale t_0 for a $20M_\odot$ ZAMS non-rotating star as a function of dimensionless tidal forcing frequency Ω/Ω_c, where Ω is the orbital angular velocity and $\Omega_c^2 = GM_s/R_s^3$ (after Savonije and Papaloizou 1985).

of the perturbed star. If the tidal frequency happens to be close to the eigenfrequency of one of the normal g-modes of non-radial oscillation of the perturbed star the (resonant) response can be large, with very small values of t_0 (see Figs. 2 and 3). Let us consider

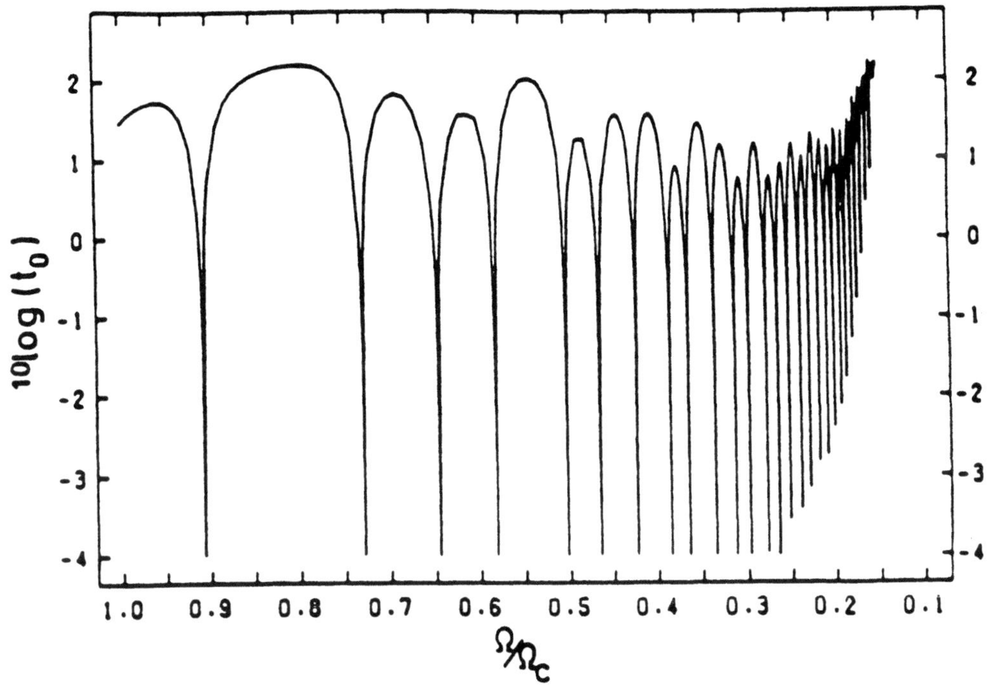

Fig. 3. Same as Fig. 2, but now for a $20M_\odot$ star near the end of core hydrogen burning ($X_c = 0.1$). Note that the maximum values of t_0 are more than an order of magnitude smaller than in Fig. 2

uniformly rotating stars and assume for simplicity that we may apply the calculated t_0 values for non-rotating stars to uniformly rotating stars (with angular velocity ω) by simply replacing $t_0(\Omega)$ by $t_0(\Omega-\omega)$. For a moderately eccentric orbit (Eqn. 3) and small tidal effects (linear approximation) the tidal torque on the uniformly rotating secondary can be expressed as (Savonije and Papaloizou , 1985):

$$\frac{dH}{dt} = -I\left(\frac{M_p}{M_s}\right)^2\left(\frac{R_s}{a}\right)^6\left[\frac{(\Omega-\omega)}{t_0(\Omega-\omega)} + \frac{e^2}{8}\frac{(\Omega-2\omega)}{t_0(\frac{1}{2}\Omega-\omega)} + \frac{147e^2}{8}\frac{(\Omega-\frac{2}{3}\omega)}{t_0(\frac{3}{2}\Omega-\omega)}\right] \quad (6)$$

Note that if a term in the tidal potential varies as $e^{i(\nu\Omega t - m\phi)}$ the rate of change of orbital energy $\frac{dE}{dt}$ is related to the rate of change of orbital angular momentum $\frac{dH}{dt}$ by:

$$m\frac{dE}{dt} = \nu\Omega\frac{dH}{dt}$$

We can now derive from (1) the tidal circularization rate for a moderately eccentric orbit as:

$$-\frac{1}{e}\frac{de}{dt} = \frac{I}{\mu a^2}\left(\frac{M_p}{M_s}\right)^2\left(\frac{R_s}{a}\right)^6\left[\frac{3}{8t_0(\frac{1}{2}\Omega)} - \frac{(1-2\frac{\omega}{\Omega})}{16t_0(\frac{1}{2}\Omega-\omega)} - \frac{(1-\frac{\omega}{\Omega})}{2t_0(\Omega-\omega)} + \frac{147(1-\frac{2\omega}{3\Omega})}{16t_0(\frac{3}{2}\Omega-\omega)}\right] \quad (7)$$

where μa_2 is the orbital moment of inertia. Note that the rate of change of the eccentricity can be either positive or negative, depending on $(\frac{\omega}{\Omega})$ and the value of t_0 for the different tidal components. In the weak friction approximation for which t_0 is a constant, independent of tidal frequency, $\frac{de}{dt} > 0$ if $\frac{\omega}{\Omega} > \frac{18}{11}$, a fact noted already by Darwin (1879).

Once we know t_0 for all possible tidal frequencies we can, for a given set of initial conditions, calculate the evolution of a binary system (adopting the assumptions mentioned above). Fig. 4 shows the tidal evolution of a binary system consisting of a 20 $M_\odot$ star and a 1 $M_\odot$ companion which is typical for massive X-ray binaries. The initial eccentricity of $e = 0.25$ appears to vanish just before the end of core hydrogen burning in the massive component if the initial orbital period is less than about 6 days. Near the end of core hydrogen burning the star accelerates its evolution, whereby the the stellar g-modes begin to shift rapidly to higher frequencies as the core contracts. This causes the system to pass through many resonances (corresponding to the wiggles in the curves) whereby the tidal evolution speeds up considerably.

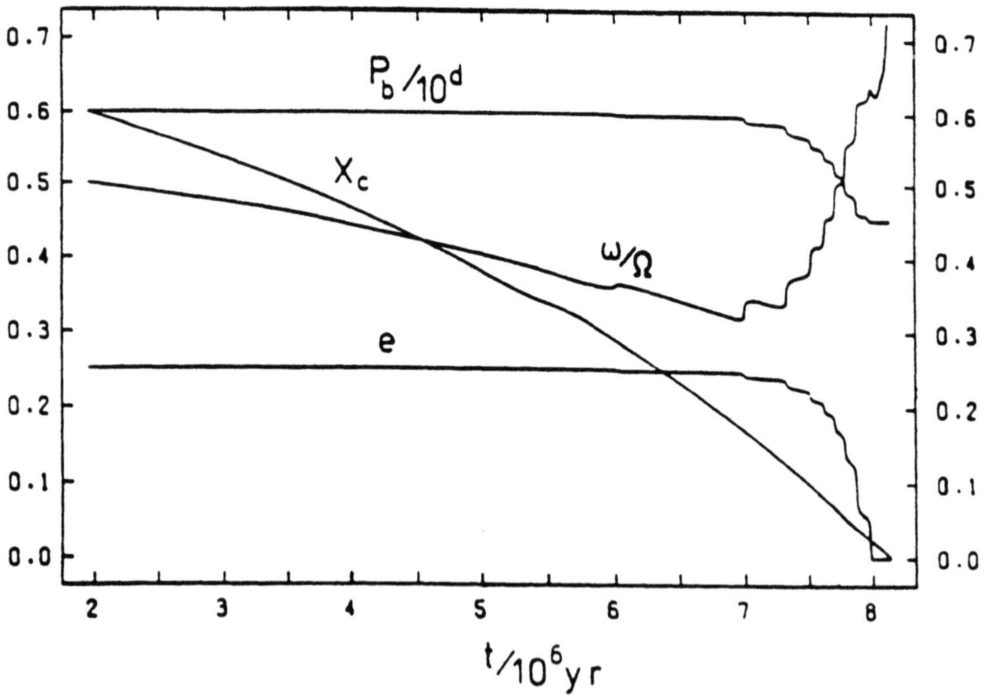

Fig. 4. Evolution of a CBS consisting of a $20M_\odot$ star and a $1M_\odot$ compact companion with initial orbital period of 6 days and orbital eccentricity of 0.25. Tidal evolution was followed using (6) and (7).

Close to a resonance the linear approximation obviously breaks down. However, close to resonance, only a limited amount of angular momentum can be exchanged anyhow, because the strong interaction shifts ω and the orbital parameters in general rapidly away from the resonance condition. Due to the expansion of the secondary (R_s/a) increases and this adds in causing enhanced tidal evolution near the end of the main sequence phase of the massive star (Eqns. 6 and 7).

Because in these systems (with an extreme mass ratio of $\frac{1}{20}$) the stellar moments of inertia are not much less than the orbital moment of inertia μa^2, the tidal evolution does not lead to synchronous rotation of the massive star. In order to spin up the massive star, so much angular momentum needs to be taken from the orbit that it shrinks noticeably. Due to the resulting increase of Ω the ratio $(\frac{\omega}{\Omega})$ hardly increases so that the massive star needs more and more angular momentum from the orbit to spin up to synchronism. Under these circumstances tidal equilibrium cannot be attained: the tidal interaction forces the low mass star to spiral down into the outer layers of its massive companion (see Sect. 5.3).

For stars with convective envelopes t_0 cannot be determined as there is no fundamental theory of turbulent convection. One can only make a crude guess of the 'eddy viscosity' $\nu = \frac{1}{3}vl$, where v and l denote the typical velocity and size of the (largest) turbulent eddies, with help of the mixing length 'theory' (see contribution by J.P. Zahn in these lecture notes). Estimating t_0 from the characteristic viscous timescale $\tau_{visc} = \frac{\nu^2}{R}$, it follows that, typically, $t_0 \simeq 1$ year for low-mass MS-stars. Therefore we expect tidal effects to be important in CBS containing such stars (Zahn 1989).

Concluding, we can say that there is observational (as well as theoretical) evidence that CBS with orbital periods less than about 6-7 days tend to evolve to a state of tidal equilibrium in which $e = 0$ and $\omega = \Omega$, unless the mass ratio is extreme. In the latter case tidal equilibrium may be unattainable. In the following discussion of mass transfer in binary systems we will in general assume that the binary is in a state of tidal equilibrium. Note that stellar evolution (expansion or contraction of the stars) can change the stellar spin rates and disturb the tidal equilibrium.

2 Roche Approximation

Before we address the problem of mass transfer in CBS's we will determine the gravitational potential Ψ in a binary system and the shape of the two stars in the so-called Roche-approximation. Hereby it is assumed that the gravitational fields of *both* stars can be approximated by that of a point mass and that the binary is in tidal equilibrium (synchronous rotation and circular orbit). Introducing a Cartesian reference frame centred at star 1 with star 2, both on the X-axis (at the orbital separation a), and rotating about the centre of mass with the binary angular velocity Ω, we can write the effective gravitational potential corrected for centrifugal effects due to the rotating binary frame as:

$$\Psi(x,y,z) = \frac{GM_1}{r_1} + \frac{GM_2}{r_2} + \frac{1}{2}\Omega^2\left[(x-x_c)^2 + y^2\right] \quad (8)$$

where $x_c = M_2/M_t a$ (with $M_t = M_1 + M_2$) is the distance between the centre of star 1 and the binary's gravity centre, and

$$r_1 = \sqrt{x^2 + y^2 + z^2}$$

$$r_2 = \sqrt{(x-a)^2 + y^2 + z^2}$$

Fig. 5 shows a cross-section of the equipotential surfaces, as defined by (8), with the orbital plane. Close to the stars the contours of constant potential are closed and almost circular. Further away from the stars the contours become more and more distorted, due to the perturbing influence of the companion. The largest closed equipotential surface around the stars is called the **Roche-lobe**. The point at which the two Roche lobes (one

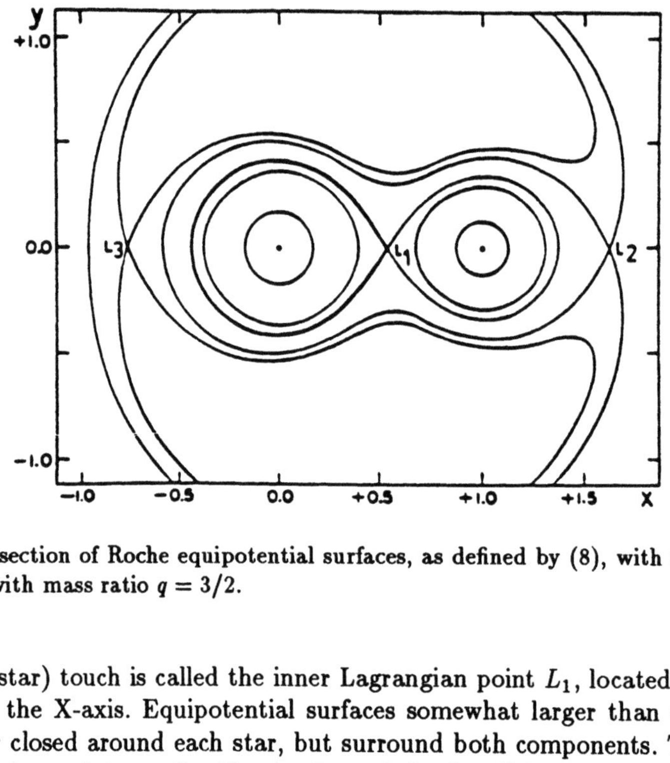

Fig. 5. Cross-section of Roche equipotential surfaces, as defined by (8), with the orbital plane for a system with mass ratio $q = 3/2$.

around each star) touch is called the inner Lagrangian point L_1, located in between the two stars on the X-axis. Equipotential surfaces somewhat larger than the Roche lobe are no longer closed around each star, but surround both components. There exist two more Lagrangian points on the X-axis: L_2 and L_3. In all Lagrangian points $\nabla\Psi = 0$, i.e. the effective force of gravity (corrected for centrifugal forces) vanishes. Check that there must indeed be three such Lagrangian points on the X-axis, and one on each side off the X-axis, in the orbital plane, i.e. five in total. Note that, due to rotation, the equipotential surfaces show no symmetry of revolution around the X-axis (flattening in the Z-direction), see Kopal 1959.

2.1 Shape of Binary Components in Roche Approximation

In the Roche potential given by (8) the gravitational field of the binary components is approximated by that of a point mass. Since all stars show a certain central condensation, this is a reasonable approximation for Ψ in the tenuous outer layers of both components.

In a corotating reference frame the equation of motion of a small gas element in the star is given by:

$$\frac{d^2\mathbf{r}}{dt^2} + 2\Omega \times \frac{d\mathbf{r}}{dt} = -\frac{1}{\rho}\nabla P + \nabla\Psi \tag{9}$$

Note that the centrifugal acceleration has been included in the gradient of the gravitational potential. For synchronous stars in a circular orbit the left hand side of (9) vanishes since the stars are stationary in a corotating frame. This implies hydrostatic equilibrium:

$$\nabla P = \rho\nabla\Psi \tag{10}$$

Because the gradient of a scalar (P) is a vector perpendicular to the contours of equal value of this scalar quantity, we deduce from (10) that surfaces of equal pressure (isobars) coincide with equipotential surfaces. Furthermore it is obvious that surfaces of equal density $\rho(\Psi) = \frac{dP(\Psi)}{d\Psi}$ also coincide with equipotential surfaces. Hence, in the Roche approximation the shape of the binary components coincide with the Roche equipotential surfaces. For later use it is convenient to introduce here the concept of the **Roche Radius** R_c. This is the radius which corresponds to a sphere with the same volume as the Roche lobe. The latter volume can be calculated by numerical integration. For a given orbital separation a the volume of the Roche lobe depends on the binary's mass ratio $q = M_1/M_2$ only. Eggleton (1983) determined an analytical fit formula for the Roche Radius, accurate to within 1 %:

$$R_c/a = 0.49\,q^{\frac{2}{3}}\left(0.6\,q^{\frac{2}{3}} + ln(1 + q^{\frac{1}{3}})\right)^{-1} \tag{11}$$

A convenient expression for the Roche radius only valid for the less massive star ($M_1/M_2 \leq 0.8$) in the binary is (Paczyński 1971):

$$R_c/a = 0.462\left(\frac{M_1}{M_1 + M_2}\right)^{\frac{1}{3}} \tag{12}$$

As an example, table 1 shows the ratio's R_c/a and r_L/a, whereby r_L denotes the distance from the centre of the primary to the L_1 point, boyh as a function of the binary's mass ratio q=M_1/M_2: Note that for $q \to \infty$ the ratio $r_L/a \to 1$, because the companion reduces in that limit to a test particle in a Keplerian orbit at distance a of the primary.

2.2 Mass Loss through L_1: Roche Lobe Overflow

Due to the nuclear fusion reactions in the stellar interior a star expands when it is in the main sequence stage or in the following phase of hydrogen shell burning. The more massive a star the faster its nuclear evolution, so that this effect will be most noticeable for the more massive component in a binary (the primary). When the primary (with mass M_1) expands it may at some moment fill its entire Roche lobe. We have seen that the net gravitational force at L_1 vanishes, whereas $\nabla P \neq 0$ at L_1 if the star fills its entire Roche lobe. Hence, there can be no hydrostatic equilibrium at L_1: the pressure gradient near L_1 will push the gas through the gravitational saddlepoint at L_1 into the Roche lobe of the the secondary (with mass M_2). This gas will move with respect to our corotating binary frame, so that it is subject to both centrifugal and Coriolis forces. The Coriolis

Table 1. The distance of L_1 from the primary's centre and the Roche radius, normalized to the orbital separation a, as a function of the binary's mass ratio q.

q	r_L/a	R_c/a
1	0.500	0.379
1.25	0.523	0.398
1.67	0.552	0.424
2.50	0.593	0.460
3.33	0.621	0.485
10.0	0.718	0.578
20.0	0.769	0.630
100.	0.859	0.720
1000.	0.932	0.785

forces deflect the gas stream from L_1 away from the attracting secondary, so that the gas swirls around this star and cannot fall straight onto it. We will discuss the qualitative features of this flow in a later section. Let us here consider the stability of the mass flow through L_1. Assuming that no mass escapes from the Roche lobe around the secondary (**Conservative Case**) we have: $\dot{M}_2 = -\dot{M}_1$, where the dot means a time derivative. By differentiating logarithmically with respect to time the expression for the orbital angular momentum in a circular orbit H:

$$H = \frac{M_1 M_2}{(M_1 + M_2)} a^2 \Omega$$

we obtain with $\Omega = \sqrt{G(M_1 + M_2)/a^3}$:

$$\frac{\dot{a}}{a} = -2\left(1 - \frac{M_1}{M_2}\right)\frac{\dot{M}_1}{M_1} + 2\frac{\dot{H}}{H} \tag{13}$$

In general the angular momentum of the infalling material (with respect to the secondary) may be neglected compared to H, so that $\dot{H} = 0$. With $\dot{M}_1 < 0$ we conclude that $\dot{a} < 0$ for $\frac{M_1}{M_2} > 1$, so that the mass exchange between the two stars makes the orbit, and thereby the Roche lobe, smaller. Therefore, more mass will tend to flow through the L_1-region. The mass transfer is unstable and increases in strength, unless the primary ('contact') star shrinks more than its Roche lobe in reaction to the mass loss. In Sect. 2.4 we will address the problem of how a star reacts to sudden mass loss. But first we will briefly discuss the dynamics of the gasstream from L_1.

2.3 Dynamics of Gas Stream from L_1

The gaselements that flow through the L_1 region into the Roche lobe of the secondary rapidly accelerate to supersonic velocities when they fall towards the secondary. One can, therefore, in first approximation neglect pressure effects and use a ballistic approximation to determine the dynamics of the gasstream. Fig. 6 shows a few trajectories of particles which originate from the L_1 point with different velocity vectors (thermal evaporation speed) for a binary with $q = \frac{2}{3}$ (from Flannery, 1975). All three particles approach the secondary at a minimum distance of $R_{\min}$ and then fly by to rebounce at the Roche lobe (no dissipation).

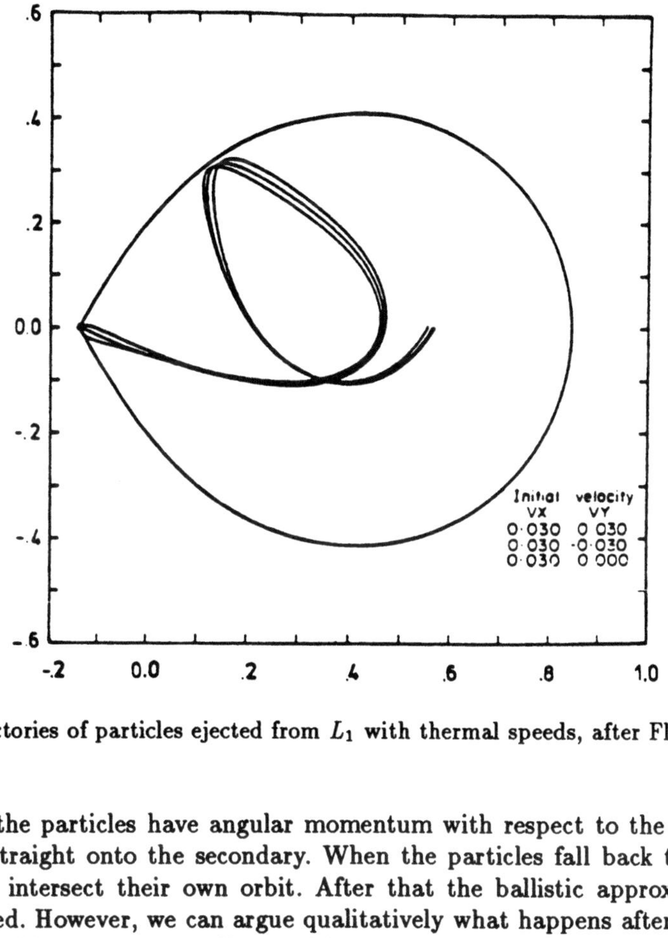

Fig. 6. Trajectories of particles ejected from L_1 with thermal speeds, after Flannery 1975.

Because the particles have angular momentum with respect to the secondary, they cannot fall straight onto the secondary. When the particles fall back towards the secondary they intersect their own orbit. After that the ballistic approximation can no longer be used. However, we can argue qualitatively what happens after the stream collides with itself. During the collission kinetic energy is dissipated, while the angular momentum with respect to the secondary is approximately conserved. The dissipation therefore tends to make the orbit circular (Sect. 1.1). We can make a crude estimate of the radius R_h of such a circular Keplerian orbit by taking for the specific orbital angular momentum $j = \sqrt{GM_2R_h}$ that of matter at rest at $L_1 : j = \Omega(a - r_L)^2$. This yields for R_h :

$$R_h \simeq \left(1 - \frac{r_L}{a}\right)^4 \left(\frac{M_2}{(M_1 + M_2)}\right)^{-1} a$$

For a system with $q = 1$, for example, this yields $R_h \simeq 0.125a$. Table 2 gives more accurate values for R_h and the distance of closest approach $R_{\min}$ for various mass ratios q, as derived by Lubow and Shu (1975) by integrating particle trajectories from L_1.

Table 2. Ballistic approximation for the gasstream from L_1. The distance of closest approach R_{min} and the radius of the circularized Kepler orbit R_h about the secondary, normalized to the orbital separation a, as a function of the binary's mass ratio (after Lubow and Shu (1975).

M_2/M_1	R_{min}/a	R_h/a
0.067	0.023	0.040
0.150	0.028	0.049
0.300	0.033	0.059
0.500	0.038	0.068
0.750	0.044	0.078
1.000	0.049	0.087
1.667	0.060	0.106
3.000	0.079	0.136
4.500	0.097	0.163
7.000	0.120	0.198
10.000	0.142	0.230
15.00	0.171	0.270

If the secondary's radius is larger than R_{min} the stream from L_1 will directly hit the star. On the other hand, if $R_2 < R_{min}$, the infalling gas will form a ring about the secondary. This will generally be the case if the secondary is a compact star (a white dwarf, neutron star or black hole). Because of viscous effects (the nature of which is still controversial) the ring spreads into a so-called accretion disc, which extends inwards either to the magnetosphere (if the secondary has a non-negligeable magnetic field) or all the way down to the surface of the secondary. At the inside of the disc the gas will settle onto the secondary, after it has transferred its excess angular momentum to exterior matter by viscous interaction in the accretion disc. Because of the outward transport of angular momentum the outer edge of the accretion disc expands, until tidal interaction with the primary becomes sufficiently efficient to transport the excess angular momentum back into the orbit (e.g. Papaloizou and Pringle 1977, Lin and Papaloizou 1979). For a review on accretion discs see, e.g. Shakura and Sunyaev 1973, Pringle 1981.

2.4 Response of the Contact Star to Mass Loss

Suppose we suddenly remove some mass from the outer layers of a star that was initially in thermal and hydrostatic equilibrium. In thermal equilibrium the structure of the star is such that the rate of energy production by nuclear fusion is precisely balanced by the outward energy transport and atmospheric radiation into space: $L_{nuc} = L_*$. The perturbed star will restore hydrostatic equilibrium on a short, dynamical timescale by expanding slightly. Inside the star the initial expansion will be nearly adiabatic since the dynamical timescale is much smaller than the timescale for thermal readjustment (the ratio τ_{dyn}/τ_{th} is very small, except close to the stellar surface). After the initial rapid expansion the star will therefore in general no longer be in thermal equilibrium. The star may be either smaller or larger than it was before the mass loss, depending on its internal structure. The star can adjust its thermal structure on a thermal timescale τ_{th}. If the star has lost only a small amount of mass, it will regain practically its original size after thermal equilibrium has been restored. If the perturbation is confined to the outer layers thermal equilibrium may be restored quite fast, but if the perturbation is more severe and affects a large fraction of the star it may take approximately the Kelvin Helmholtz timescale $\tau_{KH} = (GM^2)/(RL)$ to restore thermal equilibrium.

For mass exchange through L_1 the crucial question is whether (and how fast) the star expands or contracts in reaction to the mass loss. Let us adhere to the schematic picture discussed above and divide the stellar response to mass loss into two stages, an adiabatic expansion followed by a thermal readjustment. Fig. 7 shows schematically the run of specific entropy S as a function of radius in the outer layers of a star in thermal equilibrium, with either a radiative or a convective envelope. When the star is convectively stable, i.e. $\nabla < \nabla_{ad}$ the entropy S increases outwards, while for convective stars (especially for giants with very tenuous, cool outer layers in which convection is less effective in transporting energy) $\nabla > \nabla_{ad}$ and S decreases outwards.

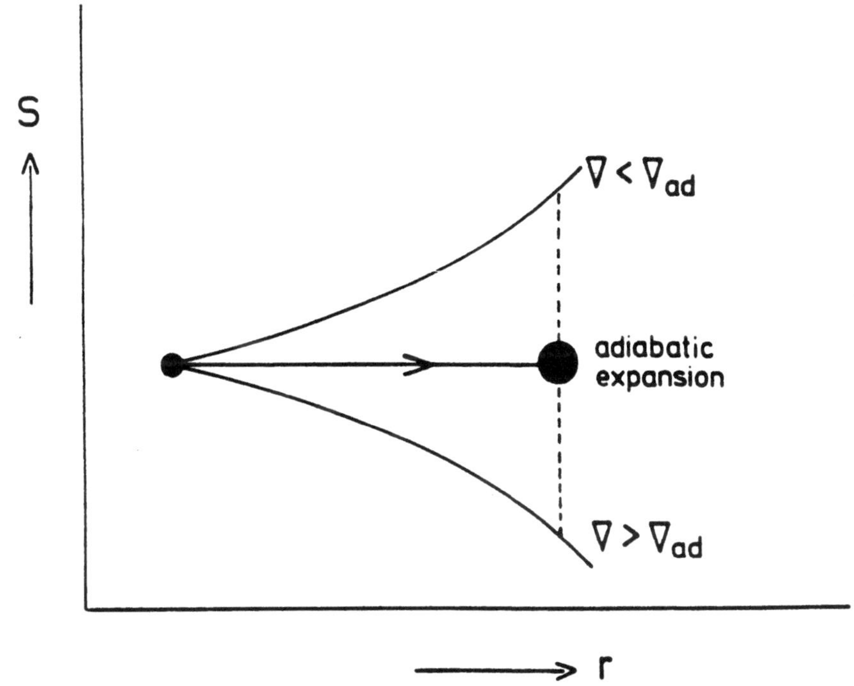

Fig. 7. Schematic picture of (adiabatically) expanding gaselements in the outer layers of a mass losing star. S is entropy of stellar gas per unit mass. See text for details.

Consider a gaselement that rises upwards and expands (adiabatically) due to the mass loss at the surface. If the star is radiative the gaselement will have in its new equilibrium position a *lack* of entropy compared to the thermal equilibrium situation, i.e. the mass losing star will tend to be cooler and *smaller* than it was before the mass loss. If on the other hand, the star has a super-adiabatic temperature gradient the gaselement will have an *excess* of entropy compared to the thermal equilibrium situation and the star will tend to be *larger* than it was before the mass loss.

Hence, if the contact star has a radiative envelope, the unstable mass flow through the L_1 region limits itself by causing the star to become smaller than its thermal equilibrium size, so that it fits more easily inside its Roche lobe. During the initial rapid phase of mass loss from such a star a substantial amount of the interior luminosity is absorbed by the expanding outer layers (in order to regain thermal equilibrium the gaselements have to increase their entropy and absorb energy from the outgoing energy-flux). However, for (more massive) contact stars with $\nabla > \nabla_{ad}$ (giants with deep convective envelopes) the star becomes larger than its equilibrium size and will not be able to adhere to the shrinking Roche lobe. Such stars undergo violent mass loss. Of course, if the contact star is sufficiently *less* massive than its companion, the Roche lobe will *expand* faster than the contact star and there is no violent mass transfer.

Figure 8 shows schematically the evolution of the stellar radius R as a function of the contact star's mass M. From point A to B the primary is detached, while in point B it has expanded to its Roche lobe.

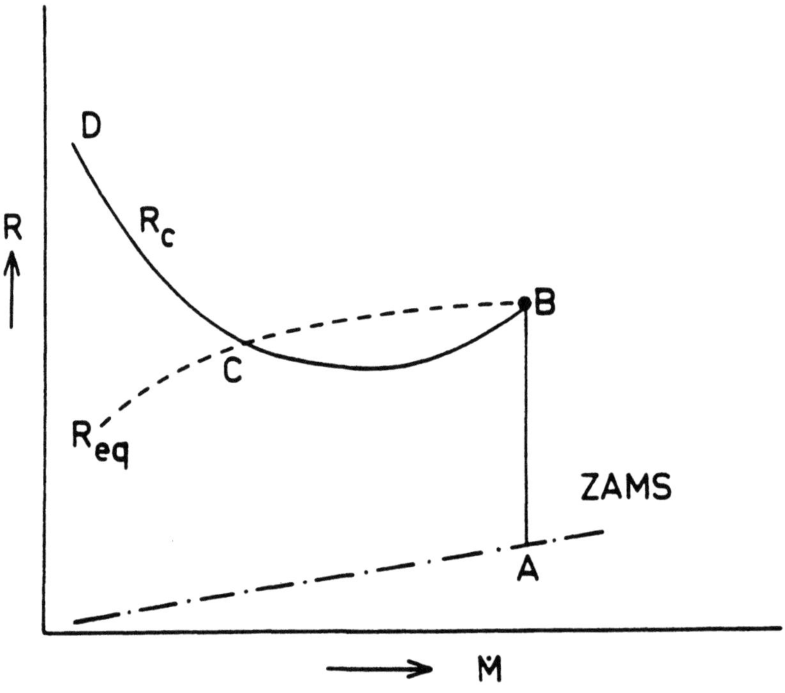

Fig. 8. Schematic picture of the Roche radius and the thermal equilibrium radius of a star as a function of its mass.

From point B on the primary's mass will decrease due to the mass transfer through

L_1. The curve labelled R_{eq} corresponds to the thermal equilibrium radii of stars with different masses, but with a chemical composition corresponding to point B. This curve is analogous to the (stripe-dotted) curve labelled ZAMS, which corresponds to a homogeneous chemical composition. From point B on the star will transfer mass at such a rate that its radius remains approximately equal to the Roche radius. Initially this requires very rapid mass loss, on a timescale of the order of the Kelvin-Helmholtz (global thermal) timescale. Once the binary's mass ratio has been reversed and the Roche radius begins to reexpand (according to (11) the latter two phenomena do not precisely coincide) the contact star has time to restore its thermal equilibrium. This occurs near point C. Note that the evolution from B to C is so fast that we may neglect the change in the star's chemical composition due to nuclear burning ($\tau_{\rm nuc} >> \tau_{\rm KH}$). When the Roche lobe expands in reaction to mass loss the mass transfer can only continue if there is a mechanism that causes the mass losing star to expand as well. Usually nuclear evolution causes a star to expand, and from about C to D (and further) the mass transfer proceeds on the much longer nuclear evolution timescale of the contact star. During the slow phase the contact star remains approximately in thermal equilibrium. The star is no longer constrained to follow the curve $R_{\rm eq}$, since its chemical composition has been changed compared to that of point B as a result of nuclear burning.

Concluding, we note that one can distinguish two phases in the mass transfer stage, namely:
i) an initial phase of rapid mass loss during which typically a peak mass transfer rate of $\dot{M} \simeq M/\tau_{\rm KH}$ is reached (for giant stars the transfer rate can be much larger), and during which the binary's mass ratio is reversed, followed by
ii) a phase of slow mass transfer, with typically $\dot{M} \simeq M/\tau_{\rm nuc}$. However, if after the phase of rapid mass loss, nuclear evolution causes the contact star to shrink (e.g. during core helium burning) then the star detaches itself from the Roche lobe and the mass transfer will cease.

3 Numerical Calculations of Close Binary Evolution

Many approximate calculations of binary evolution have been performed with 1 dimensional stellar evolution codes (see contribution by C. de Loore) in which the outer boundary conditions of a single spherical star are adapted to the possibility of mass transfer through the L_1 point, e.g. the pioneering calculations of: Morton (1960), Paczyński (1966), Kippenhahn and Weigert (1967), Plavec (1968) and many others. Usually one determines the mass transfer rate from the condition that the stellar radius R of the spherical 'contact' star equals its Roche radius R_c. One neglects the rotational and tidal deformation of the binary stars because existing stellar evolution codes are 1-dimensional and can only treat spherical stars. However, this approximate procedure does account for the basic fact that stars in CBS cannot expand beyond a certain size, approximated by the Roche-radius of a spherical star, and are forced to lose (an appreciable fraction of their) mass to their (originally) less massive companion. As a result their evolution is fundamentally different from single stars with the same initial conditions.

The flow through L_1 and the accretion of this flow by the secondary is a hydrodynamic problem that has not yet been solved in its full complexity. In most cases the evolution of the mass gaining companion in the binary has not been followed, because the problem of accretion is more complicated than that of mass loss through L_1. If the transferred mass would fall adiabatically onto the mass gaining companion, the liberated accretion energy would soon cause this star to swell up (e.g. Benson 1970, Flannery and Ulrich 1977, Packet and De Greve 1979). Very soon the secondary would expand to its Roche lobe, and the binary would become a contact system. Obviously, that would make the evolution much more complicated, and very difficult to treat, e.g. Sugimoto and Miyaji 1980, Webbink 1980, Plavec 1980. However, the accretion onto the tenuous outer layers of the secondary is presumably far from adiabatic, and the outcome is not at all clear. As long as the mass gaining star remains inside its Roche lobe we can, at least initially, ignore it and consider the primary's evolution only.

A further complication is that the binary system may not be tidally relaxed, so that the adopted Roche approximation is invalid. The stars may rotate non-synchronously, or the orbit may be elliptic, so that the tidal perturbation is time dependent and dynamical effects are important. The concept of Roche lobes makes in that case little sense, e.g. Limber (1963), Lubow (1979) so that hydrodynamic calculations of the mass transfer are required, which is technically difficult and has not yet been done.

In most calculations of binary evolution it is assumed that all the transferred mass is ultimately accreted by the companion, so that the total mass and angular momentum of the binary is conserved. Observations indicate that this may be a simplification and that some mass (and angular momentum) may be lost from the system during the mass transfer phase, e.g. Polidan and Peters 1980. In very compact binaries gravitational radiation may carry away a non- negligeable amount of angular momentum, see Sect. 6.2.

3.1 Classification of Mass Transfer: Cases A, B and C

It is common to distinguish three different cases (A,B and C) of binary evolution, depending in what phase of stellar evolution the primary is when it first expands to its Roche lobe (Kippenhahn and Weigert 1967). To illustrate this let us consider the evolution of a $5M_\odot$ star. Fig. 9 shows the stellar radius as a function of evolution time.

For a circular orbit with radius a Kepler's law can be written as:

$$\log P_b(\text{days}) = 1.5 \log\left(\frac{a}{R_\odot}\right) - 0.5 \log\left(\frac{M_1 + M_2}{M_\odot}\right) - 0.936 \tag{14}$$

Adopting $M_1 = 5M_\odot$ and $q = M_1/M_2 = 2$, we find from (11) for the Roche radius of the primary $R_c = 0.44a$. Combining this with (14), we find a relation between the orbital period and the Roche radius of the $5M_\odot$ star:

$$\log P_b(\text{days}) = 1.5 \log\left(\frac{R_c}{R_\odot}\right) - 0.84 \tag{15}$$

Fig. 9 shows three evolutionary phases for which the primary expands. The first phase corresponds to core hydrogen burning (Main Sequence phase), while the second and third phase correspond to the rapid core contraction preceding helium and carbon ignition in the core. In the latter two phases the star is not in thermal equilibrium. When the primary expands to its Roche lobe during the first expansion phase we speak of **Case A**, otherwise of **Case B** or **Case C** mass transfer, respectively. With help of (15) we can calculate the orbital periods that correspond to the limits of Case A, B and C (Fig. 9).

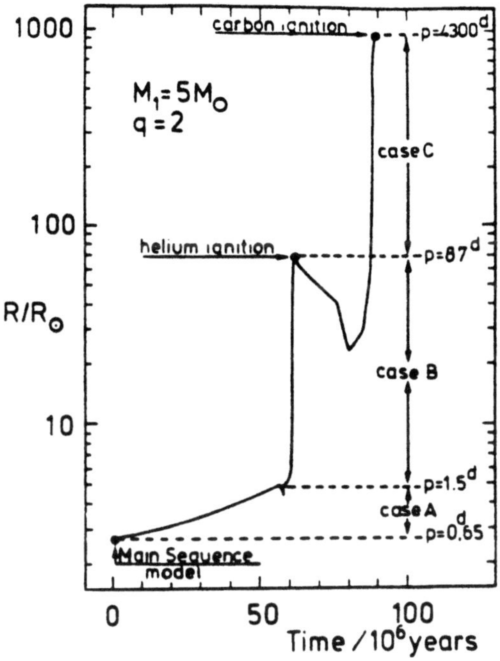

Fig. 9. Stellar radius as a function of time, after Paczyński 1971. Depending on the initial orbital period of the CBS the mass transfer will occur as Case A, B or C. See text for details.

For example, if the orbital period of our binary lies in the range 0.65 days to 1.5 days, the primary would fill its Roche lobe while still on the main sequence, and the mass transfer would be classified as Case A. Case B would correspond with an initial orbital period in the range 1.5 to 87 days, and Case C with an initial orbital period in the range of 87 to 4300 days.

Most calculations of binary evolution were performed for the cases A and B. Low mass primaries that undergo Case C mass transfer (e.g. Lauterborn 1970) have generally very extended envelopes with poorly determined radii (uncertain mixing length) and are prone to violent mass transfer (see Sect. 2.4). For massive primaries the evolution after core helium burning is not well understood because of uncertainties with (semi-convective) mixing while stellar wind losses should be taken into account. See Paczyński 1971 for an overview of standard evolution calculations for Cases A, B and C. For a recent study of Case C mass transfer, see Pastetter and Ritter 1989.

3.2 Short Overview of Binary Evolution in Case A and B

For Case A evolution the primary expands to its Roche lobe while it is in the core hydrogen burning phase. In the rapid phase of mass transfer the mass ratio q is more than reversed, while the primary's thermal equilibrium is greatly disturbed. The subsequent evolution proceeds on the primary's nuclear timescale. The primary, still in the core hydrogen burning phase, fills its Roche lobe as an overluminous subgiant. Compared with a main sequence star of the same mass its overluminosity can be as large as 3 magnitudes. The secondary, which has become the more massive component, now evolves much faster and after some time, it will also expand to its Roche lobe. Therefore Case A evolution leads in general to contact binaries.

Case B of evolution proceeds differently in massive binaries and low-mass systems. The initial rapid phase of mass transfer, whereby the mass ratio q reverses, proceeds in a qualitatively similar way. But the subsequent phase of slow mass transfer, which occurs on the intrinsic expansion timescale of the primary, differs substantially in massive and low-mass binaries.

For primaries with initial mass $M_i \leq 3M_\odot$ the phase of rapid core contraction ends when the electron gas in the helium core becomes degenerate. After that the intrinsic expansion timescale is that of shell hydrogen burning in a star with a degenerate helium core. The primary can reach very high overluminosities, up to 10 magnitudes, in such systems. The mass ratio q can become as low as 1/5 or 1/10, whereby the main sequence star is much brighter than its 'overluminous' companion. If a binary system like this happens to eclipse, it will show an Algol-type lightcurve. The mass transfer ceases when the contact star has exhausted its hydrogen-rich envelope by the combined action of shell burning and mass transfer. The final outcome is thus a detached system with a helium white dwarf. When the companion expands to its Roche lobe a second phase of mass transfer (in the reverse direction) will start.

In primaries with an initial mass larger than about $3M_\odot$ the mass exchange is terminated when the primary ignites helium in its core. After that the star starts to shrink and detaches itself from its Roche lobe. Before that the primary has lost almost its complete hydrogen-rich envelope during the mass exchange and settles itself near the helium Main Sequence. The helium star remnant may look like a Wolf Rayet star (e.g. Paczyński 1968 and Kippenhahn 1969). If the helium star is more massive than about $2.5M_\odot$ (Habets 1986) it will evolve to a supernova stage. When the (now more massive) companion expands to its Roche lobe a second phase of mass transfer will begin.

3.3 Example of Binary Evolution Calculation for Case B

As an example we consider a binary with initial parameters: $M_1 = 4M_\odot$, $M_2 = 3.2M_\odot$ and $P_b = 1.8$ days. With Kepler's law we derive for the orbital separation $a = 12.0R_\odot$. The primary's Roche radius follows as $R_c = 4.78R_\odot$. The primary attains this radius only during the core contraction phase prior to helium ignition (Case B). Fig. 10 shows the Hertzsprung-Russell (HR)diagram with the evolution tracks of a $4M_\odot$ single star and of a star with the same mass in a binary system with the above specifications.

At point 4 the primary fills its Roche lobe. From then on the primary deviates from thermal equilibrium and the (surface) luminosity shoots down because of strong energy absorption in the radiative outer layers (see Sect. 2.4). The points in Fig. 10 are defined according to the table 3, where 'conv. ++' means that the contact star's envelope becomes convective and 'conv. −−' that the convection disappears.

$L_{\max}$ is the maximum value of the internal luminosity in the star. The difference $L_{\max} - L_*$ is absorbed by the outer layers. Fig. 11 shows the calculated mass transfer

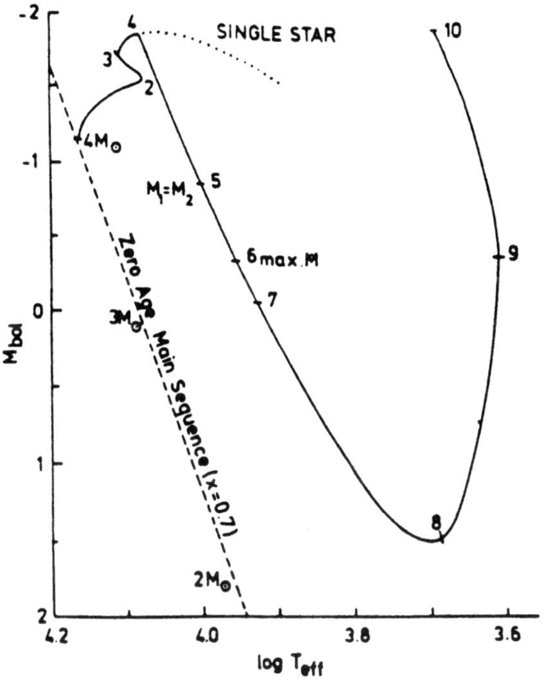

Fig. 10. Evolution of CBS with mass transfer (Case B) in HR-diagram (after van der Linden, unpublished).

Table 3. Legend to Figure 10

point	condition	$M_1/M_\odot$	$R_1/R_\odot$	$L_{max}/L_\odot$	$L_*/L_\odot$
4	$R_1 = R_c$	4.0	4.8	430	430
5	$M_1 = M_2$	3.6	4.4	350	176
6	$\dot{M}_1$ max	3.3	4.3	290	98
7	R_1 min	3.1	4.3	250	71
8	conv.++	1.7	6.4	73	20
9	conv.−−	0.60	22.4	122	121
10	He ign.	0.46	31.6	521	521

rate as a function of time, where $t = 0$ corresponds to the beginning of Roche lobe overflow.

The mass transfer rate reaches a peak of more than $6 \times 10^{-6} M_\odot$/year. The second, smaller peak, is caused by the appearance of convection in the contact star's outer layers (point 8). This is consistent with the discussion in Sect. 2.4, i.e. a contact star with a (super-)adiabatic temperature gradient in (part of) its outer layers (convection) shows less tendency to shrink when it is driven out of thermal equilibrium by mass loss. Near point 9 the contact star re-approaches the thermal equilibrium state ($L_{max} \simeq L_*$), the

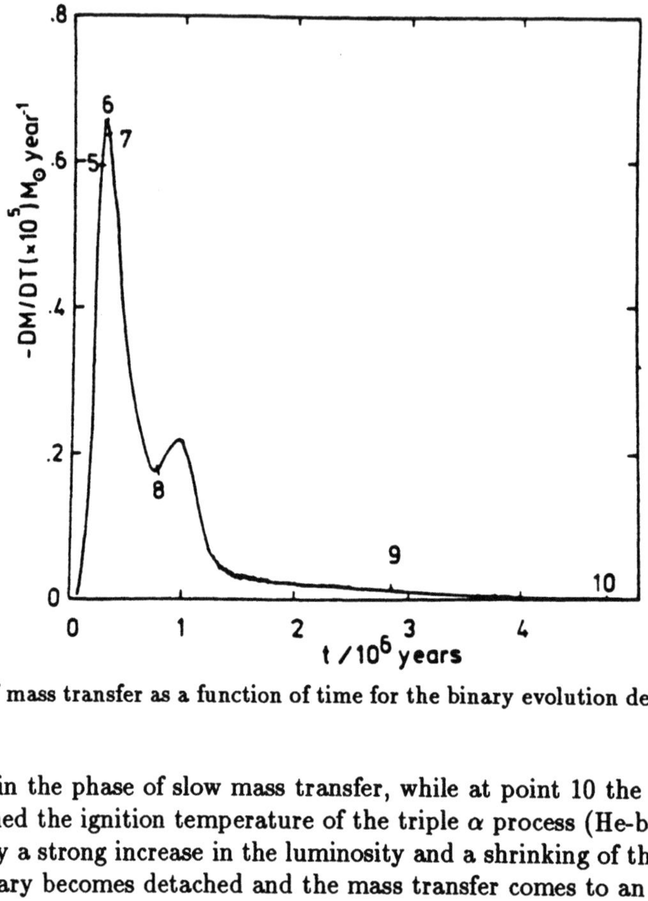

Fig. 11. Rate of mass transfer as a function of time for the binary evolution depicted in Fig. 10.

system is now in the phase of slow mass transfer, while at point 10 the central temperature has reached the ignition temperature of the triple α process (He-burning). This is accompanied by a strong increase in the luminosity and a shrinking of the stellar radius, so that the binary becomes detached and the mass transfer comes to an end.

4 Late Stages of Close Binary Evolution: X-ray Binaries

Let us now apply the above analysis to the late stages of close binary evolution in which one of the stars has finished its nuclear evolution and managed to collapse to a compact state (white dwarf, neutron star or even black hole).

There exist roughly one hundred bright ($L_x \geq 10^{36}$erg/s) X-ray sources in our galaxy, most of which are close binary systems containing a neutron star (e.g. Hutchings 1982, Long and Van Speybroeck 1983, Van Paradijs 1983, Verbunt 1989). The companion of the neutron star ejects mass, either by Roche lobe overflow, by some kind of equatorial outflow (Be stars), or by the emission of a strong stellar wind (massive OB-star). When the ejected mass is accreted by the neutronstar the gravitational potential energy of about GM_x/R_x per gram of accreted matter (where M_x and R_x are the mass and radius of the neutron star) is converted into kinetic energy and must be almost completely dissipated before the matter can settle on the neutron star's surface. Substituting the canonical neutron star values $M_x = 1.4M_\odot$ and $R_x = 10^6$ cm, one finds that about 10^{20} erg/g, (which amounts to $0.1c^2$ i.e. about 10% of the rest mass), is liberated in the accretion process. Accretion onto a neutron star is thus a few orders of magnitude more efficient in liberating energy than nuclear fusion. To power a bright X-ray source of 10^{36} erg/s would require an accretion rate of 10^{16}g/s, which is about $10^{-10}M_\odot$/year, a figure that is very well possible in close binaries.

One distinguishes low-mass X-ray binaries (LMXB's), in which the mass of the mass losing star is less than about $2M_\odot$, from massive X-ray binaries (MXB's), which harbour a massive star (say $M \geq 10M_\odot$). X-ray binaries containing mass losing stars with a mass in between these two limits seem rare. A short characteristic of both classes is given in table 4.

Table 4. Classification of X-ray Binaries

MXB's	LMXB's
massive donorstar $M \geq 10M_\odot$	low-mass donorstar $M \leq 2M_\odot$
young stellar population	old stellar population
often contain X-ray pulsar	rarely contain X-ray pulsar
often show X-ray eclipse	very rarely show X-ray eclipse
no X-ray bursts	lower luminosity sources
	often show X-ray bursts
optical light dominated	optical light dominated
by massive companion	by accretion disc
accretion by stellar wind	accretion by Roche lobe overflow
(incipient Roche lobe overflow?)	presence of accretion disc

If the accreted matter is supplied by Roche lobe overflow it forms an accretion disc around the neutron star (Sect. 2.3). The supplied matter must then first lose a large fraction of its angular momentum before it can be accreted onto the neutron star (e.g. Pringle 1981).

If the neutron star possesses a strong magnetic field the disc cannot reach the stellar surface, but is truncated on the inside at the stellar magnetosphere. Inside the magnetosphere the accreting plasma is forced to flow along magnetic fieldlines and to corotate with the central star (e.g. Lamb et al. 1973). For a surface magnetic field with a dipole character the accreting matter can only reach the stellar surface near the two magnetic poles. Two 'hot spots' will form at the magnetic poles where the accreting matter hits the stellar surface. If the neutron star spins along an axis which makes an angle with the magnetic axis, the two hot magnetic poles will act as a kind of X-ray lighthouse. We observe an X-ray pulsar if the beam strikes along the earth. Apparently, the neutron stars in LMXB's have in general no strong magnetic dipole field, in contrast to the ones in MXB's which often contain X-ray pulsars. This seems not surprising because the neutron stars in LMXB's are much older, so that the magnetic field has had time to decay (on a typical timescale of some 10^7 years as indicated by the observations of radio pulsars) from the $10^{12} - 10^{13}$ G believed to be common for young neutron stars to much smaller values (Ostriker and Gunn 1969, van den Heuvel et al. 1986).

The absence of X-ray eclipses in LMXB's is thought to be due to the relatively small solid angle subtended by the companion (Joss and Rappaport 1979) and/or to the shielding of the companion star by a thick accretion disc (Milgrom 1978).

The lower luminosity LMXB's often show X-ray bursts, short periods of much higher X-ray intensity. There appear two kind of bursts, one associated with instabilities in the accretion flow near the neutron star, the other with thermonuclear flashes of accreted material on the neutron star's surface (e.g. Lewin and Joss 1983).

While the companion of the neutron star is well visible in the optical in MXB's, and appears to be a bright early type star, it is not so in LMXB's. The companion in LMXB's is in general invisible because the accretion disc in LMXB's dominates the optical light by intercepting X-rays from the central regions of the disc and reprocessing the X-radiation into the optical part of the spectrum. Therefore we know very little about the companion stars in LMXB's. Apparently it are not luminous stars like in MXB's. However, in some systems the X-ray emission sometimes vanishes (X-ray transients like Aql X-1 or Cen X-4). During the quiescent phase one has observed the absorption spectrum of a companion, which appears to be a main sequence star of type G-K (van Paradijs 1983).

Some LMXB's show periodic X-ray absorption dips from which orbital periods have been determined. The inferred orbital periods range from a very short 41^m to a more common value of $\sim 10^h$. However, the brightest LMXB's do not show these X-ray dips and because most of them are situated in the galactic bulge and therefore not observable in the optical, we do not know their orbital periods. It has been suggested (Webbink et al. 1983) that the very bright ($L_x \simeq L_{\rm Edd}$) sources are wide binaries ($P > 1$ day) with (sub)giant contact stars. Several of these brighter sources, which do not show the phenomenon of X-ray bursts, show instead substantial changes in the shape of their X-ray power spectra. These so called **quasi periodic oscillations** (QPO) on short timescales are thought to be associated with the accretion flow close to the compact object (e.g. van der Klis 1989).

5 Massive X-ray Binaries (MXB's)

5.1 Formation of MXB's

X-ray binaries contain a collapsed star (neutron star or black hole) and must therefore be evolved systems. Observations indicate that collapsed stars can be formed by core collapse (type II SN) in the final stages of evolution in stars whose original mass was larger than about $7M_\odot$. Most of the gravitational energy liberated during core collapse is emitted in the form of neutrino's. In type II supernova explosions the stellar envelope is blown into space, for which only a small fraction of the liberated gravitational energy is needed. The precise mechanism(s) which causes the envelope ejection is not well understood (see contribution by E. Müller). Since the more massive star in the binary evolves fastest, it will reach the supernova stage first. This may seem a strong impediment to the formation of X-ray binaries, because the supernova explosion of the most massive component is likely to disrupt the binary.

Effect of Supernova Explosion: For it can be shown (e.g. Boersma 1960) that when a star in a binary system with a circular orbit undergoes instantaneous (short compared to orbital period) isotropic mass loss, the binary will be disrupted if more than half of the system's mass is ejected. For supernova explosions instantaneous mass loss seems a fair approximation. It is assumed that the impact of the exploding supernova shell on the companion can be neglected. Numerical hydrodynamic calculations (2D) suggest that this is a rather good approximation (Fryxell and Arnett 1981). However, if the explosion is non-isotropic the resulting orbit depends on the amount and direction of the non-isotropy. Depending on whether the non-isotropy would be directed in the same direction or the reverse direction of orbital motion it would favour or disfavour the disruption of the binary system.

The above statement on the disruption of the binary system can be shown easily by noting that an isotropic explosion produces no net force on the star, so that the relative velocity of the two stars remains invariant just before and after the explosion. If ΔM is the mass lost instantaneously by star 1 then the total binary energy E_f after the explosion (in a frame moving with the system's new centre of mass) can be written as the sum of the potential and kinetic energy:

$$E_{\rm f} = -G\frac{(M_1 - \Delta M)M_2}{a_{\rm i}} + \frac{1}{2}\frac{(M_1 - \Delta M)M_2}{M_1 + M_2 - \Delta M}V^2$$

where the relative velocity of the two stars is given by: $V = \sqrt{G(M_1 + M_2)/a_{\rm i}}$ with $a_{\rm i}$ the radius of the original circular orbit. This yields:

$$E_{\rm f} = -\frac{G(M_1 - \Delta M)M_2}{2a_{\rm i}}\left(\frac{M_{\rm t} - 2\Delta M}{M_{\rm f}}\right)$$

where $M_{\rm t} = M_1 + M_2$ and $M_{\rm f} = M_{\rm t} - \Delta M$ is the system's initial and final total mass.

Indeed we find a parabolic orbit with $E_{\rm f} = 0$ if $\Delta M = \frac{1}{2}M_{\rm t}$. If less than half of the total mass is ejected the orbit becomes elliptic ($0 < e < 1$). Because the radial component of relative velocity is zero before (and just after) the explosion the periastron distance of the new elliptic orbit (with semi-major axis $a_{\rm f}$) $a_{\rm p} = a_{\rm f}(1 - e)$ must equal the original orbital separation $a_{\rm i}$. By equating $E_{\rm f} = -G(M_1 - \Delta M)M_2/(2a_{\rm f})$ to the expression for $E_{\rm f}$ derived above, we find another relation between $a_{\rm i}$ and $a_{\rm f}$: $a_{\rm i} = a_{\rm f}(1 - \Delta M/M_{\rm f})$. This yields for the orbital eccentricity:

$$e = \Delta M/M_{\rm f}$$

Conservation of momentum in the frame moving with the binary's original centre of mass yields (the net momentum of the ejected matter is zero) requires:

$$M_{\rm f}\,V_{\rm kick} = M_2V_2 - (M_1 - \Delta M)V_1$$

where V_1 and V_2 are the orbital velocities of the two stars just before (or just after) the explosion. The mass loss therefore gives the binary system a kick velocity:

$$V_{\rm kick} = \frac{M_2}{M_{\rm t}}\frac{\Delta M}{M_{\rm f}}\sqrt{\frac{GM_{\rm t}}{a_{\rm i}}}$$

Since in a type II supernova only a small fraction of the stellar mass collapses (about $1.4M_\odot$) and several $M_\odot$ are ejected, the binary is likely to be disrupted.

However, the disruption is avoided if the binary is sufficiently close for *mass exchange to take place before the supernova event.* In that case the massive star will lose almost its complete hydrogen-rich envelope by Roche lobe overflow to its companion (e.g. Van den Heuvel and Heise 1972, van den Heuvel 1978) before the supernova explosion. In the progenitor system of a MXB the first phase of mass transfer presumably takes place according to Case B, whereby the remnant star is an almost pure helium star when the mass loss ceases. The helium star evolves in only a few times 10^5 years to a supernova stage and collapses to a neutron star if its mass is more than 2 - 3 $M_\odot$ (e.g. Habets 1986). Because the mass ejection of the supernova is now a small fraction of the binary's mass the explosion is not likely to be disruptive. Tidal interaction will tend to make the orbit circular again.

X-ray Stage: When the unevolved companion begins to eject a strong stellar wind the neutron star companion may capture mass at a rate sufficient to turn itself into a bright X-ray source (see Sect. 4.3). When the massive star swells to its Roche lobe and begins to transfer mass the neutron star soon accretes so much mass that the X-rays are likely to be absorbed by the accreting matter whereafter the bright X-ray source disappears from the sky.

A typical maximum X-ray luminosity of an X-ray source is given by the so called **Eddington limit**, which applies strictly only to spherical symmetric accretion. L_{Edd} is the luminosity for which the outward radiation pressure (in a hot plasma mostly due to electron scattering) balances the inward pull of gravity (which is predominantly on the protons for accreting hydrogen). The electrons and protons are of course held together by electromagnetic forces. We find (e.g. Tucker 1975):

$$\sigma_{\text{T}} \frac{L_{\text{Edd}}}{4\pi r^2 c} = \frac{G M_x m_p}{r^2}$$

where m_p is the mass of the proton, σ_T the cross section for electron scattering, G the constant of gravity, c the speed of light and M_x the mass of the accreting object. Hence,

$$L_{\text{Edd}} = \frac{4\pi G m_p c}{\sigma_T} M_x \simeq 1 \times 10^{38} \left(\frac{M_x}{M_\odot}\right) \text{erg/s}$$

If the massive star is still in the core hydrogen burning phase incipient Roche lobe overflow can power a bright X-ray source during some $10^4 - 10^5$ years before the X-ray source is quenched (Savonije 1978). This is a non-negligeable duration in terms of the estimated total lifetime of these sources. Some of the very bright X-ray sources like Cen X-3 and SMC X-1 may be in such a phase of incipient Roche lobe overflow (e.g. Lamers et al. 1976).

5.2 Accretion from a supersonic stellar wind

Let us now make a crude estimate of the accretion rate of a neutron star orbiting in the supersonic wind of a massive companion star (Davidson and Ostriker 1973). Assuming for simplicity spherically symmetric wind ejection from the massive star we can write the wind loss rate as:

$$\dot{M}_w = 4\pi r^2 \rho_w(r) V_w(r)$$

where r is the distance from the centre of the massive star, which has mass M, radius R and luminosity L. During the outflow from the stellar surface the wind is accelerated to a typical terminal velocity of a few times the escape speed at the stellar surface (e.g. Abbott 1978):

$$V_w = \beta(r)(2GM/R)^{\frac{1}{2}}$$

with $\beta \simeq 3$ attained at r of a few times R. The mechanism responsible for the wind acceleration in massive stars is thought to be resonant scattering of stellar uv-photons by ions in the stellar wind (e.g. Lucy and Solomon 1970, Castor et al. 1975, Cassinelli 1979). One can derive a crude upper limit for the mass loss rate in a radiation pressure driven stellar wind by considering the maximum photon momentum transferred to the outflowing wind material:

$$\dot{M}_w V_\infty < \frac{L}{c}$$

if we neglect multiple scatterings. This yields a maximum mass loss rate:

$$\dot{M}_w \sim \frac{L}{c}\left(\frac{R}{2GM}\right)^{\frac{1}{2}} \simeq 10^{-11}\left(\frac{L^2R}{M}\right)^{\frac{1}{2}}_{\odot} \mathrm{M_\odot/year}$$

For a somewhat evolved $20M_\odot$ core hydrogen burning star with $L \simeq 10^5 L_\odot$ and $\log(T_{\rm eff}) \simeq 4.5$, which corresponds to $R \simeq 10R_\odot$, we find a maximum wind loss rate of $\dot{M}_w \simeq 10^{-6}M_\odot$/year. Lamers (1981) derived a fit formula for the observed mass loss rate of O and B type stars (based on a sample of 53 early type stars) which yields a similar value for the stellar wind loss rate of such a star. With the expression given above we find a typical supersonic windspeed of a few thousand km/s.

Bondi Hoyle Accretion from a Supersonic Uniform Flow of Gas: Let us now estimate the efficiency by which the neutron star's gravitational field captures the wind material. For highly supersonic (negligeable thermal pressure effects) inflow of a homogeneous gas with speed V_∞ and density ρ_∞ at infinity onto a massive object with mass M_x, Bondi and Hoyle (1944) derived an expression for the rate at which matter is gravitationally captured by the massive object:

$$\dot{M}_x = \alpha\pi R_a^2 \rho_\infty V_\infty$$

where α is of order unity and the Bondi-Hoyle radius, or accretion radius, R_a is given by

$$R_a = \frac{2GM_x}{V_\infty^2}$$

Note that for matter that flows by within a radius R_a from the massive object the kinetic energy at infinity is smaller than the (absolute value of) its potential energy when it passes the object, so that R_a seems a natural size of the flow region that is severely perturbed by gravitational effects. By gravitational attraction the matter is focussed onto an 'accretion line' downstream of the object, where the focussed matter collides as a result of the cylindrical symmetry. BH assumed that the cooling of the shocked matter on the accretion line is efficient, so that the line does not blow up, but remains confined. The incoming matter loses its velocity component perpendicular to the accretion line by the collisions. After that the matter on the accretion line has only a velocity component along it and if sufficiently close (i.e. $r \simeq R_a$) to the attracting object, the gravitational attraction dominates and it will fall onto the object.

Bondi-Hoyle argued that the undetermined factor α is of order unity, which was confirmed by more recent numerical experiments (Hunt 1979, Shima et al. 1985, Livio et al. 1986 and de Kool and Savonije 1988). Applying this expression to the neutron star in a MXB, we obtain $R_a = (2GM_x)/V_{\rm rel}^2$, with $V_{\rm rel}$ the inflow velocity of the stellar wind corrected for the orbital motion of the neutron star, which has a mass M_x. Hence,

$$V_{rel}^2 = 2\beta^2(GM/R) + (GM/a) = (1 + 2\beta^2 a/R)(GM/a)$$

where a is the orbital separation and we have adopted the approximation $M_t \simeq M$. The gravitational capture efficiency of the neutron star is thus:

$$\epsilon = \frac{\dot{M}_x}{\dot{M}_w} = \left(\frac{R_a}{2a}\right)^2 \frac{V_{\rm rel}}{V_w} = \frac{1}{\beta}\left(\frac{M_x}{M}\right)^2 \left(\frac{R}{2a}\right)^{\frac{1}{2}} \left(1 + 2\beta^2\frac{a}{R}\right)^{-\frac{3}{2}}$$

Example: Let us now assume that the massive star ($M = 20M_\odot$) almost fills its Roche lobe: $R/R_c = 0.9$, so that with $M_x = 1M_\odot$ we find from (11) $R/a = 0.57$. If we adopt $\beta = 1$, the capture efficiency follows as:

$$\epsilon \simeq 2 \times 10^{-4}$$

To power an X-ray source with $L_x = GM\dot{M}_x/R_x \simeq 10^{37}$ erg/s by accretion from the massive companion's wind would require an accretion rate of $\dot{M}_x \simeq 10^{17} g/s$. With the calculated capture efficiency ϵ this would require a stellar wind loss rate of $\dot{M}_w \simeq 8 \times 10^{-6} M_\odot$/year which is a bit high compared to our crude estimate above. However, the early type stars in MXB's may be overluminous and oversized compared to their mass as a result of the substantial stellar wind losses (Ziolkowski 1977, de Loore et al. 1978), and may therefore have stronger than usual stellar winds compared to their current stellar mass. In view of the uncertainties it is generally believed that accretion from the stellar wind of the early type companion can indeed power (although perhaps not the brightest) MBX's. The brightest MXB's may be powered by incipient Roche lobe overflow from the early type star. We expect a large accretion disc around the compact companion in that case. Because of the accretion of angular momentum the neutron stars are expected to spin up to short rotational periods, i.e. X-ray pulsars in such systems are expected to have short X-ray periods. This seems consistent with observations (Bradt et al. 1979), see following table.

Table 5. Some Massive X-ray Binaries

Source	P_b(days)	P_x(sec) (pulsar)	L_x (erg/s)	Companion's Spectrum	Expected Mode
SMC X-1	3.81	0.71	6×10^{38}	B0 I	RLO
CenX-3	2.09	4.84	4×10^{37}	O6.5 II-III	RLO
LMC X-4	1.40	13.51	2×10^{38}	O8 (IV-V)e	RLO
4U1700-37	3.41	–	3×10^{36}	O6.5f	Wind
Vela X-1	8.97	283	1.4×10^{36}	B0.5 Iab	Wind
4U1538-52	3.73	529	4×10^{36}	B0 Ib	Wind

The emerging X-ray spectrum of wind-embedded MBX's has been considered by Hatchett and McCray (1977).

5.3 Final Evolution of MXB's: Spiral in Evolution of Neutron star

The X-ray phase terminates within the early type star's thermal timescale once the second phase of Roche lobe overflow has begun and the X-ray source becomes absorbed by the accumulation of mass caused by the rapid mass transfer. Tidal effects cause the neutron star to spiral down into the envelope of its massive Roche lobe filling companion (Sect. 1.4). Once the neutron star is inside the companion's envelope it is subject to a strong frictional force, which gives rise to a torque and conversion of orbital angular momentum to spin angular momentum of the massive star's envelope. The details of the dissipation process and angular momentum deposition are not well understood. Numerically it is difficult to calculate because of the very different length and timescales involved and lack of symmetry in the problem. Nevertheless, to make a start, Taam et al. (1978) treated the problem in one dimension by assuming spherical symmetry. Unfortunately, we have a poor understanding of the evolution of spin angular momentum in stars, and Taam et al. assumed that the specific angular momentum of the massive star's envelope was constant in space at the beginning of the spiral in phase. The frictional dissipation takes place in an annulus around the neutron star's orbit. Because of the adopted symmetry they had to postulate a rapid redistribution of energy and angular momentum in the z-direction (parallel to stellar rotation axis), so that they could deposit the energy and angular momentum in a spherical shell with typical radial extent:

$$R_a = \frac{GM_x}{(V - V_e)^2 + C^2}$$

This expression for the accretion radius is an ad hoc generalization of the Bondi-Hoyle radius to include both supersonic and subsonic flow around a heavy object (Bondi and Hoyle 1944). $V - V_e$ is the relative velocity of the neutron star in its orbit through the envelope and C the local sound speed. The loss rate of orbital energy can then be estimated by assuming that all envelope matter which passes through a cylinder with radius R_a around the neutron star is captured and accreted, i.e. loses its kinetic energy with respect to the neutron star. This corresponds to a frictional dissipation rate:

$$\dot{E}_{\rm dis} = \frac{1}{2}\pi R_a^2 \rho (V - V_e)^3$$

where ρ is the density of envelope matter at the position of the neutron star. Apart from the frictional energy source, energy is liberated by accretion. Taam et al. assumed that the accretion luminosity is limited by the Eddington limit (Sect. 5.1). The decay of the orbit is dictated by the orbital energy loss rate due to the dissipative processes:

$$-\frac{GM(r)M_x}{2r^2}\frac{dr}{dt} = \dot{E}_{\rm dis}$$

where $M(r)$ is the stellar mass within the orbital radius r. Adding to $E_{\rm dis}$ the dissipation due to tidal interaction (adopting a crude expression for the turbulent shear viscosity) between the neutron star and the envelope layers interior to it, Taam et al. calculated the spiral in evolution of the neutron star. In general the tidal dissipation rate appeared less important than the frictional term given above. They considered two cases:

case 1) in which the neutron star enters the envelope of the $16M_{\odot}$ star when it is at the onset of core helium burning. The envelope is not very extended and in radiative equilibrium.

case 2) in which the spiral in starts when the massive star is more evolved and has become a red giant with an extended convective envelope.

In case 1 the neutron star spends a few thousands years in the outer layers, where the frictional drag is small, before it falls down into the more dense layers and the orbital decay speeds up considerably. The large frictional luminosity of 10^{42} erg/s causes the envelope to become convective. The amount of orbital energy dissipated during the spiral in is in itself amply sufficient to eject the envelope from the star. But, because of the relatively high density of the envelope material, convection is quite efficient in transporting the large energy flux outwards, without the envelope being blown of by an accumulation of thermal energy. When the neutron star approached the helium core Taam et al. terminated their calculation.

In case 2 the density in the extended giant envelope is more homogeneous, and much lower than the average density in case 1. However, because the density in the very outer layers is now higher in comparison to case 1, the initial (slow) spiral in of the neutron star is much more rapid: the whole spiral in phase takes less than about 20 years. The frictional luminosity increases as the neutron star spirals down, but now (because of the rarified, extended envelope) the estimated timescale for convective energy transport ($\tau_{\rm conv} \simeq 10^{-1}$ years) is about two orders of magnitude longer than in case 1. By the time the orbital separation has decayed below $3.5R_{\odot}$ the orbital decay time $r/\dot{r}$ becomes shorter than τ_{conv} and a large pressure builds up near the orbit of the neutron star. This causes the matter to acquire outflow velocities in excess of the escape speed, so that the envelope is ejected from the star by hydrodynamic expansion.

In a more recent paper Bodenheimer and Taam (1984) used a two-dimensional (explicit) hydro-code to study the spiral in evolution. Because the Courant condition requires very short timesteps in the inner regions near the core of the giant star, they started with a spherical symmetric model of Taam et al. (1978), in which the neutron star was already close to the edge of the helium core, and included only the layers just outside the core in their 2D calculations of the subsequent evolution. The frictional energy is now deposited in an annulus, according to the same expressions as in Taam et al. (1978). However, for unclear reasons they now deposited the frictional energy in a region small compared to the Bondi-radius. The results are very different from the spherical case, with equatorial mass loss in both cases 1 and 2.

Concluding, one can say that the details of the common envelope evolution are far from understood. However, observations of double core planetary nebulae (e.g. Bond, 1983) indicate that spiral in evolution does indeed occur in nature and that it can lead to the ejection of the entire envelope (next section). For unknown reasons the spiral in can be interrupted before the spiral in star coalesces with the central core since the remnants are still double stars. Further evidence for the occurrence of spiral in evolution is provided by the existence of binary radio pulsars.

Binary Radio Pulsars: After the spiral in phase with envelope ejection we are left with a neutron star orbiting the helium core remnant of the massive companion star. It now depends on the mass of the helium star and possible asymmetries in the following supernova explosion whether the binary is disrupted or not. When the binary is indeed disrupted in the SN explosion, two run-away pulsars are formed, a young radio pulsar and an old one. If the binary happens to remain bound, a binary radio pulsar is formed like the observed systems PSR1913+16 and PSR2303+46 (Taylor and Weisberg 1982, Srinivashan and van den Heuvel 1982).

6 Low Mass X-ray Binaries (LMXB's)

6.1 Formation of LMXB's

The existence of low mass X-ray binaries is rather surprising, if one considers the possibility of creating a neutron star in company with a less massive star. If the neutron star was born in a typical type II supernova explosion, the binary should have been disrupted because certainly more than half of the system's mass would have been ejected. Apparently, neutron stars in LMXB's are produced by a less violent event than a standard type II supernova explosion. It is possible that the neutron star is formed by **accretion induced collapse** of a white dwarf. Chandrasekhar showed that white dwarfs can not have a mass larger than a certain limit mass $M_{\rm Ch} \simeq 1.4 M_\odot$. In more massive white dwarfs the (relativistic) degenerate electron gas is unable to balance the self-gravity and the star collapses to a neutron star. When a white dwarf receives mass from its companion it can be made more heavy than $M_{\rm Ch}$ and collapse to a neutron star. However, numerical calculations of this process (e.g. Nomoto 1982) indicate that the white dwarf will collapse only if it is already close to the Chandrasekhar limit at the beginning of mass transfer, and if the rate of mass transfer is very high ($\dot{M} > 10^{-8} M_\odot$/year). Otherwise the white dwarf will be fully disrupted in a thermonuclear explosion (type I supernova) triggered by the accretion process. However, it has been suggested that (hypothetical) separation of carbon and oxygen in a crystallizing CO-white dwarf could prevent the thermonuclear flash (Isern et al. 1983). In any case, for heavy O+Ne+Mg white dwarfs accretion leads to collapse because electron captures on Mg^{24} and Ne^{20} trigger the collapse of the core prior to the explosion (Miyaji et al. 1980). White dwarfs like this form in binaries from stars in the mass range $M = 8 - 10 M_\odot$ (Nomoto 1984) and are relatively rare, so that without separation effects accretion induced collapse would be a rare event. This seems consistent with the relatively small number of LMXB's in our galaxy ($\simeq 50$) compared to the large number of cataclysmic variables (CV's). CV's are close binaries in which mass transfer occurs from a relatively unevolved, Roche lobe filling, low mass star to a (usually more massive) white dwarf. The CV's can be divided broadly into novae and dwarf novae (e.g. Robinson 1976, Warner 1983).

Spiral in Evolution: If we adopt the possibility that accretion induced collapse can produce the neutron star in LMXB's, then we are faced with a new problem: white dwarfs are the remnants of giant stars with dimensions much larger than the average orbital separation of LMXB's or CV's. How can the binary have accomodated a giant? This is only possible if the binary system was much wider in the past. It is generally believed that this is indeed true, and that the progenitor systems of LMXB's and CV's are very wide binaries with orbital periods of order $\sim 10^2$ days and rather extreme mass ratio's. The more massive component evolves to the giant stage while its companion remains practically unevolved. When the giant approaches its Roche lobe and tidal effects become important, it absorbs so much orbital angular momentum in an attempt to keep its rotation synchronous with the orbit, that the less massive companion begins to spiral down into the giant's outer layers (Paczyński 1976). In Sect. 5.4 we have seen that the details of the spiral in process (sometimes referred to as 'common envelope' evolution) are very poorly understood, but that the dissipated orbital energy of the spiral in star is in general amply sufficient to eject the giant's envelope (see e.g. Meyer and Meyer-Hofmeister 1979 for an approach different from Taam et al. 1978).

Note that stars with common envelopes by no means necessarily undergo spiral in evolution of one star into the other. For example, in the very common W Ursae Majoris binaries both stars fill their Roche lobe, whereby the two stars are embedded in a common envelope. These binaries are rather stable and not subject to spiral in evolution. For a catastrophic spiral in the orbit must be unstable to tidal and frictional effects, which is the case only if the mass ratio (and moment of inertia of the massive component) is quite extreme (section 1.4).

Double Core Planetary Nebulae: Observations of double cores in planetary nebulae indicate that the spiral in process does indeed occur in nature. In some nebulae a hot subdwarf (the remnant core of the giant) is accompanied by a much cooler companion star (presumably the spiral in star), forming a rather close, but detached, binary. The orbital periods indicate that the cool star must have orbited inside the giant's envelope, before it was expelled. Table 6 lists some observed binaries in planetary nebulae.

Table 6. Some observed binaries in planetary nebulae, after Bond 1983

nebula	binary	orbital period
UU Sagittae (Abell 63)	sdO ($0.9 M_\odot$) dK	$0.^d465$
V477 Lyrae (Abell 46)	$R_1 \sim 0.2 R_\odot$ $R_2 \sim 0.6 R_\odot$	$11^h 19^m$
VW Pyx (Kohoutek 1-2)	ns?	$16^h 05^m$
(Abell 41)	sdO $\sim 0.6 M_\odot$ dM $\sim 0.1 - 0.3 M_\odot$	$2^h 43^m$

A number of detached binaries are known to contain white dwarfs or sub-dwarfs and to have rather long orbital periods (many days). Systems like that may become similar to the long period old nova GK Per when the unevolved star has expanded to its Roche lobe. Hence, the observations indicate that the spiral in process can result in (detached) binaries of rather different orbital periods, ranging from a few hours to more than a day. Below we will argue that this gives rise to a dichotomy in the further evolution of these systems.

Direct Collapse Scenario: There are some X-ray binaries with low mass donor stars for which the compact star cannot have been formed by accretion induced collapse. One of these is the system A0620-00, which contains a black hole with mass larger than about $5M_\odot$. May be a system like this has evolved through a spiral in phase, whereby the remnant core (helium star) collapsed to a black hole without disrupting the binary. This seems possible if the supernova occurs after, or at the end of the spiral in phase when most of the envelope has already been lost, so that the supernova explosion cannot eject much mass (e.g. de Kool et al. 1987). It is difficult to judge whether direct collapse of the remnant core after the spiral in is a viable general alternative mechanism for neutron stars in LMXB's. In this respect it is interesting to note that there is some indication that the planetary nebula VW Pyx (table 6) contains a binary with a neutron star.

Post Spiral in Binary Evolution: Bifurcation into compact and wide LMXB's. The subsequent evolution of the binary system depends critically on its orbital period. If the orbital period after the spiral in is smaller than a certain critical value (which is roughly 1 day, but depends on the stellar masses) the further evolution of a *compact* binary is dominated by orbital angular momentum losses due to gravitational radiation (e.g. Faulkner 1971, Paczyński and Sienkiewicz 1981, Rappaport et al. 1982) and possibly 'magnetic braking' (Verbunt and Zwaan 1981) by a stellar wind from the unevolved star. We will come back to these processes in the next section. The orbital angular momentum losses cause the binary to shrink on a timescale short compared to the (nuclear) expansion timescale of the unevolved, cool star. At some moment the orbital separation, and the Roche lobe, have shrunken to the point that the unevolved star fills up its Roche lobe and begins to transfer mass to the compact star. Depending on the stellar masses the system will now be a CV or a (pre-)LMXB. In the accretion induced collapse scenario the white dwarf is supposed to collapse to a neutron star in the latter case. During the collapse of a white dwarf to a much denser neutron star the liberated gravitational energy is radiated away (mostly in the form of neutrinos), which corresponds to a non-negligeable instantaneous mass loss of about:

$$\Delta M \simeq \frac{3}{5}\frac{GM_{Ch}^2}{R_x c^2} \simeq 0.18M_\odot$$

where we have assumed that the neutron star has a a radius $R_x = 10$ km (much smaller than $R_{\rm wd}$) and a uniform density. The collapsed star may also lose some mass from the rebounce shock. The mass loss will perturb the orbit, whereby the binary becomes detached. Subsequently the orbit will decay and circularize by gravitational radiation and tidal effects. At a certain moment the mass transfer restarts and a LMXB has been born.

However, if the orbital period after spiral in is larger than the critical period (of about a day), the timescale for angular momentum losses is longer than the nuclear evolution timescale of the unevolved star. This means that the latter will have time to significantly reduce or deplete the hydrogen in its core and evolve away from the main sequence. After some time it will expand to its Roche lobe and start to transfer mass to its white dwarf companion. When the companion is, or has become, the more massive star the orbital separation will increase ever more. In this case of *wide* LMXB's the mass transfer rate is expected to be higher and favourable for accretion induced collapse of a white dwarf to a neutron star. For binaries with a period longer than some 1.5 days the ($\sim 1M_\odot$) contact star develops a degenerate helium core, allowing the evolution to be followed with a set of simple algebraic equations (see next section). For binaries with smaller periods close to the bifurcation period (that is with stars that fill their Roche lobe at about the end of the core hydrogen burning phase) this is, however, not possible and one needs to use a full scale stellar evolution code to follow the evolution. This has been done by e.g. Iben and Tutukov 1984 and Pylyser and Savonije 1988.

LMXB's formed in Globular Clusters: A group of LMXB's with probably a completely different formation mechanism are the X-ray binaries in globular clusters. There have been about 10 LMXB's discovered in globular clusters, which is exceptionally high compared to the total number of LMXB's in our galaxy if one considers the relatively small amount of mass contained in globular clusters. It is believed that these X-ray binaries are formed by tidal capture of a free fossil neutron star by a main sequence star, or red giant in the crowded fields in the interior of globular clusters (e.g. Verbunt and Hut 1987).

6.2 Mass Transfer in Low Mass X-ray Binaries

We have already discussed that LMXB's can be divided into two subclasses, namely: compact systems with unevolved or degenerate contact stars, and wide systems with evolved, (sub)giant contact stars. This division corresponds to the bifurcation in the evolution after the spiral in phase into 'converging' systems dominated by angular momentum losses and 'diverging' systems driven by the nuclear evolution of the unevolved star to the giant stage. The wide systems may be related to the ~ 10 very bright ($L_x \simeq 10^{38}$erg/s) X-ray sources near the galactic centre, which are subject to heavy obscuration in the optical. Let us first discuss the compact systems and the orbital angular momentum loss mechanisms that are thought to drive the mass transfer in these systems.

Mass Transfer in compact LMXB's: The compactness of these binaries (with $P_b < 1$ day) implies that the companion of the compact star must be a small star. If it is an unevolved, low mass main sequence star which fills its Roche lobe, the mass transfer takes place on the nuclear evolution time of the star if we ignore loss of orbital angular momentum. We assume that the compact star is the more heavy component, otherwise there would be a short phase of rapid mass transfer and reversal of the mass ratio. (Accretion induced collapse might take place during this phase of rapid mass transfer). It appears that the mass transfer rate driven by the slow expansion of the main sequence star (slow phase) is too small to power a bright X-ray source. However, there is a mechanism which reduces the orbital angular momentum of a compact binary. The corresponding decay of the orbit and of the Roche lobe gives rise to an enhanced mass transfer to the compact star. **Gravitational Radiation:** According to the theory of general relativity a binary star emits gravitational waves. The energy loss rate associated with the radiation emitted by a binary with a circular orbit is given by (e.g. Landau and Lifshitz 1958):

$$\frac{dE}{dt} = -\frac{32G}{5c^5}\mu^2 a^4 \Omega^6 \tag{16}$$

where G is the constant of gravity, c the speed of light, μ the reduced mass, a the orbital separation and Ω the orbital angular velocity of the binary. The loss rate of orbital angular momentum $H = \mu a^2 \Omega$ follows from $\dot{E} = \Omega \dot{H}$. We define the characteristic timescale for gravitational radiation losses τ_g as:

$$\tau_g^{-1} = -\frac{\dot{H}}{H} = \frac{32}{5c^5} G^3 M_1 M_2 M_t a^{-4} \ s^{-1} \tag{17}$$

which can be expressed as:

$$\tau_g = 8 \times 10^7 \text{yrs} \left(\frac{M_t^{\frac{1}{3}}}{M_1 M_2} \right)_{\odot} P_b^{\frac{8}{3}}(\text{h})$$

where we have used $M_t = M_1 + M_2$.

For example, a compact binary consisting of two $1M_\odot$ stars with an orbital period of 7 hours, has $\tau_g \simeq 10^{10}$ yrs, which is roughly equal to the nuclear timescale of a $1M_\odot$ main sequence star. In a binary like this orbital decay by gravitational radiation would obviously have a significant effect on the evolution of the binary system.

Expression for the Mass Transfer Rate: Let us derive a simple recipe to determine the mass transfer rate by Roche lobe overflow under the assumption that we know the mass-radius relation for the mass losing contact star. We can only use a mass-radius relationship if the contact star remains approximately in thermal equilibrium during the mass transfer, otherwise the radius would also depend on the history of the star. This implies that we can only consider the slow stage of mass transfer from a less massive star to a more massive companion. But the slow stage is the phase where the system spends most of its time, i.e. LMXB's are probably in this stage.

If we differentiate logarithmically with respect to time the expression for the orbital angular momentum $H = \mu a^2 \Omega$, we consider circular orbits, we find (compare with (13)):

$$\frac{\dot{H}}{H} = \frac{\dot{M}_1}{M_1} + \frac{\dot{M}_2}{M_2} + 2\frac{\dot{a}}{a} + \frac{\dot{\Omega}}{\Omega} = -\tau^{-1} \tag{18}$$

where τ is the timescale for orbital angular momentum losses. By assuming conservation of mass, i.e. $\dot{M}_1 = -\dot{M}_2$, we can solve for the rate of change of the orbital separation after using Kepler's law to eliminate $\dot{\Omega}/\Omega = -\frac{3}{2}\dot{a}/a$:

$$\frac{\dot{a}}{a} = 2\left(\frac{M_2}{M_1} - 1\right)\frac{\dot{M}_2}{M_2} - \frac{2}{\tau} \tag{19}$$

The contact star's radius must equal the Roche radius R_c during mass transfer. Because the contact star (star 2) is less massive, we will use (12) for the Roche radius:

$$R_c = 0.46 \, a \left(\frac{M_2}{M_t}\right)^{\frac{1}{3}}$$

which is valid for $M_2/M_1 \leq 0.8$. Differentiating this expression with respect to time yields:

$$\frac{\dot{R}_c}{R_c} = \frac{\dot{a}}{a} + \frac{1}{3}\frac{\dot{M}_2}{M_2}$$

Let us express the general mass-radius relation as:

$$\frac{R}{R_\odot} = \alpha \left(\frac{M}{M_\odot}\right)^\beta \tag{20}$$

or

$$\frac{\dot{R}_2}{R_2} = \beta \frac{\dot{M}_2}{M_2}$$

By requiring $\dot{R}_c/R_c = \dot{R}_2/R_2$ we obtain another expression for the rate of change of the orbital separation:

$$\frac{\dot{a}}{a} = \left(\beta - \frac{1}{3}\right) \frac{\dot{M}_2}{M_2} \tag{21}$$

Note that if $\beta < \frac{1}{3}$ the orbit must necessarily expand in order to keep the the mass losing secondary inside its Roche lobe. For such low values of β the star will adjust the mass transfer (which tends to widen the orbit) to keep itself just inside its Roche lobe.

By equating (19) and (21) and eliminating $\dot{a}/a$, we obtain for the orbital angular momentum loss driven mass transfer rate:

$$\dot{M}_2 = -\left(\frac{5}{6} + \frac{\beta}{2} - \frac{M_2}{M_1}\right)^{-1} \frac{M_2}{\tau} \tag{22}$$

Obviously the mass ratio M_2/M_1 should be smaller than $\frac{1}{6}(3\beta + 5)$, otherwise the expression becomes singular.

Example: Low mass main sequence contact star. Let us apply this expression to a binary consisting of a main sequence dwarf of $0.7M_\odot$ and a $1.4M_\odot$ neutron star. The dwarf star obeys the approximate mass-radius relation (unpublished):

$$\frac{R}{R_\odot} \simeq 0.86\,(0.94 - 0.18\ln(X)) \left(\frac{M}{M_\odot}\right)^{0.9}$$

which gives for, say an hydrogen abundance of $X = 0.6$: $R \simeq 0.64R_\odot$, while from the Roche geometry we find $R/a = 0.32$, so that $a \simeq 2R_\odot$. With Kepler we find an orbital period of $P_b = 5^h.4$. The timescale for gravitational radiation losses (17) becomes $\tau_g \simeq 10^{10}$ yrs, whereas $\tau_{\rm nuc} \simeq 5 \times 10^{10}$ years.

By substituting the obtained value τ_g for τ into (22) we obtain $\dot{M} \simeq 10^{-10}M_\odot$/yr for the mass transfer rate. This corresponds to $L_x = (GM_x/R_x)\dot{M} \simeq 10^{36}$ erg/s, which is still an order of magnitude lower than the estimated typical X-ray luminosity of these sources. **Magnetic braking:** Verbunt and Zwaan (1981) suggested that additional angular momentum is lost from the binary through a weak stellar wind from the low mass star which is magnetically coupled to it out to many stellar radii. Observations of a sample of single G stars (Skumanich 1972, Smith 1979) indicate that the rotation velocity of these stars decays as a function of age t crudely as:

$$V_{\rm rot} \simeq f \times 10^{14} t^{-\frac{1}{2}} \text{ cm/s}$$

where $f \sim 1$. Presumably the magnetically forced corotation of the stellar wind out to several stellar radii exerts a braking torque on the star by which it spins down. The star's spin down rate decreases in time. This has been interpretated in terms of a decrease of the stellar magnetic field, which seems proportional to the rotation rate of the star (Zwaan 1981). For the star the corresponding magnetic braking timescale would be given by

$$\tau_{\rm mb}^{-1} = -\frac{\dot{H}}{H} \sim 5 \times 10^{-29} f^{-2} \kappa_g \left(\frac{M_t}{M_2}\right) \left(\frac{R_2}{a}\right)^4 \omega^2 a^2$$

where κ_g is the star's gyration constant and ω its spin angular velocity. Verbunt and Zwaan now assumed that a similar expression holds for the rapidly rotating contact stars in compact binaries and that tidal interaction (the braking star has a convective envelope) is sufficiently effective to keep it in synchronism with the orbit by transferring orbital angular momentum into spin momentum of the braking star. This implies a drain of orbital angular momentum and a tendency for the binary to shrink.

Adopting $\omega = \Omega$, $f \simeq 1$ and $\kappa_g \simeq 2 \times 10^{-2}$ yields for the typical timescale $\tau_{mb} = \dot{H}/H \simeq 10^9$ yrs. The above expression for magnetic braking would thus give rise to a mass transfer rate of about $10^{-9} M_\odot/yr$, which would be more consistent with observational estimates of the X-ray brightness of the LMXB's. But note that, unlike gravitational radiation, there is no sound theory of magnetic braking so that the results are very uncertain.

Example: Degenerate contact star. In ultra compact binaries with orbital periods less than say 1 hour, the contact star could be a degenerate star. If the electron gas in the star is completely degenerate, but non-relativistic, i.e. if we consider a low mass white dwarf, the mass radius relation is approximately given by (e.g. Paczyński 1967):

$$\frac{R}{R_\odot} \simeq 0.013\,(1+X)^{\frac{5}{3}} \left(\frac{M}{M_\odot}\right)^{-\frac{1}{3}}$$

We immediately see that $\beta < \frac{1}{3}$, so that a binary with a degenerate contact star must necessarily expand its orbit during mass transfer (21). Since a compact binary is subject to strong gravitational radiation losses, which tend to decay the orbit, the mass transfer rate must be rather high to accomplish this.

For a helium white dwarf (X=0) of $M = 0.16 M_\odot$ we find a radius of $R = 2.4 \times 10^{-2} R_\odot$. The Roche expression yields $R_c = 0.22a$, so that we require $a = 0.11 R_\odot$. From the expression $P_b = 10^4 \text{sec}\, (a^3/M_t)_\odot^{\frac{1}{2}}$ we obtain $P_b = 50$ min., so that $\tau_g \simeq 4 \times 10^8 (0.8)^{\frac{8}{3}} \simeq 3 \times 10^8$ yrs. Substitution into (22) yields a mass transfer rate of $\dot{M} \simeq -10^{-9} M_\odot/yr$.

Fig. 12 shows the time evolution of the mass transfer rate for a $0.16 M_\odot$ and a $0.08 M_\odot$ white dwarf, together with a $0.9 M_\odot$ main sequence contact star. The mass transfer rates from the white dwarfs rapidly decay to rather low values. This implies that if we want to explain the existence of the dozen or so very bright X-ray sources ($L_x \simeq L_{Edd}$) in terms of mass transfer from a white dwarf we would expect to observe many more weak X-ray binaries which is not the case.

Mass Transfer in wide LMXB's: If the orbital period of the progenitor LMXB is sufficiently large when it emerges from the spiral in phase, orbital angular momentum losses due to gravitational radiation or magnetic braking become unimportant: for small Ω we have $\tau_{nuc} << \tau_g$ or τ_{mb}. Before the binary has time to shrink noticeable the unevolved star will expand to its Roche lobe and begins to transfer mass. If the unevolved star is more massive than its degenerate companion it will undergo a short phase of rapid

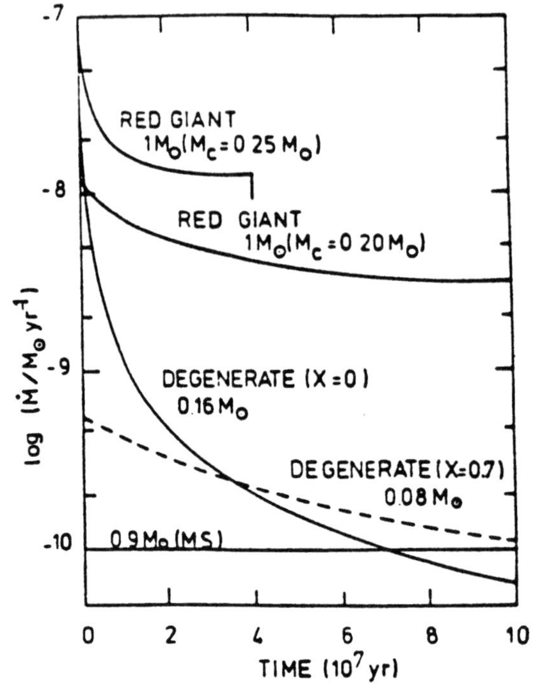

Fig. 12. Evolution of mass transfer rate for different types of contact star.

mass transfer in which the mass ratio is reversed. After that the mass transfer will drive the stars apart and it will occur on the natural expansion time of the contact star. In case this star is a (sub)giant, for which the orbital period should be somewhat larger than a day, the stellar radius and luminosity appear to be virtually independent of the mass in the envelope, but determined by the mass M_c of the degenerate helium core (Refsdal and Weigert 1970).

For $-0.2 \leq y \leq 0.4$, where $y = ln(M_c/0.25M_\odot)$, we can express the luminosity L and radius R of the (sub)giant as a simple polynomial (Webbink et al. 1983):

$$ln\left(\frac{L}{L_\odot}\right) = a_0 + a_1y + a_2y^2 + a_3y^3 \tag{23}$$

$$ln\left(\frac{R}{R_\odot}\right) = b_0 + b_1y + b_2y^2 + b_3y^3 \tag{24}$$

where for a population I composition (Z=0.02) the coefficients are given by :

$$(a_0, a_1, a_2, a_3) = (3.50, 8.11, -0.61, -2.13)$$

$$(b_0, b_1, b_2, b_3) = (2.53, 5.10, -0.05, -1.71)$$

The evolution of the giant is determined by the accretion rate of helium onto the degenerate helium core due to the H-shell burning just outside the core. The luminosity of the star is related to the accretion rate of the core by:

$$L = \epsilon_H X \dot{M}_c \tag{25}$$

where $\epsilon_H = 5.987 \times 10^{18}$ erg/g is the net amount of liberated fusion energy per gram hydrogen and X the (constant) hydrogen mass fraction of the envelope. Equations 23 to 25 form a closed set of equations from which we can determine the global evolution of the (sub)giant, once we have specified the initial core mass M_c. For if we substitute the corresponding value y into (23) and (24) we find L and R, and from this through (25), we know $\dot{M}_c$, and can step up the value for y, etc.

If the (sub)giant is situated in a close binary we can also calculate the mass transfer rate and the evolution of the binary system (Webbink et al. 1983). Because the mass transfer occurs on the nuclear evolution (expansion) timescale of the (sub)giant the latter star will remain approximately in thermal equilibrium during the mass transfer and we are allowed to use (23)-(24) for the radius and luminosity. From (24) we find after logarithmic differentiation with respect to time:

$$\frac{\dot{R}}{R} = \left(b_1 + 2b_2 y + 3b_3 y^2\right) \dot{y} \tag{26}$$

Equating this expression to that for the Roche lobe (12), and eliminating $\dot{a}/a$ with (19), in which we put $\tau = \infty$, we find for a mass ratio $M_2/M_1 < \frac{5}{6}$, the mass transfer rate as:

$$\dot{M}_2 = -\frac{\left(b_1 + 2b_2 y + 3b_3 y^2\right)}{\left(\frac{5}{3} - 2\frac{M_2}{M_1}\right)} \left(\frac{M_2}{M_c}\right) \dot{M}_c \tag{27}$$

The (sub)giant's envelope will now be slowly depleted by both the H-shell source (accretion onto the degenerate helium core) and the mass loss to the companion. As long as the envelope mass is more than a few per cent of the total stellar mass the stellar radius and luminosity are well approximated by the polynomial expressions given above. However, once the envelope mass is depleted below this limit, the expanded envelope can no longer be sustained and shrinks on a short thermal timescale to very small dimensions, whereby the mass transfer terminates quite abruptly.

We can determine the evolution of the binary during mass transfer by accounting for both the increase in $\dot{M}_c$ and the decrease of the envelope mass with help of (25) and (27). In Fig. 12 we have plotted the rate of mass transfer from a $1M_\odot$ giant with a core mass $M_c = 0.25M_\odot$ and $0.20M_\odot$, respectively. High mass transfer rates are sustained for several times 10^7 years, so that the subclass of very bright LMXB's could very well be of this type.

6.3 Final Evolution of LMXB's

The final evolution of the two subclasses of LMXB's will of course be totally different. Let us consider first the post LMXB stage of the wide, diverging binaries.

Final Evolution of Wide LMXB's: After the giant's envelope collapses the mass transfer terminates quite abruptly. The binary is now rather wide, it can have a period of a few hundred days, and consists of the degenerate helium core of the giant (which becomes a white dwarf) and a neutron star. During the previous accretion phase the neutron star has been spun up to a short spin period. This period is determined by the magnetic field of the neutron star. The larger the magnetic field the larger the magnetosphere around the star in which the accreting matter is magnetically forced to corotate with the neutron star. If, at the boundary of the corotating magnetosphere the circular velocity is larger than the local Kepler speed, matter from the accretion disc cannot accrete onto the neutron star and is ejected by centrifugal forces. Therefore the neutron star will be spun up until these two velocities become comparable (e.g. Pringle and Rees 1972, Davidson and Ostriker 1973, Gosh and Lamb 1979, Henrichs 1983). The resulting 'equilibrium' spin period is approximately:

$$P_{\mathrm{eq}} = 6\mathrm{ms}\,(B_9)^{\frac{6}{7}}(R_6)^{\frac{8}{7}}(M_p)^{-\frac{5}{7}}(\dot{M}_{17})^{-\frac{3}{7}}$$

where B_9 is the neutron star's surface (dipole) magnetic field strength in units of 10^9 G, R_6 its radius in 10^6 cm, M_p its mass in $M_\odot$ and $\dot{M}_{17}$ the accretion rate in units of 10^{17} g/s. For magnetic fields of order 10^9 G, the neutron star spins up to typically a millisecond. Even if the magnetic field is weaker it cannot spin much faster because gravitational radiation losses become significant for spin periods shorter than about a millisecond (e.g. Papaloizou and Pringle 1978).

When the accretion terminates the rapidly spinning magnetic neutron star becomes a radio pulsar. Several binary radio pulsars with wide, but almost circular orbits, and unseen (or indeed white dwarf) components have been discovered (e.g. Boriakoff et al. 1983, Dewey et al. 1986, Segelstein et al. 1986, Lyne et al. 1988). The evolution model discussed above fits these systems remarkably well (Joss and Rappaport 1983, Paczyński 1983, Savonije 1983, 1987). The circular orbits can be easily explained in the evolutionary model sketched above: the now widely separated stars formed a semi-detached binary in the past with a giant star which filled its Roche lobe, so that tidal interaction was very efficient. The short period (about 5 milliseconds) of the radio pulsars PSR1855+09 and PSR1953+29 can be understood in terms of the accretion induced spin up of a neutron star with a moderately weak (see e.g. van den Heuvel et al. 1986) magnetic field of the order of 10^9 G.

Final Evolution of Compact LMXB's: The compact LMXB's evolution is driven by angular momentum losses, whereby the binary shrinks. One may wonder whether there is a limit to the orbital decay. The answer is yes, there must be a minimum orbital period for a given binary. The reason is that when the low mass contact star loses its mass (on a timescale short compared to its nuclear timescale, so that its nuclear evolution is almost frozen) it moves down in the Hertzsprung-Russel diagram, almost parallel to the ZAMS. At a certain moment its mass has become so low that it cannot sustain nuclear burning in its interior. When the nuclear energy source vanishes the pressure balancing gravity is produced predominantly by the degenerate electron gas. The degeneracy was created by the contraction of the cooling star. However, for degenerate stars the radius increases when mass is lost and this forces the binary to expand (Sect. 6.2). Therefore, the binary must evolve through a minimum period, after which the degenerate star transfers so much mass that the binary reexpands. For hydrogen-rich contact stars this minimum period is about 1 hour (Paczyński and Sienkiewicz 1981, Rappaport et al. 1982), while for helium burning helium stars this minimum period is about 10 minutes (Savonije et al. 1986). When the binary has passed the minimum orbital period the contact star becomes fully degenerate and inert. The mass transfer is driven by the angular momentum losses due to gravitational radiation. Because the contact star's mass is now small these losses become rather weak and the mass transfer slows down appreciably. The binary evolves very slowly during this phase. When the contact star's mass has been depleted to planetary values the degeneracy is lifted and the matter becomes solid. However, it may take something of a Hubble time or longer to reach this final stage.

References

Abbott, D.C.,1978, Ap.J. **225**, 893.
Alexander, M.E.,1973, Astrophys. Sp. Sc. **23**, 459.
Benson, R.S.,1970, Ph.D. thesis, Univ. of California at Berkeley.
Bodenheimer, P. and Taam, R.E.,1984, Ap.J. **280**, 771.
Boersma, J.,1960, Bulletin of the Astron. Inst. of the Netherlands **505**, 291.
Bond, H.E.,1983, in: 'Cataclysmic Variables and Low Mass X-ray Binaries', eds. J. Patterson and D.Q. Lamb, Reidel.
Bondi, H. and Hoyle, F.,1944, Mon.Not.Roy.Astr.Soc. **104**, 273.
Boriakoff, V.,Buccheri, R. and Fucci, F.,1983, Nature **304**, 417.
Bradt, H.V.,Doxey, R.E.and Jernigan, J.G.,1979, in: 'X-ray astronomy', eds. W.A. Baity and L.E. Peterson, Oxford Univ. Press.
Cassinelli, J.P.,1979, **17**, 275.
Castor, J.I., Abbott, D.C. and Klein, R.I.,1975, Ap.J. **195**, 157.
Davidson, K. and Ostriker, J.P.,1973, Ap.J. **179**, 583.
Darwin, G.H.,1897, Phil. Trans. Roy. Soc. **170**, 1.
Dewey, R.J., Maguire, C.M.,Rawley, L.A., Stokes, G.H. and Taylor, J.H.,1986, Nature **322**, 714.
de Kool, M.,van den Heuvel, E.P.J., and Pylyser, E.,1987, Astron. and Astrophys. **183**, 47.
de Kool, M. and Savonije, G.J.,1988, Astron. and Astrophys. **189**, 331.
de Loore, C.,de Greve, J.P. and van Beveren, D.,1978, Astron. and Astrophys. **67**, 373.
Eggleton, P.P.,1983, Ap.J. **268**, 368.
Faulkner, J.,1971, Ap.J. **170**, L99.
Flannery, B.P.,1975, Mon.Not.Roy.Astr.Soc. **170**, 325.
Flannery, B.P., and Ulrich, R.K.,1977, Ap.J. **212**, 533.
Fryxell, B.A. and Arnett,W.D.,1981, Ap.J. **243**, 994.
Ghosh, P. and Lamb, F.,1979, Ap.J. **234**, 296.
Habets, G.,1986, Astron. and Astrophys. **167**, 61.
Hatchett, S. and McCray, R.,1977, Ap.J. **211**, 552.
Henrichs, H.F.,1983, in: 'Accretion Driven Stellar X-ray Sources', eds. W. Lewin and E. van den Heuvel, Cambridge Univ. Press, 393.
Hut, P.,1981, Astron. and Astrophys. **99**, 126.
Hunt, R.,1979, Mon.Not.Roy.Astr.Soc. **188**, 83.
Iben, I. and Tutukov, A.V.,1984, Ap.J. **284**, 719.
Isern, J.,Labay, J.,Hernanz, M. and Canal, R.,1983, Ap.J. **273**, 320.
Jackson, J.D.,1962, Classical Electrodynamics, J. Wiley , p67.
Joss, P.C. and Rappaport, S.A.,1979, Astron. and Astrophys. **71**, 217.
Joss, P.C. and Rappaport, S.A.,1983, Nature **304**, 419.
Kippenhahn, R.,1969, Astron. and Astrophys. **3**, 83.
Kippenhahn, R. and Weigert, A.,1967, Zeitschrift f. Astrophys. **65**, 251.

Koch, R.H. and Hrivnak, B.J.,1981, Astron. J. **86**, vol. 3,438.
Kopal, Z.,1959, Close Binary Systems, J. Wiley.
Lamb, F.K., Pethick, C.J. and Pines, D.,1973, Ap.J. **184**, 271.
Lamers, H.J.,1981, Ap.J. **245**, 593.
Lamers, H.J., van den Heuvel, E.P. and Petterson, J.A.,1976, Astron. and Astrophys. **49**, 327.
Landau L.D. and Lifshitz E.M.,1959, Classical Theory of Fields, Pergamon Press.
Landau L.D. and Lifshitz, E.M.,1959, Mechanics, Pergamon Press.
Lauterborn, D.,1970, Astron. and Astrophys. **7**, 150.
Lewin, W.H. and Joss, P.C.,1983, in: 'Accretion Driven Stellar X-ray Sources', eds. W.H. Lewin and E.P. van den Heuvel, CUP., p42.
Limber, D.N.,1963, Ap.J. **138**, 1112.
Lin, D.N.C. and Papaloizou, J.C.B.,1979, Mon. Not. Roy. Astr. Soc. **186**, 799.
Livio, M.,Soker, N.,de Kool, M. and Savonije, G.J.,1986, Mon.Not.Roy.Astr.Soc. **218**, 593.
Lubow, S.H. and Shu, F.H., Ap.J. **198**, 383.
Lubow, S.H.,1979, Ap.J. **229**, 1008.
Lucy, L.B. and Solomon, P.M.,1970, Ap.J. **159**, 879.
Lyne, A.,Biggs, J.,Brinklow, A., Ashworth, M. and McKenna, J.,1988, Nature **332**, 45.
Meyer, F. and Meyer-Hofmeister E.,1979, Astron. and Astrophys. **78**, 167.
Milgrom, M.,1978, Astron. and Astrophys. **67**, L251.
Miyaji, S.,Nomoto, K.,Yokoi, K. and Sugimoto, D.,1980, Publ. Astr. Soc. Japan **32**, 303.
Morton, D.C.,1960, Ap.J. **132**, 146.
Nomoto, K.,1982, Ap.J. **257**, 780.
Nomoto, K.,1984, Ap.J. **277**, 791.
Ostriker, J.P. and Gunn, J.R.,1969, Ap.J. **157**, 1395.
Packet, W. and de Greve, J.P.,1979, Astron. and Astrophys. **75**, 255.
Paczyński, B.,1966, Acta Astron. **16**, 231.
Paczyński, B.,1968, in: 'High Lights of Astronomy', ed. L.Perek, Reidel, p.409.
Paczyński, B.,1971, Ann. Rev. Astr. Astrophys. **9**, 183.
Paczyński, B.,1976, IAU Symp. no:73, Structure and Evolution of Close Binary Systems, eds. P.Eggleton, J. Pringle and J. Whelan, Reidel, p.75.
Paczyński, B.,1983, Nature **304**, 421.
Paczyński, B. and Sienkiewicz, R.,1981, Ap.J. **248**, L27.
Papaloizou, J.C. and Pringle, J.E.,1977, Mon.Not.Roy.Astr.Soc. **181**, 441.
Papaloizou, J.C. and Pringle, J.E.,1978, Mon.Not.Roy.Astr.Soc. **184**, 501.
Pastetter, L. and Ritter, H.,1989, Astron. and Astrophys. **214**, 186.
Plavec, M.,1968, Adv. Astron. Astrophys. **6**, 201.
Plavec, M., 1980, in: 'Close Binary Stars: Observations and Interpretations', eds. M. Plavec, D. Popper and R. Ulrich, Reidel Publ. Comp.
Plavec, M.J.,Popper, D.M. and Ulrich, R.K., IAU Symp. no: 88, Polidan, R.S. and Peters, G.J.,1980,in: 'Close Binary Stars: Observations and Interpretation', eds. M. Plavec, D. Popper and R. Ulrich, Reidel.
Pringle, J.E. and Rees, M.J.,1972, Astron. and Astrophys. **21**, 1.
Pringle, J.E.,1981,Ann. Rev. Astron. and Astrophys. **19**, 137.
Pylyser, E. and Savonije, G.J.,1988, Astron. and Astrophys. **191**, 57.
Refsdal, S. and Weigert, A.,1971, Astron. and Astrophys. **13**, 367.
Rappaport, S.A.,Joss P.C. and Webbink, R.F.,1982, Ap.J. **254**, 616.
Robinson, E.L.,1976, Ann. Rev. Astron. Astrophys. **14**, 119.
Savonije, G.J.,1978, Astron. and Astrophys. **62**, 317.
Savonije, G.J.,1983a, in: 'Accretion Driven Stellar X-ray Sources', eds. W.Lewin and E. van den Heuvel, CUP, p.342.
Savonije, G.J.,1983b, Nature **304**, 422.

Savonije, G.J.,1987, Nature **325**, 416.
Savonije, G.J. and Papaloizou, J.C.,1984, Mon.Not.Roy.Astr.Soc. **207**, 685.
Savonije, G.J. and Papaloizou, J.C.,1985, in: 'Interacting Binaries', eds. P.Eggleton and J. Pringle, Reidel, p.83.
Savonije, G.J.,de Kool, M. and van den Heuvel, E.P.J.,1986, Astron. and Astrophys. **155**, 51.
Segelstein, D.J.,Rawley, L.A.,Stinebring, D.R.,Fruchter, A.S. and Taylor, J.H.,1986, Nature **322**, 712.
Shakura, N.I. and Sunyaev, R.A., 1973, Astron. Astrophys. **24**, 337.
Shima, E.,Matsuda, T.,Takeda, H. and Sawada, K.,1985, Mon.Not.Roy.Astr.Soc. **217**, 367.
Skumanich, A.,1972, Ap.J. **171**, 565.
Smith, M.A.,1979, PASP **91**, 737.
Srinivashan, G. and van den Heuvel, E.P.J.,1982, Astron. and Astrophys. **108**, 143.
Sugimoto, D. and Miyaji, S.,1980, Proc. IAU Symp. no: 88, 'Close Binary Systems: Observations and Interpretations', Reidel, p.133.
Taam, R.E.,Bodenheimer, P. and Ostriker, J.P.,1978, Ap.J. **222**, 269.
Taylor, J.H. and Weisberg, J.M.,1982, Ap.J. **253**, 908.
Tucker, W.H.,1975, Radiation Processes in Astrophysics, MIT Press.
van den Heuvel, E.P.J.,1978, in: 'Physics and Astrophysics of Neutron stars and Black Holes', eds. R. Giacconi and R. Ruffini, North Holland Publ. Comp.
van den Heuvel, E.P.J. and Heise, J.,1972, Nature Phys. Sci. **239**, 67.
van den Heuvel, E.P.J., van Paradijs, J.A. and Taam, R.E.,1986, Nature **322**, 153.
van der Klis, M.,1989, Ann. Rev. Astron. Astrophys. **27**, 517.
van Paradijs, J.A.,1983, in: 'Accretion Driven Stellar X-ray Sources', eds. W.Lewin and E. van den Heuvel, CUP, p.189.
Verbunt, F., 1989, in: 'Neutron Stars: their Birth, Radiation and Winds', NATO ASI, Erice, September 1988, ed. W. Kundt, Kluwer, Dordrecht.
Verbunt, F. and Zwaan, C.,1981, Astron. and Astrophys. **100**, L7.
Verbunt, F. and Hut, P.,1987, in: 'The Origin and Evolution of Neutron Stars',
IAU Symp. no: 125, Reidel, p187.
Warner, B.,1983,IAU Coll. nr: 72,'Cataclysmic Variables and Related Objects', eds. M. Livio and M. Shaviv, p.155.
Webbink, R.F.,1980, IAU Symp. nr: 88,'Close Binary Stars: Observations and Interpretations', Reidel, p.127.
Webbink, R.F., Rappaport, S.A. and Savonije, G.J.,1983, Ap.J. **270**, 678.
Zahn, J.P.,1977, Astron. and Astrophys. **57**, 383.
Zahn, J.P.,1989, Astron. and Astrophys. **220**, 112.
Ziolkowski, J.,1977, Proc. 8th Texas Conf. on Relat. Astrophys., Ann. N.Y. Acad. Sci. **302**, 47.
Zwaan, C,1981, in: 'Solar Phenomena in Stars and Stellar Systems', eds. R. Bonnet and A. Dupree, Reidel, p.463.

This article was processed using the LaTeX macro package with ICM style

PART II

Computational Methods in Astrophysical Hydrodynamics

An Introduction to Computational Methods in Hydrodynamics

Willy Benz

Harvard-Smithsonian Center for Astrophysics, 60 Garden Street, Cambridge, MA 02138, USA

1 Introduction

Computational astrophysics has become over the years a more and more important part of theoretical astrophysics. This should not come as a surprise since, after all, the physics of most astronomical objects can be described by coupled (generally non-linear) systems of differential equations that, unfortunately, are often too complicated to be solved analytically and for which numerical solutions have to be found.

A typical example of such a system of coupled non-linear differential equations are the equations governing the motion of fluids. Although based on the very simple physical principles of conservation of mass, momentum and energy, analytical solutions to this system of equations can only be found under some very specific and usually restrictive assumptions (special symmetry, small perturbations, etc.). The system of equations actually solved may end up bearing only little resemblance with the original set describing the physical situation. In most cases, to obtain physically realistic solutions, the fluid equations have to be solved numerically. This is certainly unfortunate, since the insight in the physics often gained by solving analytically the problem can easily be lost by resorting to quickly to the computer. Furthermore, the solutions actually obtained by the computer are necessarily only numerical approximations to the real solutions. Determining how close these numerical solutions are to the real solutions is often very difficult and in practice relies on some convergence argument between results obtained with various resolutions as well as physical intuition on how the solution should look like. Additional intuition on the performance of a code in a given context should come from the many test calculations performed *before* attacking the real, in our case astrophysical, problem. The importance of such tests can never be stressed enough. Ideally, one would like to compare numerical results with analytical solutions to problems as close as possible to the actual problem to be solved. This is unfortunately not always possible, and one might have to resort to comparisons with observations or experimentations or, in the extreme case, with the results of other numerical simulations.

On the other hand, the advantages of a well written and designed numerical scheme is that once the program has been written to solve a particular problem, it is usually relatively easy to modify it to include additional physics. A properly written program will therefore not only solve one particular problem but rather a whole class of problems. This should already make clear that a careful planing of the overall structure of the program is essential to allow for these changes to be introduced without the necessity of a major rewriting of the code. A further incentive for a careful long-range planing resides in the fact that developing the software necessary to solve a problem of a minimum of complexity requires usually several months or even years of effort.

In this series of lectures, we shall present an introduction to some of the numerical techniques developed over the years to solve the hydrodynamics conservation equations. This presentation will necessarily be incomplete since there are almost as many different schemes as researchers in this field. In fact, faced with this impossible task, we adopted the diametrically opposed approach, focusing only on three specific but very different schemes used to solve the hydrodynamics equations in one, two and three spatial dimensions respectively. Although the choice of these three methods as prototypes is somewhat arbitrary and certainly reflects some personal bias, all three methods have been, and still are, widely used in the astrophysical community. They also have the additional advantage to be relatively simple which allows for a more pedagogical introduction to some of the specific problems encountered in numerical fluid dynamics. A good understanding of the basic principles outlined in these three schemes should provide the basis necessary to understand the many other, sometimes much more sophisticated, schemes found in the literature.

Before actually solving the hydrodynamics equations, it is necessary to begin with the discussion of some important aspects of finite-differencing such as centering of variables, accuracy of various approximations to derivatives and differential equations (section 2). Since we are concerned with time evolving schemes, we shall also discuss the important question of stability of any given algebraic representation of a differential equation (section 3). A method will be outlined, the von Neumann stability analysis, that allows to derive analytical stability criteria for numerical approximations of partial differential equations with constant (or slowly varying) coefficients.

The first example of numerical method to solve the equations of mass, momentum and energy conservation will be the familiar Lagrangian 1D, spherical geometry scheme. Its simplicity makes it a good pedagogical example that we use to introduce concepts like the sometimes confusing artificial viscosity. We move on to 2D cylindrical geometry and discuss an Eulerian scheme for which we shall make extensive use of the so-called operator splitting technique. Particular care will be taken to discuss the advection terms since they constitute the primary difficulty in an Eulerian approach to fluid dynamics. Finally, we conclude with an introduction to a new method to solve the hydrodynamics equations in 3D, the so-called Smooth Particle Hydrodynamics, or SPH, method. The originality of this approach based on a Lagrangian description of the fluid resides in the way spatial gradients are estimated without the need of a grid. Hence, this method, apart from being intrinsically simple, does not suffer from many of the problems and difficulties usually associated with grid distortions and/or motions.

Much of what follows can be found in standard text books on numerical fluid dynamics. In particular, for section 2 and 3 we used extensively "Computational Techniques for Fluid Dynamics" by C.A.J. Fletcher and the classical "Computational Fluid Dynamics" by P.J. Roache as they are excellent reviews of many schemes and their related properties. Additional (and sometimes complementary) reviews can also be found in "Numerical Simulations of Reactive Flows" by Oran (1987), in "Computational Methods for Fluid

Flows" by Peyret and Taylor (1983), in "Difference Methods for Initial Value Problems" by Richtmyer and Morton (1967). From an astronomer's point of view however, these books are maybe too much oriented to "practical" fluid dynamics like, airplane wing design or flows in pipes for example. An introduction specific to astrophysical fluid dynamics does unfortunately not exist yet, although it is in the making. Parts of the material presented in these lectures (especially in sections 4 and 5) are in fact taken from a book in preparation entitled "Computational Astrophysics" by R.L. Bowers and J.R. Wilson.

2 Finite Difference Approximations

The first step in solving the hydrodynamics equations (or any other system of differential equations) on a computer is to transform the set of differential equations describing the space-time continuum into algebraic expressions that can be solved by the machine. This transformation, or discretization, unfortunately modifies the original equations by introducing error terms. The art of numerical hydrodynamics is to choose algebraic expressions that minimize these errors.

To convert the governing partial differential equation(s) to a system of algebraic equations a number of choices are available. The most common in astrophysical fluid dynamics is the so-called finite difference method and we will essentially concentrate on this technique. In this method a grid, called the computational grid, is laid over a specified region of space. The variables are then assigned to individual cell centers, edges, or corners. The derivatives are computed by taking differences between neighboring cells.

Let us look at this discretization procedure in some more details. To make matter easier we will first consider only problems in one spatial dimension.

2.1 Variables, Functions and Derivatives

We characterize the grid (really an axis in 1D) by the cell size $\Delta x^n_{k+\frac{1}{2}}$ and along the time axis we define the time step size by $\Delta t^{n+\frac{1}{2}}$. In these definitions, the standard convention that superscript always refer to time and subscript to space coordinate has been used. This notation will be used throughout this paper. The coordinates at time t^n on the grid will then obviously be given by

$$x^n_k = \sum_{l=0}^{k-1} \Delta x^n_{l+\frac{1}{2}} \tag{2.1.1}$$

with k=2,3,4,...,K. K being the total number of cells in the grid. Similarly, the time t^n will be computed from

$$t^n = \sum_{l=0}^{n-1} \Delta t^{l+\frac{1}{2}}. \tag{2.1.2}$$

It is important to note that the coordinate themselves are defined on zone edges, whereas the increments are zone centered. As a general rule, in these lectures integer indexes always correspond to zone-edge defined variables whereas half-integer indexes correspond to zone centered variables.

We can also define finite difference approximations to physical variables. These variables will be associated with zone corners, edges, or centers depending on the nature of the variable and/or the scheme used. As an example, let us look at a function g(x,t) that is associated with zone centers. Its value at time t^{n+1} at spatial location $k+\frac{1}{2}$ will be given by

$$g(x,t) = g(x^{n+1}_{k+\frac{1}{2}}, t^{n+1}) = g^{n+1}_{k+\frac{1}{2}}.$$

This value is represented by a "o" in Fig. 1. The same function associated with zone edges at time t^n would be written as

$$g(x,t) = g(x^n_k, t^n) = g^n_k.$$

and its value at t^n is shown by a "•" in Fig. 1.

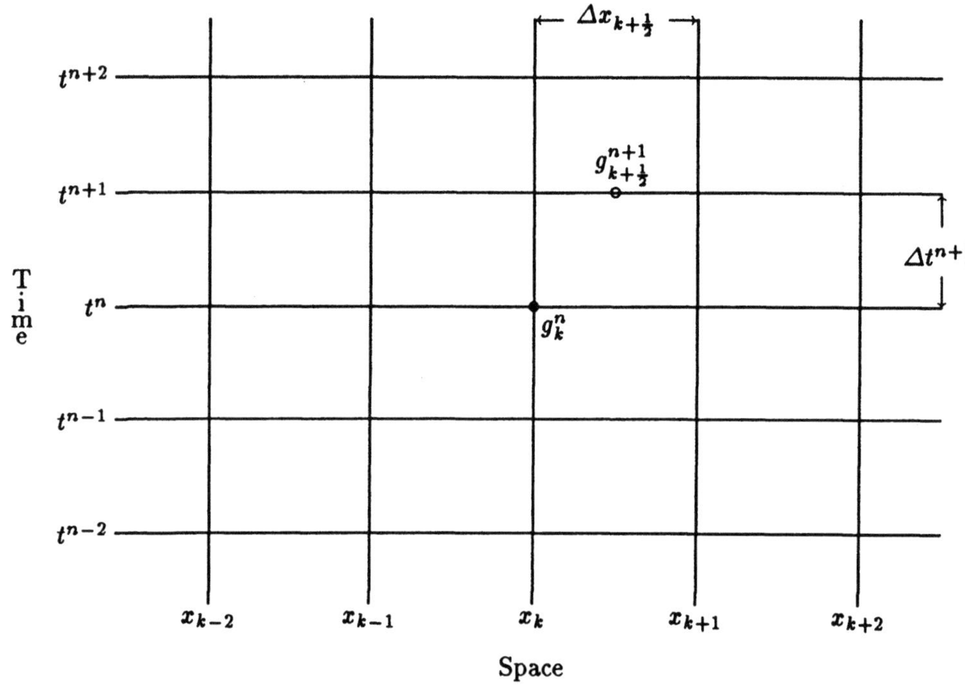

Fig. 1. Computational grid for one dimensional space-time

Applying these definitions we can write partial derivatives. Let us consider a function (still in one spatial dimension) $f(x,t)$. For a fixed time t, we can perform a Taylor expansion of f centered about x_0

$$f(x_0+\Delta x) = f(x_0) + \left(\frac{\partial f}{\partial x}\right)_{x_0} \Delta x + \frac{1}{2}\left(\frac{\partial^2 f}{\partial x^2}\right)_{x_0} \Delta x^2 + O(\Delta x^3) \tag{2.1.3}$$

where $O(\Delta x^n)$ stands for all terms of order Δx^n and higher. Using our notations we replace (2.1.3) by the following expression

$$f_{k+1}^n = f_k^n + \left(\frac{\partial f}{\partial x}\right)_k \Delta x_{k+\frac{1}{2}}^n + \frac{1}{2}\left(\frac{\partial^2 f}{\partial x^2}\right)_k (\Delta x_{k+\frac{1}{2}}^n)^2 + O(\Delta x^3). \qquad (2.1.4)$$

Solving this equation for the first order spatial derivative of f yields the following result

$$\left(\frac{\partial f}{\partial x}\right)_k = \frac{f_{k+1}^n - f_k^n}{\Delta x_{k+\frac{1}{2}}^n} - \frac{1}{2}\left(\frac{\partial^2 f}{\partial x^2}\right)_k \Delta x_{k+\frac{1}{2}}^n + O(\Delta x^2). \qquad (2.1.5)$$

For the sake of clarity, let us assume in what follows constant spacing in time and drop the superscript n on Δx. We shall also write any finite difference analog of $\frac{\partial f}{\partial x}$ as $\frac{\delta f}{\delta x}$. We obtain a first estimation to the first order spatial derivative of f called the forward-difference approximation

$$\left(\frac{\delta f}{\delta x}\right)_k = \frac{f_{k+1}^n - f_k^n}{\Delta x_{k+\frac{1}{2}}}. \qquad (2.1.6)$$

This finite difference approximation to the spatial derivative of f has a truncation error of order Δx, and is therefore only first order accurate.

By expanding $f(x,t)$ backward we obtain another finite-difference analog to $\frac{\partial f}{\partial x}$, this time called backward-difference approximation

$$\left(\frac{\delta f}{\delta x}\right)_k = \frac{f_k^n - f_{k-1}^n}{\Delta x_{k-\frac{1}{2}}} \qquad (2.1.7)$$

which is also first order accurate.

Yet another finite-difference approximation can be obtained by subtracting the forward and the backward Taylor expansions keeping terms up to third order. We find

$$\left(\frac{\partial f}{\partial x}\right)_k = \frac{f_{k+1}^n - f_{k-1}^n}{2\Delta x_k} - \frac{1}{2}\left(\frac{\partial^2 f}{\partial x^2}\right)_k \frac{(\Delta x_{k+\frac{1}{2}}^2 - \Delta x_{k-\frac{1}{2}}^2)}{2\Delta x_k} + O(\Delta x^2) \qquad (2.1.8)$$

where we have defined $\Delta x_k = \frac{1}{2}(\Delta x_{k+\frac{1}{2}} + \Delta x_{k-\frac{1}{2}})$. Now in case of uniform spatial zoning, that is, when $\Delta x_{k+\frac{1}{2}} = \Delta x_{k-\frac{1}{2}} = \Delta x$, equation (2.1.8) yields the so-called centered-difference approximation

$$\left(\frac{\delta f}{\delta x}\right)_k = \frac{f_{k+1}^n - f_{k-1}^n}{2\Delta x}. \qquad (2.1.9)$$

This expression has a truncation error of order Δx^2 and is therefore second order accurate. Notice however that this is true only for uniform zoning. In case of non-uniform zoning, a correction term involving the second derivative of f has to be taken into account for this analog to remain second order accurate. In practice, this may be a problem since second derivatives are not always known or may involve significant amount of additional computing.

Finite-difference approximations to second derivatives can be obtained in a similar manner. Consider again uniform spatial zoning, by finding first the Taylor expansions of f_{k+1}^n and f_{k-1}^n centered at f_k^n keeping terms up to fourth order, adding both expressions and finally solving for the second order derivative yields the following finite difference approximation

$$\left(\frac{\delta^2 f}{\delta x^2}\right)_k = \frac{f_{k+1}^n + f_{k-1}^n - 2f_k^n}{\Delta x^2}. \tag{2.1.10}$$

It is easy to show that the truncation error in this expression is of order Δx^2, that is (2.1.10) is second-order accurate.

Up to this point we have considered cases for which both, the value of the function and its derivatives are centered in the same way on the computational grid, which in our examples corresponded to cell boundaries. Let us now consider the case of a staggered mesh where the function itself is defined on cell edges but the derivative is cell centered. An example of such a mesh is shown in Fig. 5. To obtain a suitable expression for the derivative, we proceed again by Taylor expanding f_{k+1}^n and f_k^n but this time centered at $k+\frac{1}{2}$

$$f_{k+1}^n = f_{k+\frac{1}{2}}^n + \left(\frac{\partial f}{\partial x}\right)_{k+\frac{1}{2}} \frac{\Delta x_{k+\frac{1}{2}}}{2} + \frac{1}{2}\left(\frac{\partial^2 f}{\partial x^2}\right)_{k+\frac{1}{2}} \frac{\Delta x_{k+\frac{1}{2}}^2}{4} + O(\Delta x^3)$$

$$f_k^n = f_{k+\frac{1}{2}}^n - \left(\frac{\partial f}{\partial x}\right)_{k+\frac{1}{2}} \frac{\Delta x_{k+\frac{1}{2}}}{2} + \frac{1}{2}\left(\frac{\partial^2 f}{\partial x^2}\right)_{k+\frac{1}{2}} \frac{\Delta x_{k+\frac{1}{2}}^2}{4} + O(\Delta x^3).$$

Subtracting the second expression from the first we obtain

$$\left(\frac{\delta f}{\delta x}\right)_{k+\frac{1}{2}} = \frac{f_{k+1}^n - f_k^n}{\Delta x_{k+\frac{1}{2}}}. \tag{2.1.11}$$

This expression is easily shown to be second order accurate even for non-uniform zoning. Comparing (2.1.11) to (2.1.6),(2.1.7) or (2.1.8) we see that a staggered mesh offers the clear advantage of allowing second order accuracy without any additional computations (at least at this stage). We shall make extensive use of staggered meshes in sections 4 and 5.

2.2 Differential Equations

Using the representation of variables and derivatives introduced above, we can proceed and find the appropriate algebraic expressions for differential equations. In this section we shall consider particular examples of simple differential equations that illustrate the difficulties typically encountered in the discretization process. As our first example, we choose the following very simple differential equation

$$\frac{df}{dt} = -\frac{f}{\tau}\ , \text{with } f(t=0) = 1 \tag{2.2.1}$$

This equation has the obvious exact solution $f(t) = e^{-t/\tau}$. Using the forward, backward and centered difference approximations described in 2.1 and assuming uniform zoning for simplicity, we can derive the following three possible representations of (2.2.1)

$$\text{forward} : \frac{f^{n+1} - f^n}{\Delta t} = -\frac{f^n}{\tau} \tag{2.2.2}$$

$$\text{backward} : \frac{f^n - f^{n-1}}{\Delta t} = -\frac{f^n}{\tau} \tag{2.2.3}$$

$$\text{centered} : \frac{f^{n+1} - f^n}{\Delta t} = -\frac{1}{2\tau}(f^{n+1} + f^n) \tag{2.2.4}$$

Notice that (2.2.4) has been centered in a somewhat surprising way. Using a straightforward analog of (2.1.9) for estimating the time derivative of f would lead to a intrinsically unstable and therefore useless numerical scheme. A detailed discussion of stability conditions for various time evolving schemes will be presented in section 3. Another interesting point about (2.2.4) is that the value of the function at the new time $n+1$ (value that is still unknown at this stage) is used to evaluate the right-hand-side. In this simple example this does not pose any problems since it is still possible to solve analytically the equation for the value of the function at the new time f^{n+1}. In more complicated differential equations, however, solving analytically for the new value of the unknown is generally not possible and more complicated solution procedures have to be applied. These schemes are then called implicit as opposed to explicit schemes where the derivatives are computed using only known values of the variables. Implicit schemes despite being computationally more expensive have usually the attractive feature of having no stability limitations on the size of the time step whereas explicit schemes usually have as we shall see in section 3. It should however not be forgotten that the time step in implicit schemes is still limited by accuracy considerations. A good illustration of this is (2.2.3) which, as we shall see in the next section, has no stability limitations on the time step but being only first order accurate, errors will grow linearly with the size of the time step.

Returning to our example, it is possible for each of the three approximations to solve for its value at $t + \Delta t$ knowing the value of f at time t. It is interesting to compare the results obtained using these approximations to the exact solution. A graphical representation of the exact solution as well as the three numerical solutions is displayed in Fig. 2. Actual numerical values of the solutions at $t/\tau = 2$ are compared to the analytical solution in Table 1. All these results have been obtained taking $\Delta t/\tau = 0.2$.

Table 1. Numerical results at $t/\tau = 2$

Approximation	Value	Error	Truncation Error
exact	0.1353	0.0000	-
forward (2.2.2)	0.1074	0.0279	$O(\Delta t)$
backward (2.2.3)	0.1615	-0.0262	$O(\Delta t)$
centered (2.2.4)	0.1407	0.0054	$O(\Delta t^2)$

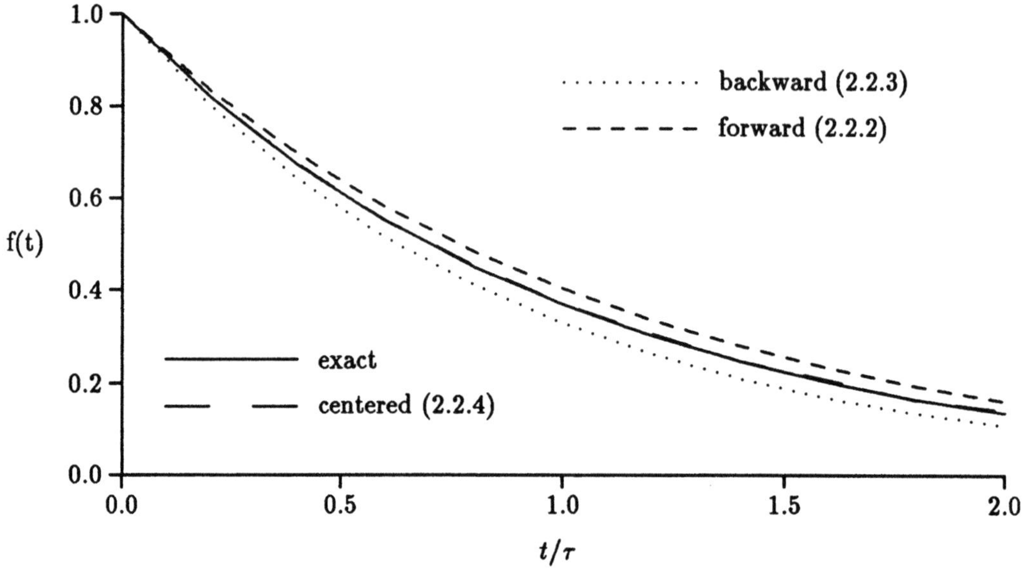

Fig. 2. Solutions to $df/dt = -f/\tau$

From an inspection of Table 1 we conclude that the truncation error gives a good indication of the accuracy of each of these approximations. The relative error at $t/\tau = 2$ is 5 times larger for our 2 first order schemes than for our second order approximation corresponding directly to the ratio of the truncation errors. The benefits of using higher order accuracy schemes becomes clearly visible.

From these results it might appear that higher-order (even higher than second) approximations should always be used. However, this is deceptive. First, higher order formulae involve more points and hence are less economical from a computational point of vue than low order formulae. From a practical perspective, the accuracy that can be achieved for a given execution time is usually more important than accuracy alone. Furthermore, the superiority of high order approximations is also dependent on the smoothness of the exact solution. Inviscid supersonic flows (a typical situation in many astrophysical flows) can produce discontinuous solutions, generally associated with shock waves. Evaluating the algebraic equations over such a discontinuity will not improve significantly by using higher order methods. In practice, second order methods work well and should be used in preference to first order methods. Beyond second order, approximations usually become barely tractable in the case of multi-dimensional fluid dynamics.

The next example of differential equation which is instructive to consider is the diffusion equation. An obvious astrophysical application of this equation would be the radiation transport in an optically thick medium. For our purpose, we simplify the problem by assuming a constant diffusion coefficient and write the diffusion equation

$$\frac{\partial f}{\partial t} = \alpha \frac{\partial^2 f}{\partial x^2}. \tag{2.2.5}$$

Applying the results of section 2.1, and assuming uniform zoning (in time and space), we can use the centered-difference approximations to obtain the following second order accurate representation of (2.2.5) called the Richardson approximation

$$\frac{f_k^{n+1} - f_k^{n-1}}{2\Delta t} = \alpha \frac{f_{k+1}^n + f_{k-1}^n - 2f_k^n}{\Delta x^2}. \tag{2.2.6}$$

This approximation is, in fact, unacceptable. As we shall see in section 3, this representation is numerically unstable for $\alpha > 0$ and $\Delta t > 0$ and thus is of no practical use. This behavior, as pointed out by Roache (1972), emphasizes the difference between an accurate finite-difference analog for a derivative, and accuracy for a differential equation. It is indeed true that both derivatives in (2.2.6) are estimated to second order accuracy and therefore are individually relatively accurate but used together they result in a fundamentally unstable scheme.

An interesting alternative is to replace the center node value of f_k^n by its time average $f_k^n = \frac{1}{2}(f_k^{n+1} + f_k^{n-1})$. The approximation (2.2.6) then becomes

$$\frac{f_k^{n+1} - f_k^{n-1}}{2\Delta t} = \alpha \frac{f_{k+1}^n + f_{k-1}^n - f_k^{n+1} - f_k^{n-1}}{\Delta x^2} \tag{2.2.7}$$

which is known as the DuFort-Frankel scheme. Although the value of f at time $n+1$ appears on both sides of the equation, it only involves the value at the same spatial location k. Equation (2.2.7) can therefore be solved explicitly for the new value of the function f_k^{n+1}. A stability analysis of (2.2.7) (see section 3) shows that this representation is unconditionally stable for $\alpha, \Delta t, \Delta x > 0$. However, there is a price to pay for this and it may come as a surprise. If we expand equation (2.2.7) in Taylor expansions centered at node n, k, we obtain the following expression neglecting terms higher than second order

$$\frac{\partial f}{\partial t} = \alpha \frac{\partial^2 f}{\partial x^2} - \alpha \frac{\Delta t^2}{\Delta x^2} \frac{\partial^2 f}{\partial t^2}. \tag{2.2.8}$$

The result is therefore a modification of the original equation. The discretization procedure has resulted in a transformation of the originally parabolic differential equation into a hyperbolic equation. From the standard theory of differential equations it is known that the solutions of these two types of equations have intrinsically different properties. From a physical point of vue, hyperbolic equations are essentially wave equations in which a perturbation introduced at time t_0 can only influence a restricted domain of space for $t > t_0$. Parabolic equations are typified by solutions that march forward in time and diffuse in space. A perturbation introduced at time t_0 can in principle affect all space for $t > t_0$ but will be strongly attenuated as it spreads.

We use this example to introduce here the important concept of *consistency*. A finite-difference approximation to a differential equation does not only have to be *accurate* but should also be *consistent* with the original differential equation. In other words, the original equation should be recovered when taking the limit of infinitely small time steps and zone sizes in the finite-difference analog. Thus, for consistency we must require in the DuFort-Frankel scheme that $\frac{\Delta t}{\Delta x} \longrightarrow 0$ as both Δt and $\Delta x \longrightarrow 0$. In practice, however, consistency becomes a vague notion since the limits can never be taken rigorously. The convergence of a solution is often checked by recomputing it with halved Δx. In our example, the time step Δt should be, in this case, reduced by more than 1/4 (Richtmyer 1957). Despite the unconditional stability of this scheme, too large a Δt can lead to large errors in the solution. The time step has in fact to be chosen so that $\alpha \Delta t^2/\Delta x^2$ remains small. The time step limit in this case comes therefore not from a stability constraint but rather from a accuracy requirement.

The above methods for solving the diffusion equation are all explicit and therefore suffer from a time step restriction that can become so limiting that the system can no longer be solved using a reasonable amount of computer time (see also the discussion on time step restrictions for the explicit diffusion equation in section 3). We outline here one implicit scheme for solving the diffusion equation which is quite popular, the Crank-Nicholson scheme. Being implicit, this scheme does not suffer from the time step limitation problems mentioned above. In this scheme, the time average is taken over the derivatives rather than over the function itself as in the DuFort-Frankel approach. Hence, we write our new finite-difference approximation to (2.2.5)

$$\frac{f_k^{n+1} - f_k^n}{\Delta t} = \alpha \left(\frac{1}{2} L_{xx} f_k^n + \frac{1}{2} L_{xx} f_k^{n+1} \right) \tag{2.2.9}$$

where L_{xx} is the second derivative operator with respect to the x coordinate, defined as in (2.1.10) by

$$L_{xx} f_k^n = \frac{f_{k-1}^n - 2f_k^n + f_{k+1}^n}{\Delta x^2}. \tag{2.2.10}$$

A Taylor expansion of equation (2.2.9) shows that this approximation is consistent with the original differential equation (2.2.5) and that the truncation error is $O(\Delta x^2, \Delta t^2)$. Rearranging slightly (2.2.9) and defining $s = \alpha \Delta t/\Delta x^2$ yields

$$-\frac{1}{2} s f_{k-1}^{n+1} + (1+s) f_k^{n+1} - \frac{1}{2} s f_{k+1}^{n+1} = \frac{1}{2} s f_{k-1}^n + (1-s) f_k^n + \frac{1}{2} s f_{k+1}^n. \tag{2.2.11}$$

This equation can not be solved explicitly for the new value of the unknown since values of f^{n+1} appear at different spatial locations. By considering all spatial nodes J in our mesh, we obtain a system of equations for the f^{n+1} that can be written in a matrix form

$$\begin{pmatrix} (1+s) & -\frac{1}{2}s & 0 & 0 & \dots \\ -\frac{1}{2}s & (1+s) & -\frac{1}{2}s & 0 & \dots \\ 0 & -\frac{1}{2}s & (1+s) & -\frac{1}{2}s & 0 \\ \vdots & \vdots & \vdots & \vdots & \vdots \\ \dots & \dots & \dots & -\frac{1}{2}s & (1+s) \end{pmatrix} \begin{pmatrix} f_2^{n+1} \\ f_3^{n+1} \\ f_4^{n+1} \\ \vdots \\ f_{J-1}^{n+1} \end{pmatrix} = \begin{pmatrix} d_2 \\ d_3 \\ d_4 \\ \vdots \\ d_{J-1} \end{pmatrix} \tag{2.2.12}$$

where the d_j are given by

$$\begin{aligned} d_2 &= \frac{1}{2} s f_1^n + (1-s) f_2^n + \frac{1}{2} s f_3^n + \frac{1}{2} s f_1^{n+1} \\ d_j &= \frac{1}{2} s f_{j-1}^n + (1-s) f_j^n + \frac{1}{2} s f_{j+1}^n \\ d_{J-1} &= \frac{1}{2} s f_{J-2}^n + (1-s) f_{J-1}^n + \frac{1}{2} s f_J^n + \frac{1}{2} s f_J^{n+1} \end{aligned} \tag{2.2.13}$$

The boundary values f_1^{n+1} and f_J^{n+1} are known from the Dirichlet boundary conditions that have to be specified to allow the problem to be solved. Since the resulting matrix is tridiagonal, this type of system of equations can be solved very efficiently using the so-called Thomas algorithm (see for example Fletcher, 1988). The Thomas algorithm consists in two parts, a forward sweep followed by a backward substitution, each taking only $O(J)$ operations. To illustrate this, let us first rewrite the above system of equations in a more general form

$$\begin{pmatrix} b_1 & c_1 & 0 & 0 & \dots \\ a_2 & b_2 & c_2 & 0 & \dots \\ 0 & a_3 & b_3 & c_3 & 0 \\ \vdots & \vdots & \vdots & \vdots & \vdots \\ \dots & \dots & \dots & a_J & b_J \end{pmatrix} \begin{pmatrix} f_1^{n+1} \\ f_2^{n+1} \\ f_3^{n+1} \\ \vdots \\ f_J^{n+1} \end{pmatrix} = \begin{pmatrix} d_1 \\ d_2 \\ d_3 \\ \vdots \\ d_J \end{pmatrix}$$

The forward sweep proceeds in eliminating the a_i coefficients and normalizing the b_i to unity. This is done for the first equation by setting

$$c_1' = \frac{c_1}{b_1} \quad , \quad d_1' = \frac{d_1}{b_1}$$

and for the following equations

$$c_i' = \frac{c_i}{b_i - a_i c_{i-1}'} \quad , d_i' = \frac{d_i - a_i d_{i-1}'}{b_i - a_i c_{i-1}'}.$$

The second stage, or backward substitution, consists in finding the new values f^{n+1}. The substitution starts with the last equation and proceeds backwards, hence $f_J^{n+1} = d_J'$ and

$$f_i^{n+1} = d_i' - f_{i+1}^{n+1} c_i'.$$

This algorithm is particularly economical, since it requires only 5J-4 operations. Furthermore, a stability analysis of the type presented in section 3 indicates that the Crank-Nicholson scheme has no restriction on the size of the time step. However, to avoid numerical problems (ill-conditioning) and round-off errors propagation it is necessary that $|b_i| > |a_i| + |c_i|$.

It is important to realize that a tridiagonal matrix is only obtained when a single governing equation is discretized using an implicit algorithm. Many flow problems are, however, governed by systems of equations (e.g. mass, momentum and energy conservation equations). Attempts to solve them in an implicit way results typically in block tridiagonal systems. The Thomas algorithm can be generalized without problems by replacing in the above description the scalar a, b, c by matrices $\mathbf{a}, \mathbf{b}, \mathbf{c}$, of size $M \times M$, with M being the number of equations formed at each grid point. The f and d become sub-vectors of length M. Since for a compressible viscous flow, M=5, the sub-matrices inversions can be done without requiring an enormous amount of computer time.

So far, we have only considered the diffusion equation in one spatial dimension. Solving it in a higher dimensional space creates substantial additional difficulties and a complete discussion would be beyond the scope of this short presentation. Nevertheless, to give a flavor of the techniques that are at hand to circumvent these difficulties we shall briefly discuss the diffusion equation in a two dimensional space. Taking again constant diffusion coefficients, we write

$$\frac{\partial f}{\partial t} = \alpha_x \frac{\partial^2 f}{\partial x^2} + \alpha_y \frac{\partial^2 f}{\partial y^2}. \tag{2.2.14}$$

A straightforward extension of the method used above, yields an implicit scheme closely resembling the one derived for the 1D diffusion equation. The resulting scheme has a truncation error of $O(\Delta t, \Delta x^2, \Delta y^2)$ and can be shown to be unconditionally stable. Unfortunately, however, the resulting matrix is not tridiagonal and the Thomas algorithm can no longer be used to solve the system. Since conventional Gauss elimination would be prohibitively expensive in computer time another method has to be found if (2.2.14) is to be solved efficiently.

The solution to this problem lies in splitting the procedure into two sub-steps, each half the length of the original time step. At each half-step, only terms associated with one particular direction are treated implicitly. Consequently, only three implicit terms appear during any of the two sub-steps and they can be grouped adjacent to the main diagonal to form again a tridiagonal matrix which can then be solved by the efficient Thomas algorithm.

An example of such a splitting technique is the so-called Alternating Direction Implicit (ADI) method of Peaceman and Rachford (1955). A normal time step is split into two sub-steps; during the first sub-step, one spatial direction is treated implicitly whereas the other direction is treated explicitly. The situation is reversed during the second sub-step. For our problem, the following x implicit discretization is used for the first half time step

$$\frac{f_{j,k}^{n+\frac{1}{2}} - f_{j,k}^{n}}{\Delta t/2} = \alpha_x L_{xx} f_{j,k}^{n+\frac{1}{2}} + \alpha_y L_{yy} f_{j,k}^{n}. \tag{2.2.15}$$

A second half-step is then taken in which the y direction is treated implicitly in a similar manner

$$\frac{f_{j,k}^{n+1} - f_{j,k}^{n+\frac{1}{2}}}{\Delta t/2} = \alpha_x L_{xx} f_{j,k}^{n+\frac{1}{2}} + \alpha_y L_{yy} f_{j,k}^{n+1}. \tag{2.2.16}$$

The operators L_{xx} and L_{yy} are defined by the relation (2.2.10). Equations (2.2.15-16) can be rewritten in a form similar to (2.2.11). Since for each of these equations only one spatial direction is treated implicitly and the other explicitly the matrices are tridiagonal and the Thomas algorithm can be applied successively to solve first (2.2.15) followed by (2.2.16). This scheme can be shown to be consistent with the original equation, unconditionally stable as well as second-order accurate in both space and time. For the latter one to be true,however, care has to be taken to treat boundary conditions appropriately at mid-time step.

The obvious generalization to three spatial dimensions of the ADI scheme is to take sub-steps at $n + \frac{1}{3}$ and $n + \frac{2}{3}$. This can be shown to lead to a method that is economical, spatially second-order accurate but only conditionally stable. For stability the following requirements have to be met $s_x, s_y, s_z \leq 1.5$, where $s_x = \alpha_x \Delta t / \Delta x^2$.

We have already mentioned several times in this section that, for some schemes, the size of the time step is subject to sometimes severe restrictions. In the following section we shall turn our attention precisely to the question of the stability of time-evolving schemes.

3 Stability

The concept of stability of a numerical scheme is concerned with the growth, or decay, of errors introduced at any stage of the computation. To study the temporal behavior of these errors we apply the standard technique used to study the stability of fluid flows, that is, small perturbations are introduced in the equations and their evolution are followed in time and space. If the perturbations are found to grow without bound, eventually overtaking the physical solution, the scheme is said to be unstable and, of course, has to be discarded since small perturbations will always be present in computer calculations due to machine round off. On the opposite, if the perturbations always vanish after some time the scheme will be declared unconditionally stable and can be used to solve the equation. A third, and most common case, are schemes for which the perturbations are found to vanish only for restricted values of the time step Δt, these will be called conditionally stable schemes.

Several methods to investigate analytically the stability of a particular scheme exist, unfortunately none is perfect. Furthermore, these stability analyses quickly become untractable even for moderately complex equations or system of equations. In practice, where we are usually concerned with actually obtaining numerical solutions, and not only with the analysis of computational methods for their own interest, numerical experimentation can not be avoided. A stability analysis in this case will merely provide helpful clues to guide the numerical experimentation. These remarks could led to the erroneous conclusion that nothing useful can be learned from performing these analyses. Quite the contrary, the detailed study of simplified equations provides invaluable insights in the discretization process, insights that will be of great help to understand (and cure) the more complex problems that will, without any doubts, arise when trying to solve much more intricate equations.

3.1 von Neumann Stability Analysis

The so-called von Neumann analysis was originated by J. von Neumann about 1944 at Los Alamos. It is the most commonly used method of determining stability criteria as it is generally the easiest to apply and the most dependable. Unfortunately, it can only be used to establish stability criteria for linear initial value problems with constant coefficients (or coefficients varying sufficiently slowly to be considered as constants). For non-linear equations, a stability criterion can only be established for the linearized equations, therefore providing only a necessary condition for stability as opposed to a necessary and sufficient condition. For a presentation of different stability analyses, see for example Roache (1972). To illustrate the von Neumann stability analysis, we shall consider in this section a number of simple differential equations, some of them already encountered in section 2, and derive explicitly the appropriate stability criteria.

Let us start with the 1D diffusion equation (2.2.5). We have seen that a second order accurate representation of the diffusion equation for uniform zoning in space and time can be written

$$\frac{f_k^{n+1}-f_k^{n-1}}{2\Delta t}=\alpha\frac{f_{k+1}^n+f_{k-1}^n-2f_k^n}{\Delta x^2}.$$

In section 2 we claimed that this equation is unconditionally unstable and therefore of no practical use. We shall now demonstrate this using the von Neumann stability analysis. We proceed by solving the above equation for f_k^{n+1}

$$f_k^{n+1}=f_k^{n-1}+d(f_{k+1}^n+f_{k-1}^n-2f_k^n) \tag{3.1.1}$$

where we have defined $d=2\alpha\Delta t/\Delta x^2$. Since for physically meaningful equations $\alpha\geq 0$, this implies that d is a positively defined quantity. The first step in a von Neumann stability analysis consists in expanding the solution in a finite Fourier series. Since the equation is linear, the growth or decay of each mode can be studied separately to determine the stability of the representation. Each Fourier component of the solution is written

$$f_k^n=V^ne^{ik\phi} \tag{3.1.2}$$

where the phase angle ϕ is defined by $\phi=k_x\Delta x$, k_x being the wave number of the component. Introducing (3.1.2) in (3.1.1) we obtain after some elementary algebra

$$V^{n+1}=V^{n-1}+V^n(2d(\cos\phi-1))$$

which can be written as

$$\begin{pmatrix}V^{n+1}\\ V^n\end{pmatrix}=G\begin{pmatrix}V^n\\ V_{n-1}\end{pmatrix}. \tag{3.1.3}$$

G is called the amplification matrix and, in our case, is given by

$$G=\begin{pmatrix}2d(\cos\phi-1) & 1\\ 1 & 0\end{pmatrix}. \tag{3.1.3}$$

If perturbations introduced in equation (3.1.1) are to remain finite, the following stability criterion has to be satisfied

$$|\lambda_m|\leq 1\ ,\ \forall m \tag{3.1.4}$$

with λ_m representing the m'th eigenvalue of the amplification matrix G. This conditions should be satisfied for all possible values of the phase angle ϕ. The eigenvalues of G are found in the usual way by setting

$$\det\begin{vmatrix}2d(\cos\phi-1)-\lambda & 1\\ 1 & -\lambda\end{vmatrix}=0$$

which yields the characteristic equation for the eigenvalues

$$\lambda^2-2d(\cos\phi-1)\lambda-1=0.$$

This equation admits two solutions

$$\lambda_\pm=d(\cos\phi-1)\pm\sqrt{d^2(\cos\phi-1)^2+1}. \tag{3.1.5}$$

It is easy to see that, for some values of ϕ (for example $\phi = 0$), λ_- violates the stability criterion (3.1.4) since d is a positively defined quantity. Therefore, we conclude that (3.1.1) is indeed unconditionally unstable and unsuitable for practical use.

Let us now consider the DuFort-Frankel scheme (2.2.7). We proceed in a similar fashion and after solving first for f_k^{n+1} and introducing the Fourier decomposition (3.1.2), we obtain the following amplification matrix

$$G = \begin{pmatrix} \frac{2d\cos\phi}{1+d} & \frac{1-d}{1+d} \\ 1 & 0 \end{pmatrix}.$$

Solving for the eigenvalues we obtain

$$\lambda_\pm = \frac{d\cos\phi}{1+d} \pm \sqrt{\left(\frac{d\cos\phi}{1+d}\right)^2 + \frac{1-d}{1+d}}. \tag{3.1.6}$$

Contrary to (3.1.5), both eigenvalues in (3.1.6) satisfy the stability requirement (3.1.4) for all positive values of d and all phase angle ϕ. The DuFort-Frankel scheme (2.2.6) is therefore unconditionally stable. As pointed out by Press *et al.* (1986), the surprising fact that the single replacement of f_k^n by $0.5(f_{k+1}^n + f_{k-1}^n)$ can have such a drastic effect on the stability of the representation shows that differencing partial differential equations can be as much an art than a science.

Another interesting analog to the diffusion equation, is the one in which the time derivative is expressed in terms of a forward-difference approximation (2.1.6). This yield an equation that is only first order accurate in time but second order accurate in space

$$f_k^{n+1} = d\,f_{k-1}^n + (1-2d)f_k^n + d\,f_{k+1}^n \tag{3.1.7}$$

where we have set $d = \alpha\Delta t/\Delta x^2$. A von Neumann stability analysis yields the following amplification factor (in this case the amplification matrix is a 1 by 1 matrix only)

$$G = 1 - 2d(1-\cos\phi). \tag{3.1.8}$$

For solutions to remain bounded we have to require that $|G| \leq 1$ for all ϕ. It is easy to see that (3.1.8) fulfills this conditions only for $d \leq 0.5$. Recalling the definition of d, we obtain the following constraint on the time-step

$$\Delta t \leq \frac{1}{2}\frac{\Delta x^2}{\alpha}. \tag{3.1.9}$$

We conclude that (3.1.7) is stable provided the time step is chosen so as to satisfy (3.1.9), hence, the representation is said to be conditionally stable. Looking at (3.1.9) it is interesting to note that the quantity $\Delta x^2/\alpha$ is actually nothing else than the diffusion time scale associated with one cell of the computational grid. The stability criterion therefore tells us that, to ensure stability, the time step should be taken smaller than half the diffusion time scale associated with a single cell of our computational mesh. In other words, stability requires that the time step be limited in such a way that no information can reach further than the neighboring cell during a single time step. This condition on the size of the time step is often called the Courant-Friedrichs-Lewy or CFL condition. It is a fundamental stability condition in all explicit approaches to numerical hydrodynamics.

It should also be realized that (3.1.9) is actually a very severe constraint on the time step. For example, halving the mesh size will require four times more time steps to reach the same physical time, but, since for each time step twice the work has to be done, the overall result is an increase of a factor 8 in computer time in one spatial dimension or 32 in three dimensions! This explains why the diffusion equation is usually solved implicitly, using a Crank-Nicholson scheme for example.

It is instructive to derive the stability condition for the three approximations to $df/dt = -f/\tau$ derived in section (2.2). These stability conditions are given in Table 2.

Table 2. Stability condition on the time step

Approximation	Stability Condition
exact	none
forward (2.2.2)	$\Delta t \leq \tau$
backward (2.2.3)	$\Delta t \geq 0$
centered (2.2.4)	$\Delta t \leq 2\tau$

Interestingly, although both the forward and the backward difference approximation have the same truncation error and hence about the same accuracy (see Table 1), we see that the backward approximation (2.2.3), as mentioned in the previous section, has in fact no limitations on the size of the time step and therefore is unconditionally stable. For the other two schemes we find a stability condition on the size of the time step which is again related to the characteristic time scale τ of the exact solution. What might happen if the stability condition is not satisfied is illustrated in Fig. 3. Plotted are the numerical solutions obtained using (2.2.2) with time steps $\Delta t/\tau = 1.5$ and 2.05 together with the analytical solution. Clearly, taking a time step larger than allowed by the stability condition can lead to quite dramatic results!

Let us now look at an equation that is more closely related to hydrodynamics, the so-called advection-diffusion equation

$$\frac{\partial f}{\partial t} = -v\frac{\partial f}{\partial x} + \alpha\frac{\partial^2 f}{\partial x^2}. \tag{3.1.10}$$

In this equation f models any advected and diffused flow quantity like vorticity (although 1D vorticity does not really exist). The flow velocity v will be assumed constant in this example. Using the first-order accurate in time forward-differencing (2.1.6) analog and the second-order accurate in space centered-differencing scheme (2.1.9) we write the following representation of (3.1.10)

$$f_k^{n+1} = f_k^n - \frac{v}{2}(f_{k+1}^n - f_{k-1}^n) + d(f_{k+1}^n + f_{k-1}^n - 2f_k^n) \tag{3.1.11}$$

where we have again defined $d = \frac{\alpha\Delta t}{\Delta x^2}$. Proceeding with a von Neumann stability analysis, we obtain the following amplification factor

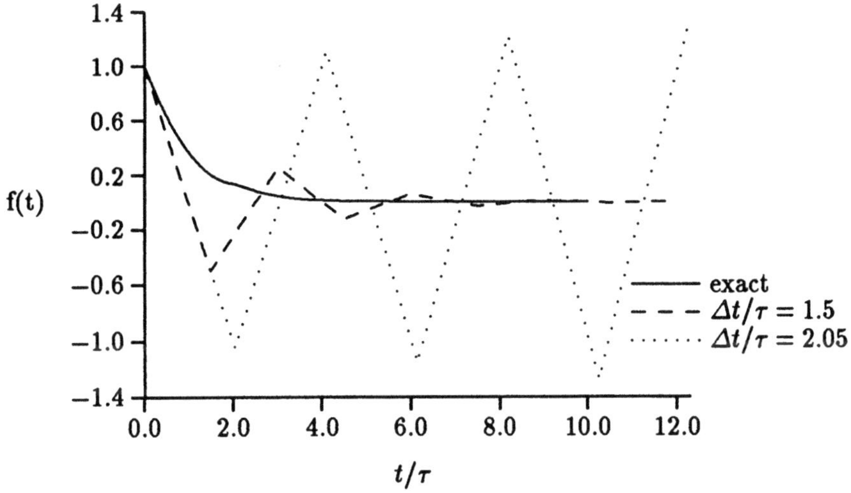

Fig. 3. Examples of numerical instabilities

$$G = 1 - 2d(1 - \cos\phi) - \mathrm{i}\, v \sin\phi. \tag{3.1.12}$$

Note that G is, in this case, a complex expression that only reduces to real for $v = 0$, that is in the case of a pure diffusion equation, limit in which we recover (3.1.8). We find the proper stability criterion by imposing $GG^* \leq 1$, with G^* being the complex conjugate of G

$$GG^* = \left(1 + 2d(\cos\phi - 1)\right)^2 + v^2(1 - \cos^2\phi) \leq 1. \tag{3.1.13}$$

Solving (3.1.3) (an easy but rather lengthy operation) yields the following two interesting constraints

$$\frac{v\Delta x}{\alpha} = R_c \leq 2 \quad \text{and} \quad d \leq \frac{1}{2}. \tag{3.1.14}$$

The second condition is nothing else than the diffusion stability condition already obtained (3.1.9). The other condition $R_c \leq 2$ implies that, for stability, the Reynolds number associated with an individual cell of our computational mesh has to be less than 2. This expression, independent of Δt, tells us that for a fluid with a given viscosity α and moving with a given velocity v, the spatial zoning is no longer a free parameter but that the cell size has to be chosen such as to meet requirement (3.1.14). For too large a velocity or too small a viscosity, this may imply too fine a zoning resulting in a number of cells growing out of proportions rendering this scheme practically useless.

Combining the two conditions yields the following constraint on the time step

$$\Delta t \leq \frac{2\alpha}{v^2}. \tag{3.1.15}$$

This relation shows that our representation (3.1.11) is unconditionally unstable for $\alpha = 0$. In other words, (3.1.11) would be completely unsuitable for treating advection alone. Another combination of the two conditions yields

$$\Delta t \leq \frac{\Delta x}{v}. \tag{3.1.16}$$

Interestingly, in this expression $\Delta x/\ v$ is the time necessary for the fluid to cross one cell. We therefore recover the fundamental stability limitation for explicit hydrodynamics, or CFL condition, namely that no information is allowed to travel more that one cell within a single time step.

Another extremely instructive example is the continuity equation. For the simple case of a constant fluid velocity we have

$$\frac{\partial \rho}{\partial t} + v_0 \frac{\partial \rho}{\partial x} = 0. \tag{3.1.17}$$

A possible finite difference approximation to (3.1.17) would be

$$\rho_{k+\frac{1}{2}}^{n+1} = -s\left(\rho_{k+\frac{1}{2}}^{n} - \rho_{k-\frac{1}{2}}^{n}\right) + \rho_{k+\frac{1}{2}}^{n} \tag{3.1.18}$$

where the fluid density ρ is now defined as a cell-centered quantity and $s = v_0 \Delta t / \Delta x$. It should be noticed that s is nothing else than the number of cells crossed by the fluid during one time step. Using our stability analysis again yields the following amplification factor

$$G = (1 - s) + s\cos\phi - \mathrm{i}\, s \sin\phi.$$

From this relation the stability condition for (3.1.18) is easily derived and found to be $0 \leq s \leq 1$. Recalling the definition of s we obtain again the stability criterion (3.1.16). Since the continuity equation physically describes mass advection it is not surprising to recover the stability criterion derived for the advection-diffusion equation.

We can deduce some more interesting properties of (3.1.18) by rewriting it in a slightly different form

$$\rho_{k+\frac{1}{2}}^{n+1} = (1 - s)\rho_{k+\frac{1}{2}}^{n} + s\, \rho_{k-\frac{1}{2}}^{n}. \tag{3.1.19}$$

Taking the largest possible value for s, namely $s = 1$, in this equation, we obtain $\rho_{k+\frac{1}{2}}^{n+1} = \rho_{k-\frac{1}{2}}^{n}$. In other words, the fluid originally in cell $k - \frac{1}{2}$ at time n has moved to cell $k + \frac{1}{2}$ at time $n + 1$, which is the exact solution. The problem arises for values of s less than 1, for which the exact solution will not be recovered. Fig. 4 illustrates the evolution of an initial spike in density at $x = 0.5$ after 2, 4, 6 and 10 time steps for a value of $s = 0.95$. Since, in case of non-uniform zoning, the criterion $s \leq 1$ has to be satisfied by all zones, s can not be equal to unity over the entire mesh. The degree to which (3.1.9) represents accurately the physical behavior of mass advection will therefore strongly depend on the location on the mesh.

Clearly such a scheme is unacceptable since any features would be spread and washed out very quickly. To understand these poor results, we expand in a Taylor series centered at $\rho_{k+\frac{1}{2}}^{n}$ all terms in (3.1.9). After some elementary algebra we obtain the following equation

$$\frac{\partial \rho}{\partial t} + v_0 \frac{\partial \rho}{\partial x} = \frac{1}{2} v_0 \Delta x \frac{\partial^2 \rho}{\partial x^2} - \frac{1}{2} \frac{\partial^2 \rho}{\partial t^2} \Delta t + O(\Delta x^2). \tag{3.1.20}$$

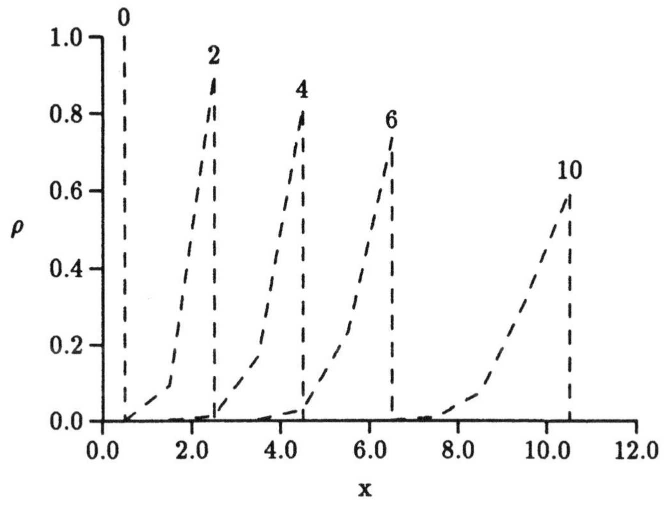

Fig. 4. First order mass advection

A comparison between (3.1.20) and the original equation (3.1.17) reveals the origin of the spreading: our representation introduces (among other terms) a spurious diffusion term not in the original equation. The diffusion coefficient $\frac{1}{2}v_0\Delta x$ depends not only on the fluid velocity but also on the mesh size, the smaller the size the smaller the coefficient. Notice that despite this serious problem, (3.1.20) is actually consistent with the original differential equation since in the limit when Δx, $\Delta t \longrightarrow 0$ the original equation is recovered. As mentioned earlier, consistency is in practice only a vague notion since the limit can actually never be taken rigorously.

This example illustrates again the important fact that the discretization procedure introduces terms that are not in the original differential equation. The exact nature of these terms and their potential effects on the solutions can usually be found by looking at the leading error terms of a Taylor expansion. A scheme for which the leading error term is made of second order spatial derivatives will essentially introduce unwanted diffusion in the numerical solution. If third order spatial derivatives are making up the leading error term, dispersion errors will dominate resulting in wave propagation speed errors that will most likely be wavelength dependent (a more detailed discussion of these points can be found in Fletcher, 1988). These problems which are most often referred to by the general denomination of *numerical viscosity*, result in considerable spreading of sharp features and can not be avoided since they are inherent to the discretization process. The art, of course, is actually to devise schemes that minimize this spreading.

As a final example, we want to investigate the stability of a numerical representation of the full hydrodynamics equation, namely the momentum conservation equation and the general continuity equation. Both equations are, however, non-linear and the von Neumann stability analysis can not be applied directly to investigate the stability of a finite-difference approximation. The procedure adopted is therefore to linearize the equations first and to derive the appropriate stability conditions for the linearized set of equations. Any finite-difference approximation to the non-linear equations should *at least* satisfy the stability condition for the linearized equations.

To derive the linearized equations, we proceed as usual in perturbation theory (see for example Chandrasekhar 1961) by writing the relevant quantities as a sum of an unperturbed term plus a term describing the perturbations, for example for the pressure $p = p_0 + p'$. We further suppose that the quantities with subscript 0 satisfy the original non-linear equations and that the perturbations are small compared to their unperturbed counterparts so that only the first order terms in the perturbations need to be considered. We obtain after some elementary algebra the standard linearized Euler equation

$$\frac{\partial \boldsymbol{v}'}{\partial t} = -\frac{1}{\rho_0}\nabla p' \tag{3.1.21}$$

and the linearized continuity equation

$$\frac{\partial \rho'}{\partial t} + \rho_0 \nabla \cdot \boldsymbol{v}' = 0. \tag{3.1.22}$$

From elementary fluid dynamics (see for example Landau and Lifshitz 1959, Courant and Friedrichs 1976), we know that these small perturbations will propagate in the fluid as a sound wave with velocity equal to the adiabatic sound speed $c = \sqrt{\gamma RT/\mu}$ in the case of a perfect gas. Assuming again for simplicity a one dimensional space and defining $\phi = p'$ and $\psi = \rho_0 c v'$ we can rewrite (3.1.21) as

$$\frac{\partial \psi}{\partial t} = -c\frac{\partial \phi}{\partial x} \tag{3.1.23}$$

and (3.1.22) as

$$\frac{\partial \phi}{\partial t} = -c\frac{\partial \psi}{\partial x} \tag{3.1.24}.$$

The spatial derivative of ϕ advances ψ in time and vice versa. This feature strongly suggest a staggered mesh approach where ϕ (or the pressure) is zone centered and ψ (or velocity) is defined on zone edges (Fig. 5).

As shown in section 2.1, such an approach results in a scheme that is second order accurate in space and time. Consequently, our choice for the finite-difference approximation (on a uniform grid in space and time) to the linearized hydrodynamics equations are for (2.1.23)

$$\frac{\psi_k^{n+\frac{1}{2}} - \psi_k^{n-\frac{1}{2}}}{\Delta t} = -c\,\frac{\phi_{k+\frac{1}{2}}^n - \phi_{k-\frac{1}{2}}^n}{\Delta x} \tag{3.1.25}$$

and for (3.1.24)

$$\frac{\phi_{k+\frac{1}{2}}^{n+1} - \phi_{k+\frac{1}{2}}^n}{\Delta t} = -c\,\frac{\psi_{k+1}^{n+\frac{1}{2}} - \psi_k^{n+\frac{1}{2}}}{\Delta x}. \tag{3.1.26}$$

Given initial conditions for ϕ and ψ, (3.1.25) can be solved for $\psi_k^{n+\frac{1}{2}}$, these values can then be used to solve (3.1.26) for $\phi_{k+\frac{1}{2}}^{n+1}$. To test the stability of our representation we use again the von Neumann stability analysis and decompose both ϕ and ψ in Fourier modes

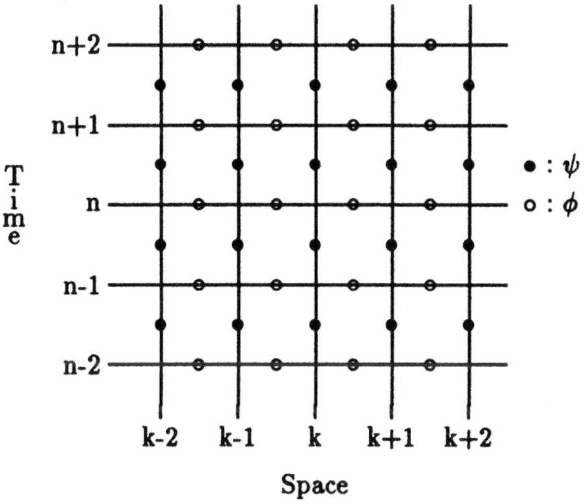

Fig. 5. Staggered mesh

$$\psi_k^{n-\frac{1}{2}} = A^{n-\frac{1}{2}} e^{ik\theta}, \quad \text{and} \quad \phi_{k+\frac{1}{2}}^n = B^n e^{i(k+\frac{1}{2})\theta}.$$

Substituting in (3.1.25) and in (3.1.26) and after some elementary algebra we find the following system of linear equations

$$\begin{pmatrix} A^{n+\frac{1}{2}} \\ B^{n+1} \end{pmatrix} = G \begin{pmatrix} A^{n-\frac{1}{2}} \\ B^n \end{pmatrix}$$

in which the amplification matrix is given by

$$G = \begin{pmatrix} 1 & -2i\, s \sin\frac{\theta}{2} \\ -2i\, s \sin\frac{\theta}{2} & 1 - 4s^2 \sin^2\frac{\theta}{2} \end{pmatrix}$$

with s defined by $s = c\Delta t/\Delta x$. Looking for the eigenvalues of G, we derive the following characteristic equation

$$\lambda^2 + \left(4s^2 \sin^2\frac{\theta}{2} - 2\right)\lambda + 1 = 0$$

and the stability condition will be determined by imposing the restriction $|\lambda_\pm| \le 1$. It is easy to see that this restriction implies that $s \le 1$, or recalling the definition of s

$$\Delta t \le \frac{\Delta x}{c}. \tag{3.1.27}$$

This expression can again be identified with the CFL condition. Its physical meaning is clear, since $\Delta x/c$ is the time needed for a sound wave to cross one cell (a quantity sometimes called the Courant number), the time step is limited (for stability) to prevent a sound wave from crossing more than one cell during one time step. In the case of a non-uniform grid, the time step would be limited by the sound crossing time of the smallest cell in the grid.

Before using this stability condition for the full non-linear hydrodynamics equations, we have to remember, however, that this condition was obtained for the linearized equations. Using it to limit the time step in a finite-difference approximation of the full non-linear hydrodynamics equations will not warrant automatically the stability of the representation especially if highly non-linear features like shocks are present in the flow. However, despite the fact that (3.1.27) is not a sufficient condition for stability it is nevertheless a necessary condition even for the non-linear equations. Therefore, in practice, (3.1.27) is actually used to limit the time step even in the non-linear regime and is often written as $\Delta t = \alpha \Delta x / c$ with α a arbitrary coefficient supposed to take care of the non-linear effects. The actual value of α can vary from one problem to another and has often to be chosen after some numerical experimentation. In practice values somewhere between 0.25 and 0.75 are usually sufficient to ensure stability.

With this example, we have completed our discussion of the question of the stability of finite-difference analog(s) to a differential equation or system of equations. We are now in a position, using the techniques outlined in sections 2 and 3, to derive finite difference analogs to the full hydrodynamics equations. We shall start with a typical astrophysical assumption and assume spherical symmetry, which reduces the hydrodynamics equations to a one dimensional space. For 1D hydrodynamics, the Lagrangian approach is the best suited and certainly the simplest from a computational point of vue.

4 1D Lagrangian Hydrodynamics

The fundamental difference between Lagrangian and Eulerian hydrodynamics lies in the choice of the coordinate system. In Eulerian hydrodynamics, the coordinate system is fixed (or moving at a prescribed velocity) whereas in Lagrangian hydrodynamics the coordinate system is attached to, and moves with the fluid. The essential advantage of attaching the coordinate system to the fluid is that the nonlinear advection terms disappear from all the equations making the system much easier to solve.

In practice, this turns out unfortunately to be true only in one dimension. In two or three dimensions, the coordinate system quickly deforms, especially if shear motions are present, and the numerical derivatives become less and less accurate if not totally wrong. At this stage, the only solution is to rezone, that is to define a new computational grid on which derivatives can again be taken accurately. These rezoning procedures are, however, usually accompanied by a significant amount of diffusion of the various quantities. This can lead to serious problems in many astrophysical situations. A typical example would be the diffusion of angular momentum in accretion disks leading to unphysical mass transport.

Over the past decade, a new type of Lagrangian hydrodynamics method has been developed, the so-called Smooth Particle Hydrodynamics (SPH) method (Lucy 1977). The interesting property of this method is to actually evaluate spatial gradients without the use of any grid. Thus, this method does not suffer from the problems associated with mesh tangling and subsequent rezoning mentioned above and can therefore be applied to multidimensional hydrodynamics. We postpone the detailed discussion of this scheme to section 6.

In this section we shall be concerned with the construction of a 1D explicit Lagrangian hydrodynamics code. For obvious astronomical reasons, we derive the appropriate equations assuming spherical symmetry. This is almost the simplest (1D cartesian would be even simpler but of much less astrophysical use), and therefore the most pedagogical, hydrodynamical code that can be built and yet it has still many possible applications in astrophysics. Late stages of stellar evolution, supernovae explosions, blast waves, non-linear (radial) pulsations can all be simulated with such types of code provided the assumption of spherical symmetry is warranted. In fact, the major difficulty in these simulations is not really the hydrodynamics itself but rather the complex physics that has to be included in the code in order to model accurately the physical system. In what follows, however, we shall be concerned only with the hydrodynamics. The code presented could be seen as a "backbone" to which complex physics can eventually be added.

4.1 1D Inviscid Lagrangian Hydrodynamics

4.1.1 Equations

As pointed out above, in Lagrangian hydrodynamics the coordinate system is attached to the fluid, consequently, the natural choice for the independent variable is mass rather than position. The transformation of variable is done easily in one spatial dimension (spherical geometry) by $dm = 4\pi r^2 \rho dr$. The relevant equations describing the motion of an inviscid fluid for which spherical symmetry can be assumed are given below. In these equations, we will chose as thermodynamical variables density and specific internal energy ϵ. Clearly, any two thermodynamical variables would do and ϵ could easily be replaced by the specific entropy s for example.

a) Momentum conservation equation

$$\frac{dv}{dt} = -4\pi r^2 \frac{\partial P}{\partial m}. \tag{4.1.1}$$

b) Mass conservation or continuity equation

$$V = \frac{1}{\rho} = 4\pi r^2 \frac{\partial r}{\partial m}. \tag{4.1.2}$$

c) Energy conservation equation

$$\frac{d\epsilon}{dt} = -P\frac{dV}{dt} = \frac{P}{\rho^2}\frac{d\rho}{dt}.$$

But recalling that the continuity equation in 1D spherical coordinate is given by

$$d\rho/dt = -\rho\nabla\cdot\boldsymbol{v} = -\rho\frac{1}{r^2}\frac{\partial}{\partial r}(r^2 v)$$

together with the definition $dm = 4\pi r^2 \rho dr$, we can rewrite the energy equation as

$$\frac{d\epsilon}{dt} = -4\pi P\frac{\partial}{\partial m}(r^2 v). \tag{4.1.3}$$

d) Equation of motion for the mass elements

$$\frac{dr}{dt} = v. \tag{4.1.4}$$

Finally, to close the system, a relation between the hydrodynamical variables is needed. This relation, or equation of state, has the general form (for our choice of thermodynamical variables) $P = f(\rho, \epsilon)$. In the simple case of a perfect gas this relation becomes $P = (\gamma - 1)\rho\epsilon$

4.1.2 Space-Time Centering

To solve the above system of equations we shall use a staggered mesh of the same type as described in section 3. As mentioned in section 2.1, the derivative estimates should be second order accurate. Unfortunately, this is not true. The scheme presented below is in fact only first order accurate in the case of non-uniform zoning but becomes second order accurate for uniform zoning in both space and time. We shall elaborate on this point below.

On such a mesh, we define the mass, radius and the velocity at zone edges whereas density, pressure and energy are defined at zone centers. Fig. 6 illustrates the space-time centering adopted. Notice that uniform zoning in mass or time is no longer assumed.

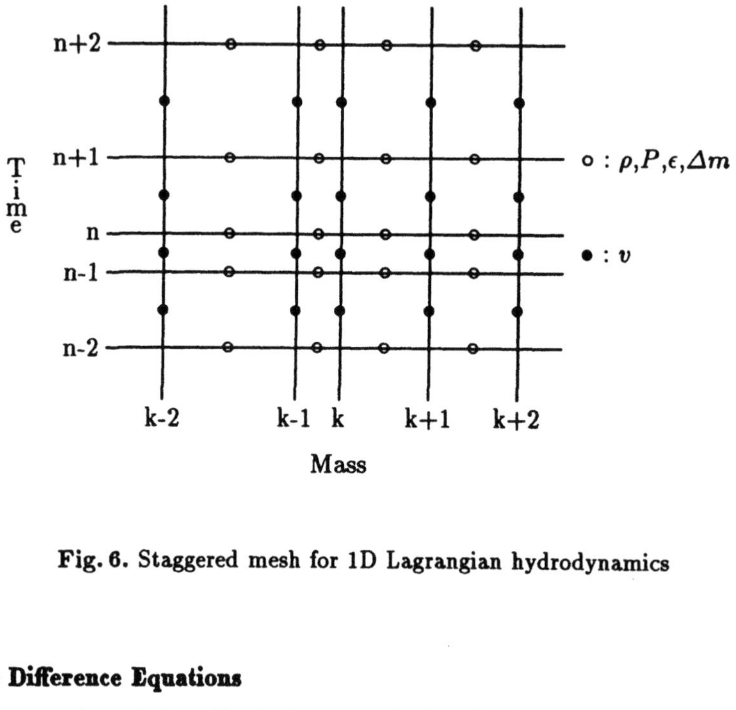

Fig. 6. Staggered mesh for 1D Lagrangian hydrodynamics

4.1.3 Finite Difference Equations

Using the staggered mesh described above, we derive the following finite-difference equations. First, the mass elements are defined by

$$\Delta m_{k+\frac{1}{2}} = \frac{4\pi}{3}\left((r^0_{k+1})^3 - (r^0_k)^3\right)\rho^0_{k+\frac{1}{2}}. \tag{4.1.5}$$

Notice that the mass elements have no time index since, by definition, Lagrangian mass elements are constant in time. To find the appropriate representation for the momentum equation (4.1.1) let us integrate it over a cell centered at k (the acceleration is centered at k)

$$\int_{\Delta m_k} \frac{dv}{dt} dm = -\int_{\Delta m_k} 4\pi r^2 \frac{\partial P}{\partial m} dm. \tag{4.1.6}$$

Let us first look at the left-hand-side of the above equation. The integration is taken over Δm_k which ranges from $r_{k-\frac{1}{2}}$ to $r_{k+\frac{1}{2}}$. Over this interval, the acceleration dv/dt is a constant. The integral therefore becomes

$$\int_{\Delta m_k} \frac{dv}{dt} dm = \left(\frac{dv}{dt}\right)_k \Delta m_k = \frac{v_k^{n+\frac{1}{2}} - v_k^{n-\frac{1}{2}}}{\Delta t^n} \Delta m_k. \tag{4.1.7}$$

The right-hand-side can be integrated by parts. Noticing further that between $r_{k-\frac{1}{2}}$ and r_k, $P = P_{k-\frac{1}{2}}$ whereas between r_k and $r_{k+\frac{1}{2}}$ we have $P = P_{k+\frac{1}{2}}$, we obtain the following expression for the right-hand-side

$$-\int_{\Delta m_k} 4\pi r^2 \frac{\partial P}{\partial m} dm = -4\pi r_k^2 (P_{k+\frac{1}{2}} - P_{k-\frac{1}{2}}). \tag{4.1.8}$$

Combining eqs. (4.1.7) and (4.1.8) and solving for the velocity at the new time step, we write the momentum conservation equation

$$v_k^{n+\frac{1}{2}} = v_k^{n-\frac{1}{2}} - A_k^n \frac{P_{k+\frac{1}{2}}^n - P_{k-\frac{1}{2}}^n}{\Delta m_k} \Delta t^n \tag{4.1.9}$$

where $A_k^n = 4\pi (r_k^n)^2$ is the surface element centered at k. Despite the appearance, the expression above is only second order accurate in case of uniform zoning in space and time. This somewhat subtle point comes from the fact that Δt^n and Δm_k are not independent quantities but are defined by $\Delta t^n = \frac{1}{2}(\Delta t^{n-\frac{1}{2}} + \Delta t^{n+\frac{1}{2}})$ and by $\Delta m_k = \frac{1}{2}(\Delta m_{k-\frac{1}{2}} + \Delta m_{k+\frac{1}{2}})$ respectively. Taylor expansions centered at node n, k in (4.1.9) (taking into consideration that between k and $k + \frac{1}{2}$ the proper mass element is $\Delta m_{k+\frac{1}{2}}$ whereas between $k - \frac{1}{2}$ and k the proper mass element is $\Delta m_{k-\frac{1}{2}}$) show that (4.1.9) is, indeed, only first order accurate becoming second order accurate in case of uniform zoning in space and time. For many applications (stellar evolution for example), a uniform zoning in mass is, however, often impossible. A more nearly uniform zoning can possibly be obtained by a judicious change of variable.

These remarks could lead the reader into questioning the use of a staggered mesh since its advantage apparently seems to be lost. It turns out that here the staggered mesh is used more for stability rather than for pure accuracy reasons. This example illustrates some of the very tricky issues involved in trying to estimate the order of an even simple numerical analog of a differential equation. A recent discussion of the order of various numerical estimates for the continuity equation can be found in Finn and Hawley (1990).

Once the velocity has been updated using (4.1.9), the new coordinates are found from equation (4.1.4) for which a suitable analog is

$$r_k^{n+1} = r_k^n + v_k^{n+\frac{1}{2}} \Delta t^{n+\frac{1}{2}}. \tag{4.1.10}$$

Notice that in (4.1.10), the time step is centered at $n + \frac{1}{2}$ whereas in (4.1.9) the time step was centered at n. As mentioned above, one is easily recovered from the other by $\Delta t^n = \frac{1}{2}(\Delta t^{n-\frac{1}{2}} + \Delta t^{n+\frac{1}{2}})$

The conservation of the mass of a Lagrangian element can be used in order to obtain the new density (or specific volume)

$$V_{k+\frac{1}{2}}^{n+1} = \frac{1}{\rho_{k+\frac{1}{2}}^{n+1}} = \frac{4\pi}{3} \frac{(r_{k+1}^{n+1})^3 - (r_k^{n+1})^3}{\Delta m_{k+\frac{1}{2}}}. \tag{4.1.11}$$

We are left with the energy equation (4.1.3). To obtain a suitable analog we proceed in a similar manner than for the momentum equation, that is we integrate (4.1.3) over a mass element $\Delta m_{k+\frac{1}{2}}$ (u is centered at $k+\frac{1}{2}$). After some elementary algebra we obtain our expression for the new specific internal energy

$$\epsilon_{k+\frac{1}{2}}^{n+1} = \epsilon_{k+\frac{1}{2}}^{n} - P_{k+\frac{1}{2}}^{n+\frac{1}{2}} \left(\frac{A_{k+1}^{n+\frac{1}{2}} v_{k+1}^{n+\frac{1}{2}} - A_k^{n+\frac{1}{2}} v_k^{n+\frac{1}{2}}}{\Delta m_{k+\frac{1}{2}}} \right) \Delta t^{n+\frac{1}{2}}. \tag{4.1.12}$$

A close examination of (4.1.12) reveals a problem since in order to compute ϵ^{n+1} we need $P^{n+\frac{1}{2}}$, the pressure being centered on integer time steps, we can only obtain the pressure at intermediate time steps by $P^{n+\frac{1}{2}} = \frac{1}{2}(P^{n+1} + P^n)$. But P^{n+1} is not yet known since it depends precisely on ϵ^{n+1}! Using P^n would not only degrade the order of the method to first order in time even in case of uniform zoning but it could also lead to large errors if the pressure is a very sensitive function of energy.

Several solutions are at hand to solve this problem but we shall only consider one that is sufficiently general not to depend on any particular form for the equation of state. The solution procedure is to obtain a first order approximation of the new internal energy by using P^n instead of $P^{n+\frac{1}{2}}$ in (4.1.12). This approximation to the new internal energy can be used to get, using the equation of state and ρ^{n+1}, an approximation to the pressure at $n+1$ that we call P'. We then compute the pressure at $n+\frac{1}{2}$ by $P^{n+\frac{1}{2}} = \frac{1}{2}(P' + P^n)$ and use (4.1.12) to obtain the final value of the specific internal energy at the new time. Notice that this procedure is very general since it does not depend on the exact nature of the equation of state. However, it requires the pressure to be evaluated twice each time step. For very complicated equations of state this can amount to a significant increase in total computing time.

The set of equations (4.1.5-12) together with an appropriate equation of state can be used to determine explicitly the evolution of the system, given suitable initial and boundary conditions. Notice that the surface elements A_k^n are the only elements specific to the spherical geometry, setting all the surface elements to 1, would be equivalent to transform our code into a 1D cartesian code.

In many astrophysically meaningful applications of a 1D spherical geometry code, self-gravity of the gas can not be neglected. In this case, it is necessary to modify the above set of equations to include self-gravity. This is done by adding an acceleration term $\nabla\phi$ to the momentum equation. The potential ϕ being given by the solution to the usual Poisson equation

$$\Delta\phi = 4\pi G\rho. \tag{4.1.13}$$

For multidimensional flows, solving this equation adds a significant amount of complexity and computer requirements to an already complicated problem. Fortunately, in 1D spherical geometry Poisson's equation can be solved analytically to yield the familiar acceleration term

$$\nabla\phi = -G\frac{M(r)}{r^2}. \tag{4.1.14}$$

The total mass interior to radius r, $M(r)$, is computed easily from

$$M(r_k) = \sum_{j=0}^{k-1} \Delta m_{j+\frac{1}{2}}. \tag{4.1.15}$$

Adding this acceleration to our momentum equation (4.1.9) we obtain the following expression for the new velocity

$$v_k^{n+\frac{1}{2}} = v_k^{n-\frac{1}{2}} - \left(A_k^n \frac{P_{k+\frac{1}{2}}^n - P_{k-\frac{1}{2}}^n}{\Delta m_k} + G\frac{\sum_{j=0}^{k-1} \Delta m_{j+\frac{1}{2}}}{(r_k^n)^2} \right) \Delta t^n. \tag{4.1.16}$$

4.2 Shock waves in 1D Lagrangian Hydrodynamics

A shock is usually defined as a discontinuity in the fluid variables. Appropriate relations between these variables on either side of the shock can be obtained from first principles like conservation of mass, momentum and energy (e.g. Landau Lifschitz 1959). Actually, a shock is not a discontinuity in a mathematical sense, molecular viscosity will always smooth the shock out to a few mean free path of the atoms or molecules making up the fluid. From a computational point of vue, this physical size is usually much smaller than any length scale we are interested in simulating. Notice that the hydrodynamics equations also have been derived assuming a continuous medium and that these equations will break down at the level of a mean free path or less.

The 1D Lagrangian code described so far is based only on Euler's equation and therefore does not include any type of physical dissipation. If the code would be applied to compute the details of a flow in a pipe ended by a piston that is gently withdrawn, the solution obtained would be a close approximation to the exact solution, a rarefaction wave propagating at sound speed. Now, if the motion of the piston would be reversed, that is the piston would be pushing on the fluid, large, non-physical, oscillations would appear in the computed solution. The reason for this behavior is simple, having not included any kinetic energy dissipation mechanism in the code, the physical temperature increase across the shock formed ahead of the piston can not be modeled appropriately. Like atoms and molecules in a real gas, the computational cells model the temperature increase by a corresponding increase in random velocities. Of course, although total energy is conserved, this is not the behavior we expect since the cells of our computational mesh do not represent atoms and/or molecules but rather macroscopic "chunks" of gas. Some form of dissipation, or viscosity, is needed to transform this microscopic kinetic energy into a macroscopic variable called temperature or, in our case, specific internal energy.

In most astronomical fluids, molecular viscosity is so small that it can be safely neglected except of course near discontinuities (or say very strong gradients) like shocks. Furthermore, the relations between the fluid variables on either side of the shock are interestingly independent of the physical mechanism providing the dissipation inside the shock itself. From these two remarks, we draw two important conclusions. First, it is not necessary to add a dissipation mechanism everywhere in the fluid but only close to discontinuities. Second, the actual implementation of the dissipation is irrelevant provided the exact physical shock jump conditions are obtained. A final requirement, of course, is that the dissipation mechanism should not broaden the shock which should remain as close as possible to a discontinuity. This is precisely what the so-called ***artificial viscosity*** was designed for by von Neumann and Richtmyer in 1950 (Richtmyer and Morton, 1967). Artificial viscosity should not be confused with numerical viscosity. The latter being usually intrinsic to a given numerical scheme and generally uncontrollable (see section 3) whereas the former is a needed addition to treat shocks.

To develop an appropriate artificial viscosity, we start with the viscous term in the general Navier-Stokes equation that is written (1D spherical geometry)

$$\text{viscous term} = \frac{1}{r^3}\frac{\partial}{\partial r}\left(r^3\eta\left(\frac{\partial v}{\partial r} - \frac{1}{3}\nabla\cdot\boldsymbol{v}\right)\right) \tag{4.2.1}$$

and define a viscous pressure by

$$Q \equiv -\eta\left(\frac{\partial v}{\partial r} - \frac{1}{3}\nabla\cdot\boldsymbol{v}\right) = -\eta\frac{2}{3}r\frac{\partial}{\partial r}\left(\frac{v}{r}\right). \tag{4.2.2}$$

We note that this expression vanishes for uniform contraction ($v \propto r$). We *define* an artificial viscosity η by the following expression

$$\eta = l^2\rho|\nabla\cdot\boldsymbol{v}| = l^2\left|\frac{d\rho}{dt}\right| \tag{4.2.3}$$

where l will turn out to be a typical length over which the shock will be smoothed. The continuity equation has been used to replace the last term. It is very important to realize that η as defined by (4.2.3) has *no* physical meaning whatsoever but that, recalling the discussion above, it is merely a numerical "trick" to provide the appropriate dissipation in shocks. Taking the artificial viscosity as a function of the divergence of the velocity ensures that this term remains small except close to shocks. Unfortunately, although small it does not completely disappear away from shocks, adding a small but unwanted viscosity to the scheme and therefore affecting to some extend the bulk of the flow.

A proper centering of the viscous pressure requires that we define Q at zone centers. A first obvious expression would be (in this expression and in what follows time superscripts have been omitted for clarity)

$$Q \propto \frac{v_{k+1} - v_k}{r_{k+1} - r_k} - \frac{v_{k+1} + v_k}{r_{k+1} + r_k}.$$

Unfortunately, since in spherical geometry usually $v_1 = r_1 = 0$ this expression will lead to problems since it vanishes for the first cell. To obtain a suitable expression we consider the volume average of both terms in Q. We obtain the following expression for the two terms

$$\int_{\Delta V_{k+\frac{1}{2}}} (\nabla\cdot\boldsymbol{v})\delta V = \frac{A_{k+1}v_{k+1} - A_k v_k}{\Delta V_{k+\frac{1}{2}}} \tag{4.2.4}$$

and

$$\int_{\Delta V_{k+\frac{1}{2}}} (\frac{\partial v}{\partial r})\delta V = \frac{A_{k+\frac{1}{2}}(v_{k+1} - v_k)}{\Delta V_{k+\frac{1}{2}}}. \tag{4.2.5}$$

In this expression we encounter for the first time the surface element at zone center $A_{k+\frac{1}{2}}$. Two expressions can be found for it depending if the cell centered radius or the cell centered surface element is defined first

$$r_{k+\frac{1}{2}} = \frac{1}{2}(r_{k+1} + r_k) \quad \text{implying} \quad A_{k+\frac{1}{2}} = 4\pi r^2_{k+\frac{1}{2}} \tag{4.2.6}$$

or

$$A_{k+\frac{1}{2}} = \frac{1}{2}(A_{k+1} + A_k) \quad \text{implying} \quad r_{k+\frac{1}{2}} = \sqrt{\frac{1}{2}(r^2_{k+1} + r^2_k)} \tag{4.2.7}$$

Both expressions are in fact equivalent as long as the definitions of the surface elements and the radii are consistent with each other throughout the entire code. In what follows, we shall use (4.2.7).

Putting (4.2.4) and (4.2.5) together and factoring out the velocities yields the expression

$$Q = -\alpha\left(v_{k+1}(A_{k+\frac{1}{2}} - A_{k+1}/3) - v_k(A_{k+\frac{1}{2}} - A_k/3)\right)\frac{1}{\Delta V_{k+\frac{1}{2}}}. \tag{4.2.8}$$

The coefficient α is best determined by taking the limit of large radius and requiring that this limit be equal to the expression derived in the simpler planar geometry. To avoid a lengthy derivation we will only give the results, dimensional arguments can be used to partially justify the expression. The final expression for the viscous pressure with proper time and spatial centering is found to be

$$Q^{n+\frac{1}{2}}_{k+\frac{1}{2}} = -\frac{3}{3}C_q^2 \rho^{n+\frac{1}{2}}_{k+\frac{1}{2}} |v^{n+\frac{1}{2}}_{k+1} - v^{n+\frac{1}{2}}_k| \left(v^{n+\frac{1}{2}}_{k+1}(1 - \frac{A^{n+\frac{1}{2}}_{k+1}}{3A^{n+\frac{1}{2}}_{k+\frac{1}{2}}}) - v^{n+\frac{1}{2}}_k(1 - \frac{A^{n+\frac{1}{2}}_k}{3A^{n+\frac{1}{2}}_{k+\frac{1}{2}}})\right) \tag{4.2.9}$$

where C_q is a numerical constant giving roughly the number of zones over which the shock will be spread. Finally, we add the supplementary constraint that the artificial viscosity should act only on shocks (compression)

$$Q^{n+\frac{1}{2}}_{k+\frac{1}{2}} = \begin{cases} Q^{n+\frac{1}{2}}_{k+\frac{1}{2}}, & \text{if } v^{n+\frac{1}{2}}_{k+1} \le v^{n+\frac{1}{2}}_k; \\ 0, & \text{otherwise.} \end{cases} \tag{4.2.10}$$

The expression for Q can be introduced into (4.2.1) and integrated over a mass element to obtain the appropriate term to introduce in the momentum equation. Since the calculation is straightforward but rather lengthy we will only give the results (omitting again the temporal index)

$$\int_{\Delta m_k} \frac{1}{\rho r^3}\frac{\partial}{\partial r}(r^3 Q)4\pi r^2 \rho dr = -\frac{1}{2}\left(Q_{k+\frac{1}{2}}(3A_{k+\frac{1}{2}} - A_k) + Q_{k-\frac{1}{2}}(3A_{k-\frac{1}{2}} - A_k)\right). \tag{4.2.11}$$

The artificial viscous stress can now be added to our previous momentum equation (4.1.16) to obtain the final form

$$v_k^{n+\frac{1}{2}} = v_k^{n-\frac{1}{2}} - \left(A_k^n \frac{P_{k+\frac{1}{2}}^n - P_{k-\frac{1}{2}}^n}{\Delta m_k} \right.$$
$$+ \frac{1}{2} \frac{Q_{k+\frac{1}{2}}^{n-\frac{1}{2}}(3A_{k+\frac{1}{2}}^n - A_k^n) - Q_{k-\frac{1}{2}}^{n-\frac{1}{2}}(3A_{k-\frac{1}{2}}^n - A_k^n)}{\Delta m_k}$$
$$\left. + G \frac{\sum_{j=0}^{k-1} \Delta m_{j+\frac{1}{2}}}{r^2} \right) \Delta t^n \qquad (4.2.12)$$

Notice that in the limit where $\Delta r/r \rightarrow 0$, all the surface elements are identical, and this expression becomes identical to the one derived for planar geometry (except for the presence of the gravity term, of course).

If the code is to conserve total energy, the kinetic energy dissipated by the artificial viscosity needs to be recovered in form of thermal energy. We need therefore to derive the appropriate expression corresponding to our artificial viscosity to be included in the energy equation (4.1.12). From standard hydrodynamics of viscous fluids (e.g. Landau Lifshitz 1959) we can write

$$\rho \frac{d\epsilon}{dt} = \frac{2}{3} \eta \left(r \frac{\partial}{\partial r} \left(\frac{v}{r} \right)^2 \right). \qquad (4.2.13)$$

By recalling the definition of Q, (4.2.2), this expression can be rewritten

$$\rho \frac{d\epsilon}{dt} = -Q \left(\frac{\partial v}{\partial r} - \frac{v}{r} \right). \qquad (4.2.14)$$

Integrating (4.2.14) in the usual manner over a cell, we obtain the contribution to the specific internal energy from the artificial viscous stress. This contribution can be added to the previous internal energy equation (4.1.12) that contained only the PdV term to obtain the final energy equation

$$\epsilon_{k+\frac{1}{2}}^{n+1} = \epsilon_{k+\frac{1}{2}}^n - \frac{\Delta t^{n+\frac{1}{2}}}{\Delta m_{k+\frac{1}{2}}} \left(P_{k+\frac{1}{2}}^{n+\frac{1}{2}} \left(A_{k+1}^{n+\frac{1}{2}} v_{k+1}^{n+\frac{1}{2}} - A_k^{n+\frac{1}{2}} v_k^{n+\frac{1}{2}} \right) \right.$$
$$\left. + \frac{1}{2} Q_{k+\frac{1}{2}}^n \left(v_{k+1}^{n+\frac{1}{2}} (3A_{k+\frac{1}{2}}^{n+\frac{1}{2}} - A_{k+1}^{n+\frac{1}{2}}) - v_k^{n+\frac{1}{2}} (3A_{k+\frac{1}{2}}^{n+\frac{1}{2}} - A_k^{n+\frac{1}{2}}) \right) \right) \qquad (4.2.15)$$

Notice again that in the limit where all surface elements are equal, that is when $\Delta r/r \rightarrow 0$ this expression becomes identical to the one derived in planar geometry.

4.3 Boundary Conditions and Stability Consideration

To be actually able to solve the system of equations described above, boundary conditions on both sides of the mesh as well as initial conditions have to be specified. In spherical geometry, if the inner boundary is at the center of the star than $r_1 = 0$ and $v_1 = 0$. If this should not be the case, than appropriate boundary conditions can be set by adding a supplementary cell, a "ghost" cell, to the grid so that the values of the various gradients for the first cell are computed correctly. At the other boundary, a free boundary can be set by imposing a zero gradient condition using again a ghost cell. A reflective boundary is obtained by reversing the sign of the velocity. If radiation transport is included, proper boundary conditions on the incoming and outgoing radiation flux will have to be specified.

The scheme described above is only conditionally stable and is therefore subject to a time step limitation condition. A necessary but not sufficient condition was derived in section 3 for the linearized hydrodynamics equations. As discussed in that section, for the full equations a modified criterion has to be used to take into account possible non-linear effects. A first condition on the time step can actually be obtained by combining the criterion for the linearized equations with the one derived for the continuity equation (3.1.17)

$$\Delta t_a = a_1 Min\{\frac{\Delta r_{k+\frac{1}{2}}}{c_{k+\frac{1}{2}} + |v_{k+\frac{1}{2}}|}\} \tag{4.3.1}$$

with $a_1 < 1$. The minimum of this expression has obviously to be taken over the whole grid. However, because of the presence of the artificial viscosity term, an additional constraint on the time step has to be added. An exact derivation of the stability criterion for the artificial viscosity term is impossible, only an approximate condition can be obtained by recalling that viscosity is essentially a diffusion process. Hence, we can use the stability condition derived for the diffusion equation (3.1.9) and write

$$\Delta t_v = a_2 Min\{\frac{1}{2}\frac{\Delta r^2_{k+\frac{1}{2}}}{\eta_{k+\frac{1}{2}}}\} \tag{4.3.2}$$

where $a_2 \leq 1$ and η is the artificial viscosity as defined by (4.2.3). To be completely rigorous, in the cases where self-gravity is important, a third constraint on the time step has to be included, Δt_g, which should include the body forces. The final time step to be used in the code can be obtained by taking the most restrictive of all individual conditions

$$\Delta t = Min\{\Delta t_a, \Delta t_b, \Delta t_g, ...\}. \tag{4.3.3}$$

In practice, to avoid divisions by zero, the inverse of all the Δt's are computed. The limitation on the time step derived above is of course not rigorous and stability is not guaranteed for all situations. Numerical experimentations will however quickly tell whether the time step has be taken too large and the free parameters $a_1, a_2, ...$ will have to be adjusted accordingly.

We these last remarks we conclude our discussion of the 1D Lagrangian, spherical geometry scheme. In the next section, we shall address the resolution of the hydrodynamics equations in 2 spatial dimensions assuming cylindrical geometry and using an Eulerian description of the fluid. Because of the increased complexity, we shall limit our presentation to some of the most important features characteristic to Eulerian hydrodynamics.

5 2D Eulerian Hydrodynamics

As mentioned in the introduction of section 4 the basic difference between Lagrangian and Eulerian hydrodynamics resides in the choice of the coordinate system. In Lagrangian hydrodynamics, the coordinate system is attached to the fluid, hence, severe distortion of the mesh can occur when the equations are solved in more than one spatial dimensions. Very quickly, especially if strong shear motions are present, the numerical derivatives become highly inaccurate. The Eulerian mesh, on the other hand, being fixed, or moved with a prescribed velocity different from the fluid velocity, does not suffer from these problems. Thus, the Eulerian formalism is usually preferred for multi-dimensional hydrodynamics despite, as we soon shall see, the additional complexity introduced by the non-linear advection terms.

In this section we shall describe in some details one of many possible 2D Eulerian approaches to solve the hydrodynamics equations the so-called donor-cell scheme. Most of the emphasize will be on the advection terms since they are the characteristics of Eulerian hydrodynamics and also the most difficult terms in the equations to discretize. We start with a first order donor-cell scheme as it has the advantage to be simple and therefore to be of pedagogical interest. In practice however, it should be avoided since it leads to an extremely diffusive scheme. Improvements to this first order scheme will be discussed, in particular, we introduce the concept of monotonic advection leading to a second order accurate in space method. Furthermore, to avoid resolution problems in situations of large changes in dimensions (as occur in collapses or explosions) we shall discretize the equations on a moving grid. We chose a cylindrical geometry since this choice of coordinate is appropriate for many astrophysical situations like the modeling of accretion disks or axisymmetric jets for example. The scheme discussed in this paper is taken from Bowers and Wilson (1990) but is also very similar to Norman and Winkler (1986) who applied it precisely to the modeling of jets in radio galaxies (Norman et al. 1981,1982).

5.1 2D Inviscid Eulerian Hydrodynamics

5.1.1 Equations

We start with the equations written in conservative form, that is integral form. The advantage of writing the equations in this somewhat unusual form is that the advection terms simply translate into fluxes of the appropriate quantities through the surface of a fluid element. Advection can therefore be computed by adding the various fluxes in and out each individual computational cell. This procedure is repeated for all advected quantities like mass, momentum, energy, etc. and for all cells in the grid. The conservation of mass, momentum and energy is therefore obviously satisfied except for the boundary terms which describe the injection of mass momentum or energy from the surroundings into the computational domain or vice-versa.

a) Mass conservation, or continuity, equation

$$\int \frac{\partial \rho}{\partial t} dV + \int \rho(\boldsymbol{v} - \boldsymbol{v}_g) d\boldsymbol{S} + \int \rho \nabla \boldsymbol{v}_g dV = 0. \tag{5.1.1}$$

b) Momentum conservation equation

$$\int \frac{\partial(\rho \boldsymbol{v})}{\partial t} dV + \int \rho \boldsymbol{v}(\boldsymbol{v} - \boldsymbol{v}_g) d\boldsymbol{S} + \int \nabla P dV + \int (\rho \boldsymbol{v}) \nabla \boldsymbol{v}_g dV = 0. \tag{5.1.2}$$

c) Energy conservation equation

$$\int \frac{\partial(\rho \epsilon)}{\partial t} dV + \int \rho \epsilon(\boldsymbol{v} - \boldsymbol{v}_g) d\boldsymbol{S} + \int P \nabla \boldsymbol{v} dV + \int (\rho \epsilon) \nabla \boldsymbol{v}_g dV = 0. \tag{5.1.3}$$

Here the time derivatives are taken with respect to a fixed observer on a grid with zones of volume dV and surface area $\boldsymbol{dS}$. The motion of the grid is described by its velocity $\boldsymbol{v}_g$ and appears in the last term of each equation. The gas pressure P is obtained from the equation of state $P = f(\rho, \epsilon)$ which does not need to be the equation of state for a perfect gas but can be much more complicated. The other symbols have the same meaning as in section 4.

Notice that if the grid velocity $\boldsymbol{v}_g = 0$, one easily recovers, using Gauss's theorem, the usual Eulerian hydrodynamics equations. In a similar way if $\boldsymbol{v}_g = \boldsymbol{v}$, that is if the grid moves with the fluid, the advection terms disappear and the Lagrangian formalism is recovered.

Before we proceed any further and start deriving finite difference approximations, it is convenient to make the following change of variables specific to our choice of cylindrical coordinates

$$u = v_z \ ; \quad v = v_r \ ; \quad S = \rho v_z \ ; \quad T = \rho v_r. \tag{5.1.4}$$

S and T are the momentum density in the z and r direction respectively. Furthermore, denoting the grid position by r_g and z_g, it follows that $v_g = dr_g/dt$ and $u_g = dz_g/dt$. We shall restrict ourselves to simple grid motions where the grid conserves its orthogonality, that is motions for which $v_g = v_g(r_g, t)$, $u_g = u_g(z_g, t)$.

5.1.2 Operator Splitting or Multi-Steps Solution Procedure

In multi-dimensional hydrodynamics operator splitting is used to avoid having to solve a set of coupled non-linear equations using a matrix inversion followed possibly by iterations to take the non-linear effects into considerations. The basic idea is the following. Consider a typical partial differential equation for the variable F

$$\frac{\partial F}{\partial t} = A + B + C + ... \tag{5.1.5}$$

where $A, B, C, ...$ are functions of $F, \boldsymbol{x}, t$. We solve (5.1.5) by the successive operations

$$\frac{\partial F}{\partial t} = A$$
$$\frac{\partial F}{\partial t} = B$$
$$\frac{\partial F}{\partial t} = C \tag{5.1.6}$$

That is, F is advanced in time in several steps, the first due to all processes noted A, the second by all processes denoted B etc. It is by no means obvious that this is justified and that the solution obtained by (5.1.6) will be identical to the one obtained solving (5.1.5). In fact, examples can be found where the operator split approach fails. On the other hand, other examples can be found where the split approach is more accurate than the non- split approach.

To split or not to split? The situation can indeed be confusing. Unfortunately, there is no definitive answer to this question. As usual in numerical hydrodynamics, experiments are the most reliable, and often the only possible, way to check whether a specific approach to a specific problem will work or not. Nevertheless, as a rough rule of thumb it is generally wise to avoid splitting two terms that nearly cancel each other. For example, one would not split the gravity force from the pressure gradients or in the case of radiation transport, the emission and absorption of photons.

Operator splitting also can significantly improve the efficiency of the code. Consider the advection-diffusion equation (3.1.10). An operator split approach to solving this equation would be to solve first the advection part explicitly followed by an implicit treatment of the diffusion part. This could lead to an important advantage since, as we have seen, the explicit diffusion equation can suffer from drastic limitations on the size of the time step for stability reasons.

In this application of 2D Eulerian hydrodynamics, we shall split the pressure gradients from the advection terms and finally the grid motion. In other words, we will solve the set of equations (5.1.1-3) in several steps. The first step can be identified as a Lagrangian-like step which we write for the Z momentum density

$$\int \frac{\partial S}{\partial t} dV = -\int \frac{\partial P}{\partial z} dV. \tag{5.1.7}$$

Notice that due to the split approach, the grid is considered fixed during this step, therefore the time derivative becomes the Eulerian derivative and commutes with the integral. Similarly, for the r momentum

$$\int \frac{\partial T}{\partial t} dV = -\int \frac{\partial P}{\partial r} dV \tag{5.1.8}$$

and finally for the energy equation

$$\int \frac{\partial(\rho\epsilon)}{\partial t} dV = -\int P\left(\frac{\partial u}{\partial z} + \frac{1}{r}\frac{\partial}{\partial r}(rv)\right) dV. \tag{5.1.9}$$

Once the system has been updated following this Lagrangian-like step, the next step, the advection step, is taken. For mass advection we have

$$\int \frac{\partial \rho}{\partial t} dV = -\int \rho(\boldsymbol{v} - \boldsymbol{v}_g) d\boldsymbol{S} \tag{5.1.10}$$

whereas momentum advection will be described by

$$\int \frac{\partial(\rho \boldsymbol{v})}{\partial t} dV = -\int \rho \boldsymbol{v}(\boldsymbol{v} - \boldsymbol{v}_g) d\boldsymbol{S} \tag{5.1.11}$$

and finally the energy advection

$$\int \frac{\partial(\rho\epsilon)}{\partial t} dV = -\int \rho\epsilon(\boldsymbol{v} - \boldsymbol{v}_g) d\boldsymbol{S} \tag{5.1.12}$$

The updated values of $\boldsymbol{v}$ and ϵ from the Lagrangian-like step are used to evaluate the right-hand-side of (5.1.10-12), and enter as the old values in the time-discretized left-hand-side of (5.1.11) and (5.1.12).

We are left with the grid motions terms. These are easily incorporated in the following way for the mass conservation equation

$$\int \frac{\partial \rho}{\partial t} dV = -\int \rho \nabla \boldsymbol{v}_g dV, \tag{5.1.13}$$

for the momentum conservation equation

$$\int \frac{\partial(\rho \boldsymbol{v})}{\partial t} dV = -\int (\rho \boldsymbol{v}) \nabla \boldsymbol{v}_g dV \tag{5.1.14}$$

and the energy equation

$$\int \frac{\partial(\rho\epsilon)}{\partial t} dV = -\int (\rho\epsilon) \nabla \boldsymbol{v}_g dV. \tag{5.1.15}$$

Consequently, a typical evolution sequence from time *old* to time *new*, for the z component of the momentum density for example, looks like

$$S^{old} \longrightarrow S^{l} \longrightarrow S^{q} \longrightarrow S^{a} \longrightarrow S^{g} = S^{new} \tag{5.1.14}$$

where the superscript l stands for Lagrangian, a for advection, g for grid compression. We have already added step q which stands for the artificial viscosity (see section 4). Notice that this step has been included between the Lagrangian and the advection term. The r component of the momentum density T will be advanced in time in a similar way. If gravity is to be included, it would have to be included in the Lagrangian step l to be solved simultaneously with the pressure gradients (see remarks on splitting above). The equivalent procedure for the energy density evolution looks like

$$(\rho\epsilon)^{old} \longrightarrow (\rho\epsilon)^{l} \longrightarrow (\rho\epsilon)^{q} \longrightarrow (\rho\epsilon)^{a} \longrightarrow (\rho\epsilon)^{g} = (\rho\epsilon)^{new}. \tag{5.1.15}$$

5.2 Numerical Equations

5.2.1 Space-Time Centering

We shall now discuss the centering of the variables on the computational mesh. We shall proceed in a similar way than in the 1D Lagrangian scheme, that is, we make use of a staggered mesh. The intensive variables (ρ, P, ϵ) will be cell centered and defined at time

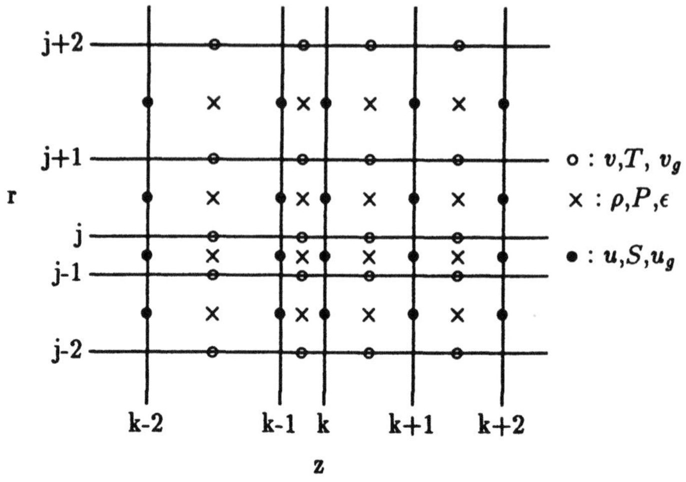

Fig. 7. Variable centering for 2D cylindrical geometry hydrodynamics

t^n. The positions, velocities and momenta will be defined at $t^{n+\frac{1}{2}}$ on the cell edge in their respective directions. This centering of the variables is illustrated in Fig. 7.

We define the following notations for the various quantities

$$q(r,z,t) \longrightarrow q^n_{k+\frac{1}{2},j+\frac{1}{2}} \tag{5.2.1}$$

where the first subscript relates to the z direction and the second one to the r direction. Superscripts, as before, refer to the time bin. Equation (5.2.1) therefore describes the centering of the intensive variables ρ, P, ϵ. For the positions, velocities and momenta we have

$$\begin{aligned} q(r,z,t) &\longrightarrow q^{n+\frac{1}{2}}_{k,j+\frac{1}{2}} \quad \text{for z component} \\ &\longrightarrow q^{n+\frac{1}{2}}_{k+\frac{1}{2},j} \quad \text{for r component} \end{aligned} \tag{5.2.2}$$

We further need to define the surface elements with normal directed in either the radial or the axial direction. The zonal surface area with its normal directed in the radial direction is written

$$A_{k+\frac{1}{2},j} = 2\pi r_j \, \Delta z_{k+\frac{1}{2}} \tag{5.2.3}$$

where we have defined

$$\Delta z_{k+\frac{1}{2}} = z_{k+1} - z_k . \tag{5.2.4}$$

The corresponding zonal area with its normal directed axially will be written as

$$A_{k,j+\frac{1}{2}} = 2\pi r_j \, \Delta r_{j+\frac{1}{2}} \tag{5.2.5}$$

where we have defined in a similar manner

$$\Delta r_{j+\frac{1}{2}} = r_{j+1} - r_j. \tag{5.2.6}$$

Finally we need to define the zone centered zonal volume which we write

$$\Delta V_{k+\frac{1}{2},j+\frac{1}{2}} = 2\pi r_{j+\frac{1}{2}} \Delta r_{j+\frac{1}{2}} \Delta z_{k+\frac{1}{2}} \tag{5.2.7}$$

with $r_{j+\frac{1}{2}} = \frac{1}{2}(r_{j+1} + r_j)$. Note that this definition of the zonal volume is exact since

$$\begin{aligned} \Delta V_{k+\frac{1}{2},j+\frac{1}{2}} &= \pi(r_{j+1}^2 - r_j^2)\Delta z_{k+\frac{1}{2}} \\ &= 2\pi \frac{1}{2}(r_{j+1} + r_j)(r_{j+1} - r_j)\Delta z_{k+\frac{1}{2}} \\ &= 2\pi r_{j+\frac{1}{2}} \Delta r_{j+\frac{1}{2}} \Delta z_{k+\frac{1}{2}} \end{aligned}$$

Similarly, we define the r edge-centered zonal volume by

$$\Delta V_{k+\frac{1}{2},j} = 2\pi \tilde{r}_j \Delta r_j \Delta z_{k+\frac{1}{2}} \tag{5.2.8}$$

Notice that this zonal volume is only equal to the exact volume if we define $\tilde{r}_j$ by $\tilde{r}_j = \frac{1}{2}(r_{j+\frac{1}{2}} + r_{j-\frac{1}{2}})$, this radius is only equal to r_j for uniform zoning.

5.2.2 Lagrangian Step

We are in a position to determine the discretized equations relevant to our Lagrangian step. Since this is achieved in a completely similar manner as discussed in section 4, we only give here the final results. Notice that the mass is no longer the independent variable. The momentum conservation equation in the z direction is given by

$$S^l_{k,j+\frac{1}{2}} = S^{n-\frac{1}{2}}_{k,j+\frac{1}{2}} - A_{k,j+\frac{1}{2}} \frac{\left(P^n_{k+\frac{1}{2},j+\frac{1}{2}} - P^n_{k-\frac{1}{2},j+\frac{1}{2}}\right)}{\Delta V_{k,j+\frac{1}{2}}} \Delta t^n, \tag{5.2.9}$$

and similarly for the radial direction

$$T^l_{k+\frac{1}{2},j} = T^{n-\frac{1}{2}}_{k+\frac{1}{2},j} - A_{k+\frac{1}{2},j} \frac{\left(P^n_{k+\frac{1}{2},j+\frac{1}{2}} - P^n_{k+\frac{1}{2},j-\frac{1}{2}}\right)}{\Delta V_{k+\frac{1}{2},j}} \Delta t^n. \tag{5.2.10}$$

Before we can compute the compressional heating we need to obtain an approximation for the velocities from the above new momenta. This is done in a obvious way by recalling the definitions of S and T. For the axial velocity

$$u^l_{k,j+\frac{1}{2}} = \frac{S^l_{k,j+\frac{1}{2}}}{\rho^n_{k,j+\frac{1}{2}}} \tag{5.2.11}$$

and for the radial velocity

$$v^l_{k+\frac{1}{2},j} = \frac{T^l_{k+\frac{1}{2},j}}{\rho^n_{k+\frac{1}{2},j}} \tag{5.2.12}$$

In the two equations above the edge-centered densities have been defined in a volume averaged form (omitting the temporal index for clarity)

$$\rho_{k,j+\frac{1}{2}} = \frac{\rho_{k+\frac{1}{2},j+\frac{1}{2}}\,\Delta V_{k+\frac{1}{2},j+\frac{1}{2}} + \rho_{k-\frac{1}{2},j+\frac{1}{2}}\,\Delta V_{k-\frac{1}{2},j+\frac{1}{2}}}{2\Delta V_{k,j+\frac{1}{2}}}$$

and

$$\rho_{k+\frac{1}{2},j} = \frac{\rho_{k+\frac{1}{2},j+\frac{1}{2}}\,\Delta V_{k+\frac{1}{2},j+\frac{1}{2}} + \rho_{k+\frac{1}{2},j-\frac{1}{2}}\,\Delta V_{k+\frac{1}{2},j-\frac{1}{2}}}{2\Delta V_{k+\frac{1}{2},j}} \tag{5.2.13}$$

The velocity divergence can be computed in a similar way as was done in the Lagrangian case described in section 4. We find

$$(\nabla v)^l_{k+\frac{1}{2},j+\frac{1}{2}} = \frac{A_{k+1,j+\frac{1}{2}}u^l_{k+1,j+\frac{1}{2}} - A_{k,j+\frac{1}{2}}u^l_{k,j+\frac{1}{2}}}{\Delta V_{k+\frac{1}{2},j+\frac{1}{2}}} + \frac{A_{k+\frac{1}{2},j+1}v^l_{k+\frac{1}{2},j+1} - A_{k+\frac{1}{2},j}v^l_{k+\frac{1}{2},j}}{\Delta V_{k+\frac{1}{2},j+\frac{1}{2}}} \tag{5.2.14}$$

where u^l and v^l are the velocities obtained from equations (5.2.11) and (5.2.12) respectively. The energy equation is solved in the following way

$$(\rho\epsilon)^c_{k+\frac{1}{2},j+\frac{1}{2}} = (\rho\epsilon)^n - \tilde{P}_{k+\frac{1}{2},j+\frac{1}{2}}(\nabla v)^l_{k+\frac{1}{2},j+\frac{1}{2}}\Delta t^{n+\frac{1}{2}} \tag{5.2.15}$$

where for the velocity divergence (5.2.14) is used. $\tilde{P}$ is defined in a similar way as in the Lagrangian scheme by $\tilde{P} = \frac{1}{2}(P^n + P^c)$. P^c is obtained by first using P^n in (5.2.15) to obtain a first order approximation of $\rho\epsilon$ at time $n+1$. From this, the new value of ϵ (a first order approximation) can be obtained by noting that

$$\rho^c = \rho^n + \Delta\rho = \frac{\rho^n}{1 + \nabla v \Delta t^{n+\frac{1}{2}}} \tag{5.2.16}$$

which can be computed since all the terms on the right-hand-side are known. Once ϵ^c is known, we proceed as already outlined in section 4.

5.2.3 Advection Step

We have to deal with the additional complexity of the advection terms. As it is apparent from the form of the equations, advection terms will all be similar whether mass, momentum or energy is advected. The advection terms are modeled in finite difference form by constructing fluxes of each quantity which are then used to compute the changes within a zone of these quantities. Over the years, many different procedures have been developed to obtain accurate estimates for the fluxes while insuring numerical stability. Since it would be hopeless here to try to present all theses procedures, we will limit ourselves to present one method only that has been widely used in numerical astrophysics, the so-called donor-cell technique. We shall discuss also some improvements to the method like the important monotonicity concept.

Let us start with mass advection as described by equation (5.1.10). This can be considered as a typical case for the advection of a scalar quantity. It will apply almost without modification to the advection of energy. The discretized form for (5.1.10) is given by

$$\rho^a_{k+\frac{1}{2},j+\frac{1}{2}} = \rho^n_{k+\frac{1}{2},j+\frac{1}{2}} - \frac{\Delta t^{n+\frac{1}{2}}}{\Delta V_{k+\frac{1}{2},j+\frac{1}{2}}} \left(F^m_{k+1,j+\frac{1}{2}} - F^m_{k,j+\frac{1}{2}} + G^m_{k+\frac{1}{2},j+1} - G^m_{k+\frac{1}{2},j} \right) \quad (5.2.17)$$

where the F^m and the G^m are the mass fluxes at the faces of the zones in the axial and radial direction respectively. Notice that these fluxes are defined at zone edges, as we will see shortly this creates some difficulties since the advected quantities, here mass, are known at cell centers. The detail expression for the flux in the axial direction is

$$F^m_{k+1,j+\frac{1}{2}} = A_{k+1,j+\frac{1}{2}} \hat{\rho}_{k+1,j+\frac{1}{2}} \left(u^l_{k+1,j+\frac{1}{2}} - u_{g,k+1} \right) \quad (5.2.18)$$

whereas in the radial direction we have

$$G^m_{k+\frac{1}{2},j+1} = A_{k+\frac{1}{2},j+1} \hat{\rho}_{k+\frac{1}{2},j+1} \left(v^l_{k+\frac{1}{2},j+1} - v_{g,j+1} \right). \quad (5.2.19)$$

In these two expressions, we have used the velocity obtained after the Lagrangian step and defined the upstream interpolated density in the axial and radial direction respectively by $\hat{\rho}_{k+1,j+\frac{1}{2}}$ and $\hat{\rho}_{k+\frac{1}{2},j+1}$. The other quantities are already defined at zone edges and pose no problems. The reason for the presence of the upstream interpolated density is clear, advection physically represents a mass flux into a zone therefore, the mass that will be transported over the cell boundary will depend on the density structure upstream of the interface. The problem is to obtain an expression for this upstream interpolated density.

The easiest approximation can be obtained by assuming the density to be constant in each cell and by recalling that the advection process physically represents mass flux into a zone. In particular, if $(u_k - u_g) > 0$, the mass flows into zone $k + \frac{1}{2}$, the upstream density to use for computing the mass flux is, in this case, $\rho_{k-\frac{1}{2}}$. The opposite being true if $(u_k - u_g) < 0$. Similar arguments hold for the radial direction. This simple scheme leads to a first order accurate advection treatment which is of course very crude and should be avoided in practice but is adequate for pedagogical purposes. Using this first order accurate method, the upstream interpolated density in the axial direction is written

$$\hat{\rho}_{k,j+\frac{1}{2}} = \begin{cases} \rho_{k-\frac{1}{2},j+\frac{1}{2}}, & \text{if } (u_{k,j+\frac{1}{2}} - u_{g,k}) > 0; \\ \rho_{k+\frac{1}{2},j+\frac{1}{2}}, & \text{otherwise.} \end{cases} \quad (5.2.20)$$

As mentioned, the advection considered here is only first order accurate and is seldom satisfactory since it leads to large diffusion of the advected quantities. An obvious improvement is to use a linear or even quadratic interpolation to obtain the upstream interpolated density. Such an interpolation is equivalent to introduce structures into the zones rather than to consider them constant and would definitively improve the order of the method. Unfortunately, it also leads to another problem which is that minima and/or maxima that were not present in the original data can be introduced. This would especially happen near strong gradients (eg. shocks) and could lead to disastrous results like negative densities. These negative densities essentially occur when the code tries to advect more mass than is actually present in a given zone. To avoid these problems we introduce the important concept of *monotonicity*. The basic principle here is that if the data are monotonic, the interpolation scheme should conserve this monotonicity. Notice that the first order scheme derived above conserves monotonicity since no interpolation is made. This property is used in practice to achieve monotonicity in higher order schemes, by simply automatically reducing to first order whenever strong discontinuities are present that lead to non-monotonic interpolations.

Using the scheme by van Leer (1977) in a similar way as in Norman and Winkler (1986) we define a quantity $d\rho_{k,j+\frac{1}{2}}$ by

$$d\rho_{k+\frac{1}{2},j+\frac{1}{2}} = \begin{cases} \frac{2\Delta\rho_{k,j+\frac{1}{2}}\Delta\rho_{k+1,j+\frac{1}{2}}}{\Delta\rho_{k,j+\frac{1}{2}}+\Delta\rho_{k+1,j+\frac{1}{2}}} & \text{if } \Delta\rho_{k,j+\frac{1}{2}}\Delta\rho_{k+1,j+\frac{1}{2}} > 0; \\ 0, & \text{otherwise.} \end{cases} \tag{5.2.21}$$

where $\Delta\rho_{k,j+\frac{1}{2}} = (\rho_{k+\frac{1}{2},j+\frac{1}{2}} - \rho_{k-\frac{1}{2},j+\frac{1}{2}})$. The upstream interpolated value for the density is taken to be

$$\hat{\rho}_{k,j+\frac{1}{2}} = \begin{cases} \rho_{k-\frac{1}{2},j+\frac{1}{2}} + \frac{(\Delta z_{k-\frac{1}{2},j+\frac{1}{2}} - \delta_{k-\frac{1}{2},j+\frac{1}{2}})d\rho_{k-\frac{1}{2},j+\frac{1}{2}}}{2\Delta z_{k-\frac{1}{2},j+\frac{1}{2}}}, & \text{if } \delta_{k-\frac{1}{2},j+\frac{1}{2}} \geq 0; \\ \rho_{k+\frac{1}{2},j+\frac{1}{2}} + \frac{(\Delta z_{k+\frac{1}{2},j+\frac{1}{2}} + \delta_{k+\frac{1}{2},j+\frac{1}{2}})d\rho_{k+\frac{1}{2},j+\frac{1}{2}}}{2\Delta z_{k+\frac{1}{2},j+\frac{1}{2}}}, & \text{otherwise.} \end{cases} \tag{5.2.22}$$

where we have defined

$$\delta_{k+\frac{1}{2},j+\frac{1}{2}} = (u_{k+\frac{1}{2},j+\frac{1}{2}} - u_{g,k+\frac{1}{2}})\Delta t/2.$$

The velocities at cell centers are defined in the usual way

$$u_{k+\frac{1}{2},j+\frac{1}{2}} - u_{g,k+\frac{1}{2}} = \frac{1}{2}\left((u_{k+1,j+\frac{1}{2}} - u_{g,k+1}) + (u_{k,j+\frac{1}{2}} - u_{g,k})\right). \tag{5.2.23}$$

Notice that the density used for the advection term is not the one interpolated at the cell edge, but the density upstream from the edge at a distance corresponding to half the distance covered by the flow in one time step. Notice also that the interpolation actually involves three cells therefore some smearing of the quantities will result from the use of this interpolation scheme. Interpolation schemes using only two zones can be devised reducing somewhat the smearing.

The derivation of the interpolated quantity in the radial direction to obtain the expression for the radial flux proceeds in the same way. For these reasons we will not derive these expressions, they can be obtained without difficulties by replacing the variations in the axial index k by the corresponding variations in the radial index j. The axial grid velocity u_g has of course also to be replaced by the radial grid velocity v_g.

As already mentioned, the advection of all scalar quantities can be treated in a similar way. In particular, the energy advection is written in a complete analog

$$(\rho\epsilon)^a_{k+\frac{1}{2},j+\frac{1}{2}} = (\rho^n\epsilon^c)_{k+\frac{1}{2},j+\frac{1}{2}} - \frac{\Delta t^{n+\frac{1}{2}}}{\Delta V_{k+\frac{1}{2},j+\frac{1}{2}}}\Big(F^e_{k+1,j+\frac{1}{2}} - F^e_{k,j+\frac{1}{2}} + G^e_{k+\frac{1}{2},j+1} - G^e_{k+\frac{1}{2},j}\Big) \tag{5.2.24}$$

where the flux in the axial direction is defined by

$$F^e_{k+1,j+\frac{1}{2}} = A_{k+1,j+\frac{1}{2}}(\hat{\rho}^n \hat{\epsilon}^c)_{k+1,j+\frac{1}{2}} \left(u^l_{k+1,j+\frac{1}{2}} - u_{g,k+1}\right) \tag{5.2.25}$$

whereas in the radial direction we have

$$G^e_{k+\frac{1}{2},j+1} = A_{k+\frac{1}{2},j+1}(\hat{\rho}^n \hat{\epsilon}^c)_{k+\frac{1}{2},j+1} \left(v^l_{k+\frac{1}{2},j+1} - v_{g,j+1}\right). \tag{5.2.26}$$

Notice that the value of the density at time n has been used to determine the interpolated quantity, whereas the internal energy resulting from the compression step is used. The first or second order interpolation scheme presented for the mass advection can be used to obtain approximations for the specific internal energy.

We are left with momentum advection. Because momentum is a vector quantity defined at cell edges, the situation is slightly different from the scalar advection we have just discussed. Each momentum component will be advected in the axial and in the radial direction but for each of these terms the centering will be different. Let us start with the axial component of the momentum. A finite-difference approximation to (5.1.11) can be written

$$S^a_{k,j+\frac{1}{2}} = S^l_{k,j+\frac{1}{2}} - \frac{\Delta t^{n+\frac{1}{2}}}{\Delta V_{k,j+\frac{1}{2}}} \left(F^S_{k+\frac{1}{2},j+\frac{1}{2}} - F^S_{k-\frac{1}{2},j+\frac{1}{2}} + G^S_{k,j+1} - G^S_{k,j}\right) \tag{5.2.27}$$

where F^S and G^S represent the flux of the z component on the momentum in the axial and radial direction respectively defined by

$$F^S_{k+\frac{1}{2},j+\frac{1}{2}} = A_{k+\frac{1}{2},j+\frac{1}{2}} \hat{S}_{k+\frac{1}{2},j+\frac{1}{2}} (u - u_g)_{k+\frac{1}{2},j+\frac{1}{2}} \tag{5.2.28}$$

and

$$G^S_{k,j+1} = A_{k,j+1} \hat{S}_{k,j+1} (v - v_g)_{k,j+1}. \tag{5.2.29}$$

In (5.2.28) the velocity at cell center $(u - u_g)_{k+\frac{1}{2},j+\frac{1}{2}}$ has been defined as in (5.2.23). In (5.2.29) the velocity is defined by

$$(v - v_g)_{k,j} = \frac{1}{2}\left(v_{k+\frac{1}{2},j} - v_{k-\frac{1}{2},j}\right) - v_{g,j} \tag{5.2.30}$$

since from our assumption, the grid velocity in the radial direction is independent of the axial coordinate. The edge-centered zonal surface area is defined by

$$A_{k,j} = 2\pi r_j \Delta z_k. \tag{5.2.31}$$

The velocities are taken after updating the momenta from pressure gradient effects and the densities from the continuity equation

$$u = \frac{S^l}{\rho^a} \quad \text{and} \quad v = \frac{T^l}{\rho^a}. \tag{5.2.32}$$

Finally, we are left with having to define the upstream interpolated quantity $\hat{S}$. The same discussion applies here as for the upstream interpolated density or energy. The easiest, first order donor cell method, where the quantities are considered constant within each cell as in (5.2.20) yields

$$\hat{S}_{k+\frac{1}{2},j+\frac{1}{2}} = \begin{cases} S_{k,j+\frac{1}{2}}, & \text{if } (u_{k+\frac{1}{2},j+\frac{1}{2}} - u_{g,k+\frac{1}{2}}) > 0; \\ S_{k+1,j+\frac{1}{2}}, & \text{otherwise}. \end{cases} \tag{5.2.33}$$

If second order momentum advection is desired (and this should be the case) an interpolation scheme similar to the one outlined for mass advection (5.2.21-23) can be used.

Of course, the momentum advection in the radial direction will be treated in exactly the same manner

$$T^a_{k+\frac{1}{2},j} = T^l_{k+\frac{1}{2},j} - \frac{\Delta t^{n+\frac{1}{2}}}{\Delta V_{k+\frac{1}{2},j}} \left(F_{k+1,j} - F_{k,j} + G_{k+\frac{1}{2},j+\frac{1}{2}} - G_{k+\frac{1}{2},j-\frac{1}{2}} \right). \tag{5.2.34}$$

The flux in the axial direction being determined by

$$F_{k+1,j} = A_{k+1,j}\, \hat{T}_{k+1,j}\, (u - u_g)_{k+1,j} \tag{5.2.35}$$

whereas the flux in the radial direction will be given by

$$G_{k+\frac{1}{2},j+\frac{1}{2}} = A_{k+\frac{1}{2},j+\frac{1}{2}}\, \hat{T}_{k+\frac{1}{2},j+\frac{1}{2}}\, (v - v_g)_{k+\frac{1}{2},j+\frac{1}{2}}. \tag{5.2.36}$$

The various surface elements and velocities in these expressions can be derived in a straightforward manner from the expressions given for the axial momentum advection.

5.2.4 Grid Motions

The final step in our Eulerian code involves the grid motion terms. These terms all involve a divergences of the grid velocity $\nabla \boldsymbol{v}_g$. Since the grid velocity is centered at exactly the same location that the fluid velocity, we can use the expression derived for the divergence of the fluid velocity, (5.2.14)

$$(\nabla \boldsymbol{v})_g = \frac{A_{k+1,j+\frac{1}{2}} u_{g,k+1} - A_{k,j+\frac{1}{2}} u_{g,k}}{\Delta V_{k+\frac{1}{2},j+\frac{1}{2}}} + \frac{A_{k+\frac{1}{2},j+1} v_{g,j+1} - A_{k+\frac{1}{2},j} v^l_{g,j}}{\Delta V_{k+\frac{1}{2},j+\frac{1}{2}}} \tag{5.2.37}$$

The grid compression terms can be solved easily using this expression. The new positions of the grid are then obtained from

$$r^{n+1}_{k+\frac{1}{2},j} = r^n_{k+\frac{1}{2},j} + v_{g,j} \Delta t^{n+\frac{1}{2}} \tag{5.2.38}$$

and

$$z^{n+1}_{k,j+\frac{1}{2}} = z^n_{k,j+\frac{1}{2}} + u_{g,k} \Delta t^{n+\frac{1}{2}} \tag{5.2.39}$$

5.2.5 Artificial Viscosity and Gravity

The Eulerian scheme as described above is not complete without the addition of an artificial viscosity term to allow for the appropriate dissipation to occur in shocks (see discussion in section 4) as well as, if needed, the addition of the gravitational force in the momentum conservation equation. A detailed discussion of both of these terms would take us much beyond the space allowed for this contribution, hence, we shall limit ourselves to the briefly outline how the appropriate expressions could be obtained.

To derive a proper expression for the artificial viscosity, it is best to follow the procedure outlined in section 4. First, it is necessary to write the viscous stress tensor in cylindrical coordinates. Each of the four terms of this tensor has then to be replaced by a corresponding artificial term, keeping in mind that viscosity should only lead to entropy increase. The appropriate artificial viscous term is then added to both the momentum and the energy equation. A more detailed discussion of the form of the artificial viscous stress tensor can be found in Norman and Winkler (1986) where they actually only consider the diagonal terms of the tensor.

Self-gravity in multi-dimensional hydrodynamics is much more complicated and time consuming that in the special 1D spherical geometry case outlined in section 4. Often, Poisson's equation is solved in Fourier space and the solution transformed back to real space using Fast Fourier Transforms (FFT) (see Press *et al.* 1986). To be applicable this procedure requires a uniform grid and periodic boundary conditions. Both conditions are unfortunately rarely met and therefore this technique is not applicable in the general case. A more general procedure is the one used by Black and Bodenheimer (1985) and also adopted by Norman and Winkler (1986) which consists of finding a steady state solution to the diffusion equation

$$\frac{\partial \phi}{\partial t} = \nabla^2 \phi - 4\pi G \rho. \qquad (5.2.40)$$

To solve this equation, the ADI method (Alternating Direction Implicit) outlined in section 2.2 can be used efficiently. Notice that the time appearing in (5.2.40) bears no relation with the physical time but only serves to find the steady state solution of the equation. Boundary conditions can be obtained from a multipole expansion. Finally, the rate of convergence is checked by monitoring the quantity

$$\chi = Max\{\frac{|\nabla^2 \phi - 4\pi G \rho|}{4\pi G \rho}\}$$

where the maximum is taken over the whole grid. As pointed out by Norman and Winkler (1986), $\chi \leq 10^{-5}$ leads to less than a 1% error in the potential.

5.3 Stability Considerations

We have to ask ourselves whether the scheme we have outlined is stable and under what conditions. Of course, a complete von Neumann stability analysis is out of the question but we can use the results from section 3 to derive an approximate criterion that can subsequently be checked by experiments. We have seen that the usual stability criterion for explicit hydrodynamics requires that no information be transported over more than one cell during one time step. We therefore use this physical interpretation to derive the following criterion

$$\Delta t^{n+\frac{1}{2}} = C_t \, Min\{\frac{\Delta r_{j+\frac{1}{2}}}{c_{k+\frac{1}{2},j} + |v_{k+\frac{1}{2},j} - v_{g,j}|} \; ; \; \frac{\Delta z_{k+\frac{1}{2},j+\frac{1}{2}}}{c_{k,j+\frac{1}{2}} + |u_{k,j+\frac{1}{2}} - u_{g,k}|}\} \quad (5.2.40)$$

where the minimum has to be taken over the entire grid and C_t is a numerical constant, sometimes called the Courant factor, and taken to be less than 1.

This concludes our description of some of the important aspects of a 2D, cylindrical coordinate Eulerian hydrodynamics code. It should become obvious that increasing the spatial dimensions lead to more and more complicated algorithms. A 3D Eulerian code becomes so complicated and confusing that any attempts to described it in the space available here would be hopeless. However, since nature is three dimensional, solving the hydrodynamics equations in three spatial dimension is essential to simulate many important phenomena in astrophysics (convection, turbulence, etc.). As announced in the introduction, there exist a simple, fully Lagrangian method to actually solve the equations in three spatial dimensions (actually this method could be trivially generalized to arbitrary dimensions) called Smooth Particle Hydrodynamics or SPH in short. This method is described in some details in the next section.

6 3D Smooth Particle Hydrodynamics

In the introduction to section 5 we mentioned that multi-dimensional Lagrangian hydrodynamics present serious difficulties as it suffers from large grid distortions as soon as some shear is present in the flow. Eulerian hydrodynamics, on the opposite, does not suffer from these difficulties since the grid is not required to move with the flow. However, as we have seen in the previous section, the additional non-linear advection terms are difficult to model and require special attention. We will now describe a new method to solve the Lagrangian hydrodynamics equations that is virtually gridless. This method, called "Smooth Particle Hydrodynamics" or SPH, combines the advantages of the simpler Lagrangian description of the fluid without suffering from mesh tangling and distortions characteristic of Lagrangian multi-dimensional hydrodynamics.

It has been now about twelve years since the first paper by Lucy (1977) describing the SPH technique appeared in the literature. During this decade SPH has changed substantially, growing from a simple intuitive scheme to a full fledged hydrodynamics method. SPH is now in a position to compete, often successfully, with most numerical schemes proposed in the literature. The main advantage of the method is, of course, to bypass the usual grid necessary to compute derivatives avoiding the severe problems always associated with mesh tangling and distortion. Since a grid is not required, and due to its Lagrangian character, SPH is especially powerful in computations were large voids are present (interactions in binary systems for example) or in which material is thrown are large distances (collisions). The astonishing success in the development of SPH can be attributed in a large part to the work of Gingold and Monaghan (1977, 1982), Monaghan and Gingold (1983) and Monaghan (1982,1985). These papers are really at the roots of the SPH technique (for a more detailed review see Benz 1989). Beside these technical papers many astrophysical problems have been attacked using this method, a review of the applications of SPH to astrophysical problems can be found in Benz (1988).

The hydrodynamics equations we are trying to solve contain both time and space derivatives. Since the originality of SPH lies in the way the space derivatives are evaluated, we will focus on this aspect. Time integration can be performed in similar ways as described in the previous sections. Current SPH codes make extensive use of the leap-frog scheme or low order Runge-Kutta integrators. Emphasis will also be given on a proper derivation of the SPH equations from the hydrodynamics equations. This will not only allow to show that SPH is indeed solving the fluid equations but also provide a natural formalism to introduce more complicated physics and possible boundary conditions. For pedagogical reasons, we shall limit ourselves to the case of constant smoothing length (a measure of the "size" of the particles, see below) in both space and time since in this case, the SPH analogs of momentum and energy conservation equations have a reasonably simple form.

6.1 Kernel Estimation

Let us start by considering the following function

$$\langle f(\boldsymbol{r})\rangle = \int W(\boldsymbol{r}-\boldsymbol{r}', h) f(\boldsymbol{r}')d\boldsymbol{r}' \tag{6.1.1}$$

where W is a function that we call the kernel and h is a measure of the width of the kernel. We further impose a normalization condition on W

$$\int W(\boldsymbol{r}, h)d\boldsymbol{r} = 1. \tag{6.1.2}$$

From (6.1.1) and (6.1.2) it immediately follows that

$$\langle f(\boldsymbol{r})\rangle \xrightarrow{h\to 0} f(\boldsymbol{r}). \tag{6.1.3}$$

Relation (6.1.1) therefore defines the kernel estimate $\langle f(\boldsymbol{r})\rangle$ of $f(\boldsymbol{r})$. In particular if $W(\boldsymbol{r}-\boldsymbol{r}', h) = \delta(\boldsymbol{r}-\boldsymbol{r}')$ than, obviously, we have the equality $\langle f(\boldsymbol{r})\rangle = f(\boldsymbol{r})$.

By considering the fact that W is a function strongly peaked at $\boldsymbol{r}' = \boldsymbol{r}$ one is allowed to expand $f(\boldsymbol{r})$ in a Taylor series about $\boldsymbol{r}$ in (6.1.1). If W is chosen to be an even function of r only (i.e. spherical geometry kernel), that is $W(\boldsymbol{r}, h) = W(|\boldsymbol{r}|, h)$, than, one can show that the following relation applies

$$\langle f(\boldsymbol{r})\rangle = f(\boldsymbol{r}) + c\,(\nabla^2 f)\,h^2 + O(h^3) \tag{6.1.4}$$

where c is independent of h. Hence, by analogy, approximating $f(\boldsymbol{r})$ by $\langle f(\boldsymbol{r})\rangle$ is said to be second order accurate in h. Notice that since the term in h^2 is multiplied by the gradient of f, this implies perfect equality for constant or linear functions. Notice also that for the term in h^2 to be small, the function f should not be discontinuous or have too large gradients over the size of W. In fact, if this should be the case, the Taylor expansion, used to derive the expression above, would be unjustified to begin with.

In a similar way one can show that the following relation also holds true

$$\langle\frac{A(\boldsymbol{r})}{B(\boldsymbol{r})}\rangle = \frac{\langle A(\boldsymbol{r})\rangle}{\langle B(\boldsymbol{r})\rangle} + O(h^2) \tag{6.1.5}$$

If we now imagine the function $f(\boldsymbol{r})$ to be known at N discrete points only, distributed in space according to the number density distribution

$$n(\boldsymbol{r}) = \sum_{j=1}^{N} \delta(\boldsymbol{r} - \boldsymbol{r}_j) \tag{6.1.6}$$

and we use (6.1.1) and (6.1.5) we can derive the following expression

$$\langle f(\boldsymbol{r})\rangle = \sum_{j=1}^{N} \frac{m_j}{\rho(\boldsymbol{r}_j)} f(\boldsymbol{r}_j) W(|\boldsymbol{r} - \boldsymbol{r}_j|, h) \tag{6.1.7}$$

where we have written the number density at $\boldsymbol{r}_j$, $\langle n(\boldsymbol{r}_j)\rangle$, in the form

$$\langle n(\boldsymbol{r}_j)\rangle = \rho(\boldsymbol{r}_j)/m_j . \tag{6.1.8}$$

Notice that this procedure formally introduces the concept of particles of mass m_j.

If, for example, we choose in (6.1.7) $f(\boldsymbol{r}) \equiv \rho(\boldsymbol{r})$ we obtain the following expression for the fluid density

$$\langle \rho(\boldsymbol{r})\rangle = \sum_{j=1}^{N} m_j W(|\boldsymbol{r} - \boldsymbol{r}'|, h) \tag{6.1.9}$$

This expression can be interpreted as follows. Every particle of mass m_j is smoothed in space according to W which represents its density distribution in space. The density at any point in space is then obtained by summing up the contributions from all particles at that point. In fact the name Smooth Particle Hydrodynamics actually derives from this interpretation. There is, however, another interpretation which, in my opinion, is better. One can look at SPH only as an interpolation technique in which particles have no real physical meaning but are just interpolation points. Contrary to the standard finite difference schemes, in SPH the interpolation points are not necessary on a lattice but can be distributed seemingly in a random manner and are moving with the fluid. The latter point need not necessarily to be true. One could easily imagine particles moving at a prescribed velocity leading to an Eulerian SPH scheme. The only (but important) advantage of the SPH approach would reside in the ease of estimating the spatial gradients.

SPH is also sometimes associated in the literature with a Monte Carlo estimation of integrals. If this were to be true, the convergence of the method would be rather poor since it would scale like $N^{\frac{1}{2}}$. Experiments have shown that the convergence rate is actually about N, much faster that expected from a pure Monte Carlo process. This difference has its origin in the fact that the interpolation points are actually not distributed in a random fashion but are in fact following the motions of the fluid (Monaghan 1985).

6.2 Hydrodynamics Equations

In SPH, all variables are centered at particle locations, therefore the space centering problems we have encountered in the previous sections do not exist. We can therefore directly proceed and derive the appropriate numerical equations.

6.2.1 Momentum Equation

We start from the usual hydrodynamical equation describing momentum conservation and derive the corresponding relation in the SPH formalism. To avoid unnecessary complications we shall consider the inviscid case first, the dissipative case being discussed later. In its Eulerian form, the momentum conservation equation is written

$$\frac{\partial \boldsymbol{v}}{\partial t} + (\boldsymbol{v} \cdot \nabla)\boldsymbol{v} = -\frac{\nabla P}{\rho} \tag{6.2.1}$$

Multiplying all terms by the kernel and integrating over the solution space V yields

$$\int_V \left(\frac{\partial \boldsymbol{v}}{\partial t} + (\boldsymbol{v} \cdot \nabla)\boldsymbol{v}\right) W(|\boldsymbol{r} - \boldsymbol{r}'|, h) d\boldsymbol{r}' = -\int_V \frac{\nabla P}{\rho} W(|\boldsymbol{r} - \boldsymbol{r}'|, h) d\boldsymbol{r}' \tag{6.2.2}$$

where all variables are taken to be functions of $\boldsymbol{r}'$.

For the sake of simplicity let us assume that the smoothing length h is constant in both space and time. Notice that the ideal case would be to allow h to vary in both space and time but this introduces complicated additional terms that will not be discussed here. With this assumption we rewrite the left-hand-side of (6.2.2) in the following way

$$\int_V \frac{\partial \boldsymbol{v}}{\partial t} W(|\boldsymbol{r} - \boldsymbol{r}'|, h) d\boldsymbol{r}' + \int_V (\boldsymbol{v} \cdot \nabla)\boldsymbol{v} W(|\boldsymbol{r} - \boldsymbol{r}'|, h) d\boldsymbol{r}' = \frac{\partial \langle \boldsymbol{v} \rangle}{\partial t} + (\langle \boldsymbol{v} \rangle \cdot \nabla)\langle \boldsymbol{v} \rangle = \frac{d\langle \boldsymbol{v} \rangle}{dt} \tag{6.2.3}$$

where any quantity (scalar or vector) $\langle f(\boldsymbol{r}) \rangle$ is defined by (6.1.1) and d/dt is the Lagrangian time derivative. The expression for the advection term above has been obtained by a Taylor expansion of the integrand similar to the one used in the derivation of (6.2.5) below.

The right hand-side of (6.2.2) needs somewhat more attention. Let us integrate by parts

$$-\int_V \frac{\nabla P}{\rho} W(|\boldsymbol{r} - \boldsymbol{r}'|, h) d\boldsymbol{r}' = -\int_S \frac{P}{\rho} W(|\boldsymbol{r} - \boldsymbol{r}'|, h) \boldsymbol{n} da + \int_V P \nabla \left(\frac{W(|\boldsymbol{r} - \boldsymbol{r}'|, h)}{\rho}\right) d\boldsymbol{r}'$$

Let us now assume that the solution space V extends far enough so that on its boundaries, S, the pressure P or W itself vanishes, the surface term is therefore identically zero.

This assumption is, of course, generally not met. In many fluid dynamics problems outside astronomy, boundaries are needed to contain the fluid. The surface integral appearing in the expression above would be the proper place to introduce boundaries in the SPH formalism. Clearly, this surface integral can be computed analytically once W

has been specified and will yield a term depending only on the distance of the particle to the boundary. A similar expression can be obtained for the energy conservation equation (see below). This form of boundary treatment, although it is the only really consistent one has unfortunately not yet been used. Calculations done so far involving boundaries have relied on "ghost" particles to contain the gas. Being a particle method, boundaries will intrinsically be more difficult to handle in SPH that they are in grid based method.

In astrophysical situations, most often the fluid is contained by gravity. In this case, no boundaries are needed and they can be assumed to be at infinity making the surface term above vanish. Expanding further the gradient in the remaining term, we get

$$-\int_V \frac{\nabla P}{\rho} W(|\boldsymbol{r}-\boldsymbol{r}'|,h)d\boldsymbol{r}' = \int_V \frac{P}{\rho}\nabla W(|\boldsymbol{r}-\boldsymbol{r}'|,h)d\boldsymbol{r}' - \int_V \frac{P}{\rho^2}\nabla\rho\, W(|\boldsymbol{r}-\boldsymbol{r}'|,h)d\boldsymbol{r}'.$$

The first term on the right hand side can be rewritten by noting that $\nabla W(|\boldsymbol{r}-\boldsymbol{r}'|,h) = -\nabla_r W(|\boldsymbol{r}-\boldsymbol{r}'|,h)$ where $\nabla_r W(|\boldsymbol{r}-\boldsymbol{r}'|,h)$ is the gradient of the kernel but with respect to $\boldsymbol{r}$ rather than $\boldsymbol{r}'$. Recalling definition (6.1.1) we rewrite (6.2.2)

$$\frac{d}{dt}\langle \boldsymbol{v}\rangle = -\nabla_r\left\langle\frac{P}{\rho}\right\rangle - \int_V \frac{P}{\rho^2}\nabla\rho\, W(|\boldsymbol{r}-\boldsymbol{r}'|,h)d\boldsymbol{r}'. \tag{6.2.4}$$

The remaining integral is treated in the same way that lead to (6.2.3). First, we note that $W(|\boldsymbol{r}-\boldsymbol{r}'|,h)$ is a strongly peaked function at $\boldsymbol{r}$. Next, suppose we develop into a Taylor expansion the function $\frac{P}{\rho^2}\nabla\rho$ about $\boldsymbol{r}'=\boldsymbol{r}$. If $W(|\boldsymbol{r}-\boldsymbol{r}'|,h)$ is an even function, the first order term will drop. So that neglecting the second order term (which is consistent with the overall order of the method) we approximate the integral above in the following way

$$\int_V \frac{P}{\rho^2}\nabla\rho\, W(|\boldsymbol{r}-\boldsymbol{r}'|,h)d\boldsymbol{r}' = \left(\frac{P}{\rho^2}\nabla\rho\right)_{\boldsymbol{r}'=\boldsymbol{r}} \tag{6.2.5}$$

in which we now replace $\nabla\rho$ by $\nabla_r\langle\rho\rangle$. Putting all terms together gives the final expression for the momentum equation (6.2.2)

$$\frac{d}{dt}\langle \boldsymbol{v}\rangle = -\nabla_r\left\langle\frac{P}{\rho}\right\rangle - \left(\frac{P}{\rho^2}\right)\nabla_r\langle\rho\rangle. \tag{6.2.6}$$

Before transforming this expression into corresponding sums using the results of section (6.1), it is important to emphasize that (6.2.6) computes the evolution of $\langle \boldsymbol{v}\rangle$ that is, the "smoothed velocity" field. This is crucial when setting up initial conditions. As pointed out in the discussion of (6.1.4), any quantity $\langle f(\boldsymbol{r})\rangle$ is formally identical to $f(\boldsymbol{r})$, provided $f(\boldsymbol{r})$ is a smooth varying function over the size of W. Discontinuous initial conditions in velocity or internal energy are therefore equivalent to large perturbations being present in the initial configuration. This obviously can have dramatic consequences if the purpose of the calculation is to check for potential instabilities in the flow for example.

We now apply the results of section 6.1 to rewrite (6.2.6) in terms of sums. The acceleration of particle i is given after regrouping all terms by

$$\frac{d\boldsymbol{v}_i}{dt} = -\sum_{j=1}^{N} m_j \left(\frac{P_i}{\rho_i^2} + \frac{P_j}{\rho_j^2}\right) \nabla_i W(|\boldsymbol{r}_i - \boldsymbol{r}_j|, h) \tag{6.2.7}$$

where $\boldsymbol{v}_i \equiv \langle \boldsymbol{v} \rangle_i$. Equation (6.2.7) describes momentum conservation in the SPH formalism.

A few comments on equation (6.2.7) can be made. First, it is easy to show that momentum is indeed conserved. We notice that forces are evaluated in pairs, so that the force due to particle j on particle i is given by

$$m_i \left(\frac{d\boldsymbol{v}_i}{dt}\right)_j = m_i m_j \left(\frac{P_i}{\rho_i^2} + \frac{P_j}{\rho_j^2}\right) \nabla_i W(|\boldsymbol{r}_i - \boldsymbol{r}_j|, h) \tag{6.2.8}$$

but from the definition of W we have $\nabla_i W(|\boldsymbol{r}_i - \boldsymbol{r}_j|, h) = -\nabla_j W(|\boldsymbol{r}_i - \boldsymbol{r}_j|, h)$ so that (6.2.8) can be rewritten as

$$m_i \left(\frac{d\boldsymbol{v}_i}{dt}\right)_j = -m_j m_i \left(\frac{P_i}{\rho_i^2} + \frac{P_j}{\rho_j^2}\right) \nabla_j W(|\boldsymbol{r}_i - \boldsymbol{r}_j|, h) = -m_j \left(\frac{d\boldsymbol{v}_j}{dt}\right)_i \tag{6.2.9}$$

which proves total momentum conservation. Notice also that angular momentum is conserved locally as well since all forces are always directed along a line joining the center of the particles.

Finally, the sum in (6.2.7) although formally extending on all particles, is in practice determined only by particles for which $\nabla_i W(|\boldsymbol{r}_i - \boldsymbol{r}_j|, h)$ has a non negligible contribution which is in turn determined by the choice of the kernel (see 6.3). This reflects the fact that pressure forces are indeed short range forces and that only neighboring particles will contribute to the local pressure gradients. The efficiency of any SPH code is therefore essentially determined by the speed at which all particles within a fixed radius around any given particle can be found. In practice, this is achieved with the use of linked lists in the case of a constant smoothing length or by hierarchical tree structures in codes allowing for local variations of the smoothing length h.

6.2.2 Energy Conservation

Most of the early SPH codes were concerned with the simulation of gravitational collapse of gas clouds (see for example Monaghan and Lattanzio 1985; Benz 1984). During the collapse, at least initially, the cloud remains very close to isothermal and therefore no energy equation is needed. For the method to simulate non-isothermal flows it is necessary to include an additional equation describing the variation of the specific internal energy or entropy of the system. Choosing density and specific internal energy, ϵ, as independent thermodynamical variables we start with the first law of thermodynamics that describes energy conservation, for adiabatic transformations

$$\frac{\partial \epsilon}{\partial t} + (\boldsymbol{v} \cdot \nabla)\epsilon = -\frac{P}{\rho} \nabla \cdot \boldsymbol{v}. \tag{6.2.10}$$

In proceeding the same way as in the derivation of the momentum conservation equation, that is by multiplying (6.2.10) by W and integrating over the solution domain which boundaries are assumed to be at infinity and transforming integrals in sums, we obtain after some algebra the following equation

$$\frac{d\epsilon_i}{dt} = \frac{P_i}{\rho_i^2}\sum_{j=1}^{N} m_j(\boldsymbol{v}_i - \boldsymbol{v}_j)\nabla_i W(|\boldsymbol{r}_i - \boldsymbol{r}_j|, h) \tag{6.2.11}$$

where as before we have defined $\epsilon_i \equiv \langle\epsilon\rangle_i$. Similar to the momentum conservation equation (6.2.7), (6.2.11) describes the evolution of the "smooth" specific internal energy and therefore the same remarks on the smoothness of initial conditions apply. The right-hand-side is nothing else, of course, than the SPH representation of the PdV/dt term. By comparing (6.2.10) and (6.2.11) one also notes for future use that

$$\rho\nabla\cdot\boldsymbol{v} = -\sum_{j=1}^{N} m_j(\boldsymbol{v}_i - \boldsymbol{v}_j)\nabla_i W(|\boldsymbol{r}_i - \boldsymbol{r}_j|, h). \tag{6.2.12}$$

As with the momentum conservation equation, the exact conservation of total energy can be easily demonstrated. We start by writing the variation of the total internal energy

$$\frac{d}{dt}\sum_{i=1}^{N} m_i\epsilon_i = \sum_{j=1}^{N} m_i\frac{d\epsilon_i}{dt} = \sum_{i=1}^{N}\sum_{j=1}^{N} m_i m_j\frac{P_i}{\rho_i^2}(\boldsymbol{v}_i - \boldsymbol{v}_j)\nabla_i W(|\boldsymbol{r}_i - \boldsymbol{r}_j|, h)$$

expanding the right-hand-side term, interchanging indexes and making use of the identity $\nabla_i W(|\boldsymbol{r}_i - \boldsymbol{r}_j|, h) = -\nabla_j W(|\boldsymbol{r}_i - \boldsymbol{r}_j|, h)$ we obtain after some rearranging of terms

$$\frac{d}{dt}\sum_{i=1}^{N} m_i\epsilon_i = \sum_{i=1}^{N}\sum_{j=1}^{N} m_i m_j\left(\frac{P_i}{\rho_i^2} + \frac{P_j}{\rho_j^2}\right)\boldsymbol{v}_i\nabla_i W(|\boldsymbol{r}_i - \boldsymbol{r}_j|, h)$$

by comparing this expression to (6.2.7) one gets :

$$\frac{d}{dt}\sum_{i=1}^{N} m_i\epsilon_i = \frac{d}{dt}\left(\frac{1}{2}\sum_{i=1}^{N} m_i v_i^2\right) \tag{6.2.13}$$

which shows that the work done by pressure forces is at the expense of internal energy and thus ensures the conservation of total energy.

6.2.3 Shocks and Artificial Viscosity

As in sections 4 and 5, a dissipation mechanism is needed to treat the shocks correctly. In fact, for particle methods, this term is even more important since without it the particles would almost free-stream past each other obviously not behaving as a fluid. In SPH, various forms for the standard bulk and/or von Neumann-Richtmyer artificial viscosity have been proposed in the literature. The straight forward expression obtained from an SPH representation of the velocity divergence was shown to give poor results, since motion of particles on very short length scales ($\leq h$) were not damped (Gingold and Monaghan 1982). An alternative, consisting in a local estimation of the velocity divergence was obtained by Gingold and Monaghan (1982) and shown to yield accurate results. The local velocity divergence at particle i due to the presence of particle j times the smoothing length is estimated by

$$\mu_{ij} = \frac{h(\boldsymbol{v}_i - \boldsymbol{v}_j)\cdot(\boldsymbol{r}_i - \boldsymbol{r}_j)}{|\boldsymbol{r}_i - \boldsymbol{r}_j|^2 + \epsilon h^2} \tag{6.2.14}$$

where the term ϵh^2 has been added to avoid divergence for small $|\boldsymbol{r}_i - \boldsymbol{r}_j|$. The corresponding pressure due to both artificial viscosities (bulk and von Neumann-Richtmyer) will be written as

$$\Pi_{ij} = \begin{cases} \frac{-\alpha c_{ij}\mu_{ij} + \beta\mu_{ij}^2}{\rho_{ij}}, & \text{if } (\boldsymbol{v}_i - \boldsymbol{v}_j)\cdot(\boldsymbol{r}_i - \boldsymbol{r}_j) \leq 0.; \\ 0. & \text{otherwise.} \end{cases} \tag{6.2.15}$$

where we have defined $c_{ij} = 0.5(c_i + c_j)$ and $\rho_{ij} = 0.5(\rho_i + \rho_j)$. This guarantees that this term remains symmetric in i and j hence ensures momentum conservation. The condition that Π_{ij} be zero for receding particles makes sure that the dissipation modeled by this artificial viscosity only adds entropy to the system.

The final conservation equations can now be written by introducing the artificial viscous pressure term in the equation of momentum conservation (6.2.6)

$$\frac{d\boldsymbol{v}_i}{dt} = -\sum_{j=1}^{N} m_j\left(\frac{P_i}{\rho_i^2} + \frac{P_j}{\rho_j^2} + \Pi_{ij}\right)\nabla_i W(|\boldsymbol{r}_i - \boldsymbol{r}_j|, h) \tag{6.2.16}$$

and the corresponding contribution to the energy equation (6.2.11)

$$\frac{du_i}{dt} = \frac{P_i}{\rho_i^2}\sum_{j=1}^{N} m_j(\boldsymbol{v}_i - \boldsymbol{v}_j)\nabla_i W(|\boldsymbol{r}_i - \boldsymbol{r}_j|, h) + \frac{1}{2}\sum_{j=1}^{N} m_j \Pi_{ij}(\boldsymbol{v}_i - \boldsymbol{v}_j)\cdot\nabla_i W(|\boldsymbol{r}_i - \boldsymbol{r}_j|, h) \tag{6.2.17}$$

6.2.4 Self-Gravity

In many astrophysical situations self-gravity can not be neglected. We will derive in this section the appropriate modifications to the momentum conservation equation due to the introduction of body forces. First, it is necessary to recall that interactions between particles are always computed in pairs, that W can be interpreted as an individual particle's density distribution and finally that particles are taken to be spherically symmetric. Having these points in mind we start with Poisson's equation

$$\Delta\phi = \frac{1}{r^2}\frac{d}{dr}(r^2\frac{d\phi}{dr}) = 4\pi G\rho \tag{6.2.18}$$

This equation can now be easily solved to give the usual solution for the forces exerted on particle i from particle j

$$-(\nabla\phi_i)_j = -G\frac{M(|\boldsymbol{r}_i - \boldsymbol{r}_j|)}{|\boldsymbol{r}_i - \boldsymbol{r}_j|^2}\frac{(\boldsymbol{r}_i - \boldsymbol{r}_j)}{|\boldsymbol{r}_i - \boldsymbol{r}_j|} \tag{6.2.19}$$

where we have defined $M(|\boldsymbol{r}_i - \boldsymbol{r}_j|)$ by the following expression

$$M(|\boldsymbol{r}_i - \boldsymbol{r}_j|) = 4\pi\int_0^{|\boldsymbol{r}_i - \boldsymbol{r}_j|} u^2\rho(u)du = 4\pi\int_0^{|\boldsymbol{r}_i - \boldsymbol{r}_j|} u^2 W(u,h)du. \tag{6.2.20}$$

This definition introduces a new constraint on the kernel, namely

$$\int_0^\infty r^2 W(r,h)dr < \infty$$

in order to avoid an infinite mass. The total gravitational force felt by particle i can be written as the sum of all the individual forces

$$-\nabla\phi_i = -G\sum_{j=1}^{N} \frac{M(|\boldsymbol{r}_i - \boldsymbol{r}_j|)}{|\boldsymbol{r}_i - \boldsymbol{r}_j|^2} \frac{(\boldsymbol{r}_i - \boldsymbol{r}_j)}{|\boldsymbol{r}_i - \boldsymbol{r}_j|} \tag{6.2.21}$$

This term can now be added to the momentum conservation equation (6.2.16) so that body forces are taken into account properly. Notice that, contrary to all previous expressions, the terms in the sum in (6.2.21) do not vanish as the distance between particles increases. This reflects the fact that gravity is a long range force. In practice, if this summation is not done in an efficient way, the method will scale like N^2 and therefore lose all its potential applications because of the tremendous amount of computer time that will be required even for a modest number of particles. The use of hierarchical tree methods is probably the most efficient way to compute the gravitational forces since they scale like $NlogN$ and allow to rapidly find all neighboring particles at the same time.

Finally, notice also that by actually replacing the density in (6.2.20) by the kernel W, we actually have given a physical meaning to this kernel. In particular, all kernels with negative values are now clearly excluded. Furthermore, by this replacement we give up the notion that SPH can be seen as an interpolation method. In fact, should we want to keep this point of view, (6.2.18) should be solved by multiplying both sides of the equation by W and integrating over the solution domain.

6.3 The Kernel

So far we have discussed the SPH formalism without specifying the kernel W. Along the way we added however, a number of constraints that the kernel should satisfy. The most frequently used kernels satisfying all these conditions that appear in the literature are listed below.

a) Exponential kernel (Wood 1981)

$$W(r,h) = \frac{1}{8\pi h^3} exp(-v) \tag{6.3.1}$$

b) Gaussian (Gingold and Monaghan 1977)

$$W(r,h) = \frac{1}{\pi^{3/2}h^3} exp(-v^2) \tag{6.3.2}$$

c) Spline (Monaghan and Lattanzio 1985)

$$W(r,h) = \frac{1}{\pi h^3} \begin{cases} 1 - \frac{3}{2}v^2 + \frac{3}{4}v^3, & \text{if } 0 \le v \le 1; \\ \frac{1}{4}(2-v)^3, & \text{if } 1 \le v \le 2; \\ 0. & \text{otherwise.} \end{cases} \tag{6.3.3}$$

where we have defined $v = \frac{r}{h}$.

The following brief comments can be made. All three kernels interpolate to second order in h, and all three are always positive in the range of interest. The advantage of (6.3.3) is that it is defined on compact support. This provides a clear limit to the number of neighbors to be considered since beyond $v = 2$ all contributions are identically zero. In the use of (6.3.1) or (6.3.2) an artificial cut-off in v has to be introduced to avoid an N^2 calculation.

As first pointed out by Gingold and Monaghan (1982) kernels can be devised that interpolate to fourth order in h. These kernels are constructed by making the quadratic error term in (6.1.4) vanish. The kernel being an even function, the third order term vanishes, resulting in a fourth order accurate interpolation. Applying this procedure to the gaussian kernel (6.3.2) the following expression called super-gaussian kernel (Gingold and Monaghan 1982) is obtained

$$W(r,h) = \frac{1}{\pi^{3/2}h^3}\left(\frac{5}{2} - v^2\right) exp(-v^2). \tag{6.3.4}$$

Although this kernel has a higher interpolation accuracy it has the problem that, unlike second order kernels, W is negative for $v > \sqrt{5/2}$. The interpretation given to W in (6.2.20) to represent an individual particle's density distribution can clearly no longer apply and therefore equation (6.2.20) can not be used. A new formalism for self-gravity has to be found. However, in the derivation of all other equations, W can be interpreted only as an interpolating kernel. In these cases the fact that W goes negative does not really matter. A word of caution is however in order, since one could imagine a particle distribution for which all the contributions from the neighbors to the density are negative. Clearly, this could lead to a situation where the particle's self-contribution (positive) will be less than the neighbors negative contribution resulting in unphysical negative densities.

Having reached the final sentences of this contribution, I would like to express my deep gratitude and my appreciation to R. Bowers and J. R. Wilson for providing me with their notes of a yet unpublished book on computational astrophysics as well as for the many enlightening discussions about what seemed to me obscure details of numerical hydrodynamics. I would also like to warmly thank E. Müller and W. Hillebrandt for their hospitality at the Max-Planck-Institut für Physik und Astrophysik where parts of these lectures were written. Partial support from the Swiss National Science Foundation is also acknowledged.

References

Anderson, D.A., Tannehill, J.C., Pletcher, R.H., (1984), "Computational Fluid Mechanics and Heat Transfer", Hemisphere Publishing Company, McGraw-Hill Book Company, New York New York.

Benz, W., (1984), *Astron. Astrophys.*, **139**, 378.

Benz, W., (1988), *Comput. Phys. Comm.*, **48**, 97.

Benz, W., (1989), in "Numerical Modeling of Non Radial Stellar Pulsation: Problems and Prospects", Ed. J.R. Buchler, Kluwer Academic Publishers, Dordrecht

Black D.C., Bodenheimer, P., (1975), *Astrophys. J.*, **199**, 619.

Bowers, R.L., Wilson, J.R., (1990), "Computational Astrophysics", in preparation.

Chandrasekhar, S., (1961), "Hydrodynamics and Hydromagnetic Stability", Clarendon Press, Oxford.
Courant, R., Friedrichs, K.O., (1976), "Supersonic Flow and Shock Waves", *Applied Mathematical Sciences*, **21**, Springer-Verlag, New York, New York.
Fletcher, C.A.J., (1988), "Computational Techniques for Fluid Dynamics", Springer Series in Computational Physics, Springer-Verlag, Berlin.
Gingold, R.A., Monaghan, J.J., (1977), *Mon. Not. R. astr. Soc.*, **181**, 375.
Gingold, R.A., and Monaghan, J.J., (1982), *J. Comput. Phys.*, **46**, 429.
Finn, L.S., Hawley, J.F., 1990, *J. Comput. Phys.*, submitted.
Landau, L.D., Lifshitz, E.M., (1959), "Fluid Mechanics", Pergamon Press, Oxford.
Lucy, L., (1977), *Astron. J.*, **82**, 1013.
Monaghan, J.J., (1982), *SIAM J. Sci. Stat. Comput.*, **3**, 422.
Monaghan, J.J., (1985), *Comput. Phys. Rep.*, **3**, 71.
Monaghan, J.J., Gingold, R.A., (1983), *J. Comput. Phys.*, **52**, 374.
Monaghan, J.J., Lattanzio, J.C., (1985), *Astron. Astrophys.*, **149**, 135.
Norman, M.L., Winkler, K.-H.A., (1986), "Astrophysical Radiation Hydrodynamics", Reidel Publishing Company, Dordrecht.
Norman, M.L., Smarr, L., Wilson, J.R., Smith, M.D., (1981), *Astrophys. J.*, **247**, 52.
Norman, M.L., Smarr, L., Winkler, K.-H.A., Smith, M.D., (1982), *Astron. Astrophys.*, **113**, 285.
Oran, E.S., Boris, J.P., (1987), "Numerical Simulation of Reactive Flow", Elsevier-Science Publishing Co., Inc., New York, New York.
Peaceman, D.W., Rachford, H.H., (1955), *SIAM J.*, **3**, 28.
Peyret, R., Taylor, T.D., (1983), "Computational Methods for Fluid Flows", Springer Verlag, New York, New York.
Press, W.H., Flannery, B.P., Teukolsky, S.A., Vetterling, W.T., (1986), "Numerical Recipes", Cambridge University Press, Cambridge.
Richtmyer, R.D., (1957), "Difference Methods for Initial Value Problems", Interscience Publishers, Inc., New York, New York.
Richtmyer, R.D., Morton K.W., (1967), "Difference Methods for Initial-Value Problems", Interscience Publishers, Inc., New York, New York.
Roache, P.J., (1972), "Computational Fluid Dynamics", Hermosa Publishers, Albuquerque, New Mexico.
van Leer, B., (1977), *J. Comput. Phys.*, **23**, 276.
Wood, D., (1981), *Mon. Not. R. astr. Soc.*, **194**, 201.

Basic Hydrodynamics

Jean Heyvaerts

Laboratoire d'Astrophysique et de Cosmologie
Université Paris 7 et Observatoire de Paris
Observatoire,92195 Meudon France

1 Why this Lecture?

The aim of this lecture is to provide an introductory exposition of basic concepts in hydrodynamics, which would help beginners in this field to approach the other, more specialized, lectures presented elsewhere in this volume. The reader already familiar with hydrodynamics and MHD could probably skip this chapter without much loss of useful information. It describes concisely the link of the hydrodynamical description to a more atomistic description of matter : this concerns both the concepts and the physical meaning of the transport coefficients. It establishes the basic equations of hydrodynamics, and quickly describes a few very classical, but important, results concerning one-dimensional flows and shocks. Though we describe, as simply as possible, some of the necessary mathematical tools , in particular tensors, the reader is still assumed to have some knowledge of vector calculus, and is also supposed to know some basic kinetic theory of gases, some thermodynamics and some electromagnetism.

2 How to Describe a Fluid?

2.1 The Distribution Function

Consider the fluid of interest to be an ensemble of particles of various species : atoms, molecules, ions, electrons. It will sometimes be useful to consider photons as constituants of the fluid too. Assume for simplicity that the translational degrees of freedom of these particles need not be described in quantum terms (this excludes solids). Then we can speak of a particle sitting at $\boldsymbol{r}$ with velocity $\boldsymbol{u}$. But of course it is impossible to describe the fate of each particle individually. The first step to be taken is to describe them statistically. Let us define the *one particle distribution function* for species A such that

$$f_A(\boldsymbol{r},\boldsymbol{u},t) \quad d^3\boldsymbol{r}\, d^3\boldsymbol{u} \tag{1}$$

is the probability, multiplied by average density, of finding a particle of type A around $\boldsymbol{r}$, within $d^3\boldsymbol{r}$, with velocity $\boldsymbol{u}$, within $d^3\boldsymbol{u}$, at time t. Later on we occasionnaly omit the time or position, or even the species index, as arguments of this function : so

$f_A(\boldsymbol{r},\boldsymbol{u})$, or $f_A(\boldsymbol{u})$, or even $f(\boldsymbol{u})$ really mean the same as (1). It should be noted that, though they do not contain all the statistical information on the system, the one-particle distribution functions still allow a fairly detailed description of it. By averaging over velocity, a number of *macroscopic* quantities can be defined, like the number density $n_A(\boldsymbol{r},t)$, the flow velocity $\boldsymbol{v}_A(\boldsymbol{r},t)$ of species A, or the average kinetic energy, $K_A(\boldsymbol{r},t)$. These are defined as follows:

$$n_A(\boldsymbol{r},t) = \iiint f_A(\boldsymbol{r},t)\ d^3\boldsymbol{u} \tag{2a}$$

$$n_A\boldsymbol{v}_A(\boldsymbol{r},t) = \iiint \boldsymbol{u} f_A(\boldsymbol{r},\boldsymbol{u},t)\ d^3\boldsymbol{u} \tag{2b}$$

$$n_A K_A(\boldsymbol{r},t) = \iiint \frac{1}{2} m_A u^2 f_A(\boldsymbol{r},\boldsymbol{u},t)\ d^3\boldsymbol{u} \tag{2c}$$

More generally, the average value at point $\boldsymbol{r}$ and time t of some function of particle velocity, $g(\boldsymbol{u})$, is defined as:

$$n_A \ < g_A > = \iiint g(\boldsymbol{u})\ f_A(\boldsymbol{r},\boldsymbol{u},t)\ d^3\boldsymbol{u} \tag{3}$$

It is sometimes useful to refer to the peculiar velocity of a particle, $\boldsymbol{w}$, defined by

$$\boldsymbol{w} = \boldsymbol{u} - \boldsymbol{v}_A(\boldsymbol{r},t) \tag{4}$$

Obviously $\boldsymbol{w}$ is a random quantity of zero average, the distribution function of which is:

$$\varphi_A(\boldsymbol{r},\boldsymbol{w},t) = f_A(\boldsymbol{r},\boldsymbol{v}_A+\boldsymbol{w},t) \tag{5}$$

For incoherent, unpolarized, radiation, it is convenient to introduce a distribution function for photons of frequency ν, propagating at point $\boldsymbol{r}$ and time t, in the direction of the unit vector $\boldsymbol{\Omega}$, such that

$$n(\nu,\boldsymbol{\Omega},\boldsymbol{r},t)\ d^3\boldsymbol{r}\ \ d\nu\ \ d\hat{\Omega} \tag{6}$$

is the probable number of photons in $d^3\boldsymbol{r}$ at time t, having a frequency in $(\nu,\nu+d\nu)$, and a direction $\boldsymbol{\Omega}$ in the small solid angle $d\hat{\Omega}$. The more commonly used specific intensity is just

$$I(\nu,\boldsymbol{\Omega},\boldsymbol{r},t) = c\ h\nu\ n(\nu,\boldsymbol{\Omega},\boldsymbol{r},t) \tag{7}$$

2.2 The Hydrodynamical Regime

Kinetic theory is the branch of physics that is concerned with the deduction of equations describing the evolution of distribution functions as a result of the interactions of particles among themselves and with other species. The nature and dynamics of these interactions is supposedly known at the microscopic level, and the so-called kinetic equations express how many particles of a given species, A, leave or enter per second a certain phase-space element $d^3\boldsymbol{r}\ d^3\boldsymbol{u}$ near position-velocity point, $\boldsymbol{r},\boldsymbol{u}$, as a result of these interactions and of all the other causes that might change their position or their velocity. Among them is the very motion of these particles, which causes changes of position, and the external forces exerted on them, which cause changes of velocity which are not due to interactions with the nearby particles. As a rule, kinetic equations have the general following form:

$$\frac{\partial f}{\partial t} + \boldsymbol{u}.\frac{\partial f}{\partial \boldsymbol{r}} + \frac{\boldsymbol{F}}{m_A}.\frac{\partial f}{\partial \boldsymbol{u}} = \mathcal{C}(f) \tag{8}$$

where the second term is due to the particle motion in real space, while the third is due to particle motion in velocity space, due to applied forces $\boldsymbol{F}$, and the r.h.s. represents the effect of interactions and mutations (the particle may be absorbed, or may decay). The explicit form of this term depends on their nature, on the regime applicable to the situation of interest (rare collisions, or not, long range of the interactions or not etc..), and may be quite complex. A well known example is the Boltzmann equation, which is relevant to gaseous media where the particles interact rarely and by short range forces, so that only binary collisions need be considered. Another famous example is the equation of radiative transfer for photons, which describes how their distribution function evolves due to emissions, absorptions and scattering by material particles. It is a property of all such equations that, in the absence of maintained gradients and external forces, their solution necessarily approaches a time independant solution which is the *thermodynamic equilibrium* distribution function. The time scale needed to reach this assymptotics is of the order of $f/\,\mathcal{C}(f)$, and the distance travelled by a particle in such a time is its mean free path, λ. So, loosely speaking, it takes one mean free path to thermalize.In practice it often happens that, even in the presence of gradients, particles have such a small mean free path that they get thermalized with their environment much before having a chance to reach places in the medium where the macroscopic parameters (density, flow velocity, average energy) are noticeably different. Let L be a characteristic gradient scale for these macroscopic parameters. Local thermalization occurs when

$$\frac{\lambda}{L} << 1 \tag{9}$$

In such situations a detailed description of the medium by distribution functions, though still correct, becomes rather superfluous, since these are necessarily very near to their thermodynamic equilibrium shape, which can be perfectly described by *a few* parameters only, namely the density, n_A, the flow velocity, $\boldsymbol{v}_A$, and the temperature, T_A which is related to the average kinetic energy by $K_A = 3/2\,k_B T_A$, where k_B is the Boltzmann constant.In a first approximation it should be enough to ignore the small difference between the actual distribution function and the perfect thermodynamic equilibrium one. We shall see below that this approximation is however not entirely sufficient, since the weak gradients still leave some trace of their existence on the distribution functions T : though small, the ratio (9) is not exactly zero! When the inequality (9) is satisfied for some species A, the latter will be said to be in the hydrodynamical regime, and the attention will in this case be focused on the macroscopic quantities mentioned above. Their evolution, of course, could, in principle, be followed by solving the appropriate kinetic equation. However, more direct reasonings, based on budget equations for mass, momentum and energy, give an easier access to the same result, if we prove able to calculate a few so-called *transport coefficients*, which relate fluxes of these quantities to their gradients ,for example heat flux to temperature gradient. Kinetic theory remains implicitly present in this determination which however has been made once and for all in a number of usual gas and plasma regimes. Note that some constituants of the medium may well be in the hydrodynamical regime while others are not, since they may have very different mean free pathes indeed. For example photons may remain in a kinetic regime, while atoms, ions and electrons fully satisfy (9).

When most constituants are in the hydrodynamical regime, it is convenient to define global quantities for the fluid as a whole, like the total mass density, ρ, the global flow velocity, $\boldsymbol{v}$, the net charge density Q, current density, $\boldsymbol{j}$, and the total pressure P :

$$\rho = \sum_A n_A m_A \tag{10}$$

$$\rho \boldsymbol{v} = \sum_A n_A \, m_A \, \boldsymbol{v}_A \tag{11}$$

$$Q = \sum_A q_A \, n_A \tag{12}$$

$$\boldsymbol{j} = \sum_A q_A \, n_A \, \boldsymbol{v}_A \tag{13}$$

$$P = \sum_A P_A \tag{14}$$

Here P_A is the partial pressure of constituant A, and q_A is its elementary charge. When all the constituants are in very close interaction with each other their temperatures T_A should be equal.

2.3 Eulerian Versus Lagrangean Description

Note that the quantities (10)-(14) are *fields*, functions of the point in space, $\boldsymbol{r}$, and time t. This is the so-called eulerian description, which concentrates on the evolution as seen at fixed points in space. It is sometimes useful to take the lagrangean point of view, and to ask about the evolution as seen by an observer which follows a given parcel of fluid. Let the quantity of interest be described the eulerian way by a field $g\,(\boldsymbol{r}, t)$. This is the value seen by an observer at $\boldsymbol{r}$ at time t. Now if our observer follows the fluid, which has velocity $\boldsymbol{v}$ at $(\boldsymbol{r}, t)$, for a small time Δt, he will observe around him, at time $t + \Delta t$, a value of g equal to

$$g\,(\Delta t) = g\,(\boldsymbol{r} + \boldsymbol{v}\Delta t, t + \Delta t) \tag{15}$$

Therefore he registers a *time derivative of g following the fluid* given by

$$\frac{Dg}{Dt} = \lim \frac{g\,(\Delta t) - g\,(0)}{\Delta t} = \frac{\partial g}{\partial t} + \boldsymbol{v}.\nabla g \tag{16}$$

where $\partial/\partial t$ denotes the eulerian time derivative, at a fixed point in space, while D/Dt denotes the lagrangean derivative "following the fluid".

A completely lagrangean description would be one in which the space variables would be so defined that they remain unchanged following the motion of the fluid . For example, if one is interested in the spherically symetric motions of a star, the mass m enclosed in a spherical shell can be used as a space variable instead of the distance to the center r. Just change variables from (r, t) to $m = \int_0^r 4\pi r^2 \rho\,(r, t)\, dr$ and t. Obviously m is a lagrangean coordinate, since by mass conservation, it does not change following the motion of any given fluid shell.

3 Eulerian Conservation Equations

Consider a certain extensive physical quantity $\mathcal{G}$, distributed in space with a density g at point $\boldsymbol{r}$. $\mathcal{G}$ can be scalar or vectorial. Let G be the total amount of it contained in an infinitesimal volume V, also near point $\boldsymbol{r}$. Let$\mathcal{B}(V)$ be the boundary of V. We can of course write :

$$G = gV \tag{17}$$

G will change in time. It is a general truth that changes, in a time dt say, can be attributed to two different causes

(a) Import-export : the quantity $\mathcal{G}$ leaks outside or inside of V through the boundary. Such exchanges are characterized, for a scalar quantity, by a flux vector ϕ, defined such that the amount of $\mathcal{G}$ which in a time dt passes through a surface element $d\boldsymbol{S}$ is

$$dG = \phi . d\boldsymbol{S} \;\; dt \tag{18}$$

This quantity is positive if there is a net outflow of $\mathcal{G}$ in the direction of $d\boldsymbol{S}$. The vector ϕ is of course a function of $\boldsymbol{r}$ and t.

(b) Creation-destruction : a number of processes can create or destroy $\mathcal{G}$inside V. For example ionization and recombinations may alter the number of electrons and ions. Such processes are characterized by a volumic rate of creation, s , defined such that in a time dt the quantity of $\mathcal{G}$ in the infinitesimal volume V increases, due to all these processes by

$$dG = s\,V \;\; dt \tag{19}$$

If s is negative, creation will indeed mean destruction; s is a function of ($\boldsymbol{r}$, t).

Explicit expressions for ϕ and s can be obtained from an analysis of the microphysics involved. Adding (18) and (19), the net change dG of G in V during dt is

$$dG = \int_{\mathcal{B}(V)} \phi . d\boldsymbol{S}_{\text{in}} dt \; + \; s\,V\,dt \tag{20}$$

In (20) the normal $d\boldsymbol{S}$ points inwards so that entries into V are counted as positive. Using the flux-divergence theorem (20) is converted into

$$dG = (-\text{div}\,\phi \; + s) \;\; V dt \tag{21}$$

Using (17) and remembering that V is fixed, we relate dG to the change in a time dt of the density g:

$$dG \;\; = \;\; \frac{\partial g}{\partial t} \;\; V\,dt \tag{22}$$

Inserting (21) into (22) gives the basic equation for the density:

$$\frac{\partial g}{\partial t} + \text{div}\,\phi \;\; = \;\; s \tag{23}$$

This is called a conservation equation, or balance equation. Note, however that specific expressions for ϕ and for s should still be obtained by a consideration of the physics involved in the creation and transport processes. Equation(23) only gives the general form of the equation obeyed by the density g, but does not contain by itself any more physical information.

4 Mass Conservation

Mass is neither created nor destroyed in newtonian physics. Then, the only term contributing to the change of mass in V is mass flux. The mass conveyed by the fluid, with velocity $\boldsymbol{v}$ through the surface $d\boldsymbol{S}$ in a time dt fills the volume of the small cylinder of base $d\boldsymbol{S}$ and generatrix $\boldsymbol{v}dt$, which has a volume $d\boldsymbol{S}.\boldsymbol{v}dt$;this can be negative, but it is a good way to keep track of the sense in which matter has gone through $d\boldsymbol{S}$. The mass that has (algebraically) gone through $d\boldsymbol{S}$ in dt seconds is then $\rho\boldsymbol{v}.d\boldsymbol{S}dt$. Comparing with (18) we identify the mass flux as being

$$\boldsymbol{\phi}_{\mathbf{mass}} = \rho\boldsymbol{v} \tag{24}$$

and the mass balance equation (23) can finally be written as:

$$\frac{\partial \rho}{\partial t} + \mathrm{div}\rho\boldsymbol{v} = 0 \tag{25}$$

5 Electric Charge Conservation

Electric charge is neither created nor destroyed, and, by definition, the electric current density $\boldsymbol{j}$ is its flux. Let Q be the charge density, then:

$$\frac{\partial Q}{\partial t} + \mathrm{div}\boldsymbol{j} = 0 \tag{26}$$

6 The Nature of the Flux of a Vector Quantity : Tensors

6.1 The Flux of Components of Vector Quantities

The next quantity we would like to write a conservation equation for is momentum. How should we define the flux of a vector quantity? Let us adopt some system of cartesian coordinates x, y, z. We also use the notation x^a for them, a being an index which can take the values 1, 2, 3. Let G be the vectorial quantity of interest and let G^a be its components. Consider the change of one component of this vector quantity. As for the scalar ones, the variation of the associated density g^a involves a rate of creation s^a, and a flux, $\boldsymbol{\phi}^a$. The equation of conservation for this component is:

$$\frac{\partial g^a}{\partial t} + \mathrm{div}\boldsymbol{\phi}^a = s^a \tag{27}$$

The flux $\boldsymbol{\phi}^a$, which is a vector, has three components, which we denote as

$$\phi^{ba} = b - component\,of\,flux\,of\,a - component\,of\,\boldsymbol{G} \tag{28}$$

Now, since there is nothing special with the coordinate system chosen, we ask what happens if we change for another one. For simplicity , let us restrict to a cartesian one, and denote the new coordinates x', y', z', by $x^{a'}$, a' being an index taking the values 1, 2, 3. The superscript "prime" on a' reminds us that this refers to new coordinates. Old and new coordinates, as well as old and new vector components, those of G in particular, are related by a linear relation represented by a transformation matrix X, with elements $X_a^{a'}$. To make our notation more efficient, we use the *Einstein dummy-index convention* which states that, if in a formula some index is repeated twice, this implicitly means a summation on the values it can take , even though the summation operator Σ is not explicitly written. For example, the relation between the old and new coordinates can equivalently be written as

$$G^{a'} = \sum_{a=1}^{3} X_a^{a'} G^a = X_a^{a'} G^a \tag{29}$$

Since the flux of a sum is obviously the sum of the fluxes, the component $G^{a'}$ has a flux $\phi^{a'}$ given by

$$\phi^{a'} = X_a^{a'} \phi^a \tag{30}$$

By the definition (28) $\phi^{b'a'}$ is the b'–component of (30), and by the general rule (29),the b'–component of ϕ^a is

$$(\phi^a)^{b'} = X_b^{b'} (\phi^a)^b = X_b^{b'} \phi^{ba} \tag{31}$$

Using (31) the b'–component of (30) can finally be written as

$$\phi^{b'a'} = X_b^{b'} X_a^{a'} \phi^{ba} \tag{32}$$

6.2 Tensors

We see that under a change of coordinates,the nine numbers ϕ^{ab} experience the specific type of transformation (32), which is rather similar to the one, (29), which holds for vector components, except that two transformation matrices instead of one are involved. Just as (29) reveals that G^a,$G^{a'}$ are the old and new components of a geometrical object,the vector$\boldsymbol{G}$, equation (32) can be regarded as evidence that ϕ^{ab}, $\phi^{a'b'}$ are the old and new components of some other kind of geometrical object, more complex than a vector,that needs two indices instead of one to label its components. Such an object is called a tensor of rank 2.The *geometrical* notation for this object will be $\overline{\overline{\phi}}$. Unlike a vector a tensor cannot be easily represented graphically.This should not obscure the fact that the tensor is a geometrical entity indeed,as distinct in nature from the matrix of its components in some reference frame, ϕ^{ab},than a vector is distinct from the column-matrix of its components. Actually, vectors are just tensors of rank 1. The basic transformation law (32) can be generalized trivially to define tensors of rank 3 and higher. It is also easy to define the tensor $\lambda\overline{\overline{A}} + \mu\overline{\overline{B}}$ as being the object of components $\lambda A^{ab} + \mu B^{ab}$ (check that the transformation law (32) holds true). Also if $\boldsymbol{A}$ is a vector and $\boldsymbol{B}$ another one, the object represented by components $C^{ab} = A^a B^b$ is indeed a second rank tensor, $\overline{\overline{C}}$, which is said to be the tensorial product of $\boldsymbol{A}$ by $\boldsymbol{B}$, noted $\overline{\overline{AB}}$ or $\overline{A} \otimes \overline{B}$. To prove this, again check that the transformation law (32) holds true. Generalization to tensorial product of higher rank tensors is trivial.

Finally consider some tensor A, of any rank,and another one, not necessarily of the same rank B. To be specific, suppose that A is second rank and B first rank (a vector). Let us define a contraction operation which associates to A and B the object of components

$$C^a = A^{ab} B^b \tag{33}$$

This object has components labeled by one index.It can be shown that it is indeed a tensor of rank 1, a vector, since (29) holds true for the components of C. This operation on tensors is called the contraction of $\overline{\overline{A}}$ by $\boldsymbol{B}$, and is denoted by a dot placed between them, like $A.B$. The contraction can apply to any pair of tensors, whatever their rank,and concerns a selected pair of indices, like

$$C^{abc} = A^{abm} B^{cm} \tag{34}$$

where the dummy summation is on m. The result of a contraction depends on the pair of indices involved in it. Contraction plays a role similar to the scalar product for vectors. The contraction of two vectors, a tensor of rank zero, i.e. a scalar, actually is just their scalar product. To prove that (33) or (34) indeed satisfy the transformation rules (29) or (32),use should be made of the fact that $X_a^{a'}$ is an orthogonal matrix, since we transform from one cartesian system to another. For such matrices,the transpose is also the inverse, so, δ_{ab} being the Kronecker symbols (equal to 1 if a=b, and zero otherwise), we have :

$$X_a^{m'} X_b^{m'} = \delta_{ab} \tag{35}$$

The object which in any frame has for components the Kronecker symbols δ^{ab} turns out to be a second rank tensor called the unit tensor, $\overline{\overline{\delta}}$.

Define the operator $\partial/\partial x_a$, noted ∇_a. These quantities transform under rule (29), and can be regarded as the components of a vector operator ∇.

Tensor components in more general coordinate systems(cylindrical,spherical, etc..) can be defined at the expense of some more mathematical complexity. We deliberately avoided this route here.This has the consequence that some of the definitions given above cannot be directly carried over to more general coordinates, the use of which would make it necessary to define for each tensor a number of different kinds of components,the "covariant" and "contravariant" ones. In general coordinates it is important to clearly distinguish between those, and this is made by placing the component indices at either upper or lower positions.However such a distinction is irrelevant to cartesian coordinates,where these different components are identical, and in this case the place of indices does not matter.

6.3 Conservation Equations for Vector Quantities

Let us now sum up the main conclusions of this long digression: The flux of a vector quantity $\boldsymbol{G}$, with vector density $\boldsymbol{g}$, is a tensor of rank 2, $\overline{\overline{\phi}}$. Its rate of creation is a vector quantity, $\boldsymbol{s}$. Equation (27), which expresses the law of evolution of the components of $\boldsymbol{g}$, can, with the definition (28) and that of the nabla operator, ∇,be rewritten as

$$\frac{\partial g^a}{\partial t} + \nabla_b \phi^{ba} = s^a \tag{36}$$

The quantities ϕ^{ba} are the components of the flux tensor $\overline{\overline{\phi}}$, while $\nabla_b \phi^{ba}$ is the a-component of the vector which results from the contraction of vector ∇ with tensor $\overline{\overline{\phi}}$. This can be noted as $\nabla.\overline{\overline{\phi}}$, or, else, as div$\overline{\overline{\phi}}$. Then,the equations of conservation (28) for the components of our vector quantity can be all summarized in the vector equation:

$$\frac{\partial g}{\partial t} + \operatorname{div} \overline{\overline{\phi}} = s \tag{37}$$

7 Microscopic Fluxes and Transport Coefficients

7.1 Convective Versus Microscopic Fluxes

Equations (23) or (37) give us only the structure of conservation equations. They would be useless if we could'nt precise fluxes and rates of creation in terms of the other basic flow variables. They are an empty structure that we have to fill with some physics!

A given quantity can leak out of a volume V by just being carried away by the fluid in which it is contained : this is a convective flux. It is possible, however, even if the medium is at rest, that individual atoms, drifting with respect to their neighbours, find their way out of the volume such that, when properly averaged out, these microscopic motions result in a certain net flux of the quantity. This is a microscopic flux also called a "diffusive" one. Think,for example, of a drop of ink injected in water. The individual ink particles random walk. Imagine a closed surface surrounding the region where ink has been injected. In the course of their random walk, more ink particles will cross it from inside to outside rather than conversely, simply because there is a slightly larger concentration of them on the inner side. So, a net flux results from these seemingly completely disordered motions, because of the concentration gradient. Let us be a bit more quantitative. In order to calculate the flux of such a minor constituant, we need to know its distribution function $f(\boldsymbol{r}, \boldsymbol{u})$, in terms of which the number flux of these particles is expressed as

$$\phi = \iiint \boldsymbol{u} f(\boldsymbol{r}, \boldsymbol{u})\, d^3u \tag{38}$$

As explained in Sect. 2, when the mean free path is very short ,the distribution function should very nearly be a maxwellian, $f_M(\boldsymbol{u})$, which should be regarded as the solution of the kinetic equation (8) to the zeroth order of approximation in the small parameter λ/L defined in (9), which is sometimes called the Knudsen number, K. If the fluid moves as a whole,this Maxwellian distribution will be drifting at the major constituant's velocity,$\boldsymbol{v}$. At zeroth order of approximation in K, the distribution function is exactly this maxwellian,and the calculation of (38) gives $\phi = n\boldsymbol{v}$, where n is the minor species density. We thus obtain, at this level of approximation, only the convective flux.It takes one order of approximation more to calculate the microscopic flux. Assume,for simplicity,that the global fluid velocity, $\boldsymbol{v}$, is zero (there is always a rest frame where this is true), and let us now take into account the fact that the Knudsen number K is not exactly zero.

7.2 Gradients Cause Distorsions with respect to the Maxwellian Shape

For simplicity, consider a one–dimensional model where particles move along an axis x, and the distribution function is $f(x, u)$. Assume, for the sake of the discussion, that at some initial time, thermalization has worked to completion everywhere, so that $f(x, u)$ is everywhere a maxwellian , with density $n(x)$ and temperature $T(x)$. These might depend on x. It could be thought that this completely thermalized situation should maintain. But, because the mean free path is not zero, this is not so. Actually, to the mean free path λ corresponds a free flight time, or collision time, τ_{coll}, equal to λ divided by the thermal velocity. Let our system, so prepared at the initial time, evolve for a time τ somewhat shorter than $\tau_{\rm coll}$. On such a short time scale, collisions can be entirely neglected, so that the class of particles with velocity in (u, u+du) , starting at the initial time, will free fly, keeping the same velocity, at x+uτ at time τ. Let $f_0(x,u)$ be the initial distribution function and $f_\tau(x,u)$ the one that establishes at time τ by this free flight process. From the preceding remark, we can write that

$$f_\tau(x,u) = f_0(x - u\tau, u) \tag{39}$$

Of course this "solution" will become totally invalid when τ will exceed $\tau_{\rm coll}$, because collisional processes, neglected in (39), will then come into play, with the effect of restauring to a large extent the thermalized situation. Note however that the distribution function at the intermediary time, f_τ, is not maxwellian, even though f_0 was assumed to be. Indeed, ignoring for simplicity temperature gradients we can write f_0 as

$$f_0(x,u) = n(x)\,\varphi_M(u) \tag{40}$$

Taking into account the fact that uτ, which is of the order of the mean free path, λ, is small as compared to the inhomogeneity scale, L, we obtain, by a first order expansion of $n(x - u\tau)$:

$$f_\tau(x,u) = n(x)\,\varphi_M(u) - u\tau\frac{dn}{dx}\varphi_M(u) \tag{41}$$

The second term of (41) represents a distorsion to the perfect maxwellian form induced by the interplay of gradient of density with the finite mean free path. Of course, for $\tau > \tau_{\rm coll}$, collision processes will struggle to restaure thermalization, but the finite mean free path effect will permanently act to rebuild the distorsion. So, in (41), τ is no larger than $\tau_{\rm coll}$, and since dn/dx is of order n/L, we see that the second term on the r. h. s. of (41) is of order λ/L as compared to the first one. An equilibrium between these conflicting processes will eventually be reached, resulting in a slightly distorted distribution, as compared to the maxwellian. This of course remains so only as long as the gradients are maintained against these effects, which tend to smoothen them.

7.3 How to Calculate Transport Coefficients

More rigourously and generally, the actual distribution function should be calculated by solving the relevant kinetic equation , assuming a stationnary state to have been reached, by an approximation procedure valid to first order in the Knudsen number. To illustrate this let us again simplify by assuming the minor species to obey a schematic kinetic equation:

$$\frac{\partial f}{\partial t} + u\frac{\partial f}{\partial x} = -\nu\,(f - \;n(x)\;\varphi_M(u)) \tag{42}$$

This is just a toy–model, which, in the absence of gradients, would restaure thermalization to $\varphi_M(u)$ in a time 1/ ν. Let us dedimensionalize (42) by adopting L, the gradient scale, as the unit of length, so that x=XL, and the thermal velocity associated to φ_M, v_T, as the unit of velocity so that $u = w\,v_T$. Inserting these in (42), the Knudsen number, v_T/νL, appears in factor of the l. h. s. of the equation:

$$K\;w\;\frac{\partial f}{\partial X} = -f + n(X)\;\varphi_M(u) \tag{43}$$

To zeroth order , that is with K=0, the solution is

$$f^{(0)} = n(X)\ \varphi_M(u) \tag{44}$$

To expand to first order write

$$f = f^{(0)} + K\ f^{(1)} \tag{45}$$

substitute in (43) and keep only first order terms in K. Then:

$$w\ \frac{\partial f^{(0)}}{\partial X} = \ -f^{(1)} \tag{46}$$

The solution correct to first order in K, reexpressed in the dimensional variables again is then:

$$f(x,u) \ = \ n(X)\ \varphi_M(u) - \frac{u}{\nu}\frac{dn}{dx}\ \varphi_M(u) \tag{47}$$

which rather expectedly is just the result given by our heuristic guess (41) , with $\tau = 1/\nu$. Inserting (47) in (38) gives the microscopic diffusion flux, to which only the "distorted" part of the distribution function contributes:

$$\phi = -D\frac{\partial n}{\partial x} \tag{48}$$

where the diffusion coefficient D is in this simple model given by:

$$D = \frac{{v_T}^2}{\nu}\int_{-\infty}^{+\infty} w^2 \exp -\frac{w^2}{2}\ \frac{dw}{\sqrt{\pi}} \ = \frac{1}{2}\frac{{v_T}^2}{\nu} \ = \frac{1}{2}\lambda v_T \tag{49}$$

The fact that microscopic fluxes are caused by distorsions to the equilibrium distribution function and are negatively proportionnal to the gradients which cause them is a general property.

7.4 Viscosity

Other types of gradients cause similar distorsions of the distribution funcion. Their form can be guessed as we did before, and a rigourous derivation follows the lines just suggested, but should use a realistic kinetic equation. Temperature gradients, for example, create a higher energy tail in the direction of decreasing temperature, and this distorsion gives rise to heat conduction.

Velocity gradients in shear flows lead to more subtle distorsions, which need at least a space with two dimensions to be properly visualized. They can easily be guessed from the heuristic "free flight in one collision time" argument used above to derive (41). Let, for example the average flow velocity at point $\boldsymbol{r}$, of coordinates (x, y), be given by

$$\boldsymbol{v}(\boldsymbol{r}) = v(y)\,\hat{\boldsymbol{x}} \tag{50}$$

The zeroth order distribution function, assuming no density or temperature gratients, is in this 2–dimensional model:

$$f_0(\boldsymbol{r},\boldsymbol{u}) = \frac{1}{\pi {v_T}^2}\ \exp -\ \frac{(\boldsymbol{u}-\boldsymbol{v}(\boldsymbol{r}))^2}{2{v_T}^2} \tag{51}$$

We guess the distorsion created by the velocity gradient by stating that it should be very nearly the one which forms from (51) when the particles are allowed to free fly for a collision time. So doing, particles at $\boldsymbol{r}$ with velocity $\boldsymbol{u}$ come at time τ at $\boldsymbol{r}+\boldsymbol{u}\tau$, keeping the same velocity. At point $\boldsymbol{r}=0$, the distribution of velocities should then have become at time τ :

$$f_\tau(0,\boldsymbol{u}) = \frac{1}{\pi {v_T}^2} \exp - \frac{(\boldsymbol{u}-\boldsymbol{v}(-\boldsymbol{u}\tau))^2}{2{v_T}^2} \tag{52}$$

Since $\boldsymbol{u}\tau$ is very small we may approximate (50) by a first order expansion

$$\boldsymbol{v}(-\boldsymbol{u}\tau) = -u_y\tau \frac{dv}{dy}\,\hat{\boldsymbol{x}} \tag{53}$$

and, since again the Knudsen number of the problem, $\tau\, dv/dy$, is small we approximate

$$(\boldsymbol{u}-\boldsymbol{v}(-\boldsymbol{u}\tau))^2 = u_x^2 + u_y^2 + 2\, u_x u_y \tau \left(\frac{dv}{dy}\right) \tag{54}$$

Then from (51) we see that the isocontours of $f_\tau(\boldsymbol{u})$ are ellipses, as shown in Fig. 1.

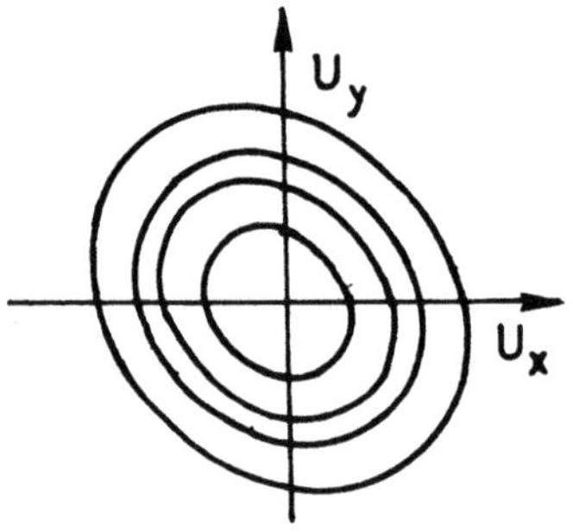

Fig. 1. Isocontours of a 2-dimensional distribution function distorted from maxwellian by a gradient of the average flow $(\partial v^x/\partial y)$.The distorsion, exaggerated in the figure, causes isocontours to become elliptical

An expansion using the smallness of $\tau dv/dy$ gives for (51)

$$f_\tau(\boldsymbol{u}) = f^{(0)}(\boldsymbol{u}) - \left(\tau\frac{dv}{dy}\right)\frac{u_x u_y}{v_T^2}\, f^{(0)}(\boldsymbol{u}) \tag{55}$$

If we now evaluate from (55) the y–x component of the momentum flux tensor, which is given in terms of the distribution function (see also (61)) by

$$\phi_{yx} = \iint m u_x u_y \; f_\tau(\boldsymbol{u})\, du_x\, du_y \tag{56}$$

we observe that only the distorsion term in (55) contributes, and yields the result

$$\phi_{yx} = -\frac{1}{4}\lambda\ v_T\ m\ dv/dy \tag{57}$$

The effect of this transport is to bring in the region $y > 0$ say, momentum of the opposite sign to the ambiant velocity, and this slows the flow there. A similar effect is produced in the region $y < 0$. Altoghether this effect of momentum transport reduces the velocity gradient. It constitutes the phenomenon of *viscosity*.

8 Momentum Conservation Equation

8.1 Rate of Creation of Momentum

The momentum density of matter is $\rho\boldsymbol{v}$. The fundamental law of newtonian mechanics says that in a time dt the momentum of a body changes by $\boldsymbol{F}dt$. Therefore, the force is the rate of change of its momentum, and the volumic rate of momentum creation $\boldsymbol{s}$ is the density of those forces which act in the volume of the fluid. These usually include gravity

$$\boldsymbol{f}_{\mathbf{grav}} = \rho\boldsymbol{g} \tag{58}$$

and electromagnetic forces associated with the macroscopic fields if the fluid is charged or conducts electricity.

$$\boldsymbol{f}_{\mathrm{e.m}} = Q\boldsymbol{E} + \boldsymbol{j} \times \boldsymbol{B} \tag{59}$$

When the fluid is in an intense radiation field, it absorbs and emit photons,exchanging momentum with the radiation. A photon of frequency ν and direction $\boldsymbol{\Omega}$ carries a momentum $(h\nu/c)\ \boldsymbol{\Omega}$. Let the rate of emission of such photons be H $(\nu, \boldsymbol{\Omega})$ photons per unit volume, per second, per unit frequency and solid angle. Let K$(\nu, \boldsymbol{\Omega})$ be the absorption coefficient and n$(\nu, \boldsymbol{\Omega})$ the photon distribution function (6). Then the rate of momentum gain per unit volume from the radiation, that is the force density exerted on the gas by the photons is

$$\boldsymbol{f}_{\mathbf{phot}} = -\iiint \frac{h\nu}{c}\boldsymbol{\Omega}\ \left(H\left(\nu, \boldsymbol{\Omega}\right) - K\left(\nu, \boldsymbol{\Omega}\right) n\left(\nu, \boldsymbol{\Omega}\right)\right) d\nu d\hat{\Omega} \tag{60}$$

8.2 Momentum Flux: Perfect Gas Part

Pressure and viscosity forces are exerted by the neighbourhood and come as flux effects. It is possible to separate two types of such effects. First particles which enter or leave the volume of interest carry with them momentum, and these import/export give rise to a net momentum flux which we refer to as the perfect gas contribution. Second, there is an average force exerted on the volume of interest by the cumulative effect of all the interaction forces between particles inside the volume and particles outside. That this net force can be expressed as the flux through the volume boundary of an "interparticular momentum flux" is not immediatly apparent, but is nevertheless true, as explained below. In gaseous media the perfect gas contribution dominates the interparticular one, and is readily evaluated. Consider a particle species of mass m and distribution function $f(\boldsymbol{u})$. The number-flux of particles having $\boldsymbol{u}$ in $d^3\boldsymbol{u}$ is $\boldsymbol{u}f\left(\boldsymbol{u}\right)d^3\boldsymbol{u}$. Each such particle carries a momentum $m\boldsymbol{u}$. Then the contribution of this category to the b–a component of the momentum flux tensor (28) is

$$d\phi^{ba} = (mu^a)\left(u^b f(\boldsymbol{u})\right) d^3\boldsymbol{u} \tag{61}$$

and the total contribution of this species to the perfect gas part is

$$\phi^{ba} = \iiint mu^a u^b f(\boldsymbol{u})\, d^3\boldsymbol{u} = nm < u^a u^b > \tag{62}$$

Note that $\phi^{ba} = \phi^{ab}$. Separating $\boldsymbol{u}$ into the average part $\boldsymbol{v}$ and a peculiar velocity $\boldsymbol{w}$, as in (4), shows that this flux is made of a convected part and a microscopic part

$$\phi^{ab} = nm < (v^a + w^a)\left(v^b + w^b\right) >= nm\, v^a v^b + nm < w^a w^b > \tag{63}$$

Now,we also know that the distribution function consists, in the hydrodynamical regime,of a (major) equilibrium part plus a small distorsion. The equilibrium part contributes to $nm < w^a w^b >$ by the so called perfect gas pressure, $P_{\rm pg}$

$$nm < \overline{\overline{ww}} >_{\rm eq} = nm \iiint \overline{\overline{ww}}\ \varphi_{\rm eq}(\boldsymbol{w})\ d^3w = P_{\rm pg}\ \overline{\overline{\delta}} \tag{64}$$

The distorsion $\delta\varphi(\boldsymbol{w})$ of the distribution function consists of terms proportionnal to $\nabla n, \nabla T$ and the various velocity gradients $\partial v^m/\partial x^n$. However detailed examination (see Huang 1976,Chap. 5) shows that only velocity gradient terms contribute to momentum flux. Scalar quantities can, in an isotropic medium, only induce vectorial effects. Gradients of vector quantities induce tensorial effects. That part of $\overline{\overline{\phi}}$ which results from this distorsion is the viscous stress tensor, usually denoted as

$$nm < \overline{\overline{ww}} >_{dist} = -\overline{\overline{\sigma}} \tag{65}$$

This tensor is linearly proportional to the velocity gradients and, quite generally its components can be written as

$$\sigma^{ij} = A^{ijkl}\ \nabla_k v_l \tag{66}$$

Where the A^{ijkl} 's are coefficients for which the linearized kinetic theory gives specific expressions. If the medium is isotropic, it is found (Landau and Lifschitz 1959, Huang 1976) that (66) must reduce to the form:

$$\sigma^{ij} = \eta\left(\frac{\partial v^i}{\partial x^j} + \frac{\partial v^j}{\partial x^i} - \frac{2}{3}\delta^{ij}\,{\rm div}\boldsymbol{v}\right) + \zeta\ \delta^{ij}\ {\rm div}\boldsymbol{v} \tag{67}$$

The coefficient η is called the dynamical shear viscosity coefficient, and ζ is the bulk viscosity coefficient or second viscosity. They have dimension $ML^{-1}T^{-1}$. For a gas of structureless particles, like billiard balls, however, the bulk viscosity coefficient is zero. Second viscosity then has to do with the internal structure of the particles. To understand how, note that the bulk viscosity term in (67) is isotropic, i.e. proportionnal to the unit tensor so that it could be added to the pressure in (64), representing a correction to it which is proportionnal to div$\boldsymbol{v}$. According to (25) and (16) div$\boldsymbol{v}$ is just the lagrangean derivative D$\log\rho$/Dt . Now, remember that the pressure in (64) is the perfect thermal equilibrium one, as given by the equation of state. This is the value actually realized if the relaxation of all the processes associated with excitation, ionization or dissociation of atoms and molecules has come to completion. But in fact this takes a finite time, $\tau_{\rm relax}$, which may well be non negligible as compared to the evolution time $\tau_{\rm ev}$ given by ${\tau_{\rm ev}}^{-1} =$

Dlogρ/Dt. Then the actual partial pressures of the constituants may slightly differ of what they ought to be at full thermodynamical equilibrium, because the chemical or ionization equilibrium have not had time enough to establish. Actual pressure then lags behind the perfect thermodynamic equilibrium value. For a constituant in the hydrodynamical regime, the ratio $\tau_{\rm relax}/\tau_{\rm ev}$ is normally small, and so is this lag too. The pressure can then be written in the form of an expansion to first order in this "relaxation Knudsen number", or equivalently as

$$P_{\rm actual} = P_{\rm th.eq} + \zeta \ Dln\rho/Dt \tag{68}$$

where ζ is the expansion coefficient. The second term of (68) turns out to be just the bulk viscosity! Often,it can be neglected.

8.3 Momentum Flux: Interparticular Interaction Forces

Let us now discuss the interparticular momentum flux. Let J and K be two particles sitting at points J and K, and let $\boldsymbol{f}$(J,K) be the force that K exerts on J. The net interparticular force exerted on a volume V is the sum of all the $\boldsymbol{f}$(J,K) 's extended to those J particles sitting in V, and to all the K particles without restriction, since the action/reaction principle causes the contribution of particles which are both in V to cancel. Let X_J be a function equal to 1 if point J is in V and zero otherwise. The net interparticular force on V is

$$\boldsymbol{F} = \sum_J \sum_K \ X_J \ \boldsymbol{f}(J,K) \tag{69}$$

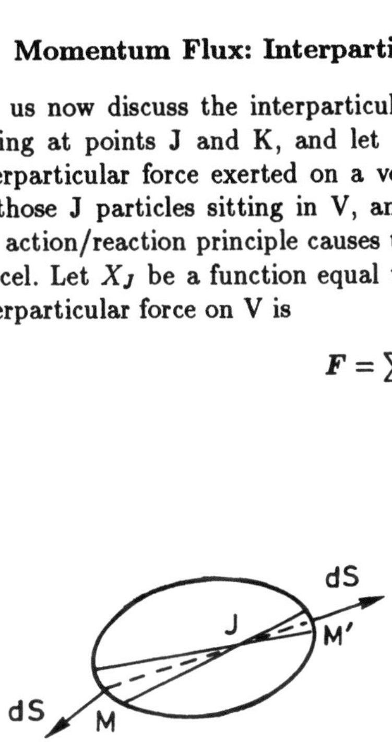

Fig. 2a. When J is inside V $\boldsymbol{JM.dS}$ is always positive and (70) is unity

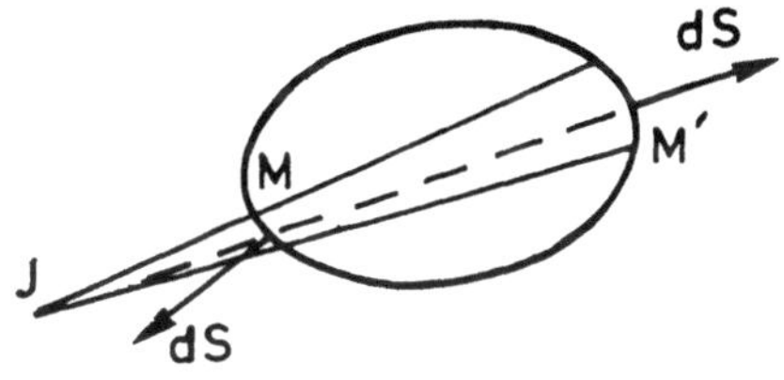

Fig. 2b. When J is outside V $\boldsymbol{JM.dS}$ has both signs and (70) is zero

To show that $\boldsymbol{F}$ can be expressd as a surface integral, note that X_J admits the following integral representation :

$$X_J = \int_{\mathcal{B}} \frac{\boldsymbol{JM.dS}}{4\pi |\ JM\ |^3} \tag{70}$$

That this integral is zero or unity can be seen by noting that, to within the sign of $\boldsymbol{JM.dS}$, the integrand is the solid angle dΩ under which the surface dS is seen from J.The arrangement of the signs when a point M varies on the boundary $\mathcal{B}$ of V is such that the integral vanishes if J is outside V, and does not otherwise(see Fig.2). Inserting

(70) in (69) and averaging by the joint probability $f_{12}(\boldsymbol{r_1},\boldsymbol{r_2})$ that particle J be at $\boldsymbol{r1}$ while particle K be at $\boldsymbol{r_2}$, we obtain :

$$< \boldsymbol{F} >= \int_{\boldsymbol{B}} \overline{\overline{\psi}}.\boldsymbol{dS} \tag{71}$$

where the interparticular momentum flux tensor $\overline{\overline{\psi}}$ is given by :

$$\overline{\overline{\psi}}(\boldsymbol{r}) = \iiint d^3r_1 \iiint d^3r_2\, f_{12}(\boldsymbol{r_1},\boldsymbol{r_2}) \frac{\boldsymbol{r}-\boldsymbol{r_1}}{4\pi|\,\boldsymbol{r}-\boldsymbol{r_1}\,|^3} \otimes \boldsymbol{f}(\boldsymbol{r_1},\boldsymbol{r_2}) \tag{72}$$

Equation (72) simplifies in the case of an homogeneous medium, because f_{12} and $\boldsymbol{f}$ are in this case functions of $\boldsymbol{R} = \boldsymbol{r_2} - \boldsymbol{r_1}$ alone, and $\overline{\overline{\psi}}$ is independant of $\boldsymbol{r}$. To avoid diverging integrals in (72) it is useful to symetrize it by exchanging the variables $\boldsymbol{r_1}, \boldsymbol{r_2}$,using the action/reaction property $\boldsymbol{f}(\boldsymbol{r_1},\boldsymbol{r_2}) = -\boldsymbol{f}(\boldsymbol{r_2},\boldsymbol{r_1})$. This gives:

$$\overline{\overline{\psi}} = -\frac{1}{8\pi} \iiint d^3r_1 \iiint d^3r_2\, f_{12}(\boldsymbol{r_1},\boldsymbol{r_2})\ [\left(\frac{\boldsymbol{r_1}}{r_1^3} - \frac{\boldsymbol{r_2}}{r_2^3}\right) \otimes \boldsymbol{f}(\boldsymbol{R})\ ] \tag{73}$$

or else

$$\overline{\overline{\psi}} = -\frac{1}{8\pi} \iiint d^3r_1 \iiint d^3R\, f_{12}(\boldsymbol{R})\ [\frac{\boldsymbol{r_1}}{r_1^3} - \frac{\boldsymbol{r_1}+\boldsymbol{R}}{|\,\boldsymbol{r_1}+\boldsymbol{R}\,|^{\,3}} \otimes \boldsymbol{f}(\boldsymbol{R})] \tag{74}$$

The integral on $\boldsymbol{r_1}$ can be performed explicitly, taking care not to separate the two terms, since each individually is divergent. By symetry, this integral on $\boldsymbol{r_1}$ has only a component along $\boldsymbol{R}$. Integration on the the components of $\boldsymbol{r_1}$ perpendicular to $\boldsymbol{R}$ is easy. The final result is :

$$\overline{\overline{\psi}} = -\frac{1}{2} \iiint d^3\,R\, f_{12}(\boldsymbol{R})\ [\boldsymbol{R} \otimes \boldsymbol{f}(\boldsymbol{R})] \tag{75}$$

If the medium is furthermore isotropic we obtain

$$\overline{\overline{\psi}} = P_{\text{int}}\, \overline{\overline{\delta}} \tag{76}$$

The interactions among particles then add an "interparticular pressure" to the "perfect gas pressure" of (64). The total pressure is the sum of these two

$$P = P_{\text{p.g}} + P_{\text{ int}} \tag{77}$$

This pressure can be calculated as a function of temperature and density by (64) and (75) or equivalently by the powerful methods of equilibrium statistical mechanics (see Balescu 1975, Chap. 6). For dilute gases the interparticular pressure is a small contribution as compared to the perfect gas part. Nevertheless it may have significant effects, on stellar structure for example.

8.4 Equation of Conservation of Momentum at last!

Summarizing, the momentum equation for a single constituant, A, can be written, now restauring the species index "A":

$$\frac{\partial}{\partial t} \rho_A \, \boldsymbol{v_A} + \mathrm{div} \left(\rho_A \, \overline{\overline{v_A v_A}} + P_A \overline{\overline{\delta}} - \overline{\overline{\sigma}}_A \right) = \boldsymbol{f_A} \tag{78}$$

where $\boldsymbol{f_A}$ incorporates all the volume forces. The general momentum equation for the fluid as a whole is obtained by simply adding the partial contributions (78) using the definition of the global quantities (10)–(14). This is all trivial except for summing the convected fluxes $\rho_A \overline{\overline{v_A v_A}}$. Since the species velocity, $\boldsymbol{v}$ may slightly differ from the mass–center velocity $\boldsymbol{v_A}$ (11) we should write$\boldsymbol{v_A} = \boldsymbol{v} + \boldsymbol{\Delta v_A}$.Note that $\sum_A \rho_A \boldsymbol{\Delta v_A}$ vanishes and obtain :

$$\textstyle\sum_A \rho_A \overline{\overline{v_A v_A}} = \left(\sum_A \rho_A\right) \overline{\overline{vv}} + \sum_A \rho_A \overline{\overline{\Delta v_A \Delta v_A}} \tag{79}$$

Now remark that the Δv_A's are quantities proportionnal to the Knudsen number. Then the second term on the r.h.s of (79) is second order in it and can be neglected. The global momentum conservation equation can then be written as

$$\frac{\partial}{\partial t} \rho \boldsymbol{v} + \mathrm{div} \left(\rho \, \overline{\overline{vv}} + P \, \overline{\overline{\delta}} - \overline{\overline{\sigma}} \right) = \rho \boldsymbol{g} + Q\boldsymbol{E} + \boldsymbol{j} \times \boldsymbol{B} + \boldsymbol{f_{\mathrm{phot}}} + other\ forces \tag{80}$$

This is a very general form of the equation. In practice, in non–relativistic fluids the net charge density Q is negligible, and some fluids will not carry electric currents. Photon forces can sometimes be ignored (see the lecture of L.Lucy to doubt it , however). Using the mass conservation equation (25) it can be shown that

$$\frac{\partial}{\partial t} \rho \boldsymbol{v} + \mathrm{div} \left(\rho \, \overline{\overline{vv}} \right) = \rho \frac{\partial \boldsymbol{v}}{\partial t} + \rho \, (\boldsymbol{v}.\nabla) \, \boldsymbol{v} \tag{81}$$

where $(\boldsymbol{v}.\nabla)$ is the operator which is written in cartesian coordinates as :

$$\boldsymbol{v}.\nabla = v_x \frac{\partial}{\partial x} + v_y \frac{\partial}{\partial y} + v_z \frac{\partial}{\partial z} \tag{82}$$

Caution must be exerted when writing it in other coordinates. Using (82),(81) becomes

$$\rho \left(\frac{\partial \boldsymbol{v}}{\partial t} + (\boldsymbol{v}.\nabla) \, \boldsymbol{v} \right) = -\nabla P + \rho \boldsymbol{g} + \boldsymbol{j} \times \boldsymbol{B} + \boldsymbol{f_{\mathrm{phot}}} + \mathrm{div} \, \overline{\overline{\sigma}} \tag{83}$$

9 Energy Conservation Equation

9.1 Kinetic and Binding Energy

Matter consists of atoms, molecules, ions, free electrons.. Most of these particles have an internal structure and can be found in a number of internal energy states. We refer below to a particle "species" as being a given kind of particle in a given internal energy state. For example we regard an hydrogen atom in the fundamental as being of a different species than one in the first excited state. A species α has its own mass m_α, and internal energy E_α , its own distribution function f_α and density n_α. The energy density associated with it consists of a kinetic energy part

$$\epsilon_\alpha^{\text{kin}} = \iiint m_\alpha \frac{u^2}{2} f_\alpha(\boldsymbol{u})\, d^3u \tag{84}$$

and an internal binding energy

$$\epsilon_\alpha^{\text{bind}} = \iiint E_\alpha f_\alpha(\boldsymbol{u})\, d^3u = E_\alpha n_\alpha \tag{85}$$

These contribute to the energy of the medium even when interaction forces between the atoms can be neglected and constitute a "perfect gas" contribution. There is also potential energy in the medium stored in interparticular potential energy. For simplicity we focus here on the perfect gas part which is more important in gaseous media. Splitting as usual the particle velocity into an average part and a peculiar velocity, as in (4), we find

$$\epsilon_\alpha^{\text{kin}} = 1/2\ \rho_\alpha {v_\alpha}^2 + 1/2\ \rho_\alpha < w^2 >_\alpha \tag{86}$$

The first part is the ordered kinetic energy, and the second part is the thermal kinetic energy. For a dilute gas:

$$\epsilon_\alpha^{\text{th}} = 3/2\ n_\alpha k_B T \tag{87}$$

The sum of (87) and (85) defines the internal energy, U_α.

9.2 Which Independant Macroscopic Variables?

In a given fluid element the number of particles of each species vary with the temperature and the mass density because of the change in the equilibrium reached by the processes of excitation, desexcitation, ionization, recombination, dissociation, ... These processes, as discussed in sect.8 relax to thermal equilibrium on different time scales. For example a Boltzmann distribution of the populations of the various energy levels is essentially established in an inelastic collision time scale, while ionization equilibrrium, as given by the Saha law, or chemical equilibrium, as given by the mass–action law, establishes in a different, usually longer time scale. For some processes , like the ones involving nuclear reactions, the time scale to establish thermodynamic equilibrium in the relative number of the different species may be so long that it is in practice never reached. For other processes, thermalization takes place rapidly, in a time shorter than τ_{ev} , defined in sect.8. Fast and slow relaxation processes , therefore, should not be treated on an equal footing: the kinetics of slow processes requires a kinetic–theory type of description, while thermodynamic equilibrium theory is essentially suitable for the others.As an illustrative example, consider an hypothetical system in which three reacting types of molecules, A, B, C are present, each having a number of energy levels, labeled by numbers i, j, or k.A "species" α is specified by the type of molecule and the energy level. Assume that, for some reason, the reaction rate of $A+B \leftrightarrows C$ is slower than τ_{ev} ,while the Boltzmann distribution establishes rapidly among the energy levels. Obviously the law of mass action will not be satisfied! Then, we should consider the number of molecules A, B or C in the system as variables which evolve "kinematically". This means that their evolution should be followed in time by solving a number–conservation equation, the right hand side of which will express the net rate of creation of molecules of the considered type as a balance of the number created and destroyed, per unit volume and time, due to all the processes we could record. By contrast, the density of molecules in a level i is simply given, in terms of their total density, by the Boltzmann law, which we can write as:

$$n(A,i) = n(A)\ \frac{g_i \exp -\frac{E_i}{k_BT}}{\sum_i g_i \exp -\frac{E_i}{k_BT}} = n(A)\, f_i(T) \tag{88}$$

Here g_i is the multiplicity of level i. If we now calculate (85) and (87) we find that the thermal kinetic energy is

$$\epsilon^{\,\mathrm{kin}} = \sum_{M=A,B,C} 3/2\ n(M)\ k_BT \tag{89}$$

while the binding energy is

$$\epsilon^{\,\mathrm{bind}} = \sum_{M=A,B,C} n(M) \sum_i E_i\ f_i(T) = \sum_{M=A,B,C}\ n(M)\ \psi_M(T) \tag{90}$$

Then the gas energy is a function of temperature and of the three molecule density, n_A, n_B, n_C, which evolve according to an equation which keeps track, by a microscopic analysis of the reaction rates, of the evolution of the chemical reaction. Of course n_A, n_B, n_C are not totally independant. Let, for example, A and B be single atoms and C the molecule AB. If we neglect diffusion the proportions of atoms A with respect to B, whether in atomic or in molecular form, would be known. With proper consideration of such constraints, independant macroscopic variables can be defined. In this case they would be ρ, T and n_{AB} say. On the other hand if the reaction had been fast enough to reach the chemical equilibrium quicker than τ_{ev} ,another relation between T, n_A, n_B and n_{AB} would have applied, namely the law of mass action, and the number of independant variables would have dropped by one as compared to the previous case, reducing to ρ and T alone.

To sum up, the number of macroscopic variables needed to describe the state of the fluid depends on which processes achieve or not thermodynamic equilibrium in the time τ_{ev}, and the evolution of some of these variables require a description by kinetic equations. Above, we considered a chemical reaction; another similar situation is when the gas is maintained by radiation in a non–LTE situation, where the populations of the atomic levels are controlled by a non–thermalized radiation field.The density of atoms in each atomic level becomes in this case itself a macroscopic variable, making the binding energy of the gas a function of these many variables! Another frequently met situation is when nuclear reactions occur, usually at such slow rates that they cannot achieve nuclear–chemical equilibrium. The nuclear binding energy, which is a part of the total binding energy, becomes a function of the many different element abundances , which themselves depend on the past history of the (stellar) gas. Again not a very simple thermodynamic function!

9.3 Energy Flux and Rate of Energy Creation

Let us now come back to the analysis of the flux terms which should appear in the energy balance equation. By a reasoning similar to the one developed in (61)–(63) we find that the energy flux is, for one species

$$\phi = n\ \ < \left(1/2\ mu^2 + e\right) \boldsymbol{u} > \tag{91}$$

This actually should be named the perfect gas part of the energy flux , but our notation ignores this for the moment. Separating as usual in $\boldsymbol{u}$ the global motion and the peculiar velocity $\boldsymbol{w}$ and proceeding as in (63) we find :

$$\phi = 1/2\ \rho v^2 \boldsymbol{v} + U\boldsymbol{v} + n < 1/2\ m w^2 \boldsymbol{w} > + n < (\boldsymbol{v}.m\boldsymbol{w})\ \boldsymbol{w} > \tag{92}$$

where U is the internal energy density

$$U = ne + 1/2\ \ nm < w^2 > \tag{93}$$

The third term in (92), though being odd in $\boldsymbol{w}$ does not vanish because the distribution function used in averaging may be distorted in such a way to be non-symetric in $\boldsymbol{w}$. The distorsion due to temperature gradients in fact causes such a skewness, and gives rise, by this term, to a heat flux proportionnal to ∇T.

$$\boldsymbol{q} = n < 1/2\ m w^2 \boldsymbol{w} > = -\kappa \nabla T \tag{94}$$

The last term of (92) can be rewritten in tensor notation as

$$< (\boldsymbol{v}.m\boldsymbol{w})\,\boldsymbol{w} > = \boldsymbol{v}. < m\,\overline{\overline{\boldsymbol{w}\boldsymbol{w}}} > = \boldsymbol{v}.\overline{\overline{\phi}}_{\mathrm{pg}} = \boldsymbol{v}.\left(P_{\mathrm{pg}}\overline{\overline{\delta}} - \overline{\overline{\sigma}}\right) \tag{95}$$

where we have identified $< m\,\overline{\overline{\boldsymbol{w}\boldsymbol{w}}} >$ as being the perfect gas part of the momentum flux tensor, which, as discussed in sect.8, separates into a pressure and a viscosity part.

The volumic rate of energy creation consists of several terms. On the one hand,volume forces perform work on the particles of the medium. Consider for example gravity and electromagnetic forces acting on a species of mass m and charge q ; they perform work on it , giving rise to an energy production rate :

$$s_{\mathrm{ext}} = n < \boldsymbol{u}.(m\boldsymbol{g} + q\boldsymbol{E} + q\boldsymbol{u} \times \boldsymbol{B}) > = nm < \boldsymbol{u}.\boldsymbol{g} > + nq < \boldsymbol{u}.\boldsymbol{E} > = \rho \boldsymbol{v}.\boldsymbol{g} + Q\boldsymbol{v}.\boldsymbol{E} \tag{96}$$

Summing on the different species we obtain :

$$s_{\mathrm{ext}} = \rho \boldsymbol{v}.\boldsymbol{g} + \boldsymbol{j}.\boldsymbol{E} \tag{97}$$

The interparticular force discussed in sect.8 also works on the particles. The total work performed on a volume V is given in the same spirit and notations as (69) by :

$$s_{\mathrm{int}} = < \sum_J \sum_K \ X_J\ \boldsymbol{u}_J.\boldsymbol{f}(J,K) > \tag{98}$$

Introducing the representation (70) of the characteristic function X_J and averaging as in (72), (98) can be transformed into a surface integral

$$s_{\mathrm{int}} = \iint d\boldsymbol{S}\,.\,\overline{\overline{\psi}}\,.\,\boldsymbol{v} \tag{99}$$

where $\overline{\overline{\psi}}$ is the interparticular momentum flux (see(71)). For homogeneous isotropic media the latter reduces to a pressure P_{int} (76). The r.h.s. of (99) is then $P_{\mathrm{int}}\,\boldsymbol{v}$ and it can be simply added to (95).

At this stage, the total pressure being as defined in(77), our expression of the energy flux is

$$\phi = 1/2\ \rho v^2 \boldsymbol{v} + U\boldsymbol{v} + \boldsymbol{q} + P\boldsymbol{v} - \boldsymbol{v}.\overline{\overline{\sigma}} \tag{100}$$

The interaction between matter and photons gives rise to a net rate of heating of the matter given by :

$$s_{\mathrm{phot}} = -\iiint h\nu \left(H(\nu, \boldsymbol{\Omega}) - K(\nu, \boldsymbol{\Omega})\, n(\nu, \boldsymbol{\Omega})\right) d\nu d\hat{\Omega} \tag{101}$$

If the medium is very opaque, the photons will constitute an hydrodynamical fluid too, and it may be useful to take advantage of the equation of transfer to express s_{phot} as

$$-s_{\mathrm{phot}} = \frac{\partial}{\partial t} U_{\mathrm{phot}} + \mathrm{div}\ \boldsymbol{\phi}_{\mathrm{phot}} \tag{102}$$

where U_{phot} and $\boldsymbol{\phi}_{\mathrm{phot}}$ are energy densities and fluxes of photons,which take a particularly simple expression for quasi–thermalized radiation. Some care must be exerted to take the fluid motion properly into account in defining these quantities (see Mihalas and Weibel-Mihalas 1984). We do not pursue more on this here.

Gathering all these results, we find from (97) and (100) that, summing over all species, the energy conservation equation can be written as :

$$\frac{\partial}{\partial t}\left(1/2\ \rho v^2 + U\right) + \mathrm{div}\left(1/2\ \rho v^2 \boldsymbol{v} + (U+P)\boldsymbol{v} + \boldsymbol{q} - \boldsymbol{v}.\overline{\overline{\sigma}}\ \right) = \rho \boldsymbol{v}.\boldsymbol{g} + \boldsymbol{j}.\boldsymbol{E} + s_{\mathrm{phot}} \tag{103}$$

9.4 More Forms of Energy in the Balance: Gravitational and Others

If the gravitational field is due to sources independant of the moving fluid,we can further manipulate this expression by noting that, G being the gravitational potential

$$\rho \boldsymbol{v}.\boldsymbol{g} = -\rho \boldsymbol{v}.\nabla G = -\mathrm{div}\,(\rho G\ \boldsymbol{v}) + G\ \mathrm{div}\,(\rho \boldsymbol{v}) = -\mathrm{div}\,(\rho G\ \boldsymbol{v}) - \frac{\partial}{\partial t}\ \rho G \tag{104}$$

Inserting this in (103) gives a conservation for kinetic, internal and gravitational energy

$$\frac{\partial}{\partial t}\left(1/2\ \rho v^2 + U + \rho G\right) + \mathrm{div}\left(1/2\ \rho v^2 \boldsymbol{v} + \rho G \boldsymbol{v} + (U+P)\boldsymbol{v} + \boldsymbol{q} - \boldsymbol{v}.\overline{\overline{\sigma}}\ \right) = \boldsymbol{j}.\boldsymbol{E} + s_{\mathrm{phot}} \tag{105}$$

Inserting (102) in (105) would extent the forms of energy included in the conservation equation to photons.Using Maxwell's equations, $\boldsymbol{j}.\boldsymbol{E}$ itself could be transformed in the famous form "$\partial/\partial t$ plus div" , allowing to merge the electromagnetic energy density and flux with the terms already present in (105).

9.5 Less Forms of Energy in the Balance: Kinetic and Internal

Conversely, instead of globalizing the forms of energy we write the balance equation for, we may wish to focus on some particular forms of it. Equation (103) is the balance equation for the kinetic and internal energy. An equation for the kinetic one alone is obtained by dotting the momentum equation (83) with $\boldsymbol{v}$. This gives, after a few algebraic manipulations :

$$\frac{\partial}{\partial t}\left(1/2\ \rho v^2\right) + \mathrm{div}\left(1/2\ \rho v^2 \boldsymbol{v}\right) = \boldsymbol{v}.\left(-\nabla P + \rho \boldsymbol{g} + Q\boldsymbol{E} + \boldsymbol{j} \times \boldsymbol{B} + \boldsymbol{f}_{\mathrm{phot}} + \mathrm{div}\overline{\overline{\sigma}}\right) \tag{106}$$

Substracting (106) from (103) we obtain, after a little more algebra, the conservation equation for internal energy :

$$\frac{\partial U}{\partial t} + \mathrm{div}\,(U\boldsymbol{v} + \boldsymbol{q}) = -P\ \mathrm{div}\boldsymbol{v} + \frac{j^2}{\sigma_e} + (\overline{\overline{\sigma}}.\nabla).\boldsymbol{v} + s_{\mathrm{phot}} - \boldsymbol{v}.\boldsymbol{f}_{\mathrm{phot}} \tag{107}$$

where σ_e is the electrical conductibility . The terms on the r.h.s of (107) are readily understood. $-P\mathrm{div}\boldsymbol{v}$, since $\mathrm{div}\boldsymbol{v}$ is $DlnV/Dt$, V being the specific volume, $1/\rho$, represents internal energy created by the work of the pressure force, j^2/σ_e is the Joule dissipation, $(\overline{\overline{\sigma}}.\nabla).\boldsymbol{v}$ represents the frictionnal heating, and $s_{\mathrm{phot}} - \boldsymbol{v}.\boldsymbol{f}_{\mathrm{phot}}$ is that part of the energy exchanged with the photons that does not go into fluid acceleration.It can be checked by direct algebra that $(\overline{\overline{\sigma}}.\nabla)\ .\boldsymbol{v}$ is always a positive quantity.

10 Second Law and Entropy Balance

The change of entropy of a definite piece of fluid is given by the first law of thermodynamics

$$d\mathcal{U} = Td\mathcal{S} - Pd\mathcal{V} + \textstyle\sum_\gamma \mu_\gamma d\mathcal{N}_\gamma \tag{108}$$

where γ is an index which refers to the various constituants of the system. For the sake of a more fluent notation, we use the dummy index rule for γ too. Following this piece of fluid for a time dt and noting that d/dt will really mean the lagrangean derivative, we find that

$$\frac{D\mathcal{U}}{Dt} = T\frac{D\mathcal{S}}{Dt} - P\frac{D\mathcal{V}}{Dt} + \mu_\gamma \frac{D\mathcal{N}_\gamma}{Dt} \tag{109}$$

Let us introduce the densities of entropy, energy and constituant γ by

$$\mathcal{S} = s\mathcal{V}; \qquad \mathcal{U} = U\mathcal{V}; \qquad \mathcal{N}_\gamma = n_\gamma \mathcal{V} \tag{110}$$

and remember that

$$\mathrm{div}\ \boldsymbol{v} = \frac{1}{\mathcal{V}}\frac{D\mathcal{V}}{Dt} \tag{111}$$

to obtain

$$\frac{DU}{Dt} + (U+P)\ \mathrm{div}\ \boldsymbol{v} = T\frac{Ds}{Dt} + Ts\ \mathrm{div}\ \boldsymbol{v} + \mu_\gamma \frac{Dn_\gamma}{Dt} + \mu_\gamma n_\gamma\ \mathrm{div}\ \boldsymbol{v} \tag{112}$$

Inserting (107) into (112) we obtain after some simple algebra and ignoring photon terms for simplicity

$$T\left(\frac{\partial s}{\partial t} + \mathrm{div}\ s\boldsymbol{v}\right) + \mu_\gamma\left(\frac{\partial n_\gamma}{\partial t} + \mathrm{div}\ n_\gamma \boldsymbol{v}\right) = -\ \mathrm{div}\ \boldsymbol{q} + \frac{j^2}{\sigma_e} + (\overline{\overline{\sigma}}.\nabla).\boldsymbol{v} \tag{113}$$

The term associated with the chemical evolution can be further expressed from a conservation equation for the density of constituant γ. The latter can be written as

$$\partial n_\gamma/\partial t + \mathrm{div}\,(n_\gamma \boldsymbol{v} + \boldsymbol{D}_\gamma) = \textstyle\sum_R w_R\ \nu_{R\gamma} \tag{114}$$

where $\boldsymbol{D}_\gamma$ is the diffusion flux of constituant γ. Such a flux is induced by concentration and thermal gradients, as shown in sect.7 and is given by a Fick's law. R is an index that labels the various reactions in which the constituant γ participates, and $\nu_{R\gamma}$ is its stoechiometric coefficient in this reaction; for example, in the reaction $2H + O \rightarrow H_2O$, one H_2O is created for one O and two H's destroyed, then $\nu_{H_2O} = 1$, $\nu_O = -1$, $\nu_H = -2$.

w_R, the velocity of the reaction is the (common) value of $\nu_{R\gamma} dn_\gamma/dt$ for all constituants participating the reaction. Substituting (114) in (113) gives a balance equation for the entropy density

$$\frac{\partial s}{\partial t} + \operatorname{div} s\boldsymbol{v} = -\frac{\operatorname{div} \boldsymbol{q}}{T} + \frac{j^2}{\sigma_e T} + \frac{(\overline{\overline{\sigma}}.\nabla).\boldsymbol{v}}{T} + \frac{\mu_\gamma}{T} \operatorname{div} \boldsymbol{D}_\gamma - \frac{\mu_\gamma}{T} \sum_R w_R \, \nu_{R\gamma} \qquad (115)$$

This can be arranged in the form

$$\frac{\partial s}{\partial t} + \operatorname{div}\left(s\boldsymbol{v} + \frac{\boldsymbol{q}}{T} - \frac{\mu_\gamma}{T}\boldsymbol{D}_\gamma\right) = \boldsymbol{q}.\nabla\frac{1}{T} + \frac{j^2}{\sigma_e T} + \frac{(\overline{\overline{\sigma}}.\nabla).\boldsymbol{v}}{T} - \boldsymbol{D}_\gamma.\nabla\frac{\mu_\gamma}{T} - \sum_R w_R \, \nu_{R\gamma} \frac{\mu_\gamma}{T} \qquad (116)$$

Equation(116) defines the flux of entropy, which contains parts due to convection, conduction and diffusion

$$\phi_S = s\boldsymbol{v} + \frac{\boldsymbol{q}}{T} - \frac{\mu_\gamma}{T}\boldsymbol{D}_\gamma \qquad (117)$$

and it defines the rate of entropy production per unit volume

$$s_S = -\frac{1}{T^2}\, \boldsymbol{q}.\nabla T + \frac{j^2}{\sigma_e T} + \frac{(\overline{\overline{\sigma}}.\nabla).\boldsymbol{v}}{T} - \boldsymbol{D}_\gamma.\nabla\frac{\mu_\gamma}{T} - \sum_R w_R \, \nu_{R\gamma}\frac{\mu_\gamma}{T} \qquad (118)$$

According to the second law of thermodynamics the change in entropy of a system should exceed the entropy it receives or ceases in exchanges with its environment. This means that the volumic ratye of entropy production must be a positive quantity. As stressed above, the positivity of most usual transport coefficients, like thermal conductivity, or viscosity coefficients ensures that this is so. Alternatively, the second law may be viewed as imposing that this must indeed be so. For example we deduce from the second law that $\sum_R w_R \, \nu_{R\gamma} \, \mu_\gamma$ be negative.

11 Summary of Fluid Dynamics Equations

The set of hydrodynamical equations provides a description of the velocity field, of the mass density and of temperature, and possibly of other variables also, such as the abundances of chemical or nuclear elements, if these are not just functions of ρ and T.If the fluid is coupled to other physical quantities, such as macroscopic electromagnetic fields or photons, the equations which control the evolution of those, such as the Maxwell equations or the transfer equation should be added to the set of hydrodynamical equations. Note that if the photons are very coupled to the gas, when the optical thickness of one gradient scale is very large, they too can be regarded as a constituant of the fluid.The hydrodynamical equations themselves comprise

(a) The conservation equations for mass,momentum and energy, which determine the evolution of $\rho, \boldsymbol{v}, T$,

(b) If ρ and T are not the only independant variables, suitable conservation equations describing the evolution of the other ones, such as abundances for example ,

(c) All the necessary equations of state, which give secundary thermodynamic quantities, such as pressure or internal energy, in terms of the independant ones, usually ρ, T and the abundances,

(d) All the necessary expressions,once derived from linearized kinetic theory, which give the transport coefficients, such as thermal conductibility or viscosity, for the type of medium considered.

To give a specific example, consider a gas moving in a supposedly known gravitational potential G, carrying no electric charge nor current Assume the nuclear composition to be known and matter to consist of neutral atomic gas only, suffering no chemical reactions. Ignore bulk viscosity. Ignore the interaction with photons . Then the state of the gas depends on ρ and T only. The hydrodynamical equations can be written as:

(a) conservation equations :

$$\frac{\partial \rho}{\partial t} + \mathrm{div}\ \rho \boldsymbol{v} = 0 \tag{119}$$

$$\rho \frac{\partial \boldsymbol{v}}{\partial t} + (\boldsymbol{v}.\nabla)\,\boldsymbol{v} = -\nabla P + \mathrm{div}\overline{\overline{\sigma}} - \rho \nabla G \tag{120}$$

$$\frac{\partial}{\partial t}\left(1/2\ \rho v^2 + U + \rho G\right) + \mathrm{div}\left(1/2\ \rho v^2 \boldsymbol{v} + (U+P)\,\boldsymbol{v} + \rho G\ \boldsymbol{v} + \boldsymbol{q} - \boldsymbol{v}.\overline{\overline{\sigma}}\right) = 0 \tag{121}$$

(b) There are no other independant variables
(c) Equations of state

$$P = P(\rho, T) \qquad U = U(\rho, T) \tag{122}$$

(d) Transport laws and coefficients

$$\boldsymbol{q} = -\kappa(\rho, T)\,\nabla T \tag{123}$$

$$\sigma_{ij} = \eta(\rho, T)\ \left(\nabla_i v_j + \nabla_j v_i - 2/3\ \delta_{ij}\ \mathrm{div}\boldsymbol{v}\right) \tag{124}$$

The functions κ and η which give the transport coefficients are supposedly known.Note that we have written (120) in the more familiar form (83) rather than in the conservation form (80). The equations of stellar structure are obtained the same way but include photons among the constituants of the medium, which contributes the dominant part of the heat conduction.The star is also self-gravitating, and its nuclear composition slowly evolves in time due to ongoing nuclear reactions. This contribute a $\partial U/\partial t$ term in the energy equation. However the evolution of the composition is very slow,so that the star essentially remains in mechanical and thermal equilibrium at all times.We can then take $\boldsymbol{v}$=0 and $\partial/\partial t$=0 in all the equations but for the $\partial U/\partial t$ of the energy equation. That term, however, can be expressed in terms of actual nuclear composition and reaction rates. Since these vary very slowly, on sufficiently short time scales $\partial U/\partial t$ can be regarded as having a constant value,given by the present rate of energy production by the nuclear reactions, ϵ . The time dependance is so concentrated in the equations, referred above under the heading (b), which describe the secular variations of the nuclear composition of the stellar material.

12 Incompressible Limit

12.1 Condition for a Fluid to be Incompressible

An incompressible fluid is one in which the relative change of density is negligible following the motion. Changes in density can be related, by the equation of state, to changes in pressure, since these are related by :

$$\Delta P \simeq \frac{\partial P}{\partial \rho} \delta\rho = c_S^2 \, \Delta\rho \tag{125}$$

where c_S is the speed of sound. On the other hand the changes in the pressure of the fluid are related to velocity changes when electrical MHD forces can be ignored. In order of magnitude

$$\rho \frac{v^2}{l} \simeq \frac{\Delta P}{l} \tag{126}$$

where l is a characteristic flow gradient length, and ΔP is a characteristic variation in space of the pressure of the fluid , which we take as representative of time variations as well. Inserting (126) in (125), we see that $\Delta\rho/\rho << 1$ indeed means

$$v \; << \; c_S \tag{127}$$

So the motion is very nearly incompressible in a fluid if the velocities involved are subsonic.

12.2 Navier Stokes Equation and Vorticity

Several simplifications result. For example, since mass conservation may be written as :

$$\frac{1}{\rho}\frac{D\rho}{Dt} \quad = \quad - \operatorname{div} \boldsymbol{v} \tag{128}$$

it results that if the density variations are negligible, the latter reduces to

$$\operatorname{div} \boldsymbol{v} \; = 0 \tag{129}$$

Also the viscous force, from (12O) and (124), simply becomes

$$\operatorname{div} \overline{\overline{\sigma}} \; = \eta \, \Delta \boldsymbol{v} \tag{130}$$

Then the equation of motion, neglecting gravity, reduces to the famous Navier-Stokes equation for viscous fluids

$$\frac{\partial \boldsymbol{v}}{\partial t} + (\boldsymbol{v}.\nabla)\,\boldsymbol{v} \;\; = \;\; - \frac{\nabla P}{\rho} + \nu \, \Delta \boldsymbol{v} \tag{131}$$

where ν is the kinematic viscosity coefficient (dimension L^2T^{-1})

$$\nu \;\; = \;\; \eta/\rho \tag{132}$$

In an incompressible fluid, since the velocity is divergenceless, its rotational

$$\boldsymbol{\omega} = \operatorname{rot} \; \boldsymbol{v} \tag{133}$$

called the vorticity field carries the same information as the velocity itself (think in terms of Fourier transforms) and therefore plays an important role. Vorticity, assuming constant kinematic viscosity, obeys in an incompressible fluid the equation

$$\frac{\partial \omega}{\partial t} = \text{rot} \quad (\boldsymbol{v} \times \omega) + \nu \Delta \omega + \frac{\nabla \rho \times \nabla P}{\rho^2} \tag{134}$$

If furthermore ρ is constant in space, not just constant following the motion, or if for some reason $\nabla \rho \times \nabla P = 0$, which happens for example if P is a function of ρ alone , then

$$\frac{\partial \omega}{\partial t} = \text{rot} \quad (\boldsymbol{v} \times \omega) + \nu \Delta \omega \tag{135}$$

This equation contains on its r.h.s. an advection term, which results from the $(\boldsymbol{v}.\nabla)\boldsymbol{v}$ term of the equation of motion and a dissipative term. The former dominates the latter if the Reynolds number of the flow, which is the following combination of typical gradient length, velocity and viscosity ν:

$$\mathcal{R} \quad = \quad \frac{vl}{\nu} \tag{136}$$

is very much larger than unity. This number is also the order of magnitude of the ratio of $(\boldsymbol{v}.\nabla)\boldsymbol{v}$ to $\nu \, \Delta \boldsymbol{v}$ in (131).

13 Gas Dynamics in One Dimension

13.1 Linear Motion in One Dimension

We now restrict to a "simple" problem, in which the motions are along and depend on a single space variable x. This one-dimensional gas dynamics is admittedly idealized, but helps to understand causality relations, non linearities and shock formation in fluid dynamics, which are important aspects of it. Viscosity, MHD effects, heat conduction, diffusion of constituants and reactions are ignored. Then (116) reduces to :

$$\frac{\partial s}{\partial t} + \text{div}\, s\boldsymbol{v} \quad = \quad 0 \tag{137}$$

Combining this whith (119) we can show that the specific entropy $\sigma = s/\rho$ is constant following the motion. The equations governing the one-dimensional flow are then :

$$\frac{\partial \rho}{\partial t} + \frac{\partial}{\partial x}\rho v \quad = \quad 0 \tag{138a}$$

$$\rho\left(\frac{\partial v}{\partial t} + v\frac{\partial v}{\partial x}\right) \quad = -\frac{\partial P}{\partial x} \tag{138b}$$

$$\frac{\partial \sigma}{\partial t} + v\frac{\partial \sigma}{\partial x} \quad = \quad 0 \tag{138c}$$

Assume moreover that at time t=0 σ has the same value everywhere, σ_0. Then (138-c) shows that this remains so later on, whith the same value. This is an isentropic flow. Consequently, here , the pressure, which generally is a function of density and specific entropy, behaves as a function of density alone .

$$P \quad = P(\rho, \sigma_0) \quad = P(\rho) \tag{139}$$

First, examine small amplitude motions around a state of rest, characterized by uniform density and pressure, ρ_0, P_0. Linearizing (138 a-b) by writing $\rho = \rho_0 + \rho_1$, $P = P_0 + P_1$ and retaining only first order terms in ρ_1, P_1 and v, we get for these quantities the equations :

$$\frac{\partial \rho_1}{\partial t} + \rho_0 \frac{\partial v}{\partial x} \quad = \quad 0 \tag{140a}$$

$$\rho_0 \frac{\partial v}{\partial t} + \frac{dP_0}{d\rho_0}\frac{\partial \rho_1}{\partial x} \quad = \quad 0 \tag{140b}$$

which can be combined to give the wave equations obeyed by ρ_1 and v :

$$\frac{\partial^2 \rho_1}{\partial t^2} - c^2 \frac{\partial^2 \rho_1}{\partial x^2} \quad = \quad 0 \tag{141a}$$

$$\frac{\partial^2 v}{\partial t^2} - c^2 \frac{\partial^2 v}{\partial x^2} \quad = \quad 0 \tag{141b}$$

where c_0, the sound speed of the medium of density ρ_0, is defined by :

$$c_0^2 = dP/d\rho_0 = P'(\rho_0) \tag{142}$$

where P' denotes the derivative of the function defined in eq (139). The general solution of eq (141) is a combination of forward and backward traveling waves.

$$\rho_1(x,t) = \rho_+(\xi_+) + \rho_-(\xi_-) \tag{143a}$$

$$v(x,t) = v_+(\xi_+) + v_-(\xi_-) \tag{143b}$$

where ρ_+, ρ_-, v_+, v_- can be arbitrary functions. Here, however they are related by (140 a-b), and are constrained by initial conditions. The variables ξ_+ and ξ_- are the following combinations of space and time variables :

$$\xi_+ = x \; - \; c_0 t \tag{144a}$$

$$\xi_- = x \; + \; c_0 t \tag{144a}$$

The choice of signs in the indexation is to remind us that ξ_+ is constant on travelling forward at the sound speed. Similarly ξ_- is constant on travelling backwards. From (140 a) ρ_+, v_+ and ρ_-, v_- are related by :

$$c_0\rho_+ = \; + \rho_0 v_+ \tag{145a}$$

$$c_0\rho_- = \; - \rho_0 v_- \tag{145b}$$

If the medium is infinite, and if we know initial conditions at time t=0, the waves v_+ and v_- can be expressed in terms of the supposedly known functions $v(x,0)$ and $\rho_1(x,0)$ Actually, at that time, ξ_+ and ξ_- are both equal to x , and we can write :

$$v(x,0) \; = \; v_+(x) + v_-(x) \tag{146a}$$

$$c_0 \frac{\rho_1(x,0)}{\rho_0} \; = \; v_+(x) - v_-(x) \tag{146b}$$

from which the functions v_+ and v_- can be expressed in terms of initial conditions. Since v_+ keeps the same value when travelling forward at speed c_0, and v_- when travelling backwards, as (144) indicates, the solution at later times can be very easily constructed from (146), (145) and (143).

13.2 Riemann Invariants and Characteristics

Consider now situations such that changes in density are not small as compared to the average and v is not small as compared to the sound speed. The latter, $c(x,t)$, is now a function of space and time defined by :

$$c^2 = P'(\rho(x,t)) \tag{147}$$

where P' is the derivative of the function defined in (139). The problem (138 a), (138b) cannot be linearized any more. Nevertheless, as we now show, a similar construction of its solution, though less explicit than (143), is still possible. It is based on the recogintion that some combination of flow properties are propagated unaltered along "lines" in the x-t plane which generalize the concept of forward and backward traveling waves defined in (144). We start from (138 a-b), taking (139) and (147) into account which gives :

$$\frac{\partial P}{\partial t} = c^2 \frac{\partial \rho}{\partial t} \qquad \frac{\partial P}{\partial x} = c^2 \frac{\partial \rho}{\partial x} \tag{148}$$

and multiply (138 a) by $c^2/\rho c$ in order to match its dimension with (138 b). This gives in place of (138a) :

$$\frac{1}{\rho c}\frac{\partial P}{\partial t} + \frac{v}{\rho c}\frac{\partial P}{\partial x} + c\frac{\partial v}{\partial x} = O \tag{149}$$

while (138 b) can be written as

$$\frac{\partial v}{\partial t} + \frac{c}{\rho c}\frac{\partial P}{\partial x} + v\frac{\partial v}{\partial x} = O \tag{150}$$

Substract and add (149) and (150) to obtain :

$$\left(\frac{\partial v}{\partial t} + \frac{1}{\rho c}\frac{\partial P}{\partial t}\right) + (v+c)\left(\frac{\partial v}{\partial x} + \frac{1}{\rho c}\frac{\partial P}{\partial x}\right) = 0 \tag{151a}$$

$$\left(\frac{\partial v}{\partial t} - \frac{1}{\rho c}\frac{\partial P}{\partial t}\right) + (v-c)\left(\frac{\partial v}{\partial x} - \frac{1}{\rho c}\frac{\partial P}{\partial x}\right) = 0 \tag{151a}$$

Note that here P and c are in fact functions of ρ so that $1/\rho c\ \partial P/\partial t$ and $1/\rho c\ \partial P/\partial x$ are in fact the time and space derivatives of a certain function of ρ, call it $H(\rho)$, the differential of which is :

$$dH = \frac{dP(\rho)}{\rho\ c(\rho)} \tag{152}$$

Since ρ, $P(\rho)$ and $c(\rho)$ are functions of eachother, we actually could consider P or c to be the independant variable on which H depends as well. For example in the case of a perfect gas of constant specific heat, we have

$$P = K\rho^\gamma \tag{153}$$

where the constant K actually depends on the specific entropy paramater. Then we have:

$$\frac{dP}{P} = \gamma\frac{d\rho}{\rho} \tag{154}$$

and then from (147) :

$$c^2 = \gamma \frac{P}{\rho} \tag{155}$$

which gives by logarithmic derivation and using (154) :

$$\frac{dP}{P} - \frac{d\rho}{\rho} = 2\frac{dc}{c} = \frac{\gamma - 1}{\gamma}\frac{dP}{P} \tag{156}$$

In this particuliar case (152) becomes using (156) and (155) :

$$dH = \frac{2}{\gamma - 1}\, dc \tag{157}$$

Note that the function H itself is defined up to an additive constant, which corresponds to the arbitrary choice of the lower limit of integration in :

$$H = \int \frac{dP}{\rho c} \tag{158}$$

Let us now define the functions J_+ and J_-, called Rieman invariants by :

$$J_+ = v + H \tag{159a}$$

$$J_- = v - H \tag{159b}$$

Equations (151 a-b) can be written in terms of them as :

$$\frac{\partial J_+}{\partial t} + (v + c)\frac{\partial J_+}{\partial x} = 0 \tag{160a}$$

$$\frac{\partial J_-}{\partial t} + (v - c)\frac{\partial J_-}{\partial x} = 0 \tag{160a}$$

This says that J_+ remains constant in the x-t plane along the line C_+ defined by the differential equation :

$$dx = (v + c) \quad dt \tag{161}$$

while J_- remains constant following the line C_- defined by the differential equation :

$$dx = (v - c) \quad dt \tag{162}$$

Lines C_+ and C_- are called forward and backward characteristics of the flow,. In the case of a flow which consists of a very small departure from a uniform rest state $c \simeq c_0$ and $v << c$,and (161) and (162) reduce to :

$$dx = \pm c_0 \; dt \tag{163}$$

which just corresponds to forward and backward travelling wave motion. In this case the function H defined by (158) with an adopted lower integration limit corresponding to the imperturbed state becomes simply :

$$H = \frac{P_1}{\rho_0 c_0} \tag{164}$$

where P_1 is the pressure perturbation, linearly related to density perturbation ρ_1 by (147), whence, in the linear approximation :

$$J_{\pm} = v \pm c_0 \frac{\rho_1}{\rho_0} \tag{165}$$

Comparing whith (146) we see that the Riemann invariants, J_+ and J_- which, as a general rule remain constant following their corresponding characteristics C_+ or C_-, reduce in the linear approximation to the functions v_+ and v_-, the forward and backward travelling parts of the perturbation, which remain constant in this approximation on travelling forward (resp. backwards) at the unperturbed sound speed c_0, as described by (163).

In the linear approximation, then, the characteristics reduce to the straight lines defined by (163), which are the trajectories of small sound signals in this approximation.

In the general case, the characteristic lines, given by (161)-(162) are not straight anymore, but they still are the path followed in the x-t plane by small localized sound signals, propagating forward and backward, at the local and actual sound speed $c(x,t)$ whith respect to a medium which is itself flowing at the local and actual velocity $v(x,t)$. There is however one important difference : though the characteristics of the linear problem can be found before solving it, this is not so for the non-linear problem, since v and c in (161) and (162) are not a priori known.They actually are the solution we are searching for!

13.3 Solutions to 1-dimensional Flows Using Characteristics

The constancy of J_+ on C_+ and J_- on C_- curves in principle provides a mean to progressively construct the actual solution. Suppose we know $v(x,t)$ and $\rho(x,t)$, that is also $c(x,t)$, at some particular time t, and for all x. Then we also know the slopes of the characteristics at all these points and we can construct at each of them a small piece of the tangents to the characteristics C_+ and C_- which pass there. On each of them we also know the value of the associated invariants J_+ andJ_- . At some point x at a slightly later time $t + dt$, we can in principle easily find, graphically or numerically say, which lines of this network of tangents intersect there , and , since each of them carries along it a definite value of the associated Riemann invariant, we so find the local values of J_+ and J_-. From those we find the local values of v and ρ. This gives in principle a mean to propagate the solution forward in time, and some numerical methods of solution actually are based on this concept .

Characteristics make it easy to understand the causal relations in the flow. This is illustrated in fig (3). Let M be an *event* , that means a point in plane x-t. The properties of the flow at M depend only on those previously taken in the cone of the past of M, which we define as the region situated "below" the two characteristics $C_+$and C_- passing at M. To see this, note that the flow at M is determined by the$J$'s brought there by the two characteristics passing at event M. There is normally only one characteristic of each set passing at any event, since otherwise the associated Riemann invariant would be double-valued. Now , since the J values are properties of these characteristics, we understand that the flow properties at M are indeed determined by anything that influences which characteristics are going to intersect at M, that is by anything that determines their actual shape. In fact the slope of a characteristic C_+ say, passing at M , at some earlier event on it like N, is determined there not only by the J_+ value common to M and N, but also by the J_- value at N, which is brought there along a J_- characteristic. However, the latter , whatever its detailed shape, is entirely below the C_- passing at M, since it cannot cross it, as noted above, and entirely on the right hand side of the C_+ passing at M : a moment's thought actually shows that two given $C_+$and C_- cannot cross more than once since at the second crossing the slope dx/dt of the C_- would be larger than that of the C_+, which is contradictory to the differential equations (161) and (162). Then anything that has some bearing on the flow at M is below C_-(M) and on the right of C_+(M), in brief, it is in the cone of the past of M. Similarly the state of the flow at M can only influence the flow at later times at events situated in the cone of the future of M, i.e. above both C_+(M) and C_-(M).

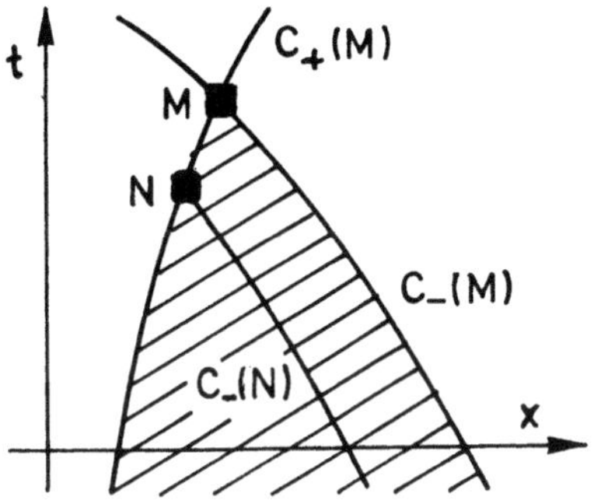

Fig. 3. The shape of the characteristics at M is under the influence of flow variables at events like N and is indirectly influenced by the shaded "cone of the past" only

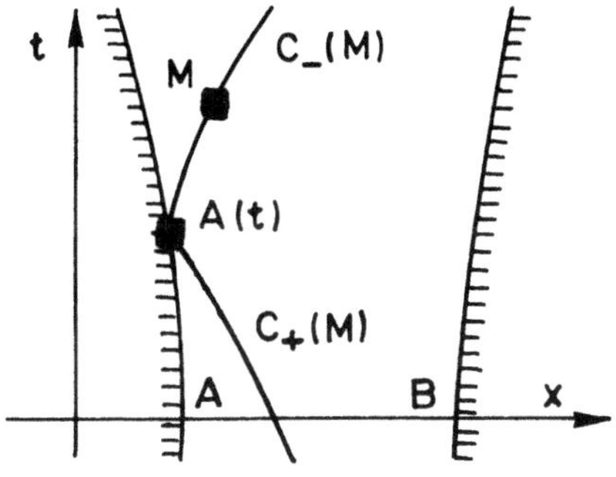

Fig. 4. The characteristic C_+ passing at event M carries a J_+ determined by boundary conditions at A(t).At this event a C_- is reflected into a C_+

Consider now boundary conditions. They have influence on the flow at M if the cone of the past of M contains part of the bondaries at times later than the initial time, as represented in Fig. (4). To be specific assume that the flow is bounded by two pistons in arbitrary motion A(t) and B(t). As shown in Fig.(4) not all the characteristics present in the flow region between the two pistons originate from the initial conditions, that is from some event situated between A and B at time zero. Only for those characteristics that can be traced back to the initial conditions are the associated J values known from the initial state of the flow. Some other characteristics, however, appear in the flow at later times at points situated on the boundary, while at the same point a characteristic of complementary nature ends. It can be said that in Fig. (4) the ending C_- characteristic is reflected into an entering C_+ characteristic. How are we going to know the J_+ value it is going to carry? The key-remark here is that the boundary condition at piston A should be fulfilled, which in this case is that the velocity of the gas there should match the independantly known velocity of the piston, v_* . This imposes that at the reflection event A(t) the following relation hold :

$$v_* = \frac{1}{2}(J_+ + J_-) \tag{166}$$

In this relation v_* is known, but J_+ is not : this then answers the question of determining the invariant attached to the newly appeared characteristic.

Interesting particular cases of one dimensional flow are provided by so called simple waves. Simple waves are the non linear generalization of linear waves travelling in only one direction. For travelling waves one of the two functions v_+ or v_- defined in (143) vanishes. As shown by (165) these functions are just linear approximations to the Riemann invariants. Then the generalization of a travelling wave is a nonlinear wave motion in which one of the invariants J_+ or J_- equals zero everywhere; by conservation on characteristics, this remains so at later times if the medium is infinite. However a Riemann invariant is defined only up to an additive constant, so it does not make any difference whether it really vanishes everywhere or is just constant. So, finally , a simple wave will be defined as one for which one of the two J's is constant in space and time. The flow quantities will in this case depend on a single variable, which is the value, at the event considered, of the other J, the "non-trivial" invariant. Since this variable will univoqually determine all the other flow variables, one could consider as well that these depend, instead of on the Riemann invariant itself, on the value assumed by any other flow variable, the density , or the flow velocity say. Thus, in a simple wave, all the flow variables must be functions of a single one of them : for example the density is a definite function of the velocity. Since along a non trivial characteristic, that is one which does not carry the"trivial"invariant, the non trivial J is, as usual, constant , so will also be all the flow variables , which, in a simple wave, are definite functions of it. This also imply that the slope of such characteristics , equal to$(v + c)$ or $(v - c)$, does not vary following them, and therefore these non trivial characteristics constitute a family of straight lines. This is in general not so for trivial characteristics. Actually, if both J_+ and J_- are uniform in space and time, the flow variables are also uniform, since they are functions of those: uniform flows are the only ones which can possibly have two families of straight characteristics.

Problems involving simple waves are amenable to analytic solutions , since advantage can be taken of the straightness property of one family of characteristics. We illustrate this by considering the following problem : at negative times a fluid has been sitting at rest in the half space x negative, bounded by a movable piston at x=0, which has been also at rest. At t=0 the piston starts to move towards positive x with a supposedly known, differentiable, motion $x = a_*(t)$, and velocity $v_*(t)$. We want to calculate the resulting gas motion. The C_- characteristic of event O (x=0 t=0), defines a region in the (x,t) plane, below it , where the fluid is still at rest, not beeing yet informed by sound signals that the piston has started its motion (see Fig.(5)). Above that line, the C_+ characteristics still carry the J_+ value which existed in the rest state, since these lines connect directly to the initial conditions. In this rest state J_+ was the same everywhere, and this property is carried over to the non-rest state. The motion for $t > 0$ then has the character of a simple wave. Non-trivial characteristics, i.e. the C_- ones since it is J_+ which remains uniform , are straight even in the region of motion but not the C_+'s, however. This is because J_- is not uniform there, since the C_- characteristics in the region of motion are formed by reflection on the moving piston , and by equation (166) each of them is going to carry a different value of J_-, namely

$$J_- = 2\, v_*\,(t_*) - J_+ \tag{167}$$

where t* is the time at which the considered C_- characteristic originate on the moving piston, at $x = a_* = a(t_*)$. The slope of this C_- is $v_* - c_*$, where c_* denotes the speed of sound in the gas at the characteristic's reflection event(a_*, t_*).This quantity can be found from the function H defined in (158) (which is a function of sound speed). H is simply given in terms of the invariants by:

$$H = \frac{1}{2}(J_+ - J-) \qquad (168)$$

For a perfect gas of constant specific heat, (157) gives H in terms of the sound speed:

$$c = \frac{\gamma - 1}{2} H \qquad (169)$$

and from (159) and (167) we obtain

$$c_* = \frac{\gamma - 1}{2} J_+ - v_* \qquad (170)$$

Then the slope of the reflected C_- is just:

$$v_* - c_* = -\frac{\gamma - 1}{2} J_+ + \frac{\gamma + 1}{2} v_* \qquad (171)$$

and the equaton of that characteristic is

$$x - a(t_*) = \left(-\frac{\gamma + 1}{2} J_+ + \frac{\gamma - 1}{2} v(t_*) \right) (t - t*) \qquad (172)$$

If we select an event P of coordinates (x,t) in the region of motion, the time t_* at which the C_- characteristic passing at this event has been reflected off the piston can be found by solving (172) for t_* , x and t being regarded as given parameters, the functions a(t_*) and v (t_*) being known, and the value of J_+ being that which prevails everywhere, in particular in the rest state at negative times, namely

$$J_+ = \frac{2}{\gamma - 1} c_0 \qquad (173)$$

where c_0 is the sound speed in rest state. Once (172) has been solved we find the value of the J_- invariant at event P, given by(167). With (173), this completely determines the invariants at P and then all the flow variables. All the mathematics of the problem then concentrates in solving the, possibly transcendental, equation (172). It can be shown that (172) has a unique solution if the piston's velocity is a monotonously increasing function of time, which remains small enough. If the piston accelerates too much the contact with the gas may be lost, that is the C_+ characteristics may become unable to reach the piston, and we meet a problem of free expansion.This happens when c_* reaches the value zero, in which case the reflected characteristic has the same slope as the incident one. So v_* should not exceed $2/(\gamma - 1)\ c_0$. We leave it to the reader to discuss for himself a few specific examples and to see in detail how the loss of contact between gas and piston may arise.

Suppose that after some time, T, corresponding to boundary event Q in Fig.5), the motion of the piston becomes uniform. Then, from (167), the value of J_- on C-'s reflected at later times will be the same for all of them. This means that a region of uniform flow, where both J_+ and J_- are uniform functions, appears above the C_- characteristic reflected at time T. In the region between that C_- and the one which passes at the event O of coordinates (0,0) is a region of unsteady, non uniform, flow called a rarefaction

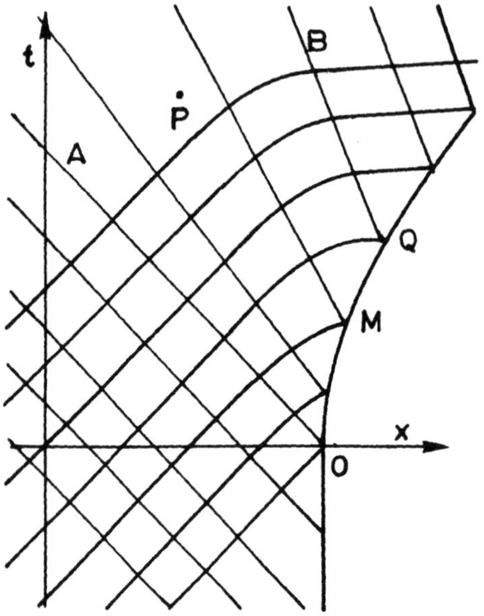

Fig. 5. A map of the characteristics of the flow generated by pulling the rigt boundary (hatched) of a 1-dimensional gas initially at rest.A nonlinear rarefaction wave forms.See text.

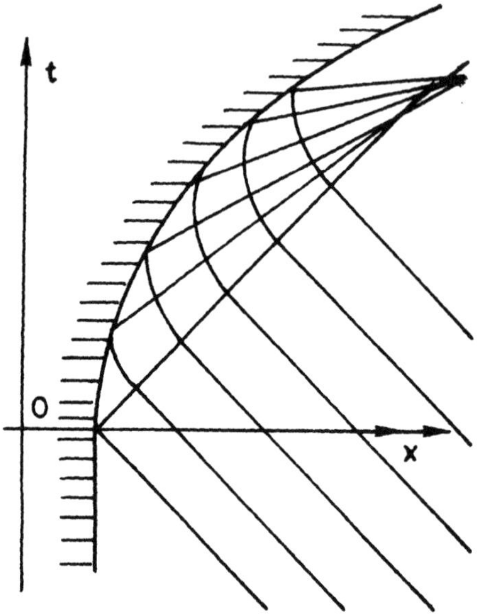

Fig. 6. A map of the characteristics of the flow generated by pushing the left boundary (hatched) of a 1-dimensional gas initially at rest. A shock forms by intersecting characteristics.See text.

wave. Consider now what happens when the period of piston acceleration, T, approaches zero, the final piston velocity, v_{**}, still remaining finite. In that case the network of C_- characteristics in the non uniform region, that is those reflected off the boundary between 0 and T, approaches, when T becomes smaller and smaller, a family of straight lines, all originating at event O, with different slopes, which forms a "centered fan". This limit is somewhat singular in the sense that the piston's motion is not twice derivable anymore, resulting in many C_-'s crossing eachother at O, but it should be remembered that this is only an idealized limit. In this limit, the C_- passing at a given event P , of coordinates x and t, in the fan is not characterized by the time t_* when it reflects off the piston, since in the fan this time is 0 for all C_-'s, but rather by its slope, equal to t/x. The invariant J_-, since it is constant on C_-'s, so appears to be a function of x/t, in the fan region, and so are all the flow variables too. The flow, depending on this unique combination of space and time variables, is said to be self-similar. Its structure can be found by the same reasoning as before, slightly adapted. One equation will express that event (x,t) sits on a C_- characteristic along which the flow velocity is v and the sound speed c, namely, since C_- 's, as a rule, have slope $v - c$, we can write:

$$\frac{x}{t} = v - c \tag{174}$$

The other equation states that J_+ has the same value it has in the rest region :

$$\frac{2}{\gamma-1} c_0 = v + \frac{2}{\gamma-1} c \tag{175}$$

Equations (174) and (175) constitute a pair of equations for v and c at event (x,t). These quantities are seen to indeed depend on (x/t) :

$$c = \frac{2c_0}{\gamma+1} - \frac{\gamma-1}{\gamma+1}\frac{x}{t} \tag{176a}$$

$$v = \frac{2}{\gamma+1}\left(c_0 + \frac{x}{t}\right) \tag{176b}$$

The fan zone, associated with the rarefaction wave is where v is between zero and the terminal value of the piston velocity v_{**}.namely

$$-c_0 < \frac{x}{t} < -c_0 + \frac{\gamma+1}{2} v_{**} \tag{177}$$

A graphical study readily shows that if instead of expanding outwards the piston would have been pushed inwards into the gas, no consistent solution would have been obtained, since for times larger than a certain value the solution of equation (172) becomes multi-valued. This is illustrated in Fig.(6). Something similar happens to a non linear simple wave in an infinite medium which we now describe in some more detail to illustrate the phenomenon of steepening of non linear disturbances.

13.4 Breaking of Non Linear Waves and Shock Formation

To be specific, we consider isothermal thermodynamics ($\gamma = 1$) for which the sound speed c is a constant and the Riemann invariants are ((158), (159))

$$J_{\pm} = v \pm c \ln(\rho) \tag{178}$$

We consider a forward propagating simple wave for which J_- is uniform. This imply that density ρ and velocity v should be related by the relation :

$$\rho = \rho_0 \exp v/c \tag{179}$$

This should in particular be so at the initial time. The C_+ are straigt and have, as always, slope $v+c$. Since these are known at the initial time, the C_+ network can be constructed from a knowledge of the initial conditions. Let us remind that in a simple wave, these lines convey constant values of all the flow variables, in particular the velocity. This property can be expressed in a simple formula, which states that some variable has at (x,t) the same value it has at the initial time on the C_+ which passes at (x,t), where it has a slope $c+v$:

$$v(x,t) = v(x - (c+v)\,t,\ 0) \tag{180}$$

Since c is assumed uniform, we can go to the rest frame which moves at velocity c.Defining $y\ (x,t)= x - ct$, we can write this relation in the equivalent and simpler form :

$$v(y,t) = v(y - vt,\ 0) \tag{181}$$

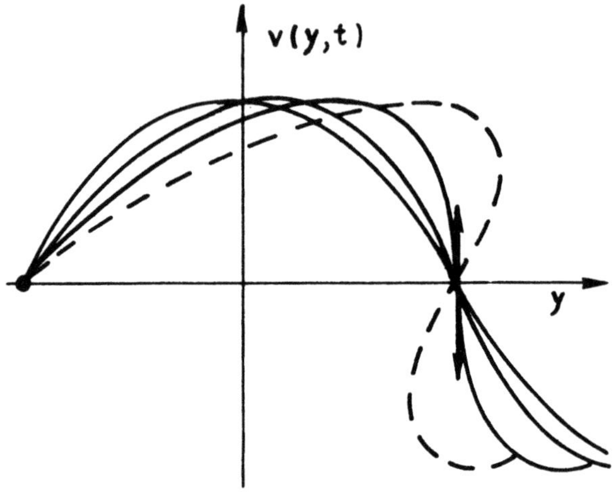

Fig. 7. The graphical construction of the evolution of the velocity profile of a simple non linear wave.Steeper profiles correspond to larger times.When the time becomes too large , the construction gives an unphysical double-valued solution.In fact, a shock has formed when the profile has steepened to vertical.

This indicates a way to construct graphically the solution : plot the initial distribution of velocities $v(y,0)$. To obtain the corresponding profile at time t transport the points which have ordinate v and abcissa x at an abcissa $x+vt$, keeping the same ordinate. This makes it possible to construct the profile at time t point by point. An example of such a construction is shown in Fig. (7). The profile steepens in the vicinity of points where the velocity equals the sound speed and has a negative slope, because faster parcels of fluid tend to overrun slower ones in front of them. For large times the solution so constructed becomes multivalued, which reflects the fact that different C_+ characteristics intersect eachother. This obviously leads to a non-physical solution.

The overturning of the solution however does not arise in reality. This is because the perfect fluid picture, which neglects viscosity, does not make sense anymore when the velocity derivative becomes very large. Neglecting second viscosity, the viscous force acting on a compressible one-dimensional fluid is

$$\mathrm{div}\overline{\overline{\sigma}}\,.\,\hat{x} = 4/3\ \eta \quad d^2v/dx^2 \tag{182}$$

It can be shown that the second derivative of the velocity diverges in the dissipationless solution at the same time as the first derivative does.One can easily check this general property on an example, like $v(y,0)$ $=(1-y^2)$, for which the graphical construction implied by (181) gives overturning at $y=1$. Equation (182) actually imply that v, y and t are related in this case by the relation

$$y = \sqrt{(1-v)} \quad + vt \tag{183}$$

Differentiating once at constant t we obtain dv/dy

$$\frac{dv}{dy} = -\left(\frac{1}{\sqrt{1-v}} - t\right)^{-1} \tag{184}$$

and differentiating once more we get the second derivative:

$$\frac{d^2v}{dy^2} = -\frac{1}{4}\left(\frac{1}{2} - t\sqrt{1-v}\right)^{-3} \tag{185}$$

Both expressions indeed diverge at t= 1/2. How small the viscosity coefficient might be, the viscous force can never be neglected near the point where the gradient diverges, and indeed the growth of the viscous force will limit the growth of dv/dx to a finite, albeit very large value. If nevertheless we want to insist on the zero viscosity limit, we must consider that this force limits the validity of the dissipation-free model when the characteristics intersect, and that its effect is exerted in a small vicinity of the point where the diverging gradients appear. The very large values to which they ultimately saturate because of these dissipative effects appear as the formation of discontinuities in the flow. So, solutions of non linear hydrodynamics problem develop discontinuities of the flow variables in a finite time. Such discontinuities obey physical laws which connect the upstream and downstream values of the velocity, density, temperature, which we establish in the next paragraph. They are the seat of dissipative phenomena. Numerically, discontinuities can be treated as such, using the laws established below, or can be simulated by somehow exaggerating the viscosity in the region of interest; this broadens the region where the gradients are large without breaking the jump relations across the discontinuity.

14 Shocks

Discontinuous features in the flow which involve dissipative effects are called shocks. Shocks also occur in multidimensional flows. The locus of the discontinuities occurring on all the possible stream lines in them traces a surface, generally non planar nor stationnary. However, since the laws which apply to these shock jumps are local relations they can be analyzed in the vicinity of any given point on the shock surface. In such a small region it is enough to represent this surface by its tangent plane, and, though it is in motion, it is always possible to find some rest frame, called a shock frame, where the little bit of surface considered is instantaneously at rest. As discussed earlier, at the shock, none of the flow variables diverges, but their spatial gradients do. Obviously their lagrangean time derivatives do too, but if we place ourselves in a rest frame where the surface of discontinuity is instantaneously at rest, the eulerian time derivatives will be regular. The shock jump relations which we are going to establish express the laws of conservation of mass, momentum and energy in a form integrated on the region of the shock, namely integrated on the volume of a flat box placed on the shock front, as represented in Fig.(8). In doing so, we shall place ourselves in the shock rest frame We cannot neglect dissipative phenomena in the shock region itself, but we shall consider them as negigible away from the shock surface, that is , they will be neglected at the two outer surfaces of the integration box. This assumption could be criticized in actual situations, but here we shall not elaborate further on this point.Shock fronts in heat conducting media, for example, are possible and obey other jump relations than those to be established below.Similarly, astrophysics offers a number of contexts where the situation may be a lot more complicated, with, for example, photons conveying energy to be dumped in the shock region (ionization fronts) .All these situations can be handled the same way as we now do for "classical" shocks.

Let us write again the basic conservation laws :

$$\frac{\partial \rho}{\partial t} + \mathrm{div} \rho \boldsymbol{v} = 0 \tag{186}$$

$$\frac{\partial}{\partial t} \rho \boldsymbol{v} + \mathrm{div} \left(\rho \overline{\overline{\boldsymbol{v}\boldsymbol{v}}} + P \,\overline{\overline{\delta}} - \overline{\overline{\sigma}} \right) = \rho \, \boldsymbol{g} \tag{187}$$

$$\frac{\partial}{\partial t} \left(U + \frac{1}{2} \rho v^2 + \rho G \right) + \mathrm{div} \left(\frac{1}{2} \rho v^2 \boldsymbol{v} + (U + P) \boldsymbol{v} + \rho G \, \boldsymbol{v} + \boldsymbol{q} - \overline{\overline{\sigma}}.\boldsymbol{v} \right) = 0 \tag{188}$$

We integrate them on the flat box of Fig.(8), and take the limit where the thickness of this box is made to vanish : all nondiverging terms, even if discontinuous, disappear in this process. This is so for the gravity force and also, because of our special choice of rest frame , for the time derivative terms. By contrast the divergence terms, which contain singular space derivatives, make a contribution which can be calculated by use of the flux-divergence theorem. Let us precise a few conventions : let $\boldsymbol{n}$ be a unit vector normal to the shock front. It has two possible orientations, at this point equivalent. The side number 2 is the one towards which $\boldsymbol{n}$ is pointing, the side 1 is the other. The small box, of negligible thickness , has two faces of surface S, with surface elements oriented outwards of it, given by:

$$d\boldsymbol{S}_1 = -S \, \boldsymbol{n} \qquad ; \qquad d\boldsymbol{S}_2 = +S \, \boldsymbol{n} \tag{189}$$

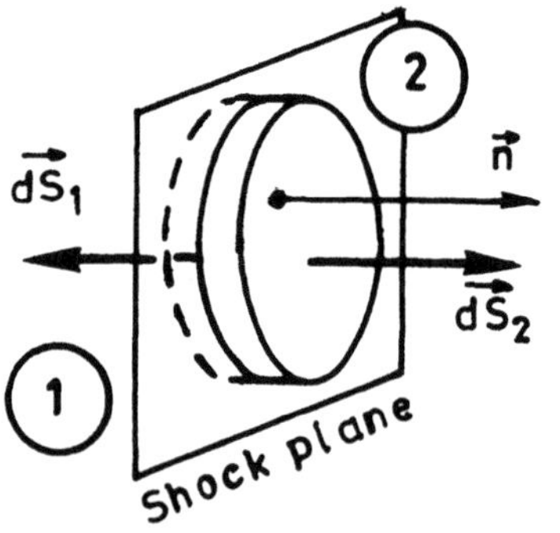

Fig. 8. An illustration of the thin box used as an integration volume to deduce the shock jump relations.The conventions for labeling the two sides and orientating the normals are also represented

Let us introduce the notation

$$\boldsymbol{v}_n \; = \; (\boldsymbol{v}.\boldsymbol{n}) \, \boldsymbol{n} = \; v_n \, \boldsymbol{n} \qquad ; \qquad \boldsymbol{v}_t \; = \; \boldsymbol{v} \; - \; \boldsymbol{v}_n \tag{190}$$

and box-integrate (186-188). We remind the reader that the flux-divergence theorem also applies to tensors. We so obtain :

$$\rho_1 v_{1n} = \rho_2 v_{2n} \tag{191}$$

$$\rho_1 v_{1n}\, \boldsymbol{v}_1 + P_1 \boldsymbol{n} = \rho_2 v_{2n}\, \boldsymbol{v}_2 + P_2 \boldsymbol{n} \tag{192}$$

$$\frac{1}{2}\rho_1 v_1^2 v_{1n} + (U_1 + P_1) v_{1n} + \rho_1 v_{1n}\, G = \frac{1}{2}\rho_2 v_2^2 v_{2n} + (U_2 + P_2) v_{2n} + \rho_2 v_{2n}\, G \tag{193}$$

We have already neglected here the viscosity and heat conduction terms away from the shock as explained above. Advantage can be taken of equation (191) to simplify somewhat the others . Equation (192) being still a vector one , it can be split into a normal and a tangential component. The latter, considering eq (191) simply states that the components of the velocity tangent to the shock front are continuous. This makes it easy to eliminate the tangential kinetic energy from equation (193).Thus:

$$\rho_1 v_{1n} = \rho_2 v_{2n} \tag{194}$$

$$\boldsymbol{v}_{1t} = \boldsymbol{v}_{2t} \tag{195}$$

$$\rho_1 {v_{1n}}^2 + P_1 = \rho_2 {v_{2n}}^2 + P_2 \tag{196}$$

$$\frac{1}{2}{v_{1n}}^2 + \frac{U_1 + P_1}{\rho_1} = \frac{1}{2}{v_{2n}}^2 + \frac{U_2 + P_2}{\rho_2} \tag{197}$$

There is a number of possible "shock-rest frames", in which the front is instantaneously at rest in the n-direction, but which differ in their state of motion in perpendicular directions. According to (195), one can be chosen such that the velocity be normal to the shock, both before and after it. Selecting this rest frame, and omitting the now superfluous subscripts "n", the set (194-197) reduces to the famous Rankine-Hugoniot jump relations, which we particularize here, as is quite usual, to the case of a perfect gas of constant adiabatic index, for which:

$$U = \frac{P}{\gamma - 1} \qquad (U + P) = \frac{\gamma}{\gamma - 1} P \qquad c^2 = (dP/d\rho)_S = \gamma P/\rho \tag{198}$$

The jump relations can then be written as :

$$\rho_1 v_{1n} = \rho_2 v_{2n} \tag{199}$$

$$\rho_1 {v_1}^2 + P_1 = \rho_2 {v_2}^2 + P_2 \tag{200}$$

$$\frac{{v_{1n}}^2}{2} + \frac{\gamma}{\gamma - 1}\frac{P_1}{\rho_1} = \frac{{v_{2n}}^2}{2} + \frac{\gamma}{\gamma - 1}\frac{P_2}{\rho_2} \tag{201}$$

From (199) we see that the normal velocity has the same sign on both sides : matter cannot converge to, nor diverge from the shock. There is then a definite side from which matter enters the shock, the preshock , or upstream, side, and one at which it leaves the shock, the postshock , or downstream, side.Though at this point the jump relations do not introduce any asymetry between these two sides, we shall see shortly that the second law of thermodynamics breaks this symetry. As a convention the side 1 will now always mean the preshock side.

The jump relations (199-201) can be manipulated and reexpressed in a number of ways. A useful one is to arrange them such that postshock flow variables be expressed explicitly in terms of preshock ones. Let us introduce the compression ratio

$$r = \frac{\rho_2}{\rho_1} \tag{202}$$

in terms of which (199) gives the postshock velocity :

$$\frac{v_2}{v_1} = \frac{1}{r} \tag{203}$$

and equation (200) and (199) give the post shock pressure :

$$\frac{P_2}{P_1} = 1 + \frac{\rho_1 v_1^2}{P_1} - \frac{\rho_2 v_2^2}{P_1} = 1 + \frac{\rho_1 v_1^2}{P_1}\left(1 - \frac{1}{r}\right) \tag{204}$$

Using (202)-(203), the energy equation (201) finally gives a relation between the compression ratio and the preshock flow variables

$$\frac{v_1^2}{2}\left(1 - \frac{1}{r}\right)\left(1 + \frac{1}{r}\right) + \frac{\gamma}{\gamma - 1}\frac{P_1}{\rho_1}\left(1 - \frac{1}{r}\right) = \frac{\gamma}{\gamma - 1}\frac{v_1^2}{r}\left(1 - \frac{1}{r}\right) \tag{205}$$

Excluding the"no change" solution r=1, this gives for $(1/r)$ the simple equation

$$\frac{v_1^2}{2}\left(1 + \frac{1}{r}\right) + \frac{\gamma}{\gamma - 1}\frac{P_1}{\rho_1} = \frac{\gamma}{\gamma - 1}\frac{v_1^2}{r} \tag{206}$$

from which r can be expressed in terms of the preshock Mach number of the flow, M_1, defined by

$$M_1^2 = v_1^2/c_1^2 \; ; \qquad c_1^2 = \gamma P_1/\rho_1 \tag{207}$$

Note that M_1 is the ratio of the upstream flow velocity evaluated in the shock frame to the sound velocity in the upstream medium.The compression ratio is given in terms of M_1 by :

$$r = \frac{(\gamma + 1)\, M_1^2}{2 + (\gamma - 1)\, M_1^2} \tag{208}$$

Substituting in (202)-(204) we obtain explicit expressions for the postshock values in terms of preshock ones :

$$\frac{\rho_2}{\rho_1} = \frac{v_1}{v_2} = \frac{(\gamma + 1)\, M_1^2}{2 + (\gamma - 1)\, M_1^2} \tag{209}$$

$$\frac{P_2}{P_1} = \frac{2\gamma M_1^2 - (\gamma - 1)}{\gamma + 1} \tag{210}$$

The postshock state of the gas (P_2, ρ_2), in particular, appears to be expressible in terms of the corresponding preshock values and of the Mach number M_1 which parametrizes the strength of the shock. For a given upstream gas state , it is customary to represent this relation, for different inflow velocities, or Mach numbers, as a locus in the specific volume-pressure plane.The specific volume is the inverse of the density. This locus is an hyperbola, as can be seen by eliminating M_1 between (209) and (210), which is called an "Hugoniot curve" or sometimes a "shock adiabatics", a rather unfortunate terminology, since of course this locus is by no means an adiabat! The Hugoniot curve passes for $M_1=1$ at the preshock state. Expanding around this point it can be seen that the Hugoniot curve osculates there to the adiabat of the preshock state . For very small M_1 the solution does not make sense, giving negative postshock pressures, indicating that not all values of the preshock Mach number are possible. For M_1 approaching infinity, the case of an infinitely strong shock, the post shock pressure approaches infinity, but the compression ratio approaches only the finite value

$$\frac{\rho_2}{\rho_1} \rightarrow \frac{\gamma+1}{\gamma-1} \tag{211}$$

From (209)-(210) we can calculate the postshock Mach number M_2 (defined by (207)) with postshock variables replacing preshock ones)in terms of the preshock one M_1 . This gives:

$$M_2^2 = \frac{2+(\gamma-1)\,M_1^2}{2\gamma M_1^2-(\gamma-1)} \tag{212}$$

Since the jump relations are symetric the relation which gives M_1 as a function of M_2 should be exactly the same, as can be checked.Then the graph of (212) in the M_1-M_2 plane is symetric to the main diagonal. It goes through the point (1, 1), and for M_1 very large M_2^2 approaches the value $(\gamma-1)/2\gamma$. It is also easy to check on this relation that the transition through a shock is necessarily transsonic, that is, if $M_1 > 1$, then $M_2 < 1$, and conversely : there is a subsonic side and a supersonic side.In fact, as we now show, it is forbidden by the second law of thermodynamics that the preshock state be the subsonic one.

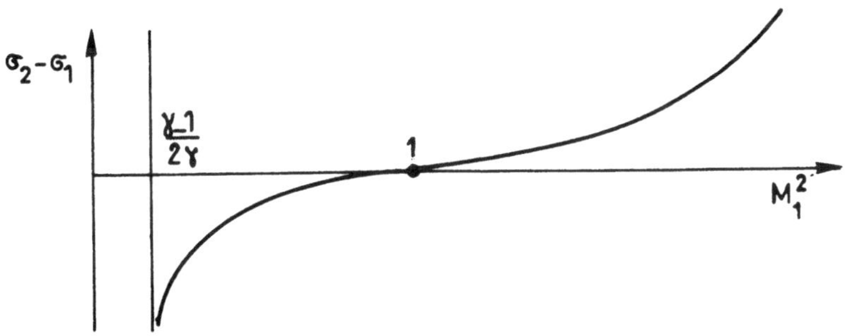

Fig. 9. A schematic representation of the variations of the jump in specific entropy, $\sigma_2 - \sigma_1$, as the Mach number M_1 of the upstream medium varies.Only for supersonic inflow in the rest frame of the shock is it possible to comply with the second law of thermodynamics.

Let us calculate the jump in specific entropy. For a perfect gas of adiabatic index γ, this quantity, σ, is proportionnal to $ln(P/\rho^\gamma)$. Then :

$$\sigma_2-\sigma_1 = \ln\left(\frac{P_2}{{\rho_2}^\gamma}\frac{{\rho_1}^\gamma}{P_1}\right) = \ln\frac{2\gamma M_1^2-(\gamma-1)}{(\gamma+1)} - \gamma\ln\frac{(\gamma+1)M_1^2}{2+(\gamma-1)M_1^2} \tag{213}$$

The second law demands that the specific entropy, that is the entropy per unit mass of material, should increase when matter passes through the shock. Note that if pre- and postshock states are exchanged, which is allowed by the jump equations themselves, the jump in specific entropy changes sign, which clearly indicates that the second law authorizes only one sense of crossing! If we now plot carefully the r.h.s. of (213) as a function of the upstream Mach number M_1, we shall discover that it is an increasing function of M_1, for$M_1 >(\gamma-1/2\gamma)$ which is necessarily so, since otherwise the downstream pressure would be negative. This function vanishes at $M_1 =1$, as well as its first and second derivative. A plot of $(\sigma_2 - \sigma_1)$ is represented in Fig.(9). This analysis shows that, in order to satisfy the second law, the upstream flow must be supersonic with respect to the upstream sound speed .Then the downstream flow is subsonic.In brief, a shock must be a transition from supersonic to subsonic.

References

Balescu,R. (1975): "Equilibrium and Non Equilibrium Statistical Mechanics", Wiley-Interscience Pub., John Wiley and Sons.

Huang, K. (1963): "Statistical Mechanics", John Wiley and Sons.

Landau,L.D. and Lifshitz,E.M. (1959): "Fluid Mechanics", Pergamon Press.

Mihalas,D. and Weibel-Mihalas, B. (1984): "Foundations of Radiation Hydrodynamics", Oxford University press.

This book was processed by the author using the TEX macro package from Springer-Verlag.

Convection and Turbulence in Stars

Jean-Paul Zahn

Observatoire Midi-Pyrénées, 31400 Toulouse, France

Columbia University, New York, NY 10027, U.S.A.

Since the nuclear reactions provide most of the energy radiated into space, the evolution of a star is determined primarily by the modification of its chemical composition, in time and in depth. And this composition is the result of both nuclear burning, which transforms the chemical species, and of various transport processes, which may either mix the chemicals, or separate them. The late stages of evolution are particularly sensitive to such processes, since the inhomogeneities are then more pronounced.

This series of lectures is devoted to the dynamical processes that mix the stellar material. Among these, the most familiar by far is thermal convection, and it is also by far the most efficient. In most stellar interiors, the temperature gradient exceeds at some depth the critical value allowed for static, radiative equilibrium. There, as predicted by linear perturbation theory, the fluid becomes unstable, and a fraction of the star adjusts into a state of turbulent motion. One of the most difficult problems of stellar structure theory is to describe such convection zones. Immediately, a few questions arise. Which is the value of the temperature gradient? What fraction of the energy flux is transported by the convective motions? How deep do these motions penetrate into the adjacent stable regions?

To answer these questions, several methods have been proposed, and our purpose is to discuss some of them. We shall start by the simplest, which is still the most commonly used, namely the mixing-length approach. Why does it perform so well, in spite of its severe shortcomings? Is it worth improving that method?

Then we shall review the more recent approaches, which attempt to solve the full hydrodynamical problem. How realistic are they? What do they teach us about stellar convection, that we did not know before?

Next we shall examine the evidence gathered by the observations, which have improved on a very fast pace in recent years. What picture do they draw of the solar convection zone?

Finally, we shall touch on a subject which is still in its exploratory phase, namely the mixing of chemicals within a radiation zone. Such mixing is suggested by many observations, although its cause is still a matter of debate. A possible explanation will be presented, which attributes that mixing to a mild turbulent diffusion due to the differential rotation of the star.

1 The mixing-length approach

The mixing-length treatment has been described in many articles, reviews and textbooks, of which we shall quote only the pionneering work of E. Vitense (1953), completed shortly after (Böhm-Vitense 1958), and the excellent reviews by E. Spiegel (1971, 1972) and by D. Gough (1977b).

Here we shall deal mainly with the basic assumptions of this treatment, and on the approximations made. But before we do so, let us first ask why a star becomes convective, after all.

1.1 Why are stars convective?

Thermal convection occurs in a compressible fluid when the temperature gradient established by conduction or radiation exceeds the critical value corresponding to an adiabatic transformation. The so-called Schwarzschild criterion for instability is conveniently expressed in terms of the logarithmic gradient of the temperature with respect to the pressure:

$$\frac{d\ln T}{d\ln P} > \left(\frac{\partial \ln T}{\partial \ln P}\right)_{ad} \quad \text{or, in shorthand} \quad \nabla_{rad} > \nabla_{ad} \,. \tag{1.1}$$

The adiabatic gradient ∇_{ad} depends on the thermodynamic properties of the stellar plasma; for a perfect (monatomic or ionized) gas $\nabla_{ad} = 0.4$. This criterion applies to an ideal fluid; the effect of viscosity and thermal diffusion is to increase slightly the instability threshold.

Let us consider the case of a convective envelope. Combining the equation of hydrostatic equilibrium

$$dP = g\rho\, dz \tag{1.2}$$

with the definition of the optical depth τ

$$d\tau = \kappa\rho\, dz \tag{1.3}$$

we find that the (total) pressure increases as

$$P = \int_0^\tau (g/\kappa)\, d\tau = g\tau < \kappa^{-1} > \,. \tag{1.4}$$

The notations are as usual for the depth z, the gravity g, the density ρ and the Rosseland mean opacity κ.

On the other hand, the radiative transfer equation has the asymptotic form (for $\tau \gg 1$)

$$T^4 = \frac{3}{4} Te^4 \left(\tau + \frac{1}{\sqrt{3}}\right) , \tag{1.5}$$

T and Te being respectively the local and the effective temperatures.

Therefore, the radiative gradient varies with optical depth as

$$\nabla_{rad} = \frac{d\ln T}{d\ln P} = \frac{1}{4}\left(\frac{\tau}{\tau + 1/\sqrt{3}}\right)\frac{<\kappa^{-1}>}{\kappa^{-1}} . \tag{1.6}$$

If the opacity were constant with depth, $\nabla_{rad} \rightarrow 0.25$, and the envelope would remain in radiative equilibrium. Moreover, in the interior of such a star, which was for a while considered as the "standard model" (Eddington 1926), the ratio between the radiation pressure (which is proportional to T^4) and the total pressure would be constant.

However, with the progress of atomic physics it became clear that, in stars like the Sun, the opacity increases sharply with depth, due to the ionization of hydrogen and helium. Thus the local value of κ^{-1} is much smaller than its average $<\kappa^{-1}>$ over $[0,\tau]$; therefore, as soon as τ becomes of order unity, the radiative gradient exceeds the adiabatic one (Biermann 1935, Cowling 1936). Note that it is not the strength, as such, of the opacity which is responsible for the convective instability, but its increase with depth.

Likewise, one can establish the expression for ∇_{rad} which is valid in the deep interior of a star:

$$\nabla_{rad} = \frac{1}{4\pi G\chi}\frac{L}{M}\frac{P}{\rho T} , \tag{1.7}$$

L and M being the luminosity and the mass at the considered depth, and G the gravitational constant. The radiative conductivity χ, whose expression is given below in (1.13), does not vary much, once hydrogen and helium are fully ionized. It is the increase of L/M, as one approaches the center of the star where the energy production culminates, which is responsible for the existence of a convective core, at least in massive stars in which the nuclear reaction rate, dominated by the CNO cycle, is very sensitive to temperature.

1.2 The temperature stratification

In a convection zone, the temperature stratification is determined by the efficiency of the turbulent motions in transporting the energy flux. Near the surface, as long as the turbulent eddies remain optically thin, they contribute very little to this transport. When they are optically thick, the convective efficiency depends on a non-dimensional parameter, the Péclet number $u\ell/K$, which compares the two competing processes: convection and radiation (u and ℓ characterize the velocity and size of the turbulent eddies, K is the radiative diffusivity).

The usual way to evaluate the convective flux is to express it by the correlation between the vertical velocity w and the enthalpy fluctuations $\delta h = \rho C_P\, \delta T$. The temperature fluctuations are then related to the velocities w, the vertical displacement of the eddies (or mixing-length) ℓ and the temperature gradient (Vitense 1953), which yields the classical expression

$$F_{conv} = cst\, \rho C_P T\, w\ell(\nabla - \nabla_{ad}) . \tag{1.8}$$

In fact, the procedure is more rigorous than it appears, and it has a firm theoretical justification. Consider the transport of a scalar s by a turbulent motion field, which has a typical scale (or correlation length) ℓ. The conservation equation

$$\frac{\partial s}{\partial t} + \mathbf{u} \cdot \nabla s = \{sources\,/\,sinks\} \tag{1.9}$$

can be averaged locally over a volume whose size is of order ℓ; the result is a series of spatial operators of increasing order which are applied to the local average $\overline{s}$ of s. To lowest order, this series reduces to the familiar diffusion equation

$$\frac{\partial \overline{s}}{\partial t} = \nabla \cdot (K_t \nabla \overline{s}) + \{sources\,/\,sinks\}\ , \tag{1.10}$$

where K_t is the turbulent diffusivity – a tensor in the general case, when the motion field is not isotropic. We refer to Knobloch (1978) for the validity of this diffusion approximation; one obvious requirement is the separation of scales, namely that $\overline{s}$ is not allowed to vary much over a distance of order ℓ.

The same treatment can be applied to the specific entropy S, which satisfies the heat conservation equation

$$\rho T \left(\frac{\partial S}{\partial t} + \mathbf{u} \cdot \nabla S \right) = -\nabla \cdot \mathbf{F_{rad}}\ , \tag{1.11}$$

with $\mathbf{F_{rad}}$ being the radiative flux

$$\mathbf{F_{rad}} = \chi \nabla T\,. \tag{1.12}$$

The radiative conductivity is related to the opacity and to the thermal diffusivity by

$$\chi = \frac{16\,\sigma T^3}{3\,\kappa \rho} = \rho C_P K\ , \tag{1.13}$$

with the classical notations for the Stefan constant σ and the specific heat at constant pressure C_P.

The advection term on the l.h.s. of (1.11) again changes into a diffusion operator on the r.h.s., where it yields the following expression for the convective flux

$$\mathbf{F_{conv}} = -\rho T K_t \nabla \overline{S}\,. \tag{1.14}$$

Here the turbulent diffusivity K_t is the vertical component of the tensor mentioned above; it is related to the r.m.s. of the vertical velocity w and to the vertical correlation length ℓ of the convective eddies: $K_t = w\ell$.

The magnitude of the total flux may be written in terms of the logarithmic gradients

$$F_{total} = \left(\frac{\rho C_P T}{H_P} \right) K \left(\nabla + \frac{K_t}{K} (\nabla - \nabla_{ad}) \right)\ , \tag{1.15}$$

$H_P = |d \ln P/dz|^{-1} = P/\rho g$ being the pressure scale-height. We see that the relative contribution of the radiative and the convective fluxes depends on the Péclet number $K_t/K = w\ell/K$ (. Introducing the radiative gradient, i.e. the gradient which would be needed to carry the total flux in the absence of convection,

$$F_{total} = \left(\frac{\rho C_P T}{H_P} \right) K\,\nabla_{rad}\ , \tag{1.16}$$

we obtain the following relation between the three logarithmic gradients

$$\frac{\nabla_{rad} - \nabla}{\nabla - \nabla_{ad}} = \frac{w\ell}{K} = Pec\,. \tag{1.17}$$

This expression is valid as soon as the turbulent eddies are optically thick; it is possible to adapt it to the optically thin case (Vitense 1953).

1.3 Prescribing the mixing-length

So far, no assumption has been made about the nature and the properties of the turbulent motions. The expression (1.15) above could be applied for instance to an accretion disk, where the instability causing the turbulence is presumably not thermal convection but is of dynamical origin.

To proceed with the description of a convective envelope, we need to specify the turbulent diffusivity, and therefore both the length ℓ and the velocity w characterizing the convective motions.

The convective velocity is estimated from the kinetic energy gained by an eddy as it travels over a vertical distance ℓ:

$$\frac{1}{2}u^2 = \int_0^\ell g\left|\frac{\delta\rho}{\rho}\right| dz\,. \tag{1.18}$$

In this picture, ℓ appears as the mean-free-path after which a turbulent eddy merges with its surroundings, hence the appellation "mixing-length". Assuming that this eddy remains in pressure equilibrium as it moves vertically, its relative buoyancy excess (or defect) $-\delta\rho/\rho$ is, to lowest order, proportional to both the distance travelled and to the superadiatic gradient. Therefore the r.m.s. of the vertical velocity scales as

$$w^2 = c\,\frac{gQ}{H_P}\,(\nabla - \nabla_{ad})\,\ell^2\,, \tag{1.19}$$

Q being the thermal expansion coefficient $Q = -(\partial \ln\rho/\partial\ln T)_P$.

The numerical factor c is assumed to take in account the anisotropy of the convective motions (horizontal versus vertical) and also the fact that a fraction of the potential energy is dissipated into heat by viscous friction before it can serve to accelerate the turbulent eddy. This constant cannot be derived from first principles, at least in the present state of the theory; its choice, which is guided by phenomenological considerations, is rather arbitrary. E. Vitense (1953) first took $c = 1/4$, as well as Biermann and Schwarzschild; later, she preferred $c = 1/8$ (Böhm-Vitense 1958), which is the value that has since been commonly used in stellar structure models.

Let us now turn to the mixing-length. This concept has been introduced by Prandtl (1925) to model the turbulent shear flows; analyzing the mean profile of boundary layer flows, he found that the theoretical predictions agree very well with the experimental measures when one chooses for ℓ the distance to the boundary:

$$\ell = z\,. \tag{1.20}$$

In the case of the Sun, Biermann (1938) proposed quite naturally to identify the mixing-length with the observed size of the solar granulation ($\approx 1\,000\,\mathrm{km}$). However, in such a highly stratified medium, ℓ presumably increases with depth, much like in Prandtl's turbulent boundary layer. For this reason, the prescription which has become widely used for a convective envelope is that of E. Vitense (1953), who took the mixing-length proportional to the pressure scale-height:

$$\ell = \alpha\, H_P\,, \tag{1.21}$$

the constant α being of order unity. This choice is made more for its convenience than for physical reasons; some authors expressed a preference for the density scale-height (Öpik 1938; Biermann 1943), and others combined Vitense's prescription (1.21) with that of Prandtl (1.20), whichever gave the lesser mixing-length (Hofmeister and Weigert 1964; Böhm and Stückl 1967). Notice that the two are almost undistinguishable for $\alpha = 1/\nabla_{ad}$, since asymptotically $H_P = \nabla_{ad} z$ at large depth.

Replacing the turbulent diffusivity $K_t = w\ell$ in terms of (1.19) and (1.21), one obtains the following expression for the convective flux

$$F_{conv} = \left(\frac{\rho C_P T}{H_P}\right) (g Q H_P^3)^{1/2} (\nabla - \nabla_{ad})^{3/2}\,. \tag{1.22}$$

It is then possible to establish between the three logarithmic gradients a relation analogous to (1.17):

$$\frac{\nabla_{rad} - \nabla}{(\nabla - \nabla_{ad})^{3/2}} = [\sqrt{c}\,\alpha^2]\,\frac{(g Q H_P^3)^{1/2}}{K}\,. \tag{1.23}$$

The transition from the radiative to the adiabatic gradient is governed mainly by the increase of the opacity (and thus of $1/K$) with depth; it takes place over an interval which is less than a pressure scale-height. Deeper down in the convection zone, the r.h.s. of (1.23) becomes very large; hence the temperature gradient is nearly adiabatic, until the radiation flux, which increases proportional to the conductivity χ, becomes capable of carrying the whole energy flux.

The situation is much simpler in a convective core. There one can dispense altogether with the mixing-length for the purpose of building a stellar model, at least if one ignores penetration: an excellent approximation is just to assume that the stratification is adiabatic.

1.4 Calibrating the mixing-length parameter α

The increase of temperature with pressure in the envelope of a star like the Sun is depicted in fig. 1. Note the sudden transition from the radiative to the nearly adiabatic slope; according to (1.23), the location of that transition depends on the actual value of $[\sqrt{c}\,\alpha^2]$, and so does also the extent of the convection zone. In other words, the structure of a star with a convective envelope is entirely determined by this mixing-length parameter α, once the constant c has been specified.

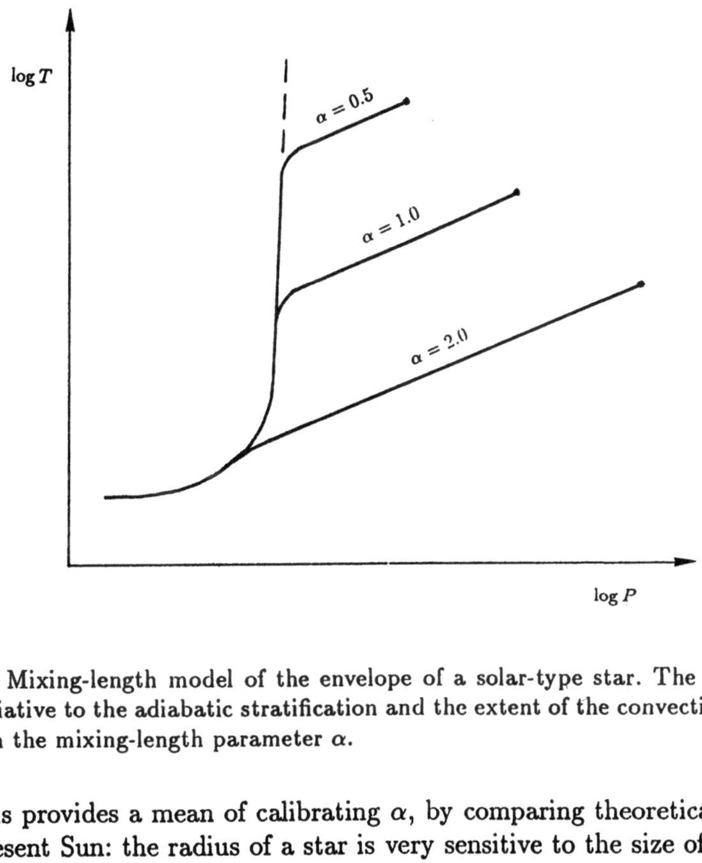

Fig. 1. Mixing-length model of the envelope of a solar-type star. The transition from the radiative to the adiabatic stratification and the extent of the convection zone depend both on the mixing-length parameter α.

This provides a mean of calibrating α, by comparing theoretical models with the present Sun: the radius of a star is very sensitive to the size of its convective envelope, and hence to the value of α. Unfortunately however, another physical quantity enters in that comparison, namely the abundance of helium, which has a direct impact on the opacity. This abundance cannot be derived with great accuracy from the solar spectrum, and the uncertainty of its value is reflected in that of α.

But the situation is improving very rapidly, thanks to the recent developments of helioseismology. This technique enables to probe the interior of the Sun, and to determine the sound speed (and thus the temperature) as a function of depth. It is now possible to locate the lower boundary of the convection zone, by detecting the change of the temperature gradient from the adiabatic to the radiative slope (Gough 1986). This direct determination of the extent of the solar convection zone (which is found to be close to 200 000 km) yields an α of ≈ 2, with the currently available opacities.

But it must be stressed that this value depends on the mixing-length formalism which has been adopted. All mixing-length treatments involve at least one other parameter, such as the constant c above. When calibrated with he actual Sun, alternate mixing-length treatments yield essentially the same model (T and P versus depth) as the classical procedure, as was demonstrated by Gough and Weiss (1976). The reason is simple. The transition from the radiative to the adiabatic slope is so abrupt that it can be described as the intersection between two lines, which requires only one parameter to be located: the product $[\sqrt{c}\,\alpha^2]$ in the classical formalism.

Apparently, the value of α deduced from the Sun also applies to other stars, at least on the main-sequence, and it has become a common practice in stellar structure theory to consider the mixing-length parameter as a physical constant. Does this reflect a universal property of the dynamics of turbulent compressible fluids, yet to be discovered? Or simply the fact that α depends on some non-dimensional parameter which does not vary much from star to star? (One could think for instance of $(MR^2/L)^{1/3}(GM/R^3)^{1/2}$, the convection time scaled by the free-fall time.)

There is no obvious reason either whether the mixing-length itself should be strictly proportional to the pressure scale-height throughout the convection zone. The ratio ℓ/H_P could well depend on other non-dimensional quantities, related to the depth, such as z/H_P, or to the efficiency of the convective transport, such as $(gQH_P^3)^{1/2}/K$, suggested by (1.23).

1.5 Extending the mixing-length treatment

The main quality of the mixing-length treatment is its simplicity, and that it offers a concrete interpretation for the single parameter required to build models of stars with a convective envelope. It is therefore questionable whether one should attempt to improve this procedure beyond what is strictly needed to predict the temperature gradient.

One justification, of course, for such improvements, or extensions, of the mixing-length treatment has been to overcome its deficiencies (cf. Gough 1977b).

It its standard form, it is a local theory: all physical quantities, the vertical velocity, the convective flux, etc. are functions of the local properties of the envelope. That assumption would be valid if the mixing-length were much smaller than all relevant scales; but this not the case with an α of order unity. One consequence of such local theory is that no motions are allowed beyond the level where the stratification becomes stable again, contrary to what is observed in the laboratory and in the Earth atmosphere. Several attempts have thus been made to render the mixing-length formalism non-local, in order to apply it to such convective overshooting (Saslaw and Schwarzschild 1965; Shaviv and Salpeter 1973; Maeder 1975).

But the predictions differ widely, depending on the assumptions made; recently, Renzini (1987) summarized the situation in a striking formula: "overshooting is found small if supposed small, large if supposed large".

The mixing-length procedure has also been stretched to treat the coupling of convection with a large scale motion field, such as the radial oscillation of a pulsating star. The modulation of the convective energy flux, which was considered by Unno (1967) and Gough (1977a), seems unable alone to explain the red edge of the Cephei instability strip observed in the Hertzsprung-Russel diagram (Gonczi and Osaki 1980). Gonczi (1981, 1982) was more successful when he added the damping due to the turbulent viscosity caused by the convective motions. This damping mechanism plays also a major role in the tidal braking of close binary stars (Zahn 1966, 1989a).

Further steps in the sophistication were taken by Xiong (1979, 1985) and Kuhfuß (1986), at the expense however of additional free parameters. At this point, it is preferable in my opinion to directly tackle the hydrodynamical problem, and that approach will be discussed in the next chapter.

2 Hydrodynamical approaches

The goal of the direct simulations of thermal convection is to solve the equations governing compressible fluid dynamics in a star with the least possible approximations. To take up this challenge without any concession, it would require a computer powerful enough, in speed and memory, to encompass all scales which are present in stellar convection, from the largest which are of the size of the star, to the smallest at which the kinetic energy is dissipated through viscous friction (a few centimeters in the Sun!). But even if such a computer were available in some distant future, would it be reasonable to spend almost all computing time with the smallest scales, i.e. those which do not participate in the heat transport? Clearly, there must be better ways of tackling the problem, even if they imply some loss of accuracy.

In any case, there is presently no other choice than to calculate only the largest scales of the convective flow, and to use some prescription to take into account the smaller, subgrid scales. Several recipes now exist, with their advantages and their disadvantages. The crudest of all is just to increase artificially the viscosity, and to pretend that it represents the effect of the Reynolds stresses due to the subgrid scales. Ideally, one should restrict such prescriptions to those scales which already participate in the inertial cascade, but even this is very rarely possible with the currently available computers.

In compressible convection, the control parameters are no longer constants, as in the Boussinesq limit, but they vary with depth. The Rayleigh number measures the degree of instability

$$Ra = \frac{g}{H_P} \frac{(\nabla - \nabla_{ad})}{K\nu} L^4 , \tag{2.1}$$

where L is the vertical extension of the convective layer. In the Sun, this number is about 10^{20} times critical, whereas in the numerical simulations it rarely exceeds 10^4 critical. The Prandtl number (ν/K), which in the solar convection zone can be as low as 10^{-9}, is also artificially increased in the calculations, with the same understanding that the viscosity is of turbulent nature.

We shall now present the methods which have been employed to simulate stellar convection, and examine their main results.

2.1 Modal expansions

Once thermal convection has reached a regime which is stationary in some statistical sense, its characteristics (the velocity field, the temperature fluctuations, etc.) are rather homogeneous over a horizontal plane; in other words, they are invariant in a horizontal displacement, and depend only on depth.

It is this property which inspired the modal approach. In this treatment, all functions are expanded in a series of prescribed horizontal planforms; for instance, the vertical velocity is given by

$$w(x,y,z,t) = \sum_k f_k(x,y)\, W_k(z,t)\,. \tag{2.2}$$

In the more general Galerkin method, the planforms $f_k(x,y)$ can be arbitrary functions; in the modal treatment, one chooses the functions that factorize the linear problem. These are periodic in space, and satisfy the orthogonality condition

$$< f_i(x,y)\, f_j(x,y) > = \delta_{ij} \tag{2.3}$$

when averaged over a horizontal plane.

The equations of motion are projected on the orthonomal base represented by these planforms, which yields a system of non-linear differential equations in time and in the vertical coordinate z (see Latour *et al.* 1976). That system not only involves the wavenumber a_k of the planforms, and but also triple correlation coefficients defined as

$$C^{ijk} = \frac{1}{2} < f_i f_j f_k >\,, \tag{2.4}$$

which govern the non-linear interactions between the modal functions $W_k(z,t)$, etc. The self-interaction coefficient C^{kkk} is a measure of the three-dimensionality of the mode of index k; it vanishes when this mode is a horizontal roll.

The series (2.2) is usually truncated after a few terms (sometimes after only one!); most of the computing effort is spent in obtaining a good resolution in the vertical direction (typically 200 mesh points, or more).

This modal method has been first applied to laboratory convection, in the so-called Boussineq approximation, where the compressibility of the fluid is neglected, except in the buoyancy force (Gough, Spiegel and Toomre 1975). The results were encouraging enough to engage in the simulation of compressible convection. The calculations were accelerated by using the anelastic approximation (Gough 1969): it filters out the sound waves, and retains only the (slower) convective modes.

A variety of situations have been considered, ideal polytropes and more realistic models, which included the physical properties of the stellar plasma.

Let us summarize the salient results which have been obtained.

i. In a strongly stratified medium, the convective flows extend over several pressure scale heights, and vigorous horizontal shears appear at their top to ensure the conservation of mass. For instance, the horizontal velocity associated with scales of the size of the solar convection zone reaches 100 m/s at a depth of 7 000 km, according to a single-mode calculation made by Latour, Toomre and Zahn (1983).

ii. The flux of kinetic energy carried by three-dimensional modes becomes comparable to the convective flux. Depending on the sign of the self-interaction coefficient (2.4), it is directed upwards or downwards; in the latter case, it neutralizes partly the convective flux.

iii. Near the top of the unstable domain, the pressure fluctuations, whose role is to deflect the vertical upwellings into horizontal streams, are responsible for density fluctuations that may be large enough to change the sign of the buoyancy force. This 'buoyancy braking' is illustrated with the solution displayed in fig. 2; it is a property of convection in a stratified medium and vanishes in the incompressible (Boussinesq) limit (Massaguer and Zahn 1980; Massaguer *et al.* 1984). The pressure fluctuations, which are ignored in the mixing length theory, are of such importance that the net work done by them is in some cases larger than that done by the buoyancy.

iv. The convective motions penetrate rather deep into the adjacent stable layers; for instance, they link together the two convection zones of an A-type star, which are due respectively to the ionization of hydrogen and to the second ionization of helium (Latour, Toomre and Zahn 1981). The extent of penetration depends sensitively on the geometry of the considered cellular flow: cells with a central upwelling penetrate much deeper than those with a central downstream, as demonstrated by the supergranular velocity field (Toomre *et al.* 1976; Massaguer *et al.* 1984).

The modal procedure is unquestionably an improvement with respect to the mixing length treatment. However, when reduced to only one or two planforms, as they have usually been because of the limited computer power, the modal calculations suffer from the fact that the horizontal size of the cells, and also their geometry, is arbitrarily imposed. One consequence of such a severe truncation is that the smallest scales, at which kinetic energy is dissipated through viscous friction, have a very large aspect ratio, leading to artificial horizontal boundary layers when the viscosity becomes too small. Another is that those truncated solutions are stationary at Rayleigh numbers where they should be time dependent; the time dependence can be recovered when including modes of vertical vorticity, as shown recently by Massaguer and Mercader (1988).

However, in spite of all these shortcomings, the modal method has permitted to establish the rather important results presented above, and those have been confirmed since by the more realistic calculations to be presented next.

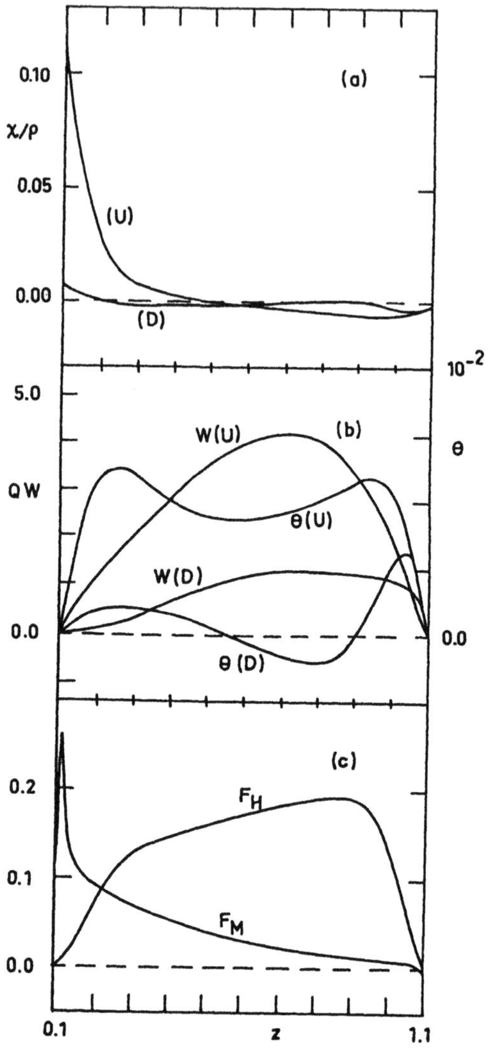

Fig. 2. A modal solution with two hexagonal planforms of the same size (Massaguer and Zahn 1980). (a) Relative density fluctuation χ/ρ of respectively the upwards (U) and the downwards (D) directed mode; note the inversion of the density perturbation in the upper part of the domain, responsible for the 'buoyancy braking'. (b) Temperature fluctuation θ and vertical component W of the momentum density, again for the two modes. (c) The convective flux F_H and the kinetic energy flux F_M.

2.2 Two-dimensional simulations

Such simulations assume also a periodic pattern in the horizontal direction, but they allow for much better resolution than the severely truncated modal series. Both finite differences and spectral methods can be used to discretize the equations; the latter are particularly well adapted to the present vector computers. Run with such powerful computers, the two-dimensional codes enable one to explore the full effects of compressibility, without being restricted to the anelastic approximation.

The two-dimensional solutions have most properties of the modal solutions presented above. But they are all time-dependent, once the Rayleigh number becomes larger than about hundred times critical. Furthermore, they are characterized by strong, narrow, unstable streams, which are directed downwards; these extend over the whole unstable domain and penetrate deeply into the stable region below (Hurlburt *et al.* 1986). The result of such a simulation is displayed in fig. 3. Such an asymmetry between upward and downward velocities is noticeable already at lower Rayleigh number, when the solutions are still stationary (Graham 1975).

The great advantage of such two-dimensional calculations is that they enable to resolve much finer features, such as the downdrafts mentioned above. Unfortunately, they are not typical of real, three-dimensional convection, the reason being that, in the inviscid limit, they not only conserve energy (i.e. the square of the velocity), but also enstrophy (the square of the vorticity). This additional constraint completely alters the coupling between the motions of different scales; it produces an inverse cascade of energy to the large scales (which in the simulations can be observed as the pairing of vortices), and a steep enstrophy cascade to the smallest scales.

In spite of this serious bias, two-dimensional calculations serve as a powerful first exploration of a given problem, provided that energy is injected at the largest scales, and that the Reynolds number is kept reasonably small (in order to avoid the cascades mentioned above). They have permitted the discovery of one of the major properties of compressible convection, namely the existence of these strong downflows that seem to play a prominent role in a stratified fluid.

2.3 Three-dimensional simulations

Relatively few simulations have been performed so far in three dimensions, but they will soon eclipse all others. Their main limitations are their cost and their low spatial resolution (rarely over 64^3). The first of such calculations was accomplished by E. Graham (1977), truly the pionneer in that field; it demonstrated the three-dimensional nature of convection, and the important function of the vertical vorticity in the nonlinear coupling of the various modes.

Å. Nordlund (1982) took boldly the Sun as the target of his calculations, even though he could only treat the uppermost 4 000 km of the envelope. He was therefore obliged to impose somewhat arbitrary conditions at the lower boundary of its computational domain. But he treated with great care the radiative transfer of energy, and was able to simulate the granular motions with striking realism: exploding granules, sharp intergranular lanes, time dependence, etc. The only serious

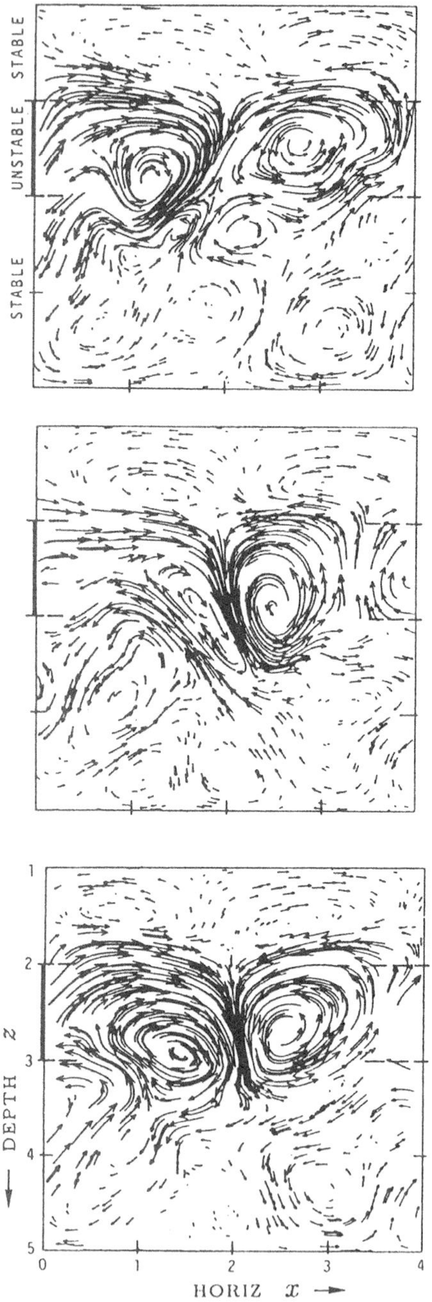

Fig. 3. A time sequence of a two-dimensional simulation of thermal convection in a stratified medium (Hurlburt *et al.* 1986). Note the strong, unsteady, downwards directed flows.

discrepancy was the large temperature contrast needed to carry the convective flux, which was higher than the observed one, even after correction from instrumental diffusion and poor seeing.

Similar calculations were performed by Chan and Sofia (1986), with different boundary conditions. These authors found that the vertical correlation of the motions is only a little more than a pressure scale height and they considered this as a justification for the usual mixing-length procedure.

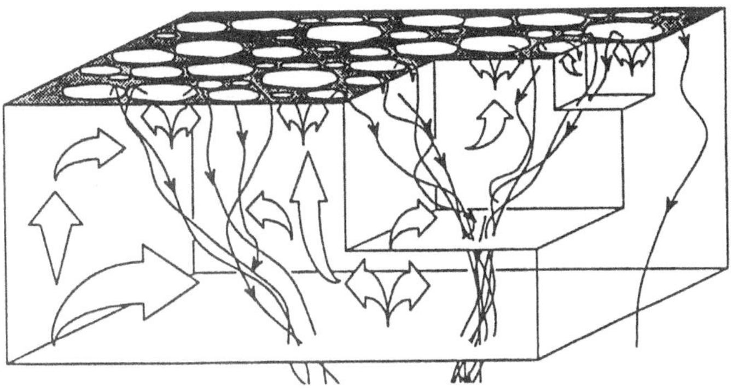

Fig. 4. Sketch of a three-dimensional simulation of thermal convection in a stratified medium, showing the successive merging of downdrafts (Spruit *et al.* 1990).

Quite on the contrary, both Graham (1977) and Nordlund (1984) obtained strong concentrated downdrafts which extend over the whole computational domain, much like those displayed by the two-dimensional calculations. This behavior has been confirmed since by Stein and Nordlund (1989) and independently by Cattaneo, Hurlburt and Toomre (1989), in a polytropic envelope and with different boundary conditions. These downflows seem to represent a genuine and important feature of stratified convection (see fig. 4); for a recent description of their properties, see Spruit *et al.* (1990). They behave more like long-lived plumes than as thermals which would just traverse the domain. They originate in the upper thermal boundary layer, which is strongly super-adiabatic, and they merge together as the depth increases. Furthermore, they carry most of the convective flux. It may well be that such plumes extend over the whole convection zone and that they are responsible for the penetration below.

While all simulations we have reported so far deal only with the convective motions of granular size that transport most of energy, some others are aimed at the large-scale flows which might be driven through buoyancy. Such are the calculations performed by Marcus (1980), Glatzmaier and Gilman (1982), Glatzmaier

(1984, 1985) and by Gilman and Miller (1986), who take the whole convection zone as their computational domain. Since these larger scales contribute little to the convective energy transport, an eddy diffusivity of heat is required to establish the nearly adiabatic stratification; it is assumed to represent the effect of the small scales: the granulation, and even the supergranulation. With such a large eddy diffusivity, the effective Rayleigh number is quite low (50 times critical), and the solutions display a fairly regular cellular pattern. At these global scales, the motions are sensitive to the Coriolis force, and they organize themselves in long cells aligned parallel to the rotation axis (they have been nicknamed 'banana cells').

This geometry was already predicted quite a while before by Busse (1970) and by Durney (1970), who found that those modes are the most unstable and who postulated that they would therefore dominate in the nonlinear regime. This trend has been confirmed in a cleverly designed microgravity experiment, which was performed on board of Spacelab 3. There, a spherical shell of fluid was heated from inside and was submitted to a radial electrostatic force mimicking the gravity; when rotation was imposed, the degeneracy was lifted again in the way predicted by Busse (Hart *et al.* 1986; Toomre, Hart and Glatzmaier 1987). We shall see in the next chapter whether this behavior has actually been observed on the Sun.

What can be concluded from the numerical simulations that have been carried out so far? Clearly, they draw a picture which is quite different from the mixing-length model. The pressure fluctuations are far from negligible, the vertical excursion of the motions is much larger than a pressure scale height, and the kinetic energy flux cannot be ignored, since it may be comparable to the convective flux.

This does not mean that we should abandon the mixing-length procedure, but rather that we should restrict its use to the purpose for which it has been designed, namely to fit the radiative temperature stratification in the atmosphere of a star with the adiabatic envelope below. Beyond that, the mixing-length model is of little help in describing the actual properties of a convective envelope, especially the dynamical ones, and it can even lead to erroneous results.

3 What the observations tell us about convection

In recent years, significant progress has been achieved in observing convective motions on the surface of stars. It is true that the information must be extracted out of the scrambled signal originating from the whole disk of the star, and that not many photons are available. But careful analysis of line shifts and line asymmetries is providing more and more clues about stellar photospheric motions and their vertical distribution (Gray 1980, 1981, 1982, 1986; Dravins, Lindegren and Nordlund 1981; Dravins 1982; Gray and Toner 1986; Dravins, Larsson and Nordlund 1986).

However, understandably most observations concerning stellar convection are provided by the Sun, whose proximity permits both extremely good spectral resolution and very high spatial definition. Let us review the most striking results that have been obtained.

3.1 Progress in the observational techniques

Unquestionably, the most spectacular results have been obtained through helioseismology, and no wonder that this technique has given a new and strong impetus to solar research. The acoustic modes of the Sun are detected by high-resolution spectroscopy, and their frequency spectrum reveals the variation with depth of the sound speed. Helioseismolgy thus permits to probe the vertical stratification of the convection zone, and to locate its bottom through a direct method. Its rotation state can also be investigated now, even though the signal to noise ratio still limits the vertical resolution. A series of colloquia, workshops, etc. have been devoted to this exciting research, and therefore I shall not expand on it.

Observation from space, free from atmospheric effects, is the other modern technique which has contributed enormously to solar research. It has been initiated by the celebrated Stratoscope experiment performed in 1959 on a stratospheric balloon (Bahng and Schwarzschild 1961). Some pictures and spectra are still taken by rockets, but only satellites allow observations of long enough duration to study the photospheric motions. The last magnificent results bearing directly on convection are those obtained on Spacelab 2 (Title *et al.* 1986; Title *et al.* 1987). The granular field has been filmed with an instrument called the Solar Optical Universal Polarimeter (SOUP), and the pictures have been processed by a very clever method. The five-minute oscillations have been subtracted, to leave only the granular motions, which are of convective origin.

But the more classical solar observers have also been extremely active, and the results they have gathered can easily compete with those obtained by the more recent techniques. Pieced all together, they unravel a new picture of the solar convection zone, which we shall now describe.

3.2 Granulation

Over more than a decade, the solar refractor of the Pic-du-Midi has yielded images with the nominal resolution of its 50 centimeter objective, i.e. 0.25 second of arc; a sample is given in fig. 5. Since 1987, the Swedish Vacuum Solar Telescope located at La Palma in the Canary Islands has produced pictures of comparable quality. These images taken by ground-based instruments are still of better resolution than those obtained in space, where smaller apertures have been used until now; the distorsion caused by the Earth atmosphere can be partly removed through sophisticated processing.

Thorough analysis of the best Pic-du-Midi data has revealed that granulation consists of a continuum of sizes, with no dominant scale, contrary to previous claims based on images of poorer resolution. And this distribution has been shown by Macris and Rösch (1984) to vary with the solar cycle: the granules are somewhat smaller at solar maximum than at the minimum.

To render the analysis more quantitative, an attempt has been made, following a suggestion by E. Spiegel, to characterize the form of the granules by their fractal dimension. This parameter can be readily deduced from a logarithmic graph of the area versus the perimeter of the granules; the result is shown on fig. 6 (Muller and

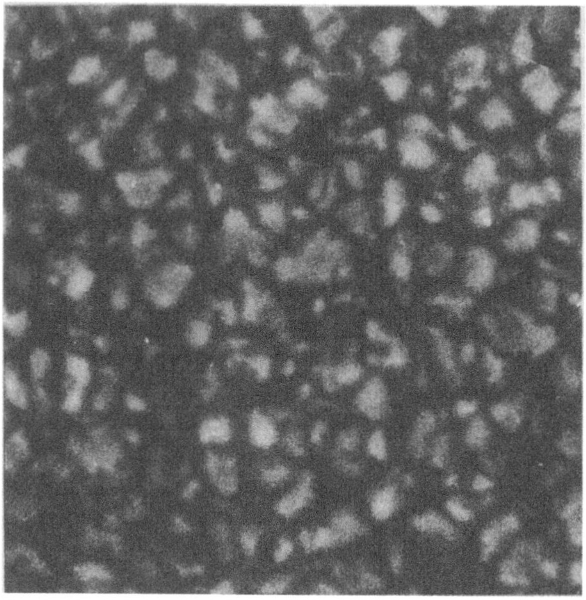

Fig. 5. A photograph of the solar granulation taken at Pic-du-Midi, with a resolution of 0.25 arcsec (R. Muller; 9 July 1978).

Roudier 1985; Roudier and Muller 1986). It confirms the visual impression that the small granules are much more regular than the large ones, but the distinction between the two is surprisingly sharp with this diagnostic tool. The transition occurs for a size of 1.4 second of arc, which corresponds to $\ell \approx 10^8$ cm, or 1 Mm. That scale corresponds to a Péclet number wl/K of order 1, if one evaluates it with the vertical velocity ($w \approx 1\,\mathrm{km/s}$); at optical depth $\tau = 1$, the thermal diffusivity K is about 10^{13} in cgs units.

This dimensionless number measures the strength of the coupling between the velocity field and the temperature field in a conducting fluid: when this number is larger than unity, heat is advected by the motions; when it is smaller, diffusion dominates. Therefore the interpretation of the observed discontinuity in the fractal dimension is that the large eddies are irregular because they advect the temperature field, whereas the small ones are regular since they let the heat diffuse more easily. There is another distinction between the large and the small eddies: they tend to segregate in different clusters, in a pattern that has the scale of the mesogranulation (Muller *et al.* 1990).

Not much more can be deduced from the images alone. That is why the Pic-du-Midi refractor has been equipped with the MDPS (Multichannel Double Pass Subtractive) instrument built by the Meudon observatory; it gives nine images of

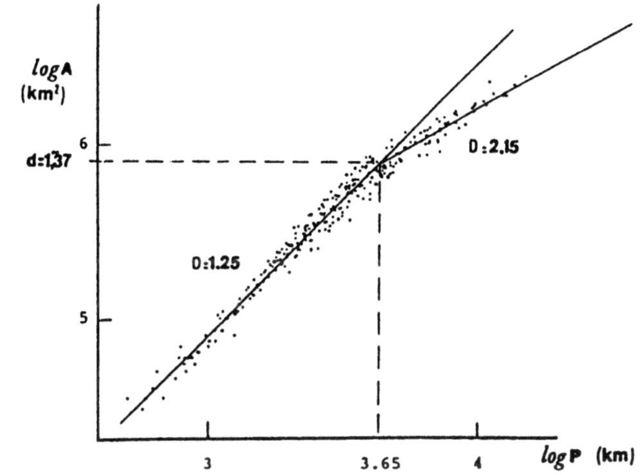

Fig. 6. Area-perimeter relation of the solar granulation. Each point represents a granule measurement; D is the fractal dimension (Roudier and Muller 1986).

the same field in different wavelengths scanning a spectral line. The first observation campaign was extremely successful: the best Doppler images obtained with this instrument, on photographic film, have a definition of about half a second, and the vertical velocity field can now be determined with that high resolution (Muller *et al.* 1987).

As is well known, this velocity field is made of two components: the waves due to the five minute oscillations, and the granular motions in which we are interested here. Ideally, to disantangle the two fields, the pictures ought to be processed in the same way as the SOUP images, and this will be done in the near future. In the meanwhile, a much cruder method has been used to filter out the acoustic waves: it consists simply in superposing two Doppler images taken two and half minutes apart (after careful recentering). When examining the kinetic energy spectra derived before and after this filtering, one verifies that it affects principally the supergranular and mesogranular scales, as expected, and that it barely modifies the small scale region. This is fortunate, since it is in that region of the spectrum that the most interesting property has been found.

The result is shown in fig. 7, where the spectral energy $E(k)$ is plotted as usual versus the wavenumber k in logarithmic coordinates; $E(k)dk$ is the kinetic energy present in the scales between k and $k+dk$. Two straight lines can be drawn through the data points: for the scales larger than 3 seconds of arc ($\approx 2\,\mathrm{Mm}$), the slope is about -0.70, whereas below that scale, and down to the resolution limit, it is -1.70.

Within the error bars of the crude treatment which has been used, this latter value is compatible with the famous $-5/3$ of the Kolmogorov law. That slope

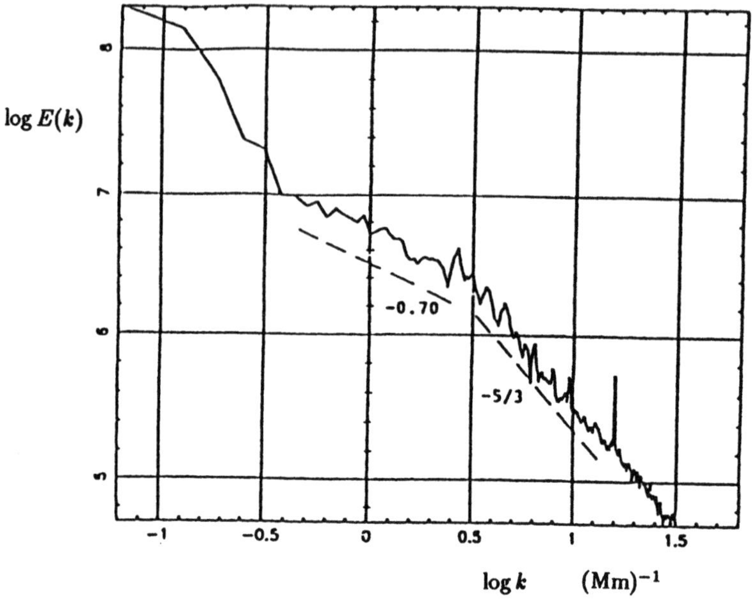

Fig. 7. Power spectrum of the solar photospheric motions, derived from two superposed Doppler images taken two and half minutes apart (Muller *et al.* 1987).

characterizes the energy spectrum of homogeneous isotropic turbulence in the so-called inertial domain, where the kinetic energy cascades down to the smallest scales, to be dissipated there through viscous friction (see for instance Landau and Lifschitz 1953). Therefore the observations above must be interpreted as a strong indication that the granular motions are of turbulent origin. They do advect heat in their vertical displacement, as revealed by the well known correlation between temperature and velocity, but according to this result the buoyancy force plays only a minor role in their dynamics, and most of their energy originates from scales larger than 2 Mm.

Let us stress that this is the first time that such a slope has been reported. Earlier work by Aime, Ricort and Harvey (1978), with a different technique (speckle-interferometry through photoelectric scans), revealed instead an exponential tail of the kinetic energy spectrum. And recent observations by Nesis *et al.* (1987), likewise recorded on a photographic film but apparently carried out with a somewhat poorer seeing, also show a steeper slope. More recent observations by the French team however confirm their earlier result.

Another indication of the turbulent character of the granular motions is the fact that their life time (about 10 mn) is of the order of their turn-over time ℓ/w. Furthermore, the pictures shot in space by the SOUP instrument seem also to confirm the existence of a turbulent cascade, with the large eddies breaking into smaller ones, whereas the small eddies never merge into bigger ones; the evidence is striking just by looking at the film, and playing it back and forth.

3.3 Large scale motions

Several attempts have been made in the last few years to detect a large scale pattern in the photospheric motions, which could be related to the global convection of the Sun and its dynamo; the subject has been recently reviewed by Schröter (1985).

The first direct, kinematic evidence for such motions was the modulation of the differential rotation discovered by Howard and LaBonte (1980) on the Mount Wilson full-disk velocity data. They interpreted it as a torsional wave synchronous with the 11 year cycle, which travels from pole to equator in about twice that time.

The second discovery was made by Ribes, Mein and Mangeney (1985) in analyzing the spectroheliograms recorded for almost a century at the Meudon observatory, which are now being digitalized. They found that the motions of the newly-born sunspots are organized in several latitudinal bands, as if they were the tracers of a meridional circulation with several counter-rotating, azimuthal rolls (3 or 4 in each hemisphere). The velocities average about 30 m/s, which leads to a turn-over time, for the rolls, of the order of one month.

The authors also found that the mean rotation rate of these young spots is nearly constant with latitude, and thus does not reflect the differential rotation observed in the photosphere. Moreover, they showed that this meridional circulation is closely related to the large scale, long-lived magnetic structures outlined by the Hα filaments: these structures coincide with the regions of convergence and divergence (and hence of downflow or upflow) of the azimuthal rolls. Finally, the authors found that this roll system migrates towards the pole in each hemisphere, with a new pair of rolls emerging at the equator every solar cycle, as illustrated in fig. 8.

This last property seems to contradict that of the torsional wave of Howard and LaBonte, which moves in the opposite direction, from pole to equator. Moreover, those authors have found no indication of meridional flows with velocities larger than 10 m/s (their detection limit). But the tracers used in the two cases are not the same: Howard and LaBonte measure their Doppler velocities in the photospheric plasma, whereas the young sunspots of Ribes *et al.* are probably anchored at the bottom of the convection zone, judging from their uniform rotation rate.

At first, the theoricians were puzzled by the fact that these roll-like cells are parallel to the equator, since it had been proved that convection in a rotating shell should proceed in so-called 'banana-cells', aligned with the rotation axis (cf. section 2.3 above). But this discrepancy can be explained: the toroidal (azimuthal) magnetic field, which in the classical dynamo theory is generated by the differential rotation, may be strong enough to lift the degeneracy due to the spherical symmetry in a way opposite to that of the Coriolis force, namely in azimuthal rolls. Further work is in progress to clarify this point. Of course, one could also

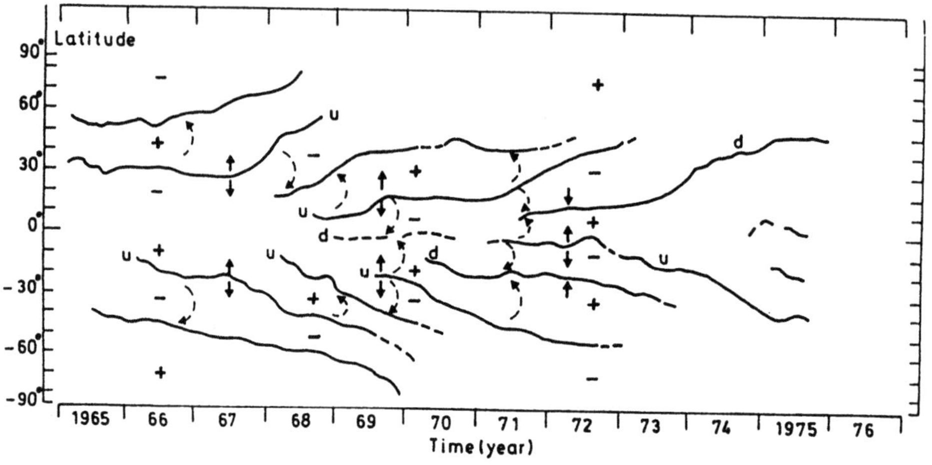

Fig. 8. Large-scale meridional circulation in the convection zone, detected through the migration of young sunspots (Ribes *et al.* 1985).

argue that the drift of the magnetic pattern, as visualized by the young sunspots, is not necessarily connected with convective motions.

Other magnetic tracers have been used in the past with no definite conclusion (see Schröter 1985). For instance, Ambrož (1987) derived a circulation pattern from the Hα Synoptic Charts, using a somewhat subjective method. He concluded that the displacements of these structures cannot be interpreted by purely zonal flows, and detected even higher velocities in the meridional direction (of order 150 m/s). Unfortunately, a confrontation with the results of Ribes *et al.* (1985) is not possible, since they concern different epochs.

3.4 Discussion

What can be learned, from the observations that have been presented above, about the nature and the mechanism of solar convection?

The result obtained by Muller *et al.* (1987), which apparently proves the turbulent nature of the granular motions, is of such importance that one is tempted to seek a theoretical interpretation before its validation by other observers. As already mentioned, another indication of such a turbulent cascade is brought by the SOUP experiment. What is surprising is the almost perfect $-5/3$ slope, which is commonly regarded as a signature of homogeneous isotropic turbulence. Such ideal turbulence is not likely to exist in this strongly stratified medium, where the density scale height is about 250 km, i.e. much less than the horizontal size of the observed eddies. Furthermore, the vertical motions are hindered by the stable stratification above optical depth $\tau = 1$.

But perfect isotropy is not required to establish the Kolmogorov law: it suffices that the anisotropy be independent of scale. And homogeneity is clearly achieved in the horizontal directions, which are those where most dynamical interactions occur in the photosphere.

At first glance, this result of a turbulent cascade, which implies that the kinetic energy of the granules proceeds from larger scales, is in contradiction with the numerical simulations performed so far. For instance, in the calculations made by Nordlund (1982) the granules live by themselves, from the work done by the buoyancy and pressure forces alone, and they need no other source of energy. But this result is hardly surprising, since the horizontal dimension of the computational domain implicitly imposes that the convective heat flux be carried only through the eddies of granular size. As already mentioned, Nordlund predicts a temperature contrast which is larger than that allowed by the observations; this may be taken as an indication that other scales, presumably larger ones, participate in the convective transport. If these scales were included in the simulation, they would probably lower the superadiabatic gradient, and hence the temperature fluctuations and the velocities associated with the granular motions.

If we accept the idea that the granules are not the major actors in the solar convection, we must identify the scales which are the dominant ones, namely those at which the major part of the kinetic energy is injected into the turbulence. In all likelihood, these are the supergranuler eddies, which are known to be convectively unstable (Van der Borght 1979; Antia, Chitre, and Pandey 1981; Antia, Chitre and Narasimha 1983).

To confirm this property, it would suffice to check that the energy spectrum is indeed maximum for the corresponding wavenumbers. Unfortunately, what seems the best spectrum obtained so far, that of Muller *et al.* (1987), is not reliable enough in that wavenumber range, mainly because of the rough method used to filter out the five minute oscillations. Nevertheless, we can use the best established part of this spectrum, namely the inertial range, and compare it with the energy of the supergranular motions, using the information which is available on sizes and velocities.

We shall proceed as follows (see Zahn 1987). First we schematically represent the spectrum of Muller *et al.* by two straight lines (fig. 9); the knee (k_G, $E(k_G)$), where the slope steepens from -0.70 to $-5/3$, is located at 2 Mm, hence $2\pi/k_G = \lambda_G = 2\,\text{Mm}$. There

$$\frac{1}{2}V_G^2 = \int_{k_G}^{\infty} E(k)dk = \frac{3}{2}k_G E(k_G), \tag{3.1}$$

V_G being the r.m.s. velocity characterizing the granular motions. We take for V_G the value 1 km/s which may be inferred from the vertical velocity measurements (see for instance Muller 1985); we neglect the contribution of scales larger than λ_G.

Likewise, for the supergranules we write

$$\frac{1}{2}V_S^2 = \int_{k_S+\Delta k}^{k_S-\Delta k} E(k)dk = \eta_S k_S E(k_S), \tag{3.2}$$

where V_S is the typical velocity (250 m/s) deduced from the lifetime of these eddies (1-2 days), and k_S the typical wavenumber ($2\pi/k_S = \lambda_S \approx 30$ Mm) of the supergranules; the integral is taken over the wavenumber range $2\Delta k$ which characterizes them. The coefficient η_S is proportional to the bandwith $\Delta k/k$, and it also depends somewhat on the slope of the energy spectrum in that domain; a conservative assumption is to take $\eta_S \approx 0.5$.

We thus obtain the ratio

$$\frac{E(k_G)}{E(k_S)} = \frac{2\eta_S}{3}\frac{\lambda_G}{\lambda_S}\left(\frac{V_G}{V_S}\right)^2 \approx 0.35\,. \tag{3.3}$$

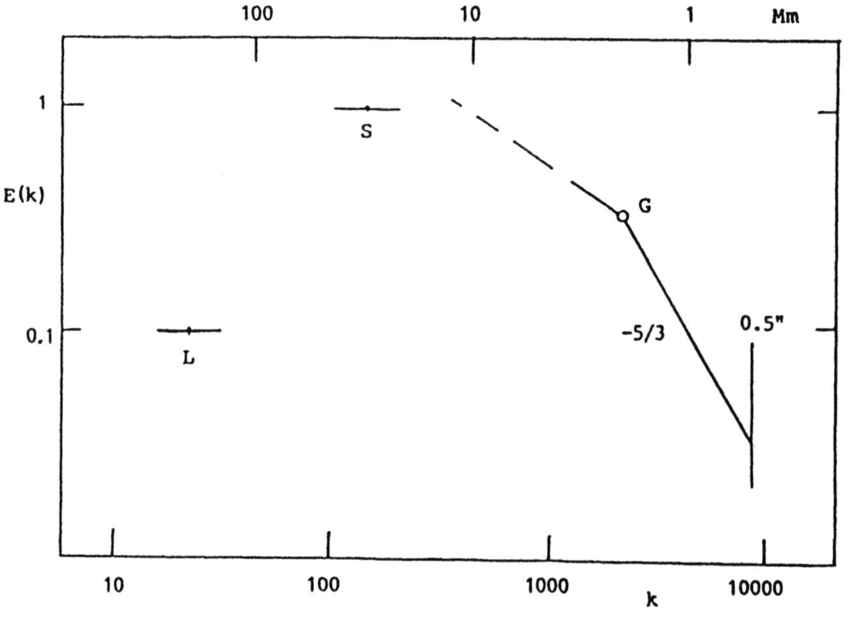

Fig. 9. Sketch of the power spectrum of the solar convection, showing the portion (starting at G) determined by Muller *et al.* (1987), the supergranular motions (S), and the large scale motions (L) discovered by Ribes *et al.* (1985).

The result is shown on fig. 9: $E(k_S)$ is definitely located above the inertial range, and this strengthens our claim that the supergranular eddies are indeed the dominant ones in solar convection, at least near the surface.

It is tempting to use the same method to plot on this graph the large scale motions reported by Ribes *et al.* (1985); we shall use the following parameters: $V_L = 30\,\mathrm{m/s}$ and $2\pi/k_L = \lambda_L \approx 200\,\mathrm{Mm}$, and again $\eta_L \approx 0.5$. Here we get

$$\frac{E(k_L)}{E(k_S)} = \frac{\eta_S}{\eta_L}\frac{\lambda_L}{\lambda_S}\left(\frac{V_L}{V_S}\right)^2 \approx 0.10\,, \tag{3.4}$$

and we see that the energy level of these motions is also located below the maximum reached for the supergranular scales, which could indicate that these large eddies are likewise fed by the supergranular motions. But it may well be that they are unstable enough to be driven by buoyancy.

Let us summarize. The picture of the solar convection zone which emerges from the most recent observations is that of a classical turbulent fluid, with energy injected at the supergranular scales and cascading down to the smallest scales. However, according to the numerical simulations reported in chap. 2, this turbulent regime presumably exists only in the uppermost part of the convective envelope; below, the flow appears to be organized in strong, concentrated downdrafts. It remains to be seen how these will interact with the large-scale rolls detected through the migration of the young sunspots.

4 Turbulence and mixing in radiation zones

Much attention has been paid in recent years to the transport processes that are likely to occur in the radiative interior of stars. The main reason is that the abundance of some elements at the surface of the Sun and solar-like stars can only be interpreted by invoking a mild transport of matter below their convective envelope (Schatzman 1969, 1977; Schatzman and Maeder 1981). For instance, if there were no mixing at all in the radiative zone, helium would drift to the interior due to microscopic diffusion, and none would be left in the atmosphere of the present Sun. Such transport is required also to explain the presence of ^{3}He in the solar wind (Geiss *et al.* 1972), since this element is produced rather deep in the radiative core. Moreover, the depletion of Li observed in solar-like stars (Boesgard 1976, Duncan 1981, Cayrel *et al.* 1984) can only occur if there is a link between the bottom of the convection zone (which is at $10^6\,^0$K) and the depth where this fragile element is destroyed (around $2.5\;10^6\,^0$K).

There are indications of such transport in other stars, for instance the presence of ^{13}C in the giant stars of the first ascending branch (Lambert *et al.* 1980). It may also affect the abundance of Li in the old halo stars, which is taken as a test for the validation of cosmological theories (Spite and Spite 1982). More importantly, if such mixing takes place in the radiation zones, it can alter the chemical composition within a star to a point where it modifies the late stages of its evolution.

4.1 Large scale circulations

Such transport of chemicals can be achieved through large-scale circulations, for which two causes have been identified. One has been thoroughly studied, since its discovery by von Zeipel (1924): a rotating star cannot achieve radiative equilibrium, and that thermal imbalance induces a meridional advection of heat (Eddington 1925; Vogt 1925; Sweet 1950; Mestel 1953; Kippenhahn 1958; McDonald 1972; Pavlov and Yakovlev 1978; Tassoul and Tassoul 1982), which has been called the Eddington-Sweet circulation. This circulation proceeds on a time-scale which is of order

$$t_{ES} \approx t_{KH} \left(\frac{\Omega^2 R}{g}\right)^{-1} = \left(\frac{GM^2}{RL}\right)\left(\frac{\Omega^2 R^3}{GM}\right)^{-1}, \tag{4.1}$$

i.e. the Kelvin-Helmholtz time divided by the oblateness due to the rotational velocity Ω. The circulation time largely exceeds the nuclear life time in slowly rotating stars, like the present Sun, but in fast rotators, such as the massive main-sequence stars, they can be of the same order.

It is relatively easy to calculate the meridional circulation for a given angular velocity. Assuming that the rotation remains uniform, the circulation is stationary, and its streamlines are depicted in fig. 10. But such a circulation advects angular momentum, and thus modifies the rotation law. Moreover, as we shall see below, non-uniform rotation generates instabilities which also take part in the redistribution of angular momentum. The situation becomes then rather complex, and no satisfactory self-consistent solution has been obtained yet.

Even less is known about the circulations which arise when the convective envelope rotates at a different regime as the radiation zone below. This seems to be the case of the Sun, where a strong differential rotation is maintained by the convective motions, whereas the radiation zone below appears to rotate much more uniformly (Brown *et al.* 1989). Such circulations are similar to the well-known Ekman flow observed when a boundary rotates at a different speed then the bulk of the fluid (Ekman 1905). A viscous layers then forms, where the excess Coriolis force is balanced by the viscous drag; its thickness is of order $(\nu/\Omega)^{1/2}$. In a star, however, the fluid is compressible, and the temperature field also adjusts to the imposed differential rotation. Even in the absence of dissipation, a circulation is established below the convection zone over a fraction (Ω/N) of the stellar radius (Bretherton and Spiegel 1968), N being the buoyancy frequency defined next (4.2).

Beside these large-scale circulations, another transport mechanism probably operates in stellar radiation zones, namely turbulent diffusion. Schatzman (1977) was the first to propose such a mechanism to explain the depletion of Li on the surface of the Sun, and he gave the first estimate of the turbulent diffusivity. It remained to identify the cause of such turbulence.

The principal candidate is the differential rotation that is generated by the large-scale circulations: it is known to be liable to various instabilities (cf. Spiegel and Zahn 1970; Knobloch and Spruit 1982, 1983; Zahn 1983). Among those instabilities, and given the rotation laws that are likely to occur in stars, the most

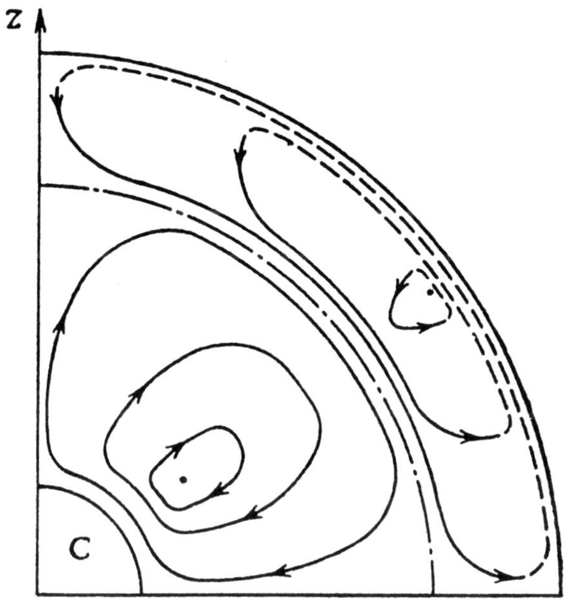

Fig. 10. Meridional circulation in a uniformly rotating star with a radiative envelope (Pavlov and Yakovlev 1978).

powerful appear to be the shear instabilities, whose properties we shall examine next.

4.2 Shear instabilities in parallel flows

The simplest shear flow is a plane-parallel flow whose velocity U is constant in the downstream direction, but varies, across the stream, in the z-direction. The intensity of the shear is measured by the gradient dU/dz, which is also the magnitude of the vorticity. Locally, any flow may be approximated, to lowest order, by a plane-parallel shear of this kind.

In the absence of other restoring forces, all shear flows are unstable, provided the Reynolds number which characterizes them is large enough. The detailed properties of the instability depend on the profile $U(z)$. When U has an inflexion point somewhere in the domain, corresponding to an extremum of the vorticity, the flow is unstable to infinitesimal perturbations (Rayleigh 1880; Fjørtoft 1950); it is then a dynamical instability, with a growth-rate of order $|dU/dz|$.

When the flow has no such inflexion point, it generally requires a perturbation of finite amplitude to become unstable. The theoretical study of that case is much more intricate; it is found that the instability threshold depends not only on the Reynolds number, but also on the shape of the perturbation (Gill 1965, Lerner and Knobloch 1988).

Such shear flows can be stabilized by a restoring force, and the best example of this is provided by the buoyancy. In a stratified medium, when a parcel of matter is perturbed from its equilibrium position, it oscillates with the buoyancy frequency N, which in a homogeneous star is given by

$$N^2 = \frac{g}{H_P}(\nabla_{ad} - \nabla). \tag{4.2}$$

If the perturbations associated with the motions were adiabatic, the shear instability would be prevented as soon as

$$\frac{N^2}{(dU/dz)^2} > 1/4, \tag{4.3}$$

a condition known as the Richardson criterion. In a star, however, the buoyancy force is weakened through radiative damping, and the stability criterion above takes the form (Zahn 1974)

$$\frac{N^2}{(dU/dz)^2} > (\sigma R_c)^{-1}, \tag{4.4}$$

where R_c is a critical Reynolds number (of order 10^3) depending on the flow profile, and σ the Prandtl number (ratio between the viscosity ν and the thermal diffusivity K, a very small quantity in stellar interiors).

But if there is a (stable) stratification of molecular weight, due for instance to nuclear reactions, then the buoyancy force is only partly reduced by radiative damping, and one recovers the original Richardson condition, the characteristic frequency now being

$$(N_\mu)^2 = \frac{g}{H_P}\frac{d\ln\mu}{d\ln P}. \tag{4.5}$$

One may wonder whether the Coriolis force, in a rotating fluid, could not play a role similar to the buoyancy. This possibility has been examined by Johnson (1963), who found that the most unstable modes are not affected by the rotation. Another proof is provided by the very nice experiment performed by Rabaud and Couder (1983), which clearly demonstrates that the onset of the shear instability is insensitive to the Coriolis force.

Therefore, the results of plane-parallel shear flow can be transposed as such to differential rotation. For convenience, let us distinguish between vertical differential rotation, in which the angular velocity Ω varies with depth r, and horizontal differential rotation, where it varies with latitude. In the first case, the instability criterion is

$$s\frac{\partial\Omega}{\partial r} > N\,(\sigma R_c)^{1/2}, \tag{4.6}$$

(s being the distance to the rotation axis), which holds for a homogeneous star. When there is a gradient of molecular weight, one recovers the Richardson criterion

$$s\frac{\partial\Omega}{\partial r} > N_\mu . \tag{4.7}$$

In the second case, that of horizontal differential rotation, the instability cannot be hindered by the stratification, since the buoyancy force acts only in the vertical direction. Therefore, due to the large Reynolds number characterizing such shear flow in stars, a differential rotation in latitude is always unstable (Zahn 1975). Since the velocity profile has then probably no inflexion point, it is linearly stable, but it will give rise to finite amplitude instabilities, as mentioned above.

Another restoring force may be provided by a magnetic field. Such a field is progressively wound up in a differentially rotating star until its azimuthal component becomes strong enough to react back on the rotation (Mestel 1953; Mestel and Moss 1986; Mestel and Weiss 1987). That interaction between shear and magnetic field certainly occurs in the regions where the magnetic field is generated through a dynamo mechanism, presumably just below the convection zone of a solar-type star (Spiegel and Weiss 1980). But if the field alternates its polarity in time, as in the Sun, it will not penetrate deep into the radiation zone, and it should not interfer much with the differential rotation (and with the shear instabilities mentioned above). The possibility of a fossile field pervading the star is not very likely either; it is hard to see how it would survive the pre–main-sequence phase and the various instabilities which have been identified so far (see Tayler 1982).

4.2 Turbulence in differentially rotating stars

When a horizontal shear flow becomes unstable, it generates eddies which have the same vorticity as the mean flow, and which are therefore horizontal, and two-dimensional. Those billows, in turn, undergo a three-dimensional instability, unless the vertical motions are hindered by some restoring force. If that is the case, the turbulent motions are anisotropic, and one has to take this into account when estimating the transport efficiency of such turbulence. as will be done next.

But before we examine in more detail the properties of the turbulence arising in differentially rotating stars, let us recall how one estimates the transport efficiency of such turbulence. To first approximation, the transport in a turbulent medium of a scalar quantity (such as temperature, the concentration of chemical species, etc.) can be described as a diffusion process, as we have seen already in chap. 1. If the velocity field is isotropic, the turbulent diffusivity is simply a scalar, which is given by

$$D_t = \frac{1}{3}\, u\ell\,, \tag{4.8}$$

where u is the r.m.s. velocity and ℓ the correlation length of the turbulent velocity field (see Knobloch 1978).

We have seen in chap. 1 how one can estimate these quantities in a stellar convection zone. Here the problem is different, since the instability is of dynamical origin. Moreover, that turbulence is strongly anisotropic: the motions are more vigorous in the horizontal than in the vertical direction. The diffusion coefficient is thus no longer a scalar, but a tensor (which is diagonal if one coordinate axis is chosen vertical).

To determine the turbulent diffusivity in the vertical direction, we may proceed as follows. We have seen that the horizontal shear is always unstable; it gives rise to horizontal vortices, which themselves are unstable and split into three-dimensional eddies. But this transition from 2 to 3 dimensions is hindered in the presence of a restoring force.

Among such forces, the Coriolis force plays a major role: it governs the dynamics of all eddies for which

$$u/\ell < \Omega\,, \tag{4.9}$$

a property which is well known in geophysical fluids, and which it is illustrated by various laboratory experiments (see for instance Hopfinger *et al.* 1982). Those large-scale motions will remain horizontal and two-dimensional, and they may also take the form of inertial waves; in any case, they will not contribute to the turbulent transport in the vertical direction. But the smallest eddies, i.e. those which do not obey the inequality above, are not sensitive to the Coriolis force; they are three-dimensional, and they follow to first approximation the Kolmogorov law

$$\varepsilon_t \approx u^3/\ell, \tag{4.10}$$

where ε_t is the rate at which kinetic energy is injected into the turbulence (see Landau and Lifshitz 1987).

The distribution of these three-dimensional eddies thus begins at the scale which verifies both

$$u'/\ell' \approx \Omega \quad \text{and} \quad (u')^3/\ell' \approx \varepsilon_t\,, \tag{4.11}$$

and therefore the vertical turbulent diffusivity is given by

$$D_v \approx u'\,\ell' \approx \varepsilon_t/\Omega^2\,. \tag{4.12}$$

The other restoring force present in a star is the buoyancy, but it is weakened through thermal dissipation, as explained in the preceding section. In practice, it only operates when there is a vertical gradient of molecular weight, which inhibits three-dimensional turbulence for the eddies whose turn-over rate is less than the residual buoyancy frequency (4.5)

$$u/\ell < N_\mu\,. \tag{4.13}$$

Vertical diffusion is suppressed entirely when (Zahn 1983)

$$(N_\mu)^2 > \varepsilon_t/\nu\,, \tag{4.14}$$

where ν is the kinematic viscosity (which fixes the size of the smallest scales, where the kinetic energy is dissipated through viscous friction). Even a rather small gradient of molecular weight suffices to prevent turbulent diffusion in the vertical direction (for instance, that due to the varying composition of ^{3}He in the

Sun). However, such a "μ-barrier" can still allow the transport of momentum, which will then be accomplished through isothermal gravity waves (Zahn 1989b).

To proceed further, one needs an estimate of ε_t, the generation rate of turbulent kinetic energy. This quantity is determined by the very cause of the differential rotation: meridional circulation, contraction or expansion of the star while it evolves, angular momentum loss through a wind, coupling with a differentially rotating convection zone, tidal braking in a binary star. As a typical example, we shall estimate this rate in the case where the differential rotation is produced by a meridional circulation.

Such a circulation advects angular momentum, whose conservation is expressed by

$$\frac{\partial}{\partial t}\left(s^2\Omega\right) + \mathbf{U}\cdot\nabla\left(s^2\Omega\right) = \Gamma, \tag{4.15}$$

where $\mathbf{U}$ is the meridional velocity and Γ the torque exerted per unit mass by the turbulent motions (as above, s is the distance to the rotation axis). Likewise, we may express the rate of variation of the rotational kinetic energy:

$$\frac{\partial}{\partial t}\frac{1}{2}(s\Omega)^2 + \Omega\,\mathbf{U}\cdot\nabla\left(s^2\Omega\right) = \Omega\,\Gamma. \tag{4.16}$$

In a stationary state, the advection term balances the right hand side, which is the work done by the turbulent torque, and therefore also the injection rate of kinetic energy into the turbulence. If we now split the angular velocity into its mean and fluctuating parts $\Omega_0(r)+\delta\Omega(r,\theta)$ (over a level surface, θ being the colatitude), and subtract the kinetic energy of the mean flow from that which is advected into the layer, we obtain the following expression for the turbulent energy input, averaged over a the whole horizontal layer (Zahn 1987):

$$\varepsilon_t(r) = -\int_0^1 \delta\Omega(r,\theta)\,\mathbf{U}\cdot\nabla\left(\Omega_0(r)\sin^2\theta\right)d(\cos\theta). \tag{4.17}$$

Let us stress that this expression is valid for any type of meridional flow; it applies either to the Eddington-Sweet or to the Ekman-like circulation.

To proceed further, one needs a prescription to determine the strength of the horizontal differential rotation $\delta\Omega(r,\theta)$. This is presently the weakest point of the theory, for we do not know that actually governs the distribution of angular momentum over a level surface. At large scale, the dynamical state is presumably similar to the geostrophic turbulence observed on Earth, in the atmosphere and in the oceans. Such turbulence is essentially two-dimensional, due to the limited vertical extent, and waves play an important role in the transport of momentum. It should be possible to transpose most of the results obtained in geophysics to the stellar radiation zones, and some attempts are being made in this spirit.

In the meanwhile, we may take $\delta\Omega$ as the first term of an expansion in spherical function, $\delta\Omega = -C\Omega_0 P_2(\cos\theta)$, introducing a parameter C which we can hope to calibrate against the observations. Replacing the circulation velocity U by its value, it is then possible to give an estimate of the energy generation rate ε_t and of the vertical diffusivity D_v.

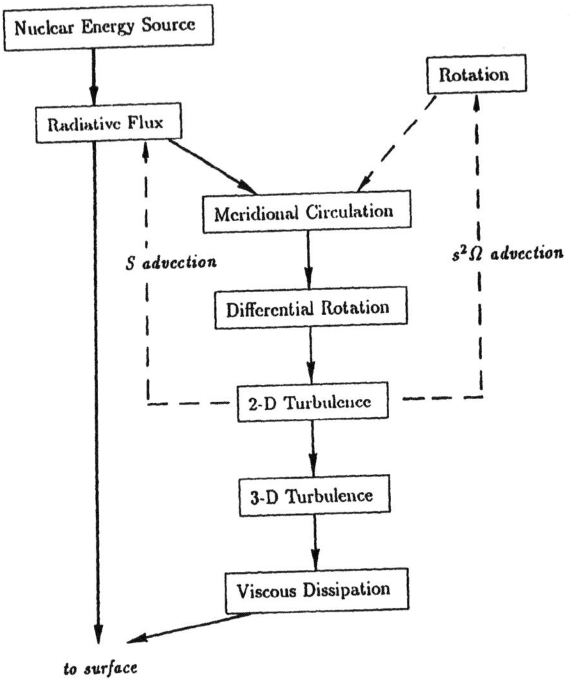

Fig. 11. Flow diagram of the energy, also showing the main feedback mechanisms (transport of entropy S and of angular momentum $s^2\Omega$).

We shall only consider the case of the Eddington-Sweet circulation; the whole scenario is represented schematically in fig. 11. The circulation is induced by the thermal umbalance of the rotating star, and through advection of angular momentum, it generates non-uniform rotation. The horizontal shear undergoes finite-amplitude instabilities; the largest scales behave as two-dimensional turbulence, and only the smallest scales have a three-dimensional component which contributes to the vertical diffusion of chemicals.

Two feed-back loops are drawn in fig. 11. One has already been mentioned, it is the redistribution of angular momentum through the two-dimensional turbulence. Another could well exist also: the horizontal transport of entropy, which would react back on the very cause of the circulation, namely the thermal imbalance; its efficiency depends on the Péclet number characterizing the largest scales, whose size is still not known (Zahn 1987).

Let us estimate the vertical turbulent diffusivity produced in this scheme. Assuming that the departures from uniform rotation are small enough, we have

$$\varepsilon_t \approx C \frac{L}{M} \left(\frac{\Omega^2 R^3}{GM} \right)^2 \quad \text{and} \quad D_v = \frac{\varepsilon}{\Omega^2} \approx C < K > \frac{\Omega^2 R^3}{GM} , \tag{4.18}$$

where $< K >= LR^3/GM^2$ is then mean thermal diffusivity of the star.

Since these lectures have been given in Punte de Lima, another process has been analyzed: it the inhibition of the large-scale advection of a chemical through the two-dimensional turbulence. Under some circumstances, it also results in a vertical diffusion, with a comparable strength (Chaboyer 1990).

To conclude, the problem of turbulent mixing in stellar radiation zones is rather intricate. Its theory is still in infancy, and one may complain that it relies too much on phenomenological arguments. That is certainly true, but cannot the same be stated about the mixing-length approach to stellar convection, whose shortcomings are well known, but which has not yet been replaced by a better procedure?

Progress will certainly be achieved by implementing the few prescriptions which are available in stellar structure codes, and to confront the theoretical models with the observations. This was done by Schatzman and Maeder (1981), Baglin *et al.* (1985), and more recently by Vauclair (1988), Charbonneau and Michaud (1989), and by Deliyannis *et al.* (1989). Some discrepancies which have been noticed may be ascribed to physical processes that have not been included yet, such as the stabilizing effect of a μ-gradient, or the momentum transport by waves. Work is in progress to render the theory more realistic, and to include additional effects which have been overlooked so far.

Acknowledgments

I am grateful to the organizers of the summer school, and especially to T. Lago, for this wonderful stay in Ponte de Lima, and to the students for their kind and responsive participation. This work was partly supported by the U.S. Air Force through grant AFOSR 89-0012 to Columbia University.

References

Aime, C., Ricort, G., Harvey, J.: 1978, *Astrophys. J.* **221**, 362

Ambrož, P.: 1987, *Bull. Astron. Inst. Czechosl.* **38**, 1

Antia, H.M., Chitre, S.M., Pandey, S. D.: 1981, *Solar Phys.* **70**, 67

Antia, H.M., Chitre, S.M., Narasimha, D.: 1983, *Monthly Not. Roy. Astron. Soc.* **204**, 865

Baglin, A., Morel, P., Schatzman, E.: 1985, *Astron. Astrophys.* **149**, 309

Bahng, J., Schwarzschild, M.: 1961, *Astrophys. J.* **134**, 312

Biermann, L.: 1935, *Astron. Nachr.* **257**, 269

Biermann, L.: 1938, *Astron. Nachr.* **264**, 395

Biermann, L.: 1943, *Z. Astrophys.* **22**, 244

Boesgard, A.: 1976, *Publ. Astron. Soc. Pacific* **88**, 353

Böhm-Vitense, E.: 1958, *Z. Astrophys.* **46**, 108

Böhm, K.-H., Stückl, E.: 1967, *Z. Astrophys.* **66**, 487

Bretherton, F.P., Spiegel, E.A.: 1968, *Astrophys. J.* **153**, 277
Brown, T.M., Christensen-Dalsgaard, J., Dziembowski, W.A., Goode, P., Gough, D.O., Morrow, C.A.: 1989, *Astrophys. J.* **343**, 526
Busse, F.H.: 1970, *Astrophys. J.* **159**, 629
Cattaneo, F., Hurlburt, N.E., Toomre, J. 1989, 'Stellar and Solar Granulation' (ed. R.J. Rutten and G. Sverino; Kluwer), 415
Cayrel, R., Cayrel de Strobel, G., Campbell, B., Däppen, W.: 1984, *Astron. Astrophys.* **283**, 205
Chaboyer, B.: 1990, 'Geophys. Fluid Dynamics' (Woods Hole Oceanograpic Institution Rep.)
Chan, K.L., Sofia, S.: 1986, *Astrophys. J.* **307**, 222
Charbonneau, P., Michaud, G.: 1988, *Astrophys. J.* **334**, 746
Cowling, T.G.: 1936, *Astron. Nachr.* **258**, 133
Deliyannis, C.P., Demarque, P., Kawaler, S.D.: 1989, *Phys. Rev. Lett.* **62**, 1583
Dravins, D.: 1982, *Ann. Rev. Astron. Astrophys.* **20**, 61
Dravins, D., Lindegren, L., Nordlund, Å.: 1981, *Astron. Astrophys.* **96**, 345
Dravins, D., Larsson, B., Nordlund, Å.: 1986, *Astron. Astrophys.* **188**, 83
Duncan, D.K.: 1981, *Astrophys. J.* **248**, 651
Durney, B.R.: 1970, *Astrophys. J.* **161**, 1115
Eddington, A.S.: 1925, *Observatory* **48**, 78
Eddington, A.S.: 1926 'The Internal Constitution of Stars' (Cambridge Univ. Press; reprinted by Dover, 1959)
Ekman, V.W.: 1905, *Arkiv Matem. Astron. Fysik* **2**, 11
Fjørtoft, R.: 1950, *Geofys. Publ.* **17**, 52
Geiss J., Buhler, F., Cerutti, H., Eberhardt, P., Filleux, C.H.: 1972, Apollo 16 Prel. Sci. Rep. NASA (SP 315)
Gelly, B., Grec, G., Fossat, E.: 1986, *Astron. Astrophys.* **164**, 383
Gill, A.E.: 1965, *J. Fluid Mech.* **21**, 503
Gilman, P.A., Miller, J.: 1986, *Astrophys. J. Suppl.* **61**, 585
Glatzmaier, G.A.: 1984, *J. Comput. Phys.* **55**, 461
Glatzmaier, G.A.: 1985, *Astrophys. J.* **291**, 300
Glatzmaier, G.A., Gilman, P.A.: 1982, *Astrophys. J.* **256**, 316
Gonczi, G.: 1981, *Astron. Astrophys.* **96**, 138
Gonczi, G.: 1982, *Astron. Astrophys.* **110**, 1
Gonczi, G., Osaki, Y.: 1980, *Astron. Astrophys.* **84**, 304
Gough, D.O.: 1969, *J. Atmos. Sci.* **26**, 448
Gough, D.O.: 1977a, *Astrophys. J.* **214**, 196
Gough, D.O.: 1977b, 'Problems of Stellar Convection' (eds. E.A. Spiegel and J.-P. Zahn; Springer), *Lecture Notes in Physics* **161**, 15
Gough, D.O.: 1986, 'Seismology of the Sun and Distant Stars' (eds. D.O. Gough; Reidel), 125
Gough, D.O., Weiss, N.O.: 1976, *Monthly Notices Roy. Astron. Soc.* **176**, 589
Gough, D.O., Spiegel, E.A., Toomre, J.: 1975, *J. Fluid Mech.* **68**, 695
Graham, E.: 1975, *J. Fluid Mech.* **70**, 689
Graham, E.: 1977, 'Problems of Stellar Convection' (eds. E.A. Spiegel and J.-P. Zahn; Springer), *Lecture Notes in Physics* **161**, 151
Gray, D.F.: 1980, *Astrophys. J.* **235**, 508
Gray, D.F.: 1981, *Astrophys. J.* **251**, 583
Gray, D.F.: 1982, *Astrophys. J.* **255**, 200

Gray, D.F.: 1986, *Publ. Astron. Soc. Pacific* **98**, 319
Gray, D.F., Toner, C.G.: 1986, *Publ. Astron. Soc. Pacific* **98**, 499
Hart, J.E., Toomre, J., Deane, A.E., Hurlburt, N.E., Glatzmaier, G.A., Fichtl, G.H., Leslie, F., Fowlis, W.W., Gilman, P.A.: 1986, *Science* **234**, 1
Hofmeister, E., Weigert, A.: 1964, *Z. Astrophys.* **59**, 119
Hopfinger, E.J., Browand, F.K., Gagne, Y.: 1982, *J. Fluid Mech.* **125**, 505
Howard, R., LaBonte, B.J.: 1980, *Astrophys. J.* **239**, 738
Hurlburt, N.E., Toomre, J., Massaguer, J.M.: 1986, *Astrophys. J.* **311**, 563
Johnson, J.A.: 1963, *J. Fluid Mech.* **17**, 337
Kippenhahn, R.: 1958, *Z. Astrophys.* **46**, 26
Knobloch, E.: 1978, *Astrophys. J.* **225**, 1050
Knobloch, E., Spruit, H.C.: 1982, *Astron. Astrophys.* **113**, 261
Knobloch, E., Spruit, H.C.: 1983, *Astron. Astrophys.* **125**, 59
Kuhfuß R.: 1986, *Astron. Astrophys.* **160**, 116
Lambert, D.L., Dominy, J.F., Sivertsen, S.: 1980, *Astron. Astrophys.* **235**, 114
Landau, L., Lifschitz, E.: 1987, Fluid Mechanics (English translation, 2nd edition; Pergamon)
Latour, J., Spiegel, E.A., Toomre, J., Zahn, J.-P.: 1976, *Astrophys. J.* **207**, 233
Latour, J., Toomre, J., Zahn, J.-P.: 1981, *Astrophys. J.* **248**, 1081
Latour, J., Toomre, J., Zahn, J.-P.: 1983, *Solar Phys.* **82**, 387
Lerner, J., Knobloch, E.: 1988, *J. Fluid Mech.* **189**, 117
Macris, C.J., Rösch,: 1984, *C. R. Acad. Sci., série II* **296**, 265
Maeder, A.: 1975, *Astron. Astrophys.* **40**, 303
Marcus, Ph. S.: 1980, *Astrophys. J.* **240**, 203
Massaguer, J.M., Latour, J., Toomre, J., Zahn, J.-P.: 1984, *Astron. Astrophys.* **140**, 1
Massaguer, J.M., Mercader, I: 1988, *J. Fluid Mech.* **189**, 367
Massaguer, J.M., Zahn, J.-P.: 1980, *Astron. Astrophys.* **87**, 315
McDonald, B.E.: 1972, *Astrophys. Space Sci.* **19**, 309
Mestel, L.: 1953, *Monthly Not. Roy. Astron. Soc.* **113**, 716
Mestel, L., Moss, D.L.: 1986, *Monthly Not. Roy. Astron. Soc.* **221**, 25
Mestel, L., Weiss, N.O.: 1987, *Monthly Not. Roy. Astron. Soc.* **226**, 123
Muller, R.: 1985, *Solar Phys.* **100**, 237
Muller, R., Roudier, Th.: 1985, 'High Resolution in Solar Physics' (ed. R. Muller; Springer), *Lecture Notes in Physics* **233**, 242
Muller, R., Roudier, Th., Malherbe, J.M., Mein, P.: 1987, 'Solar and Stellar Physics' (eds. E.-H. Schröter and M. Schüssler; Spinger) **292**, 218 (poster)
Muller, R., Roudier, Th., Vigneau, J.: 1990, *Solar Phys.* **126**, 53
Nesis, A., Mattig, W., Fleig, K.H., Wiehr, E.: 1987, *Astron. Astrophys.* **182**, L5
Nordlund, Å.: 1982, *Astron. Astrophys.* **107**, 1
Nordlund, Å: 1984, 'Small Scale Dynamical Processes in Quiet Stellar Atmospheres', (ed. S.L. Keil; National Solar Observatory, Sunspot), p. 181
Nordlund, Å.: 1985, *Solar Phys.* **100**, 209
Öpik, E.J.: 1938, *Publ. Obs. Astron. Univ. Tartu* **30**, no. 3
Pavlov, G.G., Yakovlev, D.G.: 1978 *Sov. Astron.* **22**, 595
Prandtl, L.: 1925, *Z. angew. Math. Mech.* **5**, 136
Rabaud, M., Couder, Y.: 1983, *J. Fluid Mech.* **136**, 291
Rayleigh, Lord: 1880, 'Scientific Papers' (Cambridge Univ. Press), **1**, 474
Renzini, A.: 1987, *Astron. Astrophys.* **188**, 49
Ribes, E., Mein, P., Mangeney, A.: 1985, *Nature* **318**, 170

Roudier, Th., Muller, R.: 1986, *Solar Phys.* **107**, 11
Saslaw, W.C., Schwarzschild, M.: 1965, *Astrophys. J.* **141**, 1468
Schatzman, E.: 1969, *Astrophys. Lett.* **3**, 139
Schatzman, E.: 1977, *Astron. Astrophys.* **56**, 211
Schatzman, E., Maeder, A.: 1981, *Astron. Astrophys.* **96**, 1
Schröter, E.H.: 1985, *Solar Phys.* **100**, 141
Shaviv, G., Salpeter, E.E.: 1973, *Astrophys. J.* **184**, 191
Spiegel, E.A.: 1971, *Ann. Rev. Astron. Astrophys.* **9**, 313
Spiegel, E.A.: 1972, *Ann. Rev. Astron. Astrophys.* **10**, 261
Spiegel, E.A., Zahn, J.-P.: 1970, *Comm. Astrophys. Space Sci.* **2**, 178
Spite, F., Spite, M.: 1982, *Astron. Astrophys.* **115**, 357
Spruit, H.C., Nordlund, Å, Title, A.M.: 1990, *Ann. Rev. Astron. Astrophys.* **28**, 263
Stein, R., Nordlund, Å: 1989, *Astrophys. J. Letters* **342**, L95
Sweet, P.A.: 1950, *Monthly Not. Roy. Astron. Soc.* **110**, 548
Tassoul, J.-L., Tassoul, M.: 1982, *Astrophys. J. Suppl.* **49**, 317
Tayler, R. J.: 1982, *Monthly Not. Roy. Astron. Soc.* **198**, 811
Title, A.M., Tarbell, T.D., Simon, G.W., and SOUP Team: 1986, *Proc. COSPAR meeting, Toulouse*
Title, A.M., Tarbell, T.D., Topka, K.P., Shine, R.A., Simon, G.W., Zirin, H. and SOUP Team: 1987, 'Solar and Stellar Physics' (eds. E.-H. Schröter and M. Schüssler; Spinger) **292**, 173
Toomre, J., Hart, J.E., Glatzmaier, G.A.: 1987, 'The Internal Solar Angular Velocity: Theory and Observations' (eds. B. Durney and S. Sofia; Reidel), 27
Toomre, J., Zahn, J.-P., Latour, J., Spiegel, E.A.: 1976, *Astrophys. J.* **207**, 545
Unno, W.: 1967, *Publ. Astron. Soc. Japan* **18**, 85
Van der Borght, R.: 1979, *New Zealand J. Sci.* **22**, 575
Vauclair, S.: 1988, *Astron. Astrophys.* **335**, 971
Veronis, G.: 1963, *Astrophys. J.* **137**, 641
Vitense, E.: 1953 *Z. Astrophys.* **32**, 135
Vogt, H.: 1925, *Astron. Nachr.* **223**, 229
Von Zeipel, H.: 1924, *Monthly Not. Roy. Astron. Soc.* **84**, 665
Xiong, D.-R.: 1979, *Acta Astron. Sinica* **20**, 238
Xiong, D.-R.: 1985, *Astron. Astrophys.* **150**, 133
Zahn, J.-P.: 1966, *Ann. Astrophys.* **29**, 489
Zahn, J.-P.: 1974, 'Stellar Instability and Evolution' (eds. P. Ledoux, A. Noels and R.W. Rodgers; Reidel), 185
Zahn, J.-P.: 1975, *Mém. Soc. Roy. Sci. Liège, 6e série* **8**, 31
Zahn, J.-P.: 1983, Astrophys. Processes in Upper Main Sequence Stars (Publ. Observatoire Genève), 253
Zahn, J.-P.: 1987, 'The Internal Solar Angular Velocity' (eds. B.R. Durney, S. Sofia; Reidel), 201
Zahn, J.-P.: 1989a, *Astron. Astrophys.* **220**, 112
Zahn, J.-P.: 1989b, 'Inside the Sun' (eds. G. Berthomieu and M. Cribier; Kluwer), 425
Zahn, J.-P., Toomre, J., Latour, J.: 1982, *Geophys. Astrophys. Fluid Dyn.* **22**, 159

GPSR Compliance

The European Union's (EU) General Product Safety Regulation (GPSR) is a set of rules that requires consumer products to be safe and our obligations to ensure this.

If you have any concerns about our products, you can contact us on ProductSafety@springernature.com

In case Publisher is established outside the EU, the EU authorized representative is:

Springer Nature Customer Service Center GmbH
Europaplatz 3
69115 Heidelberg, Germany

Batch number: 10838062

Printed by Printforce, the Netherlands